# Intelligent Materials

This book is dedicated to the memory of P.-G. de Gennes who passed away in May 2007

Intelligent Materials

Edited by

Mohsen Shahinpoor
Department of Mechanical Engineering, University of Maine, Orono, Maine, USA

Hans-Jörg Schneider
FR Organische Chemie, Universität des Saarlandes, Saarbrücken, Germany

RSCPublishing

ISBN: 978-085404-335-4

A catalogue record for this book is available from the British Library

© The Royal Society of Chemistry 2008

Reprinted 2009

All rights reserved

Apart from fair dealing for the purposes of research for non-commercial purposes or for private study, criticism or review, as permitted under the Copyright, Designs and Patents Act 1988 and the Copyright and Related Rights Regulations 2003, this publication may not be reproduced, stored or transmitted, in any form or by any means, without the prior permission in writing of The Royal Society of Chemistry, or in the case of reproduction in accordance with the terms of licences issued by the Copyright Licensing Agency in the UK, or in accordance with the terms of the licences issued by the appropriate Reproduction Rights Organization outside the UK. Enquiries concerning reproduction outside the terms stated here should be sent to The Royal Society of Chemistry at the address printed on this page.

Published by The Royal Society of Chemistry,
Thomas Graham House, Science Park, Milton Road,
Cambridge CB4 0WF, UK

Registered Charity Number 207890

For further information see our web site at www.rsc.org

Foreword

Claiming to be intelligent is bold. The materials described in this book are actually somewhat less ambitious; they respond in an interesting way to an external stimulus: a field, a change in pH, a pulse of light, a temperature jump. I would tend to call them "sensitive" rather than "intelligent". But they have been invented, and improved, by highly intelligent scientists, and this shows in this book.

Assume, for instance, that we wish to make an artificial muscle from a chemosensitive gel. A great man (A. Katschalsky) started this game, exposing carboxylic acids to protons, converting the gel to a neutral, more compact form. But he immediately realised that cycles (hydrochloric acid/soda) would accumulate counterions, screen out the electrical interactions, and kill the muscle. He then used a clever ion exchange ($Na^+ \rightarrow Ca^{++}$) that eliminates this form of fatigue. But he still could not solve the problem of time constants: bringing an ion, and contracting a gel, requires diffusion processes that are very slow for macroscopic samples. This is lucidly discerned in the present book. One way out is to operate with units of diameter below one micrometre. Another approach would be based on two interpenetrated networks – one elastic, and the other electrically conducting, bringing the required ions locally. But this is science fiction: to have a second network both flexible and chemically robust seems extremely difficult.

This "local feeding" problem is the heart of the matter: what the striated muscles of vertebrates achieved is local feeding of energy by ATP *all along* a myosin/actin structure.

In fact, this observation makes me doubtful about most artificial muscles based on nafions + water + ions: the ions are driven by an external voltage V (~ 1 V); the energy gain that this gives them when they move by one pore size d is only eVd/L (where e is the unit charge, L is the sample thickness). Because of the factor $d/L \sim 10^{-5}$, this can never compete with myosin, which gains $\sim 10^{-2}$ eV per jump by burning ATP. But, again, the best approach to these difficult questions is to read the present book.

There is also an interesting hope with gels doped by ferrofluids, and the corresponding experiments are nicely described here. I am not sure, however, that these gels can compete with a naïve system, where a small cylinder of permanent magnet is attached to a rubber stem. (The permanent magnet leads to much higher forces than ferrofluids.)

These random remarks illustrate (I think) the difficulty in achieving industrial materials that can produce large stresses (1 atm) with a relatively fast response (1 s): furthermore they must not heat up too much. And they must be robust (chemically and mechanically); ultimately they should be not too expensive. This list of duties is somewhat frightening. But I am convinced that some young researchers will find answers – and that they will previously read this book, as a starting point. Thus, I do congratulate the editors for their vast collection – covering so many sectors of physics and chemistry.

P.-G. de Gennes
November 2006, Paris, France
Prix Nobel de Physique 1991
Membre de l'Institut (Académie des Sciences)
Professeur Honoraire au Collège de France
Chaire de Physique de la Matière Condensée

Preface

Intelligent materials, which respond to external signals by a distinct reaction to the outside world, are the basis of promising new technologies, extending from artificial muscles, nanoscale motors to new drug-delivery devices. It may take scientists centuries to reach the complexity and efficiency of biological systems; on the other hand synthetic devices encompass a broader scope than natural functions, which are limited, *e.g.*, by the lack of protein stability and the restriction of metabolic pathways. At the same time one can hope to better understand with the help of biomimetic approaches the fundamental mechanisms of biological systems. As stated by Richard P. Feynman in 1959 in a famous address, there is plenty of room at the bottom for miniaturization. Miniaturization of technical devices, possibly down to molecular scale, is an important strategy for material and energy-saving technologies, and thus can be a significant contribution towards a sustainable development. The broad field of very different materials and external stimuli, which reach from electrical and magnetic input through effects of temperature and light to selective molecular recognition of effector compounds, and their application for new devices call for joint efforts of many chemists, physicists and engineers. Fortunately, we were able convince leading experts of the different fields to contribute their views and results to the present monograph. Our sincere thanks go to them – they deserve the merit of the book, while the editors must bear responsibility for any shortcomings. As is often the case not all of the planned chapters did materialize, and several topics that are already covered in established monographs had to be omitted in view in order to keep the book within an affordable scope. Last but not least, we also thank the Nobel Prize winner Professor Pierre-Gilles de Gennes for his support and encouragement towards publishing this book and for his enlightening, thorough and thoughtful foreword for this book.

Mohsen Shahinpoor, Hans-Jörg Schneider
Albuquerque/Saarbrücken

Contents

Introduction xxi

Chapter 1 Chemically Driven Artificial Molecular Machines
J.D. Crowley, E.R. Kay and D.A. Leigh

    1.1 Design Principles for Molecular-Level Motors and Machines   1
        1.1.1 The Effects of Scale   2
        1.1.2 Machines that Operate at Low Reynolds Number   3
        1.1.3 Lessons to Learn from Biological Motors and Machines   3
    1.2 Controlling Motion in Covalently Bonded Molecular Systems   4
        1.2.1 Controlling Conformational Changes   4
        1.2.2 Controlling Configurational Changes   9
    1.3 Controlling Motion in Mechanically Bonded Molecular Systems   12
        1.3.1 Basic Features   13
        1.3.2 Translational Molecular Switches: Stimuli-Responsive Molecular Shuttles   13
        1.3.3 Controlling Rotational Motion in Catenanes   24
    1.4 From Laboratory to Technology: Towards Useful Molecular Machines   29
        1.4.1 The Current Challenges: Constraining, Communicating, Correlating   30
        1.4.2 Reporting Controlled Motion in Solution   32
        1.4.3 Reporting Controlled Motion on Surfaces, in Solids and Other Condensed Phases   32
    1.5 Summary and Outlook   38
    References   40

Chapter 2 Photochemically Controlled Molecular Devices and Machines
V. Balzani, G. Bergamini, P. Ceroni, A. Credi and M. Venturi

2.1	Introduction	48
2.2	Molecular-level Devices for Processing Light Signals	50
	2.2.1 Wires	50
	2.2.2 Switching Devices	51
	2.2.3 Plug/socket Systems	52
	2.2.4 Molecular Extension Cables	53
	2.2.5 Antenna Systems for Light Harvesting	53
	2.2.6 Molecular Lenses Capable of Tuning the Colour of Light	56
	2.2.7 Fluorescent Sensors with Signal Amplification	56
	2.2.8 Dendrimers for a Multiple Use of Light Signals	59
	2.2.9 Logic Gates	62
2.3	Light-driven Molecular Machines	65
	2.3.1 Dethreading/rethreading of Pseudorotaxanes	65
	2.3.2 A Sunlight-Powered Nanomotor	66
2.4	Conclusions	69
	Acknowledgements	70
	References	70

Chapter 3 Transition-Metal Complex-Based Molecular Machines
B. Champin, U. Létinois-Halbes and J.-P. Sauvage

3.1	Introduction	76
3.2	Molecular Motions Driven by a Chemical Reaction	77
	3.2.1 Use of a Chemical Reaction to Induce the Contraction/Stretching Process of a Muscle-like Rotaxane Dimer	77
	3.2.2 Intramolecular Complexation/Decomplexation Processes as a Means to Make an Intermittent Degenerate Molecular Shuttle	78
	3.2.3 Molecular Machines Based on Metal-ion Translocation	82
3.3	Electrochemically Induced Motions	83
	3.3.1 Transition-metal-complexed Catenanes and Rotaxanes	83
	3.3.2 Other Related Noninterlocking Systems	89
3.4	Light-fuelled Molecular Machines	91
	3.4.1 Photoinduced Decoordination and Thermal Recoordination of a Ring in a Ruthenium(II)-containing 2catenane	91

Contents

              3.4.2   A Photochemically Driven Molecular-level Abacus   92
     3.5   Conclusion and Prospective   96
     Acknowledegments   96
     References   97

Chapter 4 Chemomechanical Polymers
H.-J. Schneider and K. Kato

    4.1   Introduction   100
    4.2   Chemomechanical Polymers Triggered by pH   101
    4.3   Particle-size Effects and Kinetics   107
    4.4   Water Uptake and Release   108
    4.5   Concentration Profiles   110
    4.6   Cooperativity and Logical Gate Functions   111
    4.7   Selectivity with Organic Effector Molecules   114
    4.8   Ternary Complex Formation for Amino Acids and Peptides as Effectors   117
    4.9   Selectivity by Covalent Interactions/Glucose-triggered Size Changes   118
    4.10 Conclusions   119
    References   120

Chapter 5 Ionic Polymer Metal Nanocomposites as Intelligent Materials and Artificial Muscles
M. Shahinpoor

    5.1   Summary   126
    5.2   Introduction   127
    5.3   Three-dimensional Fabrication of IPMNCs   128
    5.4   Manfacturing Methodologies   128
    5.5   Manufacturing Steps   129
    5.6   Electrically Induced Robotic Actuation   130
    5.7   Distributed Nanosensing and Transduction   132
    5.8   Modeling and Simulation   136
    5.9   Smart-Product Development   138
    5.10 Medical, Engineering and Industrial Applications   139
    References   140

Chapter 6 Artificial Muscles, Sensing and Multifunctionality
T.F. Otero

    6.1   Introduction   142
    6.2   Materials   143

6.3	Electrochemical Behaviour of Conducting Polymers in Aqueous Solution		143
6.4	Nonstoichiometric, Soft, and Wet Materials		147
6.5	Electrochemical Properties		149
	6.5.1	Electrochemomechanical Properties	150
	6.5.2	Electrochromic Properties	150
	6.5.3	Charge Storage	150
	6.5.4	Porosity	150
	6.5.5	Electron/Chemical Transduction	151
	6.5.6	Unparalleled Simultaneous Sensing Possibilities	151
6.6	Multifunctional and Biomimicking Properties		151
6.7	Natural Muscles		152
6.8	Devices based on the Electrochemical Properties of Conducting Polymers		153
	6.8.1	Artificial Muscles	153
	6.8.2	Other Electrochemically based Properties and Devices: Electrochromic Devices	170
	6.8.3	Batteries	172
	6.8.4	Membranes and Electron/Ion (or Electron/Chemical) Transducers	174
6.9	Theoretical Models		174
	6.9.1	Elastic Models	177
	6.9.2	Electrochemical Models	177
	6.9.3	Relaxation Models	178
	6.9.4	Molecular Dynamics Treatment	179
6.10	Final Remarks		179
References			182

Chapter 7 Electrochemically Controllable Polyacrylonitrile-Derived Artificial Muscle as an Intelligent Material
K.J. Kim and K. Choe

7.1	Polyacrylonitrile in General		191
7.2	Force-Strain Behaviour of Modified PAN		194
7.3	Actuation Properties of Modified PAN		194
	7.3.1	Length-change Characteristics of Modified PAN: Effect of pH Variation	194
	7.3.2	Generative Force Characteristics: pH-driven and/or Electrically Driven PAN Actuator	194
	7.3.3	Generative Force Characteristics: Effect of Different Anions	196
	7.3.4	Generative Force Characteristics: Effect of Acidity	197
7.4	Performance of PAN Bundle Artificial Muscle		198

	7.4.1 Electric-current Effect on Force Generation	199
	7.4.2 Work Performance	201
7.5	Summary of Performance Capability of PAN Artificial Muscle	201
References		203

Chapter 8 Unimolecular Electronic Devices
R.M. Metzger

8.1	Introduction	205
8.2	Donors and Acceptors; HOMOs and LUMOs	206
8.3	Contacts	207
8.4	Two-probe, Three-probe and Four-probe Electrical Measurements	209
8.5	Resistors	210
8.6	Rectifiers or Diodes	212
8.7	Switches	221
8.8	Capacitors	221
8.9	Future Flash Memories	222
8.10	Field Effect Transistors	222
8.11	Negative Differential Resistance Devices	222
8.12	Coulomb-blockade Device and Single-electron Transistor	222
8.13	Future Unimolecular Amplifiers	223
8.14	Future Organic Interconnects	223
Acknowledgements		223
References		223

Chapter 9 Piezoelectric Ceramics as Intelligent Multifunctional Materials
A. Yousefi-Koma

9.1	Introduction		231
9.2	Piezoelectricity		232
9.3	Piezoelectric Ceramics		232
9.4	Piezoelectric Ceramic Actuators		233
9.5	Modeling		235
	9.5.1	Sensors	239
	9.5.2	Actuators	242
9.6	Applications		242
	9.6.1	Vibration/Acoustic Control	243
	9.6.2	Rotor-blade Flap	245
	9.6.3	Adaptive Structural Shape Control	246
	9.6.4	Structural Health Monitoring	246
	9.6.5	Compact Hybrid Actuators	247

	9.7 Commercial Products	247
	References	252

Chapter 10 Ferroelectric Relaxor Polymers as Intelligent Soft Actuators and Artificial Muscles
Q. M. Zhang, B. Chu and Z.-Y. Cheng

10.1	Introduction	256
10.2	High-energy Electron-irradiated Copolymer (HEEIP)	258
	10.2.1 Microstructures of HEEIP	258
	10.2.2 Electromechanical Responses of HEEIP	262
10.3	Electrostrictive Responses and Relaxor Ferroelectric Behaviour in P(VDF-TrFE)-based Terpolymers	266
	10.3.1 The Electromechanical Response in P(VDF-TrFE)-based Terpolymers	266
	10.3.2 The Microstructure and Ferroelectric Relaxor Behaviour of P(VDF-TrFE-CFE) Terpolymers	268
10.4	Performance of Microelectromechanical Devices	273
10.5	Summary	278
Acknowledgement		279
References		279

Chapter 11 Magnetic Polymeric Gels as Intelligent Artificial Muscles
M. Zrínyi

11.1	Introduction	282
11.2	Ferrogel as a New Type of Responsive Gel	283
11.3	Interpretation of the Abrupt Shape Transition	288
11.4	Nonhomogeneous Deformation of Ferrogels	290
11.5	Muscle-like Contraction Mimicked by Ferrogels	294
11.6	Control of Pseudomuscular Contraction	295
11.7	Future Aspects	299
Acknowledgements		299
References		299

Chapter 12 Intelligent Materials: Shape-Memory Polymers
M. Behl, R. Langer and A. Lendlein

12.1	Introduction	301
12.2	Thermally Induced Shape-memory Polymers	303
	12.2.1 General Concept and Characterisation of Shape-memory Effect	303

Contents xv

        12.2.2 Thermoplastic Shape-memory Polymers 304
        12.2.3 Covalently Crosslinked Shape-memory
                Polymers 306
        12.2.4 Composites from Shape-memory Polymers
                and Particles 308
        12.2.5 Indirect Actuation of Thermally Induced
                Shape-memory Effect in Polymers 308
  12.3 Light-induced Shape-memory Polymers 311
  12.4 Multifunctional Polymers with Shape-memory Effect 312
  12.5 Conclusion and Outlook 313
  References 314

Chapter 13 Shape-Memory Alloys as Multifunctional Materials
L. McDonald Schetky

  13.1 Introduction to Shape-memory Alloys 317
  13.2 Shape-memory Alloy Applications 320
        13.2.1 Couplings 320
        13.2.2 Seals 321
        13.2.3 Electrical Connectors 322
        13.2.4 Virtual Two-way Actuation Using One-way
                NiTi Shape-memory Alloys 323
        13.2.5 Nonbiased Safety Devices 324
        13.2.6 Thermal Interrupter 325
        13.2.7 Eyeglass Frames 326
        13.2.8 Cellular-phone Antennas 327
        13.2.9 Home Appliances 327
  13.3 Medical Applications 327
        13.3.1 Orthodontics and Dental Procedures 328
        13.3.2 Superelastic Medical Devices 328
        13.3.3 Cardiovascular Stents 329
  13.4 Engineering Applications 331
        13.4.1 Adaptive Structures 331
        13.4.2 Structural Damping 333
        13.4.3 High-force Devices 334
        13.4.4 Jet-engine and Other Aeronautical Applications 334
  13.5 Thin-film and Porous Devices 336
  References 338

Chapter 14 Magnetorheological Materials and their Applications
X. Wang and F. Gordaninejad

  14.1 Introduction 339
  14.2 Historical Perspective 340
  14.3 Magnetorheological Materials 341

		14.3.1	Magnetorheological Fluids	341
		14.3.2	Magnetorheological Elastomers	344
		14.3.3	Rheological Behaviour of MR Fluids	348
		14.3.4	Models for Shear-yield Stress	351
		14.3.5	Field-induced Microstructures	353
		14.3.6	Rheometry of MR Fluids	354
		14.3.7	Effects of Surface Roughness	357
	14.4	Magnetorheological Fluid Devices		363
		14.4.1	Magnetorheological Fluid Dampers	363
		14.4.2	Modeling of Magnetorheological-Fluid Dampers	365
		14.4.3	Effect of Temperature	369
		14.4.4	Other Applications	373
	14.5	Summary		376
	Acknowledgements			376
	References			376

Chapter 15 Metal Hydrides as Intelligent Materials and Artificial Muscles
K.J. Kim, G. Lloyd and M. Shahinpoor

	15.1	Metal Hydrides in General		386
	15.2	Metal-hydride-actuation Principle		387
		15.2.1	Modeling	390
		15.2.2	Experiments	393
	15.3	Summary		394
	References			394

Chapter 16 Dielectric Elastomer Actuators as Intelligent Materials for Actuation, Sensing and Generation
G. Kofod and R. Kornbluh

	16.1	Introduction		396
	16.2	Actuation Basics		397
	16.3	Pre-stress Bias		399
	16.4	Compliant Electrodes		400
		16.4.1	Percolating Conductive Particle Networks	400
		16.4.2	Structured Metal Electrodes	400
	16.5	Theory and Modeling		401
	16.6	Actuator Design: Geometry and Structure		405
	16.7	Applications		406
		16.7.1	Artificial Muscles for Biomimetic Robots	409
		16.7.2	Linear Actuators for Industrial Applications	411
		16.7.3	Diaphragm Actuators for Pumps and Arrays	411
		16.7.4	Enhanced-thickness Mode Arrays	412

		16.7.5	Framed Actuator for Optics	414
		16.7.6	Sensors	415
		16.7.7	Generators	416
	16.8	Implementation Challenges for Dielectric Elastomers		417
	16.9	The Future: Materials Development for New Elastomers		418
		16.9.1	Improving Elastic Properties	419
		16.9.2	Improving Dielectric Properties	420
		16.9.3	Improving Breakdown Properties	420
	16.10	Conclusion		421
	References			421

Chapter 17 Azobenzene Polymers as Photomechanical and Multifunctional Smart Materials
K.G. Yager and C.J. Barrett

	17.1	Introduction		424
	17.2	Azobenzenes		425
	17.3	Azobenzene Systems		427
	17.4	Photoswitchable Azo Materials		430
	17.5	Photoresponsive Azo Materials		432
		17.5.1	Photo-orientation	432
		17.5.2	Surface Properties	434
	17.6	Photodeformable Azo Materials		434
		17.6.1	Surface Mass Transport	434
		17.6.2	Photomechanical Effects	437
	17.7	Conclusion		437
	References			438

Chapter 18 Intelligent Chitosan-based Hydrogels as Multifunctional Materials
A.F.T. Mak and S. Sun

	18.1	Introduction		447
	18.2	Characteristics of Chitosan		448
		18.2.1	Physical and Chemical Properties of Chitosan	448
		18.2.2	Biological Properties of Chitosan	449
		18.2.3	Solvent and Solubility	449
	18.3	Intelligent Properties		450
		18.3.1	pH Sensitivity	450
		18.3.2	Ionic Strength Sensitivity	452
		18.3.3	Organic Effectors Sensitivity	453
		18.3.4	Electrosensitivity	453
		18.3.5	Thermosensitivity	455

18.4	Chitosan-based Intelligent Materials		456
	18.4.1	pH-Responsive Hydrogels	456
	18.4.2	Thermoresponsive and Dual Stimuli-responsive Polymers	456
	18.4.3	Magnetic Chitosan Microsphere	457
	18.4.4	Electrical Responsive Polymers	458
18.5	Biomedical Applications		458
	18.5.1	Drug-delivery and Drug-release Systems	458
	18.5.2	Injectable Gels for Tissue Engineering	460
	18.5.3	Artificial Actuators and Muscles	460
18.6	Conclusions		461
References			461

Chapter 19   Polymer-Protein Complexation and its Application as ATP-driven Gel Machine
R. Kawamura, A. Kakugo, Y. Osada and J.P. Gong

19.1	Introduction	464
19.2	Actin Gel formed from Polymer–Actin Complexes	465
19.3	Polymorphism of Actin Complexes	467
19.4	Oriented Myosin Gel Formed under Shear Flow	469
19.5	Motility Assay of F-actin on Oriented Myosin Gel	470
19.6	Motility Assay of Polymer-Actin Complex Gel	471
19.7	Polarity of the Actin in Complexes	472
19.8	Conclusions	474
References		475

Chapter 20   Intelligent Composite Materials Having Capabilities of Sensing, Health Monitoring, Actuation, Self-Repair and Multifunctionality
H. Asanuma

20.1	Introduction	478
20.2	A New Route to Develop Intelligent Composite Materials	479
20.3	Composite Materials Fabricated by the New Route	481
20.4	A New Category of Composite Materials Having Liquid Phases for Self-repair and Other Capabilities	485
20.5	Summary and Outlook	489
References		490

Chapter 21 Overview of Liquid-crystal Elastomers, Magnetic Shape-memory Materials, Fullerenes, Carbon Nanotubes, Nonionic Smart Polymers and Electrorheological Fluids as Other Intelligent and Multifunctional Materials
M. Shahinpoor and H.-J. Schneider

21.1	Liquid-crystal Elastomers as Multifunctional Materials	491
21.2	Magnetic Shape-memory (MSM) Materials	493
	21.2.1 MSM Alloy Actuators	496
	21.2.2 Sensing and Multifunctionality Properties of MSM Materials	496
21.3	Fullerenes and Carbon Nanotubes as Multifunctional Intelligent Materials	497
21.4	Nonionic Polymer Gels/EAPs	500
21.5	Electrorheological (ER) Fluids as Multifunctional Smart Materials	500
	21.5.1 Other Applications of ER Fluids	501
References		501

Chapter 22 Overview on Biogenic and Bioinspired Intelligent Materials – from DNA-based Devices to Biochips and Drug-delivery Systems
H.-J. Schneider

22.1	Introduction	506
22.2	Biological Materials: Nucleic Acids as an Example	507
22.3	Biosensors and Biochips	508
22.4	Intelligent Bionanoparticles	509
22.5	Nanobiosensors	511
22.6	Drug-delivery and Related Systems	512
References		517

Subject Index   522

Introduction

There are two ways to study intelligent materials – the more traditional top-down approach, starting often with available macroscopic materials, and the bottom-up approach, starting from molecules or their assemblies. The first approach is mostly in the hands of engineers and physicists, the second one more in the hands of chemists. The present monograph tries to bridge a gap between these communities, with the aim of a better understanding the almost unlimited opportunities with new smart materials, also for students interested in this necessarily highly interdisciplinary field. Leading experts contributing to this book will illustrate the fundamentals and the present stage of a field, which extends over many areas of science and technology. Difficult and painful decisions were necessary in order to restrict the volume of this book to a manageable size, and to avoid too much overlap with already existing monographs. Certain areas such as liquid crystals, or piezoelectric materials, which are already very much developed, and for which excellent books are available, were deliberately confined to more recent developments, such as the use of liquid-crystal elastomers for actuators. We define intelligent materials as those that are multifunctional due to their unique molecular structure and respond to external stimuli by a characteristic behaviour to the outside world. Thus, we also can restrict the volume to fundamental events, which then can be the basis for new technologies. In this sense, technical devices, which are engineered on the basis of intelligent materials, are not emphasised within this book.

It is hoped that a condensed outline of the essentials of intelligent materials will open the doors also for newcomers to a field that experiences a very rapid development in quite different directions, and holds promise for many possible applications. These involve fields such as medicine, nanoscience and nanotechnology, engineering, biotechnology, pharmaceutical, and food industries, process control, agriculture, as well as new communication and memory devices. Many of such systems are biomimetic, and can be looked at as first steps to make intelligent use of principles invented by nature over billions of years of evolution. At the same time, such smart materials can greatly contribute to improve or repair functions and communication in complex biological systems. We sincerely hope that this book will be a valuable addition to the field of smart/intelligent materials and their applications.

Mohsen Shahinpoor
Hans-Jörg Schneider
Albuquerque/Saarbrücken

CHAPTER 1
Chemically Driven Artificial Molecular Machines

JAMES D. CROWLEY, EUAN R. KAY AND
DAVID A. LEIGH

School of Chemistry, University of Edinburgh, The King's Buildings, West Mains Road, Edinburgh EH9 3JJ, UK

1.1 Design Principles for Molecular-Level Motors and Machines

The widespread use of molecular-level motion in key natural processes suggests that great rewards could come from bridging the gap between the present generation of synthetic molecular systems – which by and large rely upon electronic and chemical effects to carry out their functions – and the machines of the macroscopic world, which utilise the synchronised movements of smaller parts to perform tasks. It is only in the last few years that it has become feasible to design and synthesise molecules in which well-defined large-amplitude or directional stimuli-induced positional changes of submolecular components can occur. Even so, all but the simplest questions remain unanswered. What are the structural features necessary for molecules to use directional displacements to repetitively do work? How can we make a synthetic molecular machine that pumps ions to reverse a concentration gradient, say, or moves itself energetically uphill? How can we make nanoscale structures that traverse a predefined path across a surface or down a track, responding to the nature of their environment to change direction?

Artificial compounds that can do such things have yet to be realised. Synthetic molecular machines remain very much in their infancy in terms of experimental systems and only the most basic types – mechanical switches, memories and slightly more sophisticated, but still rudimentary, motors – have been made thus far.[1–10] Here, we outline the early successes in taming molecular-level movement, the underlying principles that experimental designs must follow, and the progress made towards utilising synthetic molecular structures to perform tasks using mechanical motion. We also highlight some of the issues and challenges that still need to be overcome.

1.1.1 The Effects of Scale

The path towards synthetic molecular machines can be traced back nearly two centuries to the observation of effects that pointed directly to the random motion experienced by all molecular-scale objects. In 1827, the Scottish botanist Robert Brown noted through his microscope the incessant, haphazard motion of tiny particles within translucent pollen grains suspended in water.[11,12] An explanation of the phenomenon – now known as Brownian motion or movement – was provided by Einstein in one[13] of his three celebrated papers of 1905 and was proven experimentally[14] by Perrin over the next decade.[15–17] Scientists have been fascinated by the implications of the stochastic nature of molecular-level motion ever since. The random thermal fluctuations experienced by molecules dominate mechanical behaviour in the molecular world. Even the most efficient nanoscale machines are swamped by its effect. A typical motor protein consumes ATP fuel at a rate of 100–1000 molecules every second, corresponding to a maximum possible power output in the region 10^{-16} to 10^{-17} W per molecule.[18] When compared with the random environmental buffeting of $\sim 10^{-8}$ W experienced by molecules in solution at room temperature, it seems remarkable that *any* form of controlled motion is possible![19]

When designing molecular machines it is important to remember that the presence of Brownian motion is a consequence of scale, not of the nature of the surroundings. It cannot be avoided by putting a molecular-level structure in a near-vacuum for example. Although there would be few random collisions to set such a Brownian particle in motion, equally there would be little viscosity to slow it down. These effects always cancel each other out and as long as a temperature for an object can be defined, it will undergo Brownian motion appropriate to that temperature (which determines the kinetic energy of the particle). In the absence of any other molecules, heat would still be transmitted from the hot walls of the container to the particle by electromagnetic radiation, the random emission and absorption of the photons producing the Brownian motion. In fact, even temperature is not a particularly effective modulator of Brownian motion since the velocity of the particles depends on the square root of the temperature. So to reduce random thermal fluctuations to 10% of the amount present at room temperature, one would have to drop the temperature from 300 K to 3 K.[19,20] It seems sensible, therefore, to try to utilise Brownian motion when designing molecular machines rather than make structures that have to fight against it. Indeed, the question of how to (and whether it is even possible to) harness the inherent random motion present at small length scales to generate motion and do work at larger length scales has vexed scientists for some considerable time.[1]

1.1.2 Machines that Operate at Low Reynolds Number

Whilst rectifying Brownian motion may provide the key to powering molecular-level machines, it tells us nothing about how that power can be used to perform tasks at the nanoscale and what tiny mechanical machines can and cannot be expected to do. The constant presence of Brownian motion is not the only distinction between motion at the molecular level and in the macroscopic world. In the macroscopic world, the equations of motion are governed by inertial terms (dependent on mass). Viscous forces (dependent on particle dimensions) dampen motion by converting kinetic energy into heat, and objects do not move until provided with specific energy to do so. In a macroscopic machine this is often provided through a directional force when work is done to move mechanical components in a particular way. As objects become less massive and smaller in dimension, inertial terms decrease in importance and viscosity begins to dominate. A parameter that quantifies this effect is Reynolds number (R) – essentially the ratio of inertial to viscous forces – given by equation (1.1) for a particle of length dimension a, moving at velocity v, in a medium with viscosity η and density ρ.[21]

$$R = \frac{av\rho}{\eta} \qquad (1.1)$$

Size affects modes of motion long before we reach the nanoscale. Even at the mesoscopic level of bacteria (length dimensions $\sim 10^{-5}$ m), viscous forces dominate. At the molecular level, the Reynolds number is extremely low (except at low pressures in the gas phase or, possibly, in the free volume within rigid frameworks in the solid state) and the result is that molecules, or their components, cannot be given a one-off "push" in the macroscopic sense – momentum is irrelevant. The motion of a molecular-level object is determined entirely by the forces acting on it at that particular instant – whether they be externally applied forces, viscosity or random thermal perturbations and Brownian motion. Since the physics that governs mechanical dynamic processes in the two size regimes is completely different, macroscopic and nanoscale motors require fundamentally different mechanisms for controlled transport or propulsion. Moreover, the high surface area:volume ratios of molecules mean they are inherently sticky and this will have a profound effect on how molecular-sized machines are organised and interact with one another. In general terms, this analysis leads to a central tenet: while the macroscopic machines we encounter in everyday life may provide the inspiration for what we might like molecular machines to achieve, drawing too close an analogy for how they might do it is likely to be a poor design strategy. The "rules of the game" at large and small length scales are simply too different.

1.1.3 Lessons to Learn from Biological Motors and Machines

Help is at hand, however, because despite all these problems we know that motors and machines at the molecular level are conceptually feasible – they are already all around us. Nature has developed a working molecular nanotechnology that it employs to astonishing effect in virtually every significant biological process.[18] Appreciating in general terms how nature has overcome the issues of scale, environment, equilibrium, Brownian motion and viscosity is extremely useful for indicating general design traits for synthetic molecular machine systems and how they might be used.

There are many important differences between biological molecular machines and the man-made machines of the macroscopic world: Biological machines are soft, not rigid; they work at ambient temperatures (heat is dissipated almost instantaneously at small length scales so one cannot exploit temperature gradients); biological motors utilise chemical energy, often in the form of ATP hydrolysis or chemical gradients; they work in solution or at surfaces and operate under conditions of intrinsically high viscosity; they rely on and utilise – rather than oppose – Brownian motion; since their components are constantly in motion, biomolecular machines need to control their directionality of movement not power their movement; the molecular machine and the substrate(s) it is acting upon are kinetically associated during the operation of the machine; biological machines are made by a combination of multiple parallel synthesis and self-assembly; their operation is governed by noncovalent interactions; and, finally, they utilise architectures (*e.g.* tracks) which restrict most of the degrees of freedom of the machine components and/or the substrate(s) it is acting upon.

If biology and physics provide the inspiration and strategies for controlling molecular-level motion, it is through chemistry that artificial molecular-level machine mechanisms must be designed, constructed and made to work. The minimum requirements for such systems must be the restriction of the 3D motion of the machine components and/or the substrate and a change in their relative positions induced by an input of energy. Methods for achieving this are described in the following sections.

1.2 Controlling Motion in Covalently Bonded Molecular Systems

1.2.1 Controlling Conformational Changes

1.2.1.1 Stimuli-induced Conformational Control around a Single Covalent Bond

As a first step towards achieving controlled and externally initiated rotation around C–C single bonds,[22,23] Kelly combined triptycene structures with a molecular recognition event. In the resulting "molecular brake", **1** (Scheme 1.1),[24] free rotation of a triptycyl group is halted by the conformational change brought about by complexation of the appended bipyridyl unit with Hg^{2+} ions – effectively putting a "stick" in the "spokes".

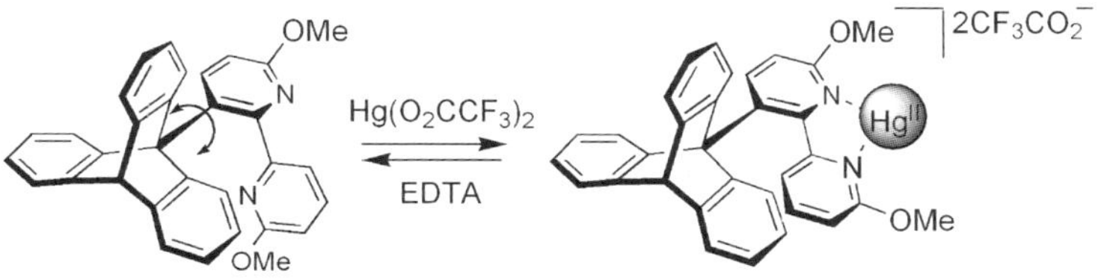

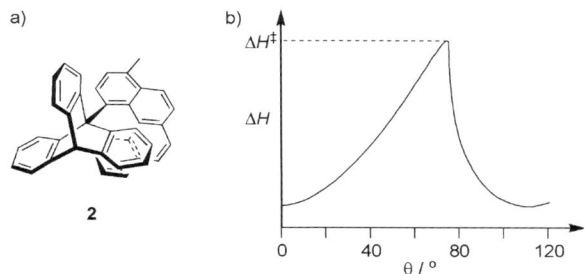

Scheme 1.1 A chemically switchable "molecular brake" induced by metal-ion binding.

Figure 1.1 (a) Kelly's realisation of Feynman's adiabatic ratchet-and-pawl in molecular form, **2**. (b) Schematic representation of the calculated enthalpy changes for rotation around the single degree of internal rotational freedom in **2**.

Kelly and coworkers then extended their investigation of restricted Brownian rotary motion to a molecular realisation (**2**) of the Feynman adiabatic Ratchet-and-Pawl[25,26] in which a helicene plays the role of the pawl in attempting to direct the rotation of the attached triptycene "cog-wheel" in one direction owing to the chiral helical structure. Although the calculated energetics for rotation showed an asymmetric potential energy profile (Figure 1.1(b)), ^{1}H nuclear magnetic resonance (NMR) experiments confirmed that rotation occurred with equal frequency in both directions. This result is, of course, in line with the conclusions of the famous Feynman thought experiment.[27] The rate of a molecular transformation – clockwise and anticlockwise rotation in **2** included – depends on the energy of the transition state (and the temperature), not the shape of the energy barrier – state functions such as enthalpy and free energy do not depend on a system's history.[28] Thus, although rotation in **2** follows an asymmetric potential energy surface, at equilibrium, the principle of detailed balance[29] requires that transitions in each direction occur at equal rates.[30]

The essential element missing from **2** needed to turn the triptycene directionally is some form of energy input to drive it away from equilibrium and break detailed balance. Kelly proposed a modified version of the ratchet structure, **3a** (Scheme 1.2), in which a chemical reaction is used as the source

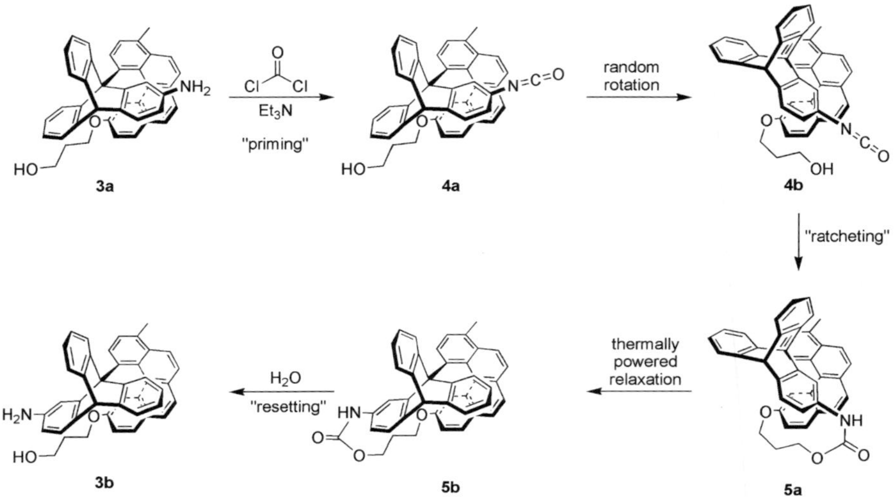

Scheme 1.2 A chemically powered unidirectional rotor. Priming of the rotor in its initial state with phosgene (**3a** → **4a**) allows a chemical reaction to take place when the helicene rotates far enough up its potential well towards the blocking triptycene arm (**4b**). This gives a tethered state, **5a**, for which rotation over the barrier to **5b** is an exergonic process that occurs under thermal activation. Finally, the urethane linker can be cleaved to give the original molecule with the rotor rotated by 120° (**3b**).

of energy.[31,32] Ignoring the amino group, all three energy minima for the position of the helicene with respect to the triptycene "teeth" are identical – the energy profile for 360° rotation would appear as three equal energy minima, separated by equal barriers. As the helicene oscillates back and forth in a trough, however, sometimes it will come close enough to the amine for a chemical reaction to occur (as in **4b**). Priming the system with a chemical "fuel" (phosgene in this case to give the isocyanate **4a**), results in "ratcheting" of the motion some way up the energy barrier (**5a**). Continuation of the rotation in the same direction, over the energy barrier can occur under thermal control and is now an exergonic process (giving **5b**) before cleavage of the urethane gives the 120° rotated system (**3b**). Although the current system can only carry out one third of a full rotation, it demonstrates the principles required for a fully operating and cyclable rotary system under chemical control and represents a major advance in the experimental realisation of molecular-level machines.

Feringa and coworkers have successfully adopted a strategy based on the stereoselective ring-opening of racemic biaryl-lactones using chiral reagents to obtain a full 360° unidirection rotation around a C–C single bond.[33] The process involves four intermediates (**A–D**, Figure 1.2(a)), in each of which rotation around the biaryl bond is restricted: by covalent attachment in **A** and **C**; and through nonbonded interactions in **B** and **D**. Directional rotation to interchange these intermediates requires a stereoselective bond-breaking reaction in steps (i) and (iii) and a regioselective bond-formation reaction in steps

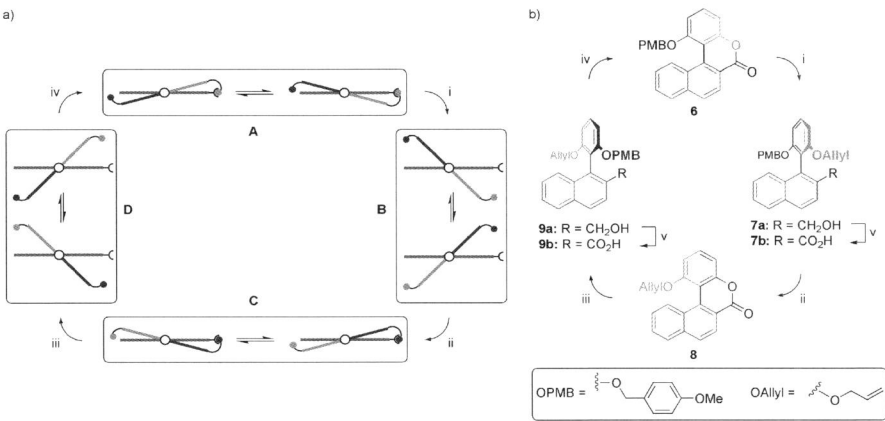

Figure 1.2 (a) Schematic representation of unidirectional rotation around a single bond, through four states. In states **A** and **C** rotation is restricted by a covalent linkage, but the allowed motion results in helix inversion. In states **B** and **D** rotation is restricted by nonbonded interactions between the two halves of the system. These forms are configurationally stable. The rotation relies on stereospecific cleavage of the covalent linkages in steps (i) and (iii), then regiospecific formation of covalent linkages in steps (ii) and (iv). (b) Structure and chemical transformations of a unidirectional rotor. Reactions: i) Stereoselective reduction with (*S*)-CBS then allyl protection. ii) Chemoselective PMB removal resulting in spontaneous lactonisation. iii) Stereoselective reduction with (*S*)-CBS then PMB protection. iv) Chemoselective allyl removal resulting in spontaneous lactonisation. v) Oxidation to carboxylic acid.

(ii) and (iv). The lactones **6** and **8** (Figure 1.2(b)) exist as racemic mixtures due to a low barrier for small amplitude rotations around the aryl–aryl bond. Reductive ring opening with high enantioselectivity is, however, achievable for either lactone, using a homochiral borolidine catalyst and the released phenol can subsequently be orthogonally protected to give **7a** or **9a**. The ring-open compounds are produced in near-enantiopure form in a process that involves directional rotation of 90° around the biaryl bond, governed by the chirality of the catalyst, and powered by consumption of borane. The *ortho*-substitution of these species results in a high barrier to axial rotation. Oxidation of the benzylic alcohol (**7a** → **7b** or **9a** → **9b**) primes the motor for the next rotational step. Selective removal of one of the protecting groups on the enantiotopic phenols results in spontaneous lactonisation when thermally driven axial rotation brings the two reactive groups together – again probably a net directional process owing to the steric hindrance of the *ortho*-substituents (although this is not demonstrated because the chirality is destroyed in this step). Figure 1.2 illustrates the unidirectional process achieved using the (*S*)-CBS catalyst, rotation in the opposite sense can be achieved by employing the opposite borolidine enantiomer and swapping the order of phenol protection and deprotection steps.

1.2.1.2 Stimuli-induced Conformational Control in Organometallic Systems

Controlling the facile rotary motion of ligands in metal sandwich or double-decker complexes is conceptually similar to controlling rotation around covalent single bonds and stimuli-induced control in such metal complexes has also been demonstrated. Ferrocene has long been known to exhibit a low barrier to Cp-ring rotation.[34] Recently, a gas-phase experimental and theoretical study of dianionic 10^{2-} and anionic $[10 \cdot H]^-$ indicated that a two-state rotary switch could be operated by monoprotonation and deprotonation (Scheme 1.3).[35] The conformation of the dianion is governed by coulombic repulsions, while an intramolecular hydrogen bond stabilises the observed conformation in the monoanion. In a related example Crowley et al.[36] have shown that 1,1′-di-(3-pyridyl)ferrocene, 11, adopts an eclipsed, π-stacked rotameric conformation both in solution and in the solid state. Quaternisation of both of the pyridine substituents, either by protonation $[11 \cdot 2H]^{2+}$ or by alkylation, results in rotation away from the fully π-stacked conformation due to the coulombic repulsion (Scheme 1.3).

In contrast to these simpler ferrocene examples, metallacarboranes tend to have rather high barriers to rotation around the metal–ligand axis. Dicarbollide ligand 12^{2-} features two adjacent carbon atoms on its bonding face that imparts a dipole moment perpendicular to the metal–ligand axis. In order to minimise electrostatic repulsion, transoid complexes are normally preferred.[37] Two exceptions however are the nickel(IV) or palladium(IV) complexes, which are cisoid.[38,39] Complex Ni · 12_2 can therefore operate as a reversible rotary switch on manipulation of the oxidation state or by photoexcitation (Figure 1.3).[40]

1.2.1.3 Stimuli-induced Conformational Control Around Several Covalent Bonds

With an external stimulus, it is possible to control conformation around not just one bond, but several all at once. Conformational control in biopolymers is of great importance in structural biology and a vast array of methods for artificially triggering, altering and reversing folding in polypeptides and

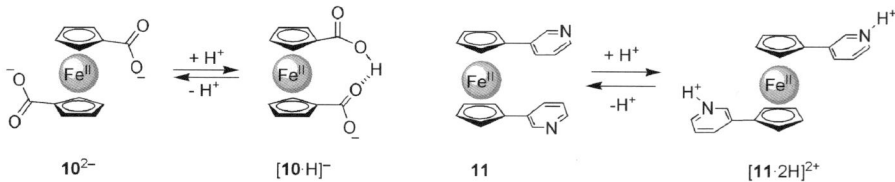

Scheme 1.3 Proton-switched intramolecular rotation in ferrocene derivatives. The amplitude of relative rotary motion of the two cyclopentadienyl ligands in **10** is ∼112° and in **11** is ∼37°.

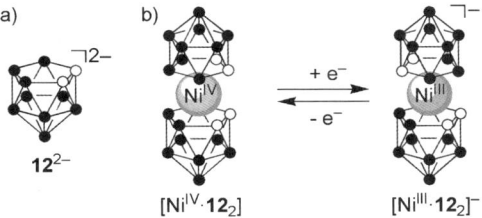

Figure 1.3 (a) Dicarbollide ligand **12**$^{2-}$. (b) Metallacarborane Ni · **12**$_2$ as an electrochemically controlled rotary switch. The cisoid–transoid exchange involves a rotation of 144°. Black spheres represent boron atoms, white spheres carbon atoms.

polysaccharides now exists.[41–43] In nature, as in artificial systems, host–guest binding is often used to trigger complex conformational changes in one or both of the interacting species. In some cases this involves no more than minor bond deformations or the restriction of a few rotational degrees of freedom. In others, molecular recognition results in significant changes to a single degree of freedom yet binding of one chemical species to another can also often result in significant changes to conformation and internal motion, affecting several bonds. Such "induced fit" mechanisms are the basis of many allosteric systems, whereby recognition of one species affects the binding properties or enzymatic activity at a remote site in the same molecule. Allosterism, cooperativity and feedback are central to many functional biological systems[44–46] and synthetic approaches towards achieving similar effects have been extensively reviewed elsewhere.[47,48]

1.2.2 Controlling Configurational Changes

Changes in configuration – in particular *cis–trans* isomerisation of double bonds – have been widely studied from theoretical, chemical and biological perspectives.[49,50] Although the small-amplitude motion involved is not generally sufficient for direct machine-like exploitation, it can provide an extremely useful photoswitchable control mechanism for more complex systems (*vide infra*) and, in some instances, can even be harnessed to perform a significant mechanical task. Such systems represent some of the first examples of molecular-level motion that can be controlled by the application of an external stimulus.

In small-molecule systems, converting a configurational change around a double bond into significant mechanical motion requires considerable ingenuity. Combining the reversible photoisomerisation of an azobenzene with the "molecular bearing" attributes of metallocene complexes, Aida has created a pair of "molecular scissors" **13** (Scheme 1.4).[51] The optically triggered change in double-bond configuration brings about an angular change of position about the ferrocene "pivot" and results in opening and closing of the phenyl "blades".

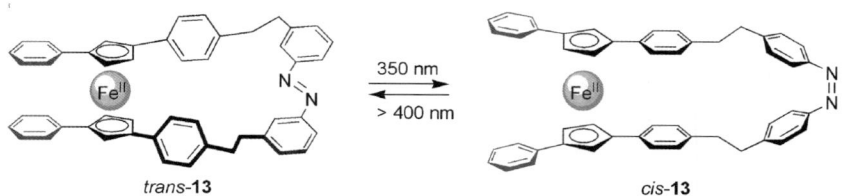

Scheme 1.4 "Molecular scissors" **13**, in which photoisomerisation of an azobenzene is converted into a 49° rotary motion around a metallocene bearing.

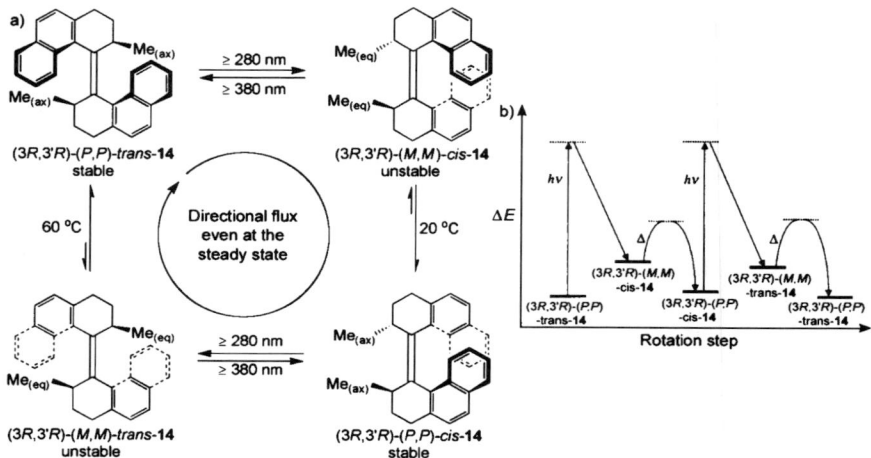

Scheme 1.5 Operational sequence (a) and potential energy profile (b) of the first, continuously operating, unidirectional 360° rotor, (3R,3R)-**14**. Note, the labels "stable" and "unstable" in (a) refer to thermodynamic stability and the reaction coordinate in (b) does not correspond directly to the angle of rotation around the central bond.

Reversible switching is possible over a number of cycles, altering the bite angle between the blades from ~9° when closed to >58° when open.

The chiral helicity in molecules such as **14** causes the photochemically induced *trans–cis* isomerisation to occur unidirectionally according to the handedness of the helix. Taking advantage of this, Feringa developed the first synthetic molecular rotor capable of achieving a full and repetitive 360° unidirectional rotation (Scheme 1.5(a)).[52–56] Irradiation ($\lambda > 280$ nm) of this extraordinary molecule, (3R,3'R)-(P,P)-*trans*-**14**, causes chiral helicity-directed clockwise rotation of the upper half relative to the lower portion (as drawn), at the same time switching configuration of the double bond and inverting the helicity to give (3R,3'R)-(M,M)-*cis*-**14**. However, this form is not stable at temperatures above –55°C as the cyclohexyl ring methyl substituents are placed in unfavourable equatorial positions. At ambient temperatures, therefore, the system relaxes via a second, thermally activated helix inversion to give (3R,3'R)-(P,P)-*cis*-**14**, the substituents at either end of the double bond continuing to rotate in the same direction with respect to each other. Irradiation of (3R,3'R)-(P,P)-cis-**14** ($\lambda > 280$ nm) results in a second photoisomerisation and helix inversion to give (3R,3'R)-(M,M)-trans-**14**. The methyl substituents are placed in an energetically unfavourable position by the photoisomerisation reaction and thermal relaxation (temperatures $>60°C$ are required) completes the 360° rotation of the olefin substituents, regenerating the starting species (3R,3'R)-(P,P)-trans-**14**. The four states can be differently populated depending on the precise choice of wavelength and temperatures used, while irradiation at 280 nm at $>60°C$ results in continuous 360° rotation.[52]

Note that if **14** is irradiated at >280 nm and at $>60°C$ the system will reach a steady state at which point the bulk distribution of diastereomers no longer changes. Crucially, however, this steady state is NOT an adiabatic equilibrium. The steady state is maintained by a process (Scheme 1.5) that features nonzero fluxes (corresponding to directional rotation of one component with respect to the other) between various pairs of diastereomers. In other words, as long as an external source of light and heat is supplied, the steady state in Scheme 1.5 is effectively maintained by a cyclic process A → B → C → D → A, with an arrow indicating a net flux. Thus, even at the steady state, **14**, and other molecules of this type, behave as directional rotary motors.

Because the photoisomerisation process in such systems is extremely fast (<300 ps),[57] the rate-limiting step in the operation of **14** is the slowest of the thermally activated isomerisations. The effect of molecular structure on this rate has been investigated in a series of second-generation motors (**15–17**, Figure 1.4).[58–62] In all the cases shown, the slowest thermal isomerisation step has a lower kinetic barrier than in **14**. However, not all the kinetic effects of the structural changes turned out to be intuitive, it appears that electronic effects also play an important role in the mechanism of these systems.[62] This was empasised by derivative **17** in which an amine in the upper half of the molecule is directly conjugated with a ketone in the lower half. The consequent increase in single-bond character of the central olefin results in greatly increased rates for the thermal isomerisation steps. Fast rotation rates have also been achieved

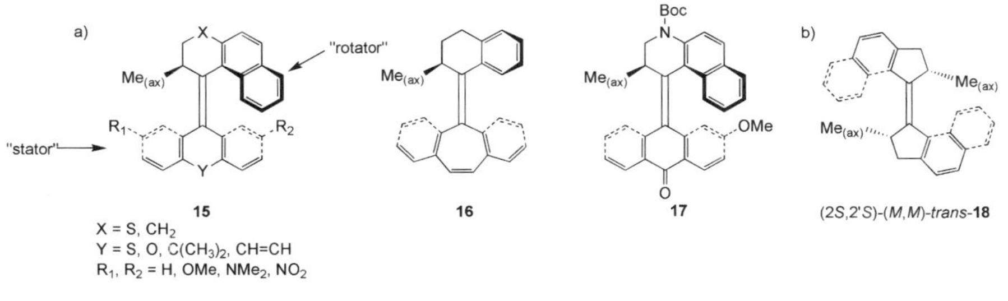

Figure 1.4 (a) General structure of second-generation light-driven unidirectional rotors **15**, **16** and the fastest member of this series, **17**. (b) Third-generation unidirectional rotor, **18**. In this case "ax" refers to the pseudoaxial orientation for substituents on the cyclopentyl ring.

with cyclopentane analogue **18** that can be accessed in higher synthetic yields than the 6-membered ring compounds.[63–65] Despite the greater conformational flexibility of the 5-membered ring, a significant energy difference between pseudoequatorial and pseudoaxial positions of the appended methyl groups still exists and unidirectional rotation occurs.

1.3 Controlling Motion in Mechanically Bonded Molecular Systems

1.3.1 Basic Features

Catenanes are chemical structures in which two or more macrocycles are interlocked, while in rotaxanes one or more macrocycles are mechanically prevented from dethreading from a linear unit by bulky "stoppers" (Figure 1.5).[66–70] Even though their components are not covalently connected, catenanes and rotaxanes are molecules – not supramolecular complexes – as covalent bonds must be broken in order to separate the constituent parts. In these structures, the mechanical bond severely restricts the relative degrees of freedom of the components in several directions, while often permitting extraordinarily large amplitude motion in an allowed vector. This is in many ways analogous to the restriction of movement imposed on biological motors by a track[18] and is one reason why interlocked structures continue to play a central role in the development of synthetic molecular machines.[71–92]

The large-amplitude submolecular motions particular to catenanes and rotaxanes can be divided into two classes (Figure 1.5): pirouetting of the macrocycle around the thread (rotaxanes) or the other ring (catenanes) and translation of the macrocycle along the thread (rotaxanes) or around the other ring (catenanes). By analogy to the stereochemical term "conformation", which refers to geometries that can formally be interconverted by rotating about covalent bonds, the relative positioning of the components in interlocked molecules (and supramolecular complexes) is often referred to as a "co-conformation".[93]

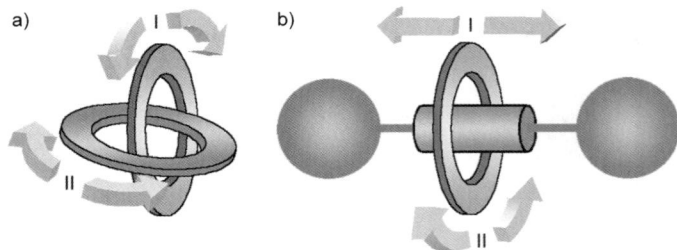

Figure 1.5 Schematic representations of (a) a [2]catenane and (b) a [2]rotaxane. Arrows show possible large-amplitude modes of movement for one component relative to another.

1.3.2 Translational Molecular Switches: Stimuli-Responsive Molecular Shuttles

The rate at which the macrocycle moves between two binding sites (stations) by Brownian motion in molecular shuttles can be regulated by temperature and, in some cases, solvent composition or binding events. However, the Principle of Detailed Balance[29] tells us that no useful task can be performed by such a process, even if the two stations are different. A more important kind of control is one in which the net location of the macrocycle is changed in response to an applied stimulus. This breaks detailed balance and, as the system relaxes back to equilibrium, the biased Brownian motion of the macrocycle can be used to perform a mechanical task.

1.3.2.1 Adding and Removing Protons to Induce Net Positional Change

The first bistable switchable molecular shuttle, reported by Stoddart and Kaifer in 1994 and arguably the first true example of a molecular Brownian motion machine, employed a two-station design (Scheme 1.6).[94] The biphenol and benzidine units in the thread of [2]rotaxane **19**$^{4+}$ are both potential π-electron donor stations for the CBPQT^{4+} cyclophane and at room temperature rapid shuttling of the macrocycle occurs as in related degenerate shuttles. Cooling to 229 K allowed observation (by NMR and UV-visible absorption spectroscopy) of the two translational isomers in a ratio of 21:4 in favour of encapsulation of the benzidine station. Protonation of the benzidine residue with CF$_3$CO$_2$D

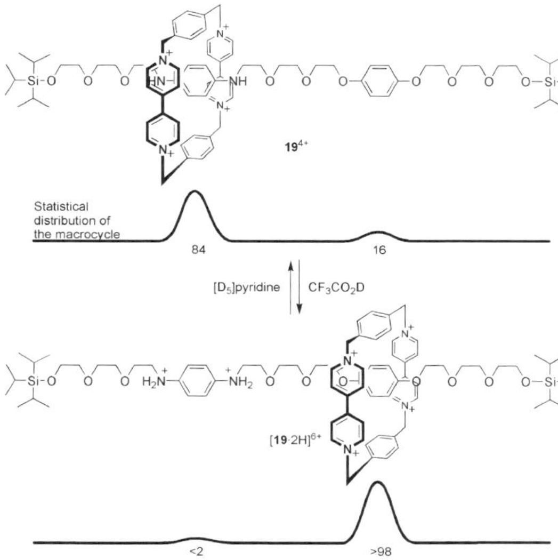

Scheme 1.6 The first switchable molecular shuttle **19**$^{4+}$.

destabilises its interaction with the macrocycle so that it now resides overwhelmingly on the biphenol station. The system can be restored to its original state by neutralisation with [D$_5$]pyridine.

Although this system is a controllable molecular shuttle, it exhibits modest positional integrity in the nonprotonated state – a much larger binding energy difference between the two stations is required. Inspired by the complexation of ammonium ions by crown ethers, a new series of rotaxanes based on threads containing secondary alkylammonium species and crown ether macrocycles was developed.[95–97] [2]Rotaxane [**20** · H]$^{3+}$ (Scheme 1.7(a)) was the first switchable molecular shuttle reported using these units.[98–100] The dibenzo-crown ether is bound to the ammonium cation by means of [N$^+$–H···O] hydrogen bonds as well as weaker [C–H···O] bonds from methylene groups in the α-position to the nitrogen. ^{1}H NMR spectroscopy at room temperature in [D$_6$]acetone shows the crown ether to be sitting mainly (but not exclusively) over the cation. Deprotonation of the ammonium centre with diisopropylethylamine turns off the interactions holding the macrocycle to this station so it resides on the alternative bipyridinium station (**20**$^{2+}$).

Stoddart and Balzani have combined three of these shuttle units in parallel to generate a so called "molecular elevator".[101,102] The "elevator" molecule, [**21** · 3H]$^{9+}$, consists of three thread-like components connected at one end and encircled by three catechol-polyether macrocycles connected in a platform-like fashion (Scheme 1.7(b)). Essentially, the shuttling action works as in **20**, just over three equivalents of base are required to move the platform from

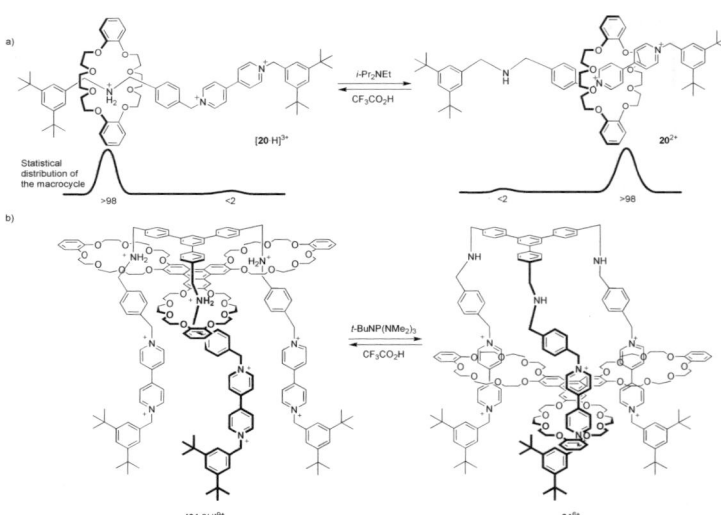

Scheme 1.7 (a) A pH-responsive bistable molecular shuttle, **20** displaying good positional integrity in both chemical states. (b) Chemical structure of a "molecular elevator" [**21** · 3H]$^{9+}$/**21**$^{6+}$.

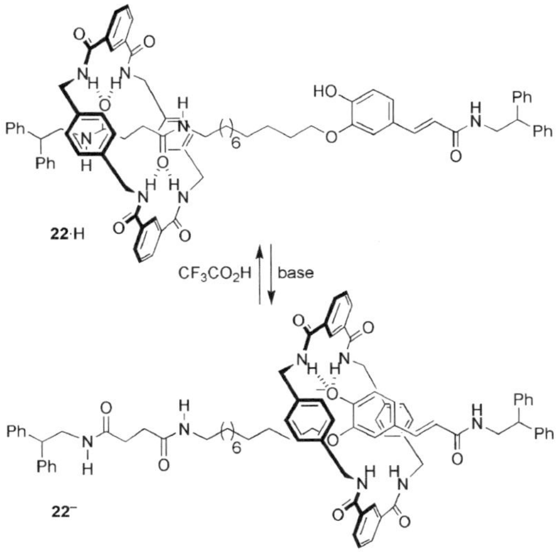

Scheme 1.8 pH-switched anion-induced shuttling in hydrogen bonded [2]rotaxane, **22**, in [D$_7$]DMF at rt. Bases that are effective include LiOH, NaOH, KOH, CsOH, Bu$_4$NOH, tBuOK, DBU, Schwesinger's phosphazine P$_1$ base, illustrating the lack of influence on shuttling of the accompanying cation.

ammonium ("top") to bipyridinium ("bottom") positions. Titration experiments and molecular modeling indicate that the shuttling motion proceeds in a stepwise fashion with each polyether macrocycle stepping individually from station to station. By combining the three shuttles in parallel excellent positional integrity is obtained in both fully protonated and fully deprotonated states.

The first pH-switchable shuttle to exploit anion hydrogen bonding interactions was recently reported.[103] In [2]rotaxane **22·H** (Scheme 1.8), formation of the benzylic amide macrocycle is templated by a succinamide station in the thread. However, the thread also contains a cinnamate derivative. In the neutral form the macrocycle resides on the succinamide station >95% of the time because the phenol is a poor hydrogen-bonding group. Deprotonation in [D$_7$]DMF to give **22$^-$** results in the macrocycle changing position to bind to the strongly hydrogen bond-basic phenolate anion. Reprotonation returns the system to its original state and the macrocycle to its original position.

1.3.2.2 Adding and Removing Electrons to Induce Net Positional Change

In the original switchable shuttle **19^{4+}** (Scheme 1.6), the change of position could also be achieved in a reagent-free manner by electrochemical oxidation of the benzidine station – shuttling away from this station occurring after

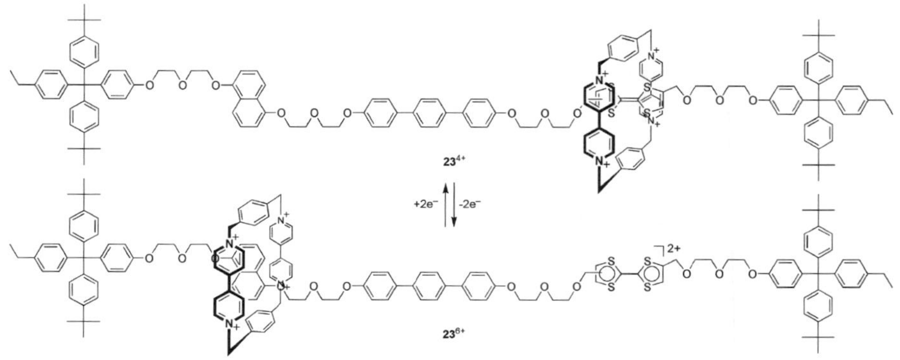

Scheme 1.9 Redox-switchable molecular shuttle **23**. The redox reactions may be carried out either chemically or electrochemically.

oxidation to the radical cation. A number of attempts to create related redox-switched shuttles with greater positional integrity failed to give an improved distribution of translational isomers in the ground state.[104–108] Incorporation of redox-active stations based on tetrathiafulvalene (TTF), however, has given rise to a whole series of redox-active shuttles, a typical example of which is illustrated in Scheme 1.9.[109] In the ground state of **23**$^{4+}$ the tetracationic cyclophane mostly sits over the more electron-rich TTF station. Oxidation (via chemical or electrochemical means) of the TTF to either its radical cation or dication (**23**$^{5+}$ or **23**$^{6+}$) results in a shift of the cyclophane to the dihydroxy-naphthalene unit (DNP). This switching of the CBPQT^{4+} cyclophane between TTF (or closely related derivatives) and dioxyarene units has since been extensively studied in both rotaxane and catenane architectures.[110–117]

One of the highest positional fidelity electrochemical switches (∼ 10^6:1 in one state; ∼ 1:500 in the other) has been demonstrated with amide-based molecular shuttles.[118,119] [2]Rotaxane **24** contains two potential hydrogen-bonding stations for the benzylic amide macrocycle – a succinamide (*succ*) station and a redox-active 3,6-di-*tert*-butyl-1,8-naphthalimide (*ni*) station – separated by a C$_{12}$ aliphatic spacer (Scheme 1.10).[118,119] While the ability of the *succ* station to template formation of the macrocycle is well established, the neutral naphthalimide moiety is a poor hydrogen-bond acceptor. In fact, the difference in macrocycle binding affinities is so great that *succ*-**24** is the only translational isomer detectable by ^{1}H NMR in CDCl$_3$, [D$_3$]acetonitrile and [D$_8$]THF, while even in the strongly hydrogen-bond-disrupting [D$_6$]DMSO, the macrocycle resides over the *succ* station about half of the time. One-electron reduction of naphthalimide to the corresponding radical anion, however, results in a substantial increase in electron charge density on the imide carbonyls and a concomitant increase in hydrogen-bond accepting ability. In **24**, this reverses the relative hydrogen bonding abilities of the two thread stations so that co-conformation *ni*-**24**$^{-\bullet}$ is preferred. Subsequent reoxidation to the neutral state restores the original binding affinities and the shuttle returns to its initial state.

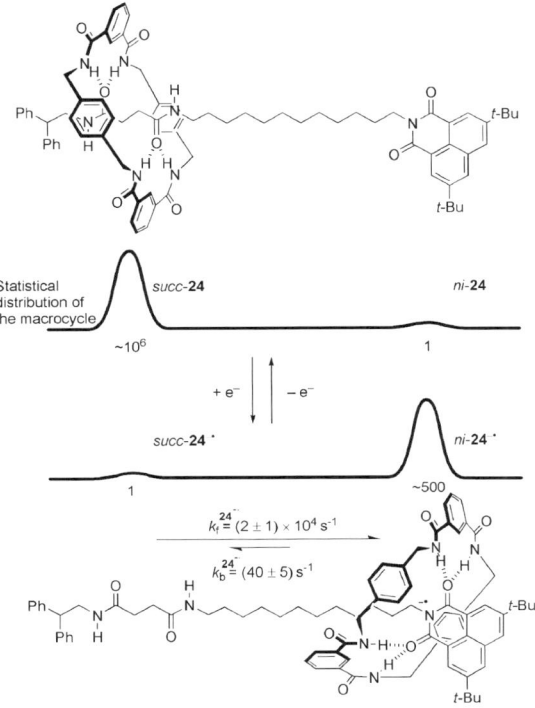

Scheme 1.10 A photochemically and electrochemically switchable, hydrogen-bonded molecular shuttle, **24**. In the neutral state, the translational co-conformation *succ*-**24** is predominant as the *ni* station is a poor hydrogen-bond acceptor ($K_n = (1.2 \pm 1) \times 10^{-6}$). Upon reduction, the equilibrium between *succ*-**24**$^{-\cdot}$ and *ni*-**24**$^{-\cdot}$ is altered ($K_{red} = (5 \pm 1) \times 10^2$) because *ni*$^{-\cdot}$ is a powerful hydrogen-bond acceptor and the macrocycle moves through biased Brownian motion. Upon reoxidation, the macrocycle shuttles back to the succinamide station. Repeated reduction and oxidation causes the macrocycle to shuttle forwards and backwards between the two stations. All the values shown refer to cyclic voltammetry experiments in anhydrous THF at 298 K with tetrabutylammonium hexafluorophosphate as the supporting electrolyte. Similar values were determined on photoexcitation and reduction of the ensuing triplet excited state by an external electron donor.

1.3.2.3 Adding and Removing Metal Ions to Induce Net Positional Change

The groups of Sanders and Stoddart have collaborated to produce another class of molecular shuttle that can be switched upon the addition of lithium ions. In the ground state, the co-conformation of [2]rotaxane **25** with the macrocycle sitting over the naphthaldiimide station is dominant, but addition of lithium ions results in the movement of the macrocycle onto the pyromellitic diimide station. Two of the small metal ions can be complexed between the

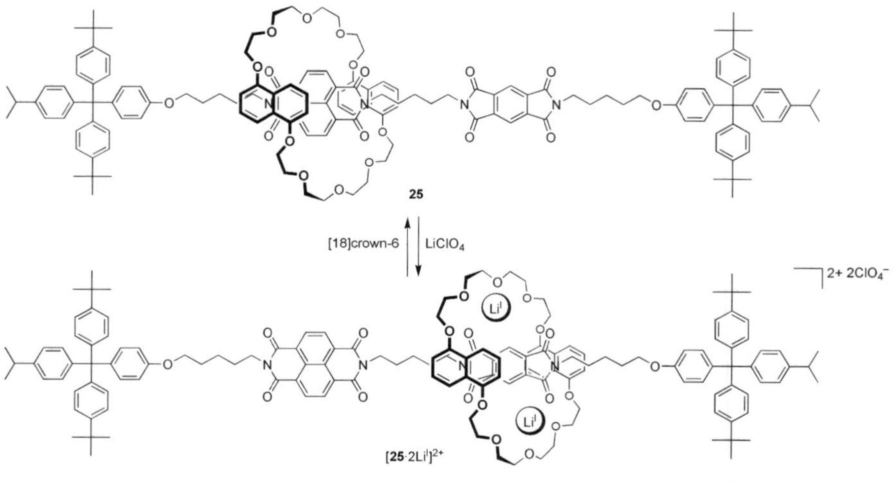

Scheme 1.11 Cation-induced shuttling based on lithium-ion complexation and decomplexation.

crown ether macrocycle and the carbonyls of the diimide moieties and this interaction is significantly stronger at the pyromellitic station (Scheme 1.11). Addition of a large excess of [18]crown-6 sequesters the lithium ions returning the system to its initial state.[120,121]

Rather than serving to enhance intercomponent interactions at a specific station, metal ion binding can be used to destabilise macrocycle binding at one site, causing it to translocate to another, unaffected unit. This is possible through two distinct mechanisms, illustrated in Scheme 1.12.[122,123] In [2]rotaxane **26** the macrocycle preferentially sits over the bis(2-picolyl)amino (BPA)-derivatised glycylglycine station. Addition of one equivalent of $Cd(NO_3)_2 \cdot 4H_2O$ generates a complex in which the metal ion is bound to the first carboxamide carbonyl oxygen as well as the three nitrogen donors of the BPA ligand, and the preferred position of the macrocycle remains essentially unchanged. However, deprotonation of the first carboxamide moiety results in coordination of the nitrogen anion, as well as the carbonyl oxygen of the second carboxamide group. The cadmium ion essentially wraps itself up in the deprotonated glycylglycine residue, switching off any intercomponent interactions with the macrocycle that now occupies the succinic amide ester station. The shuttling process is fully reversible – removal of the Cd^{II} ion with excess cyanide and reprotonation of the amide nitrogen atom with NH_4Cl quantitatively regenerates **26**.[122] In [2]rotaxane **27**, again the macrocycle preferentially resides on the carboxamide-based station adjacent to a BPA ligand. In this instance, divalent metal ions such as Cd^{II} are only able to chelate to the three BPA nitrogen atoms. In order to accommodate such a binding mode, however, the two pyridyl arms must adopt a coplanar conformation, sterically destabilising macrocycle binding to the succinamide station, causing it to move to the inherently weaker succinic amide ester unit. Rather than competing for the

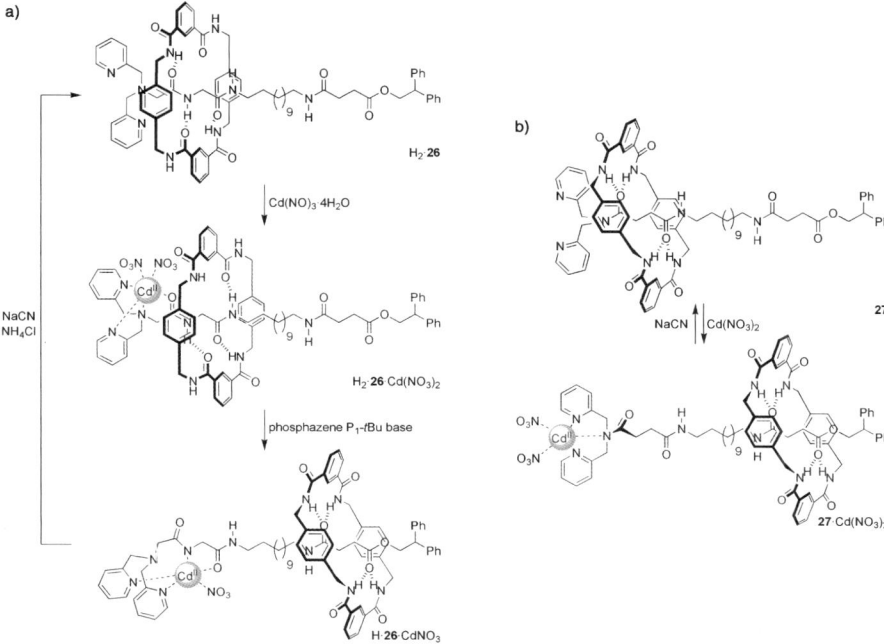

Scheme 1.12 (a) Shuttling through stepwise competitive binding. A similar shuttling mechanism can be observed using Cu^{II} ions, where the deprotonation results in a colour change. (b) An allosterically regulated molecular shuttle.

same donor atoms, as in **26**, the metal- and macrocycle-binding modes in **27** compete for the same 3D space so that this mechanism corresponds to a negative heterotropic allosteric binding event. The shuttling is fully reversible on demetallation of **27** · $Cd(NO_3)_2$ with cyanide.[123]

1.3.2.4 Adding and Removing Covalent Bonds to Induce Net Positional Change

Perhaps surprisingly, the use of covalent-bond-forming reactions to bring about positional change in molecular shuttles has not yet been extensively explored. In one successful example, the formation (and breaking) of C–C bonds through Diels–Alder ("DA") and retro-Diels–Alder ("r-DA") reactions of rotaxane **28** (Scheme 1.13) can control shuttling with excellent positional discrimination – the steric bulk of the DA-adduct displacing the macrocycle to the succinic amide ester station in *Cp*-**28**.[124]

1.3.2.5 Changing Configuration to Induce Net Positional Change

As with controlling motion in covalently bonded systems, isomerisation processes – particularly photoisomerisation processes – are a very attractive means

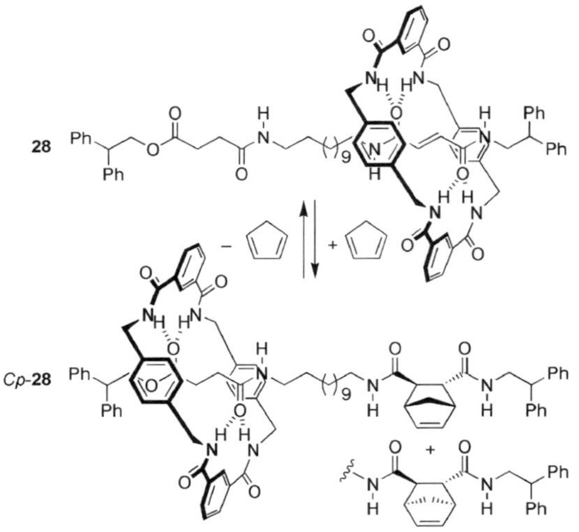

Scheme 1.13 Shuttling through reversible covalent bond formation. Absolute stereochemistry for *Cp*-**28** is depicted arbitrarily.

of inducing shuttling in rotaxanes. Shuttle E/Z-**29** (Scheme 1.14) utilises the interconversion of fumaramide (*trans*) and maleamide (*cis*) isomers of the olefinic unit.[125] Fumaramide groups are excellent binding sites for benzylic amide macrocycles:[126] the *trans*-olefin holds the two strongly hydrogen-bond-accepting amide carbonyls in a preorganised close-to-ideal spatial arrangement for interaction with the amide groups of the macrocycle. Although a similar hydrogen-bonding surface is presented to the host, macrocycle binding to the succinamide station in *E*-**29** would result in a loss of entropy (loss of bond rotation in the succinamide group) as well as one less intracomponent hydrogen bond than would be present in the fumaramide-occupied postional isomer. The result is that only one major positional isomer of *E*-**29** (shown in Scheme 1.14) is observed at room temperature in $CDCl_3$. Photoisomerisation (254 nm) reduces the number of possible intercomponent hydrogen bonds at the olefin station from four to two and so the macrocycle changes position to the succinamide station (*Z*-**29**, Scheme 1.14). Unlike many other light-switchable shuttles, this new state is essentially indefinitely stable until a further stimulus is applied to reisomerise the maleamide unit back to fumaramide.

1.3.2.6 Shuttling via Excited States

When operating in light-stimulated mode, [2]rotaxane **24** employs a highly oxidising photochemically excited state to trigger the electron-transfer process that results in macrocycle shuttling. After about 100 μs, the reduced rotaxane undergoes charge recombination with the radical cation of the external electron

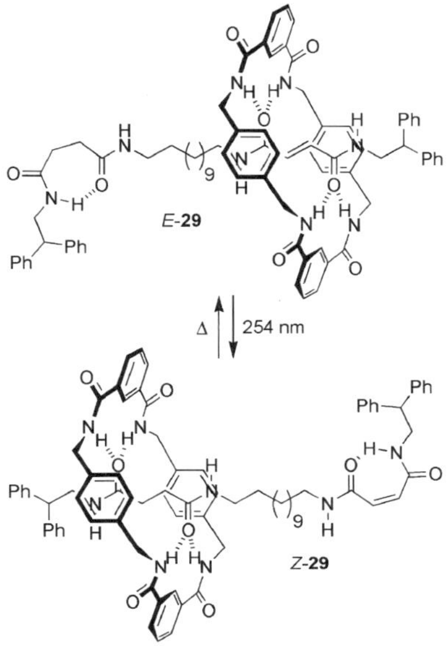

Scheme 1.14 Bistable molecular shuttle *E/Z*-**29** in which self-binding of the low affinity station in each state is a major factor in producing excellent positional discrimination. Similar results are achieved when the intermediate affinity station is a succinic amide ester.

donor, regenerating the starting state. As long as photons of a suitable wavelength are supplied, this process will continue to occur indefinitely – the photochemical process can be said to be autonomous. Attempts to induce shuttling using intramolecular electron transfer from photoexcited states have in the past been plagued by the problem of back electron transfer being fast compared to the desired nuclear motion.[127,128] Recently, however, it has been demonstrated that a previously studied [2]rotaxane, **30**, does indeed undergo shuttling in an intramolecular charge-separated state (Scheme 1.15).[129] Irradiation of the ruthenium trisbipyridine complex generates a highly reducing excited state. An intramolecular electron transfer then occurs between the excited metal centre and the most easily reduced bipyridinium station, on which the macrocycle prefers to sit. The result is destabilisation of the macrocycle–station interactions so that the alternative bipyridinium is now preferred. Remarkably, in this system the back electron-transfer process is slow enough to allow shuttling of the ring towards the other station in ∼10% of the molecules. Naturally, charge recombination quickly restores the initial state, but if photons are continually supplied, this cycle is followed indefinitely.

This shuttle operates without the consumption of chemical fuels or the formation of waste products and it automatically resets so that it operates autonomously. It must be noted, however, that if a continuous supply of photons is provided the distribution of the rings between the two stations

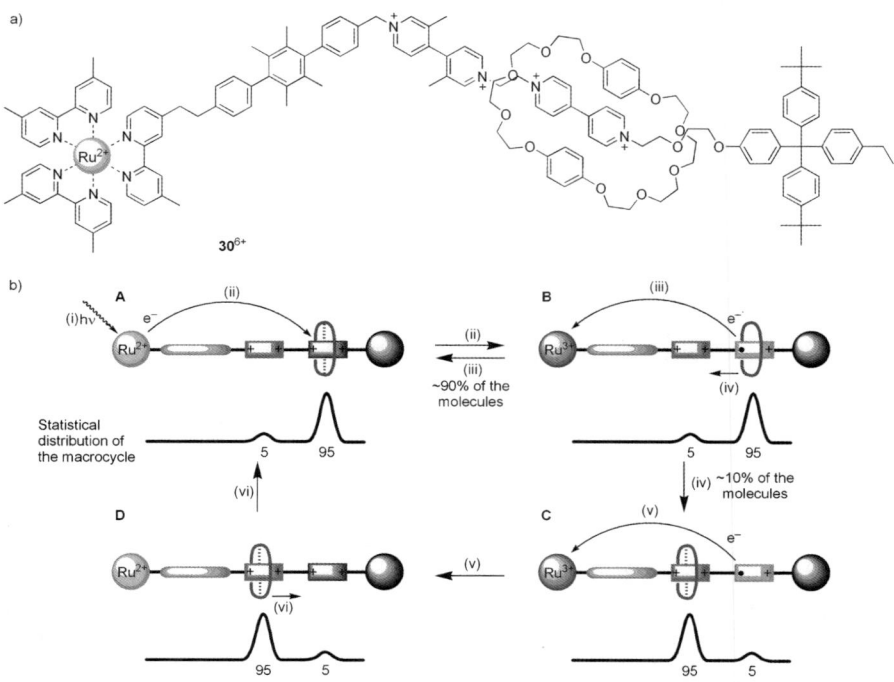

Scheme 1.15 Chemical structure (a) and operating cycle in schematic form (b) of a molecular shuttle **30⁶⁺** operating through a photoinduced internal charge-separated state. (**A**) At equilibrium in the ground state, the ring spends most of the time over the unsubstituted bipyridinium station. Irradiation (i) of the ruthenium complex generates a highly reducing excited state, resulting in electron transfer (ii) to the bipyridinium station, weakening its electrostatic interactions with the ring (**B**). Normally charge recombination processes such as (iii) are fast in comparison with nuclear motions, but here it is slow enough to allow approximately 10% of the molecules to undergo significant Brownian motion (iv) shifting the statistical distribution in this portion of the ensemble to favour the dimethylbipyridinium station (**C**). When charge recombination (v) eventually does take place, the higher binding affinity of the bipyridinium station is restored (**D**). The system relaxes (vi) to restore the original statistical distribution of rings (**A**).

would reach a steady state (the exact ratio depending on the intensity of the light) within a few milliseconds. For an autonomous molecular motor (*e.g.* **14**, Section 1.2.2) a steady state of isomer populations can still correspond to a net flux in a particular direction through these forms. In switches such as these shuttles, however, after the steady state has been reached, there is no subsequent net flux of rings between the two stations at any point in time. The only way to generate net fluxes of macrocycles between the stations in these types of systems would be to rapidly switch the photon source on and off.[130]

1.3.2.7 Entropy-driven Shuttling

Most of the shuttles that exhibit excellent positional discrimination between two stations are switched using stimuli to modify the enthalpy of macrocycle binding to one or both stations. Generally, the effect of temperature is only to alter the degree of discrimination, not to alter the station preference. In [2]rotaxane **31**, however, the macrocycle can be switched between stations simply by changing the temperature.[131] In fact, **31** is a *tristable* molecular shuttle; the ring can be switched between three different positions on the thread (Scheme 1.16).

Structurally, **31** is closely related to **29**, the differences being substitution of the isophthaloyl unit in the macrocycle for pyridine-2,6-dicarbonyl groups and replacement of the succinamide station with a succinic amide ester. In the *E*-**31** form, the macrocycle resides over the fumaramide station in CDCl$_3$ at all temperatures investigated. However, although the ^{1}H NMR spectrum of the maleamide isomer, *Z*-**31**, in CDCl$_3$ showed that the macrocycle was no longer positioned over the olefin station, the spectrum was highly temperature dependent. At elevated temperatures (308 K) the expected *succ-Z*-**31** co-conformation was observed but at lower temperatures it is the alkyl chain that exhibits the spectroscopic shifts indicative of encapsulation by the macrocycle (*dodec-Z*-**31**, Scheme 1.16). The origin of this temperature-switchable effect is the large difference in entropy of binding ($\Delta S_{binding}$) to the succinamide and alkyl chain stations, which allows the $T\Delta S_{binding}$ term to have a significant

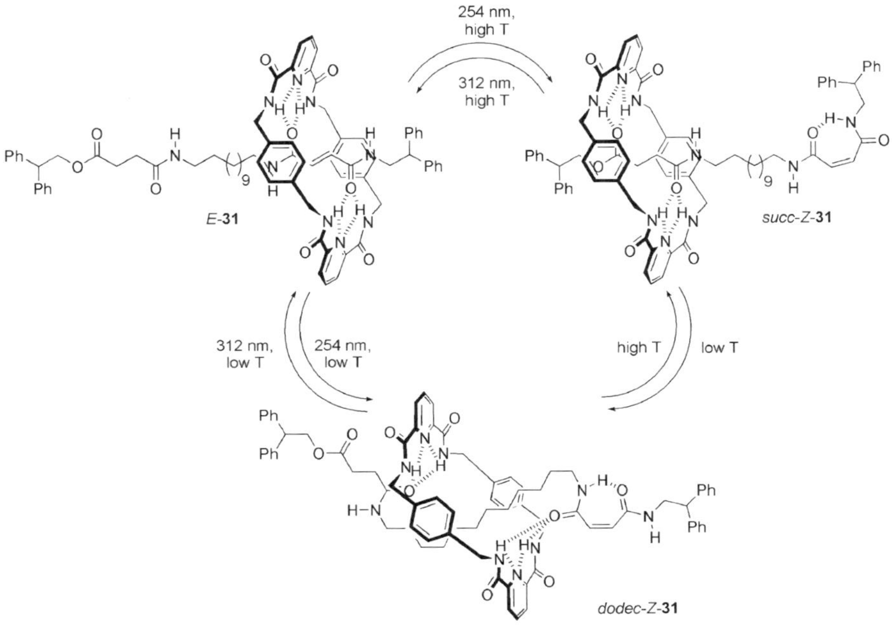

Scheme 1.16 A tristable molecular shuttle **31**.

impact on $\Delta G_{binding}$ as temperature is varied. In the *succ-Z-***31** co-conformation, the macrocycle forms two strong hydrogen bonds with an amide carbonyl and two, significantly weaker, bonds to the ester carbonyl. The *dodec-Z-***31** co-conformation allows four strong hydrogen bonds to amide carbonyls making it enthalpically favoured by $\sim 2\,\text{kcal}\,\text{mol}^{-1}$. At low temperatures, where the effects of entropy are less significant, the molecule adopts the *dodec-Z-***31** co-conformation. At higher temperatures, the increased contribution from the $T\Delta S_{binding}$ term favours the entropically preferred *succ-Z-***31** co-conformation.

1.3.2.8 Shuttling Through a Change in the Nature of the Environment

A significant advantage of using temperature to induce a change in net position of the macrocycle in a molecular shuttle is that no chemical reaction is involved and any property change associated with the new state cannot be due to a change in the covalent structure of the molecule. The same holds true for switching induced by a change in the nature (solvation or polarity) of the environment of the shuttle. The benzylic amide macrocycle in amphiphilic rotaxanes can be shuttled between various hydrogen-bonding stations ($CDCl_3$ and most nonpolar solvents) and an alkyl chain ($[D_6]DMSO$, H_2NCHO).[132,133] A similar effect has been observed in a range of poly(urethane/crown ether rotaxane)s.[134,135]

Many of the examples of stimuli-induced shuttling described in Sections 1.3.2.1–1.3.2.8 display remarkable degrees of control over submolecular fragment positioning and dynamics. They utilise a number of different stimuli-induced processes to trigger changes in the net position of macrocycles over large distances (up to $\sim 15\,\text{Å}$ in the case of **22**, **24** and **29**) and operate over a range of timescales (a complete switching cycle in **24** is over in $\sim 100\,\mu s$ while, in **29**, both states are effectively indefinitely stable). Note, however, that all these shuttles exist as an equilibrium of co-conformation and it is just the position of the equilibrium that is being varied. As the position of equilibrium changes, detailed balance is broken and it is this that can allow a useful mechanical task to be performed by Brownian motion of the macrocycle.

1.3.3 Controlling Rotational Motion in Catenanes

1.3.3.1 Two-way and Three-way Catenane Positional Switches

The fundamental principles for controlling the position of a macrocycle on a thread in a rotaxane and the relative positions and orientations of the rings in a catenane are the same. For example, the behaviour of amphiphilic homocircuit [2]catenane **32** is governed by the same driving forces that cause solvent-induced shuttling in amphiphilic shuttles.[132–133,136] In halogenated solvents such as $CDCl_3$, the two macrocycles of **32** interact through hydrogen bonds (**34**-a, Scheme 1.17, also the structure observed in the solid state) whereby each

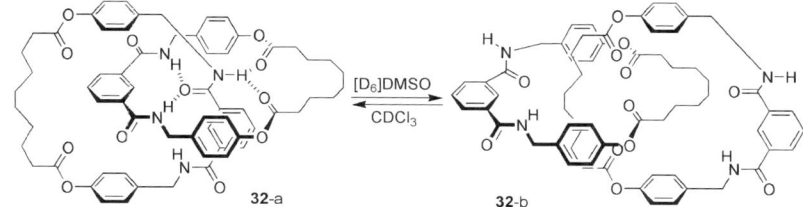

Scheme 1.17 Translational isomerism in an amphiphilic benzylic amide [2]catenane **32**.

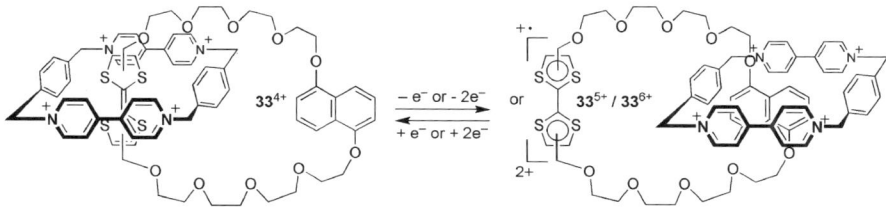

Scheme 1.18 Chemically or electrochemically driven translational isomer switching of [2]catenane **33**.

constitutionally identical ring adopts a different conformation, one effectively acting as the host (convergent H-bond sites) and the other the guest (divergent H-bond sites). In a hydrogen-bond-disrupting solvent such as [D$_6$]DMSO, however, the preferred co-conformation has the amides exposed on the surface where they can interact with the surrounding medium, while the hydrophobic alkyl chains are buried in the middle of the molecule to avoid disrupting the structure of the polar solvent (**32-b**, Scheme 1.17).[136]

In heterocircuit [2]catenanes such as **33**$^{4+}$ (Scheme 1.18) the change of position of one macrocycle with respect to the other is even more reminiscent of shuttling in the previously desribed [2]rotaxanes (such as **23**$^{4+}$, Scheme 1.9).[137] Oxidation (either chemically or electrochemically) of the TTF to either its radical cation or dication (**33**$^{5+}$ or **33**$^{6+}$) results in the CBPQT^{4+} cyclophane switching from the TTF station to the DNP unit. The movement can be reversed by reduction of the TTF unit back to its neutral state. There is, of course, no control over which direction the motion occurs; the cyclophane has a choice of two routes between the stations in either direction, half the molecules will change position in one direction, the other half in the other direction.

In a series of [2]catenates formed around square planar templates, demetallation does not result in a significant co-conformational change.[138] [2]Catenane **34** was generated in this fashion, but it was then found that the templating metal (Pd^{II}) can either be reinserted along with concomitant deprotonation of two amide nitrogens to give the original catenate; or else complexation to only one ring can be achieved, producing a half-turn in the relative orientation of the components.[139] All three forms are fully

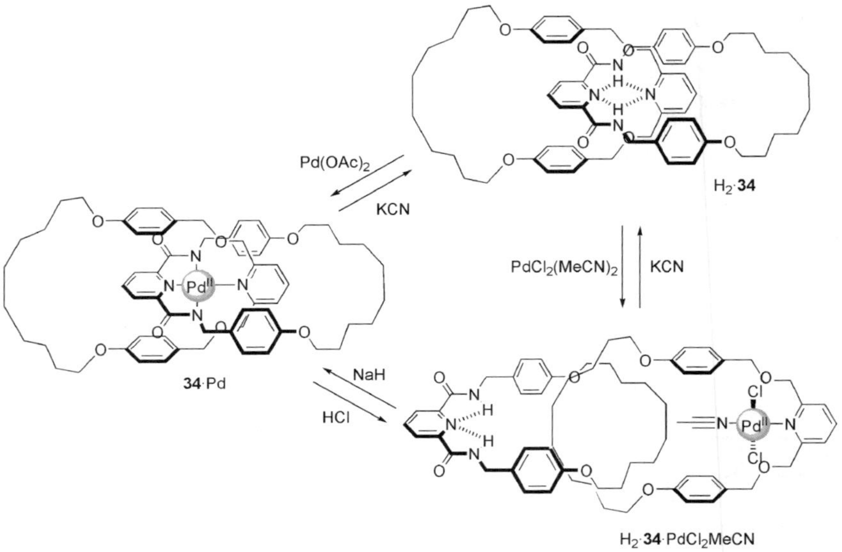

Scheme 1.19 Half-rotation in the [2]catenane **34** via interconvertible Pd^{II} coordination modes.

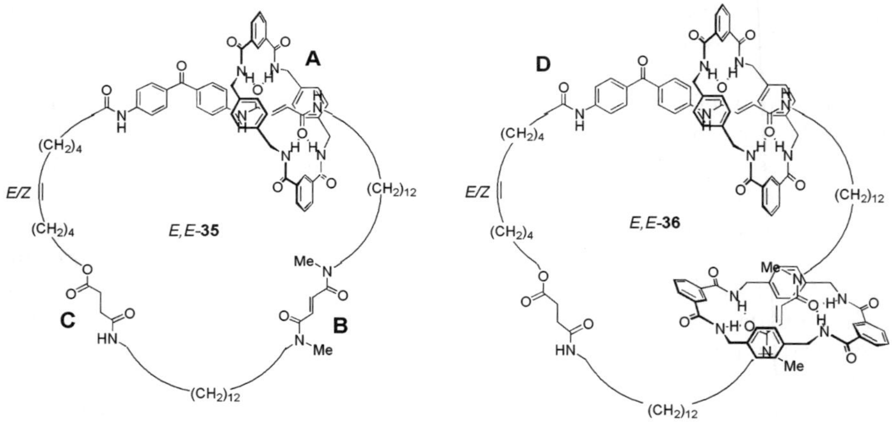

Figure 1.6 [2]Catenane **35** and [3]catenane **36**, shown as their *E,E*-isomers.

interconvertible (Scheme 1.19) and can be observed in both solution and the solid state.

The sequential movement of one macrocycle between three stations on a second ring requires independent switching of the affinities for two of the units so as to change the relative order of binding affinities.[140] In [2]catenane **35** this is achieved (Figure 1.6) by employing two fumaramide stations with differing macrocycle binding affinities (steric mismatching of some tertiary amide rotamers disfavour the methylated station, **B**), one of which (station **A**) is located next to a benzophenone unit. This allows selective photosensitised isomerisation of station **A** at 350 nm, before photoisomerisation of the other fumaramide station (station **B**) at 254 nm. The third station (station **C**) – a succinic amide ester – is not photoactive and is intermediate in macrocycle binding affinity between the two fumaramide stations and their maleamide counterparts. A fourth station, an isolated amide group (shown as **D** in *E,E*-**36**) which can make fewer intercomponent hydrogen-bonding contacts than **A**, **B** or **C**, is also present but only plays a significant role in the behaviour of the [3]catenane, **36**.

Consequently, in the initial state, the small macrocycle resides on the non-methylated fumaramide station, **A**, of [2]catenane **35**. Isomerisation of this station (irradiation at 350 nm) puts the system out of co-conformational equilibrium and the macrocycle changes its net position to the new energy minimum on the station, **B**. Subsequent photoisomerisation of this station (irradiation at 254 nm) displaces the macrocycle to the succinic amide ester unit **C**. Finally, heating the catenane (or treating it with photogenerated bromine radicals or piperidine) results in isomerisation of both the *Z*-olefins back to their *E*-forms so that the original order of binding affinities is restored and the macrocycle returns to its original position, **A**. The ^{1}H NMR spectra of each diastereomer show excellent positional integrity of the small macrocycle at all stages of the process, but the rotation is not directional – over the complete sequence of reactions, an equal number of macrocycles go from **A**, through **B** and **C**, back to **A** again in each direction.

1.3.3.2 Directional Circumrotation: A [3]Catenane Rotary Motor

In order to bias the direction the macrocycle takes from station to station in a catenane such as **35**, kinetic barriers are required to restrict Brownian motion in one direction at each stage and bias the path taken by the macrocycle. Such a situation is intrinsically present in [3]catenane **36** (Figure 1.6).[140] Irradiation of *E,E*-**36** at 350 nm causes counterclockwise (as drawn) rotation of the small macrocycle to the succinic amide ester station to give *Z,E*-**36**. Isomerisation (254 nm) of the remaining fumaramide group causes the other small macrocycle to relocate to the single amide station (*Z,Z*-**36**) and, again, this occurs counterclockwise because the clockwise route is blocked by the other macrocycle. This "follow-the-leader" process, each macrocycle in turn moving and then blocking a direction of passage for the other macrocycle, is repeated throughout the sequence of transformations shown in Scheme 1.20. After three diastereomer interconversions, *E,E*-**36** is again formed but 360° rotation of each of the small rings has not yet occurred, they have only swapped places. Complete unidirectional rotation of both small rings occurs only after the synthetic sequence (i)–(iii) has been completed twice.

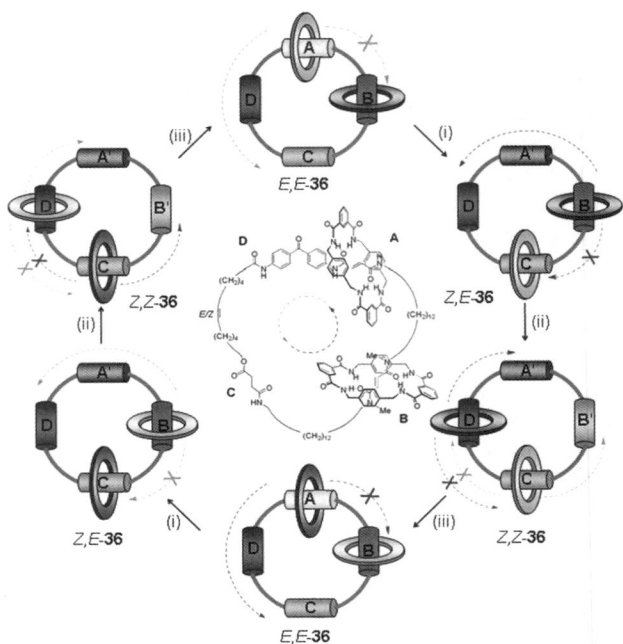

Scheme 1.20 Stimuli-induced unidirectional rotation in a four station [3]catenane, **36**. Conditions: (i) 350 nm; (ii) 254 nm; (iii) Δ; or catalytic ethylenediamine, Δ; or catalytic Br$_2$, 400–670 nm.

1.3.3.3 Selective Rotation in either Direction: A [2]Catenane Reversible Rotary Motor

A reversible molecular motor was realised in chemical terms through the synthesis and operation of catenane *fum-E*-**37** (Scheme 1.21).[141] Net changes in the position or potential energy of the smaller ring were sequentially achieved by: (i) photoisomerisation to maleamide (→*mal-Z*-**37**); (ii) desilylation/resilylation (→*succ-Z*-**37**); (iii) reisomerisation to fumaramide (→*succ-E*-**37**); and finally, (iv) detritylation/retritylation to regenerate *fum-E*-**37**, the whole reaction sequence producing a net clockwise (as drawn in Scheme 1.21) circumrotation of the small ring about the larger one. Exchanging the order of steps (ii) and (iv) produced an equivalent counterclockwise rotation of the small ring.

The simplicity of **37**, together with the minimalist nature of its design allow insight into the fundamental role each part of the structure plays in the operation of the rotary machine. The various chemical transformations perform two different functions: one pair (the linking/unlinking reactions – desilylation/resilylation or detritylation/retritylation) modulates whether the small macrocycles can be exchanged between the two binding sites on the big ring or not (*i.e.* allow the small macrocycle to reach positional equilibrium and become statistically balanced between the two binding sites); the second pair (balance-breaking reactions – either $E \rightarrow Z$ or $Z \rightarrow E$ olefin isomerisations) switch the binding affinity of the olefin station for the small macrocycle either

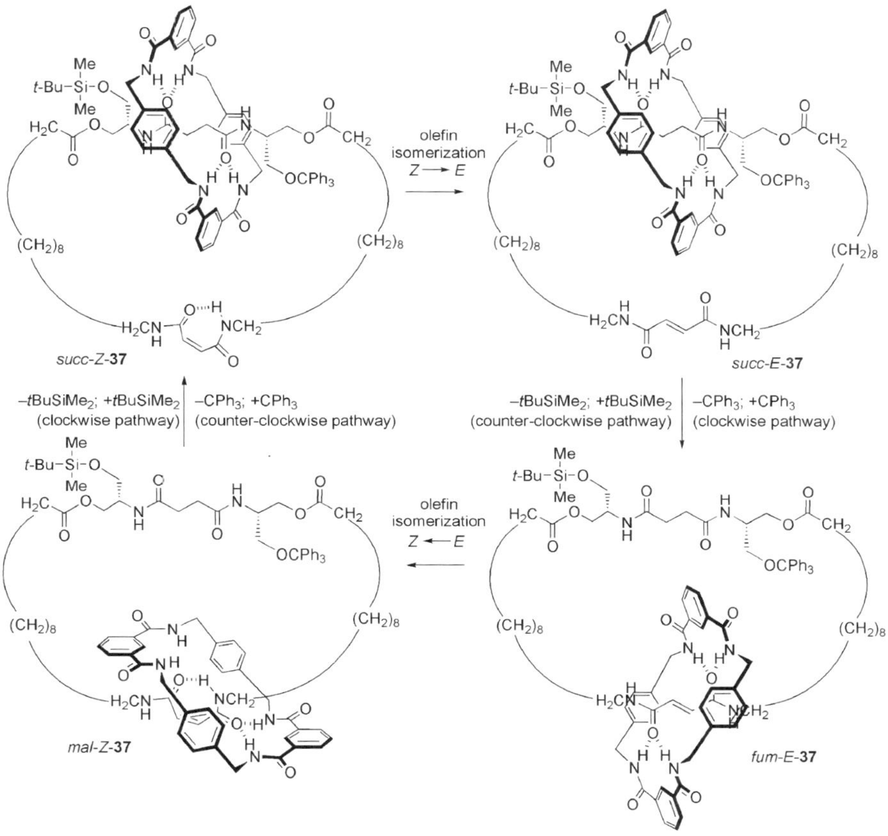

Scheme 1.21 A reversible [2]catenane rotary motor, **37**.

"on" or "off". In other words, the balance-breaking reactions control the *thermodynamics* and *impetus for net transport* by biased Brownian motion; the linking/unlinking reactions control the relative *kinetics* and *ability to exchange*. Raising the kinetic barriers "ratchets" transportation, allowing the statistical balance of the small ring to be subsequently broken without reversing the preceding net transportation sequence. Lowering the kinetic barrier allows "escapement" of a ratcheted quantity of rings in a particular direction.

1.4 From Laboratory to Technology: Towards Useful Molecular Machines

1.4.1 The Current Challenges: Constraining, Communicating, Correlating

Whatever the application or the precise mode of action it is clear any practical device requires that a molecular machine is able to interact with the outside world, either directly through macroscopic or nanoscopic property changes or through further interactions with other molecular-scale devices. The problem of how to "wire" molecular machines together and to the outside also has implications for the physical construction of the devices that, in turn, puts further demands on the fidelity of the molecular-level mechanical processes over a range of conditions. In the following sections, we examine research that addresses these challenges by focusing on using molecular-level motion to produce a functional property change or perform a physical task.

1.4.2 Reporting Controlled Motion in Solution

Simple stimuli-responsive molecular shuttles offer a generic approach to mechanical molecular switches for distance-dependent properties by suitable functionalisation of the macrocycle and one end of the thread (Figure 1.7).[142]

Following the observation that the positioning of an intrinsically achiral benzylic amide macrocycle in relation to a chiral centre can result in an induced circular dichroism (ICD) effect,[143] the chiroptical molecular shuttle E/Z-**38** was prepared (Scheme 1.22).[144] In the E-**38** form, the macrocycle is held over the excellent fumaramide template and thus far from the chiral centre of the peptidic station. Correspondingly the circular dichroism response is zero, as observed for the free thread. In the Z-**38** maleamide isomer, however, the olefin hydrogen-bonding station is "switched off", the macrocycle resides on the chiral peptide station and a strong ($-13\,\mathrm{k\,deg\,cm^2\,dmol^{-1}}$) negative ICD response is observed. Although only a modest difference in photostationary states is achieved by irradiation at 254 nm (photostationary state 56:44 $Z:E$) and 312 nm (photostationary state 49:51 $Z:E$), a large net change ($>1500\,\mathrm{deg\,cm^2\,dmol^{-1}}$) in the elliptical polarisation response is still observed (Scheme 1.22(b)) and this is reproducible over several cycles without addition of any external chemical reagents.[144]

A similar approach has been used to create a molecular shuttle switch for fluorescence, E/Z-**39** (Scheme 1.23).[142] This system also relies on the

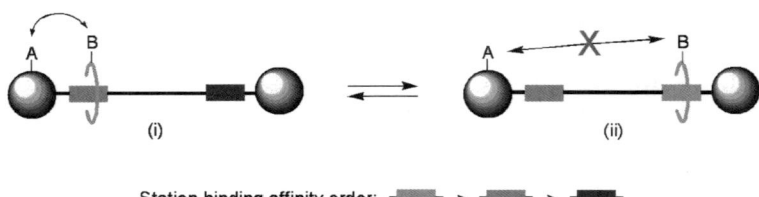

Figure 1.7 Exploiting a well-defined, large-amplitude positional change to trigger property changes. (i) A and B interact to produce a physical response (fluorescence quenching, specific dipole or magnetic moment, NLO properties, colour, creation/concealment of a binding site or reactive/catalytic group, hydrophobic/hydrophilic region, *etc.*); (ii) moving A and B far apart mechanically switches off the interaction and the corresponding property effect.

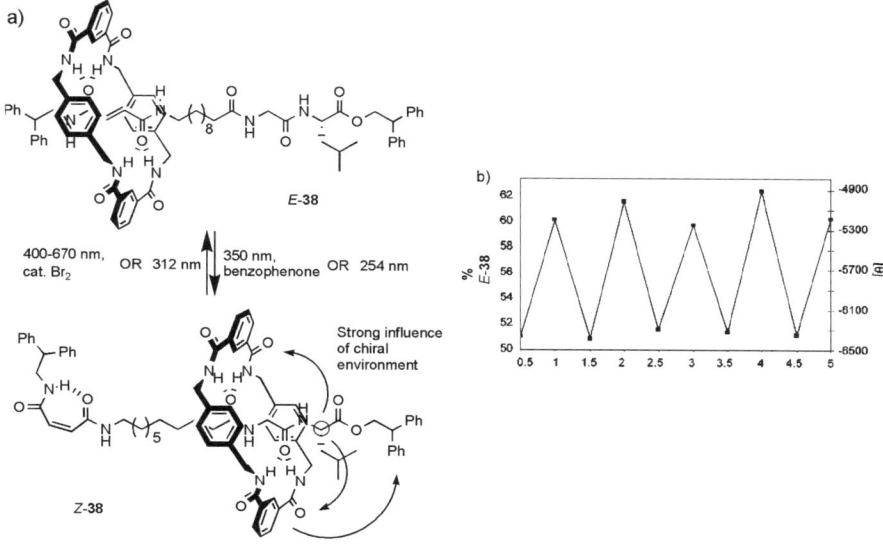

Scheme 1.22 (a) Chiroptical switching in [2]rotaxane-based molecular shuttle E/Z-**38**. (b) Percentage of E-**38** in the photostationary state (from ^{1}H NMR data) after alternating irradiation at 254 nm (half-integers) and 312 nm (integers) for five complete cycles. The right-hand Y-axis shows the CD absorption at 246 nm.

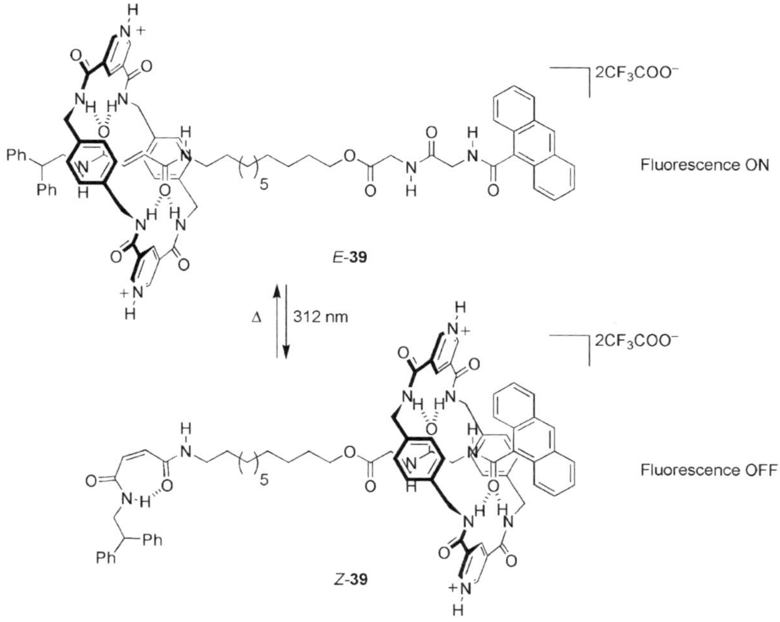

Scheme 1.23 A fluorescent molecular switch based on [2]rotaxane molecular shuttle E/Z-**39**.

photoswitchable fumaramide/maleamide station, but attached to the intermediate-strength dipeptide station is an anthracene fluorophore, while the macrocycle now contains pyridinium units – known to quench anthracene fluorescence by electron transfer. In both the free thread and *E*-**39** strong fluorescence ($\lambda_{exc} = 365$ nm) is observed, while shuttling of the macrocycle onto the glycylglycine station in *Z*-**39** almost completely quenches this emission. At the maximum of *E*-**39** emission ($\lambda_{max} = 417$ nm) there is a remarkable 200:1 difference in intensity between the two states – strikingly visible to the naked eye.

1.4.3 Reporting Controlled Motion on Surfaces, in Solids and Other Condensed Phases

1.4.3.1 Using Mechanical Switches to Affect the Optical Properties of Materials

The use of controlled molecular motion in polymer films to generate patterns visible to the naked eye (Figure 1.8) has been demonstrated with a molecular shuttle-based fluorescent switch derivatised with poly(methyl methacrylate) (**40**, Scheme 1.24).[145] A polymer film INHIBIT logic gate based on a combination of stimuli-controlled submolecular positioning of the ring and protonation was also demonstrated (**41/41** · 2H$^+$, Scheme 1.24, Figure 1.9).[145]

1.4.3.2 Solid-state Molecular Electronic Devices

In a series of ground-breaking experiments that interface molecular-level machines with silicon-based electronics, molecular shuttles have been employed as an active molecular component in solid-state molecular electronic devices.[73,84,85] A range of bistable interlocked molecules were employed and reversible switching achieved. Structures containing the tetracationic cyclophane CBPQT^{4+} can be ordered in Langmuir films using an amphiphilic

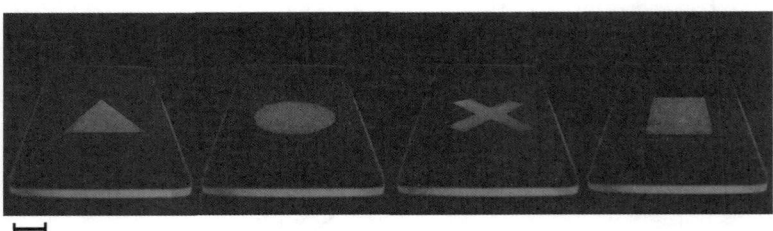

Figure 1.8 Images obtained by casting films of polymer **40** on quartz slides, covering them with an aluminium mask and exposing the unmasked area to dimethyl sulfoxide vapours for 5 minutes. The photographs were taken while illuminating the slides with UV light (254–350 nm). The symbols of Sony Playstation^{TM} are illustrative of the types of patterns that can be created.

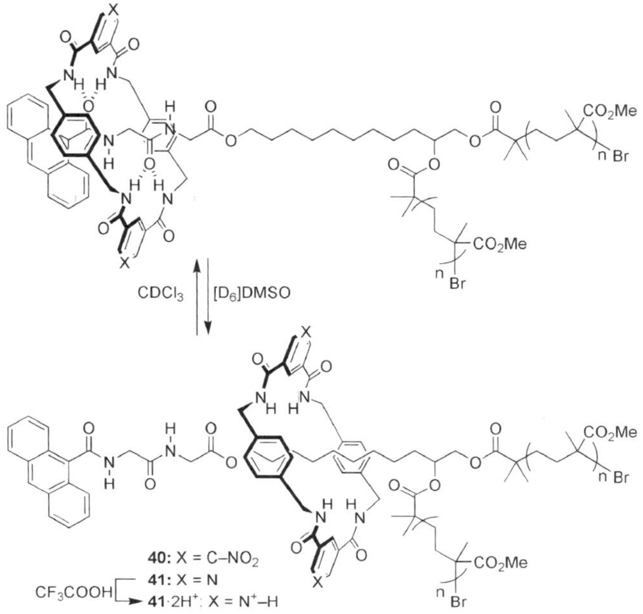

Scheme 1.24 Polymeric environment-switchable molecular shuttles **40** and **41**/**41** $\cdot$ 2H$^+$.

counterion and the monolayers subsequently transferred onto solid substrates using the Langmuir-Blodgett (LB) technique.[146–148] The process was used to create a molecular-switch tunnel junction (MSTJ) in which a monolayer of bistable [2]catenane **33**$^{4+}$ (Scheme 1.18, Section 1.3.3.1) was sandwiched between silicon and titanium/aluminium electrodes.[149] A rational design process then led to devices made from amphiphilic bistable [2]rotaxanes **42**$^{4+}$ and **43**$^{4+}$ (Figure 1.10) in which hydrophobic and hydrophilic regions are now directly incorporated into the molecular structure to allow self-organisation.[150] MSTJs of these rotaxanes possessed stable switching voltages around -2 and $+2$ V with reasonable on/off ratios and switch-closed currents.[151] These favourable characteristics allowed preparation of nanometer-scale devices that displayed properties similar to the original micrometer analogues, suggesting a molecular-level mechanism for the MSTJ operation. Furthermore, these devices could be successfully connected to form a 2D crossbar-circuit architecture. The circuit could be used as a reliable 64-bit random access memory (RAM). The more demanding task of creating a logic circuit was also demonstrated by hard-wiring 1D circuits.

Extensive control experiments on a number of noninterlocked and non-switchable molecules (for example, a simple alkyl chain carboxylic acid, the free CBPQT^{4+} ring and related degenerate catenanes) have been carried out to provide supporting evidence that a molecular-level electromechanical mechanism is responsible for the switching observed in the rotaxane-based

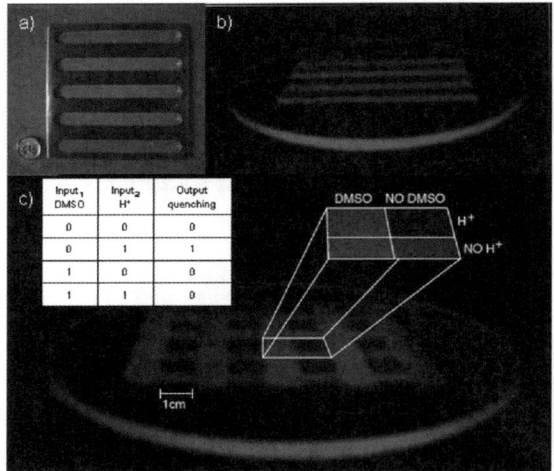

Figure 1.9 A molecular shuttle Boolean logic gate that functions in a polymer film. (a) Aluminum grid used in the experiment. The coin shown for scale is a UK 5p piece. (b) Pattern generated when films of **41** were exposed to trifluoroacetic acid vapour for 5 min through the aluminium grid mask. (c) Crisscross pattern obtained by rotating the aluminium grid 90° and exposing the film shown in (b) to DMSO vapour for a further 5 minutes. Only regions exposed to trifluoroacetic acid but not to DMSO are quenched. The truth table for an INHIBIT logic gate is shown in the inset. The photographs of the slides were taken in the dark while illuminating with UV light (254–350 nm).

devices.[149–151] The studies suggest that some form of bistability is necessary for the switching properties, consistent with a mechanism whereby electronically stimulated shuttling alters the junction conductance. The proposed mechanism for operation of the solid-state electronic devices is summarised in Figure 1.11.

Although these devices show potential as possible components for genuine single-molecule electromechanical switches, a significant barrier to achieving this goal is the nature of the physical connections used to wire the device. At very small dimensions, it is expected that metal wires will exhibit the best conductance characteristics, but unfortunately a range of molecular shuttle devices with two metal electrodes failed to demonstrate switching properties dependent on the nature of the organic monolayer.[152–155] Indeed, the polarity of molecule/metal interfaces and the potential for metal-filament formation within molecular monolayers are emerging as recurring issues in metal-contacted organic molecular electronics.[156,157] However, catenane-based MSTJs that utilise single-walled carbon nanotubes in place of the silicon electrode have been successfully created and this may provide a valuable alternative in the creation of real-world devices.[158] This interfacing of molecular machines with electronics opens up the possibility of many types of practical hybrid devices and will doubtless prove seminal in getting molecular motors and machines out of solutions in laboratories and into technological devices.

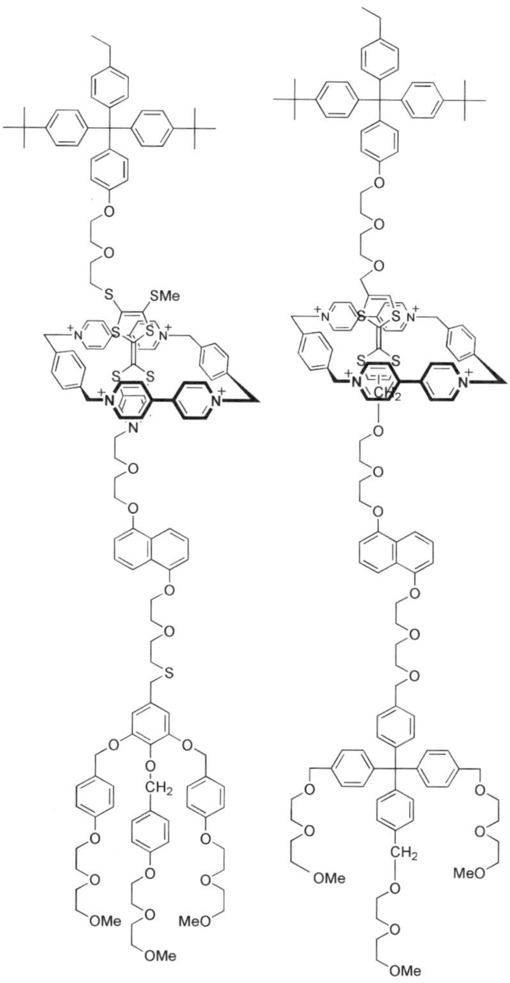

Figure 1.10 Chemical structures of amphiphilic bistable [2]rotaxanes 42^{4+} and 43^{4+} used as the active components in molecular-switch tunnel junctions (MSTJs). Hydrophobic and hydrophilic regions are incorporated directly into the structures to allow self-assembly in Langmuir films without requiring additives.

1.4.3.3 Using Mechanical Switches to Affect the Mechanical Properties of Materials

Using a biomimetic approach based on the operation of the actin–myosin complex (the basis of natural muscle), Stoddart has generated macroscopic mechanical motion using electrochemically induced shuttling in the [3]rotaxane

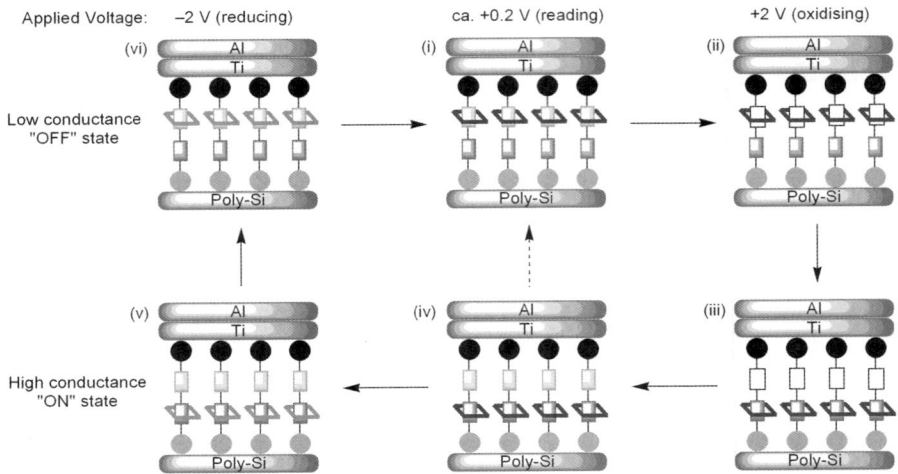

Figure 1.11 Proposed mechanism for the operation of rotaxane-based MSTJs. (i) In the ground state, the tetracationic cyclophane mainly encircles the TTF station and the junction exhibits low conductance (the precise ratio of co-conformers in this state is dependent on the exact chemical structure of the [2]rotaxane and in some cases can also be temperature dependent). (ii) Application of a positive bias results in one- or two-electron oxidation of the TTF units and the resulting electrostatic repulsion causes, (iii), shuttling of the ring onto the DNP station. (iv) Returning the bias to near 0 V reveals a high conductance state in which the TTF units have been returned to neutral but translocation of the cyclophane has not yet occurred, owing to a significant activation barrier. Thermally activated decay of this metastable state may occur slowly, (iv) → (i), over time (dependant on temperature) or can be triggered by application of a negative voltage, (v), which temporarily reduces the cyclophane to its diradical dication form, allowing facile recovery of the thermodynamically favoured co-conformation (vi). A similar mechanism is applicable to the operation of devices based on analogous [2]catenanes.

44^{8+}.[159,160] Oxidation of the TTF stations results in shuttling of the cyclophanes onto the hydroxynaphthalene units, significantly shortening the interring separation (Scheme 1.25). A self-assembled monolayer of **44^{8+}** was deposited on an array of microcantilever beams that had been coated on one side with a gold film and the setup then inserted in a fluid cell. The chemical oxidant Fe(ClO$_4$)$_3$ was added to the solution and the combined effect of co-conformational changes in approximately 6 billion randomly oriented rotaxanes on each cantilever was an upward bending of the beam of about 35 nm. Reduction with ascorbic acid returned the cantilever to its starting position. The process could be repeated over several cycles, albeit with gradually decreasing amplitude.

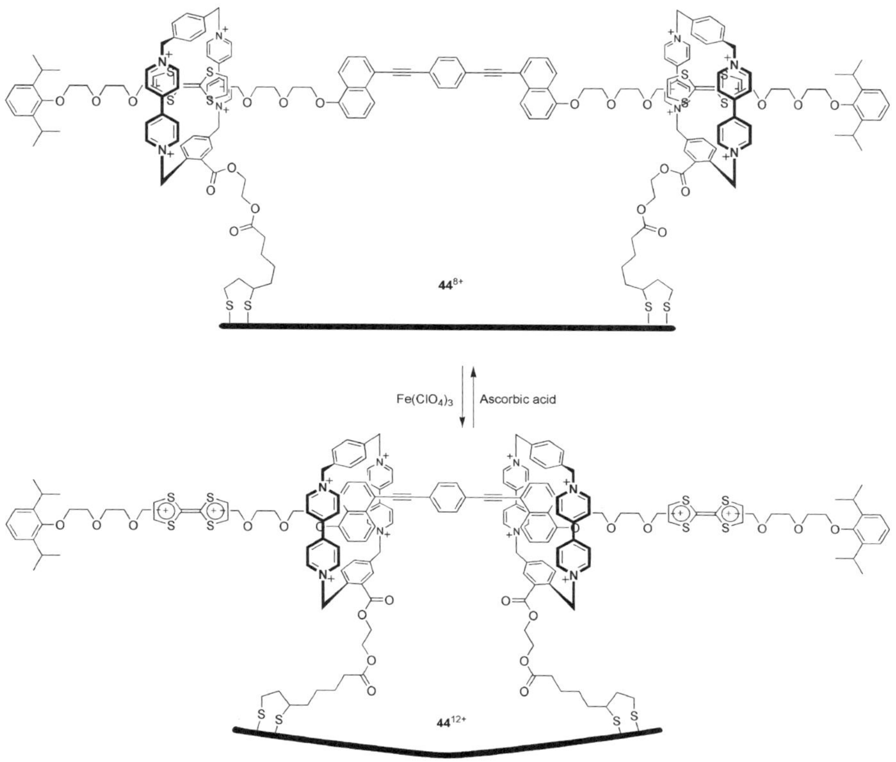

Scheme 1.25 Chemical structure of [3]rotaxane **44**$^{8+}$ and schematic representation of its operation as a molecular actuator.

1.4.3.4 Using Mechanical Switches to Affect Interfacial Properties

A final dramatic illustration of the power of controlled molecular-level motion is the use of co-conformationally switchable monolayers to control macroscopic liquid transport across surfaces.[161] The millimeter-scale directional transport of diiodomethane drops across a surface (Figure 1.12) was achieved using the biased Brownian motion of the components of stimuli-responsive rotaxane **45** (Scheme 1.26) to expose or conceal fluoroalkane residues and thereby modify the surface tension. The collective operation of a monolayer of the molecular shuttles attached to a SAM of 11-mercaptoundecanoic acid (11-MUA) on Au(111) (Scheme 1.27) was sufficient to power the movement of the microlitre droplet up a twelve degree incline (Figure 1.12(e)–(h)). In doing this, the molecular machines effectively employ the energy of the isomerising photon to do work on the drop against gravity. In this experiment, approximately 50% of the light energy absorbed by the rotaxanes was used to

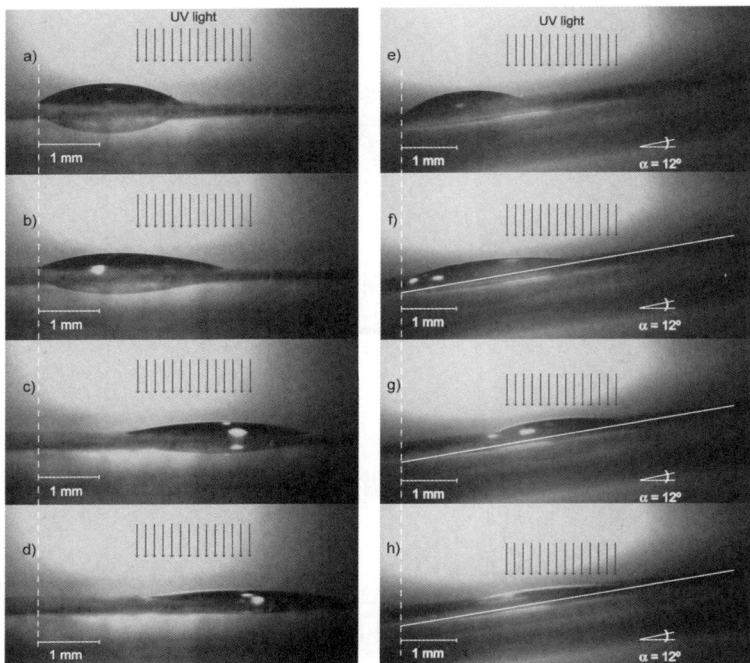

Figure 1.12 Lateral photographs of light-driven directional transport of a 1.25 μl diiodomethane drop across the surface of a E-**45**·11-MUA·Au(111) substrate on mica arranged flat (a)–(d) and up a twelve degree incline (e)–(h). (a) Before irradiation (pristine E-**45**). (b) After 215 s of irradiation (20 s prior to transport) with UV light in the position shown (the right edge of the droplet and the adjacent surface). (c) After 370 s of irradiation (just after transport). (d) After 580 s of irradiation (at the photostationary state). (e) Before irradiation (pristine E-**45**). (f) After 160 s of irradiation (just prior to transport) with UV light in the position shown (the right edge of the droplet and the adjacent surface). (g) After 245 s of irradiation (just after transport). (h) After 640 s irradiation (at the photostationary state). For clarity, on photographs (f)–(h) a white line is used to indicate the surface of the substrate.

overcome the effect of gravity, with the work done stored as potential energy (the raised position of the droplet up the incline).

1.5 Summary and Outlook

In this chapter we have outlined the current state-of-the-art with regards to how the relative positioning of the components of molecular-level structures can be switched, rotated and directionally driven in response to stimuli. In doing so they can affect the nanoscopic and macroscopic properties of the system to which they belong. Whether one chooses to call such systems "motors" and

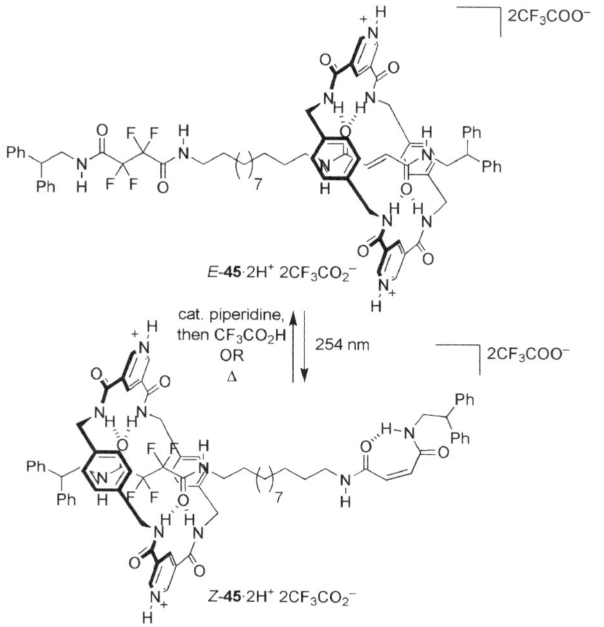

Scheme 1.26 Stimuli-induced positional change of the macrocycle in a fluorinated molecular shuttle, **45**·2H$^+$.

"machines", or prefers to consider them more classically as specific triggered large-amplitude conformational, configurational and structural changes, is irrelevant. The use of controlled molecular-level motion to bring about property changes is fundamentally different from methods that rely solely upon variations in the nature of functional groups or electrostatics.

As with many fundamental developments in science – from electricity to manned flight, the computer and the internet – it is not clear at the very beginning (and make no mistake that we are, indeed, at the very beginning in terms of experimental systems) in exactly what ways synthetic molecular-level motors and machines are going to change technology. In the short term, changing surface properties, switching and memory devices look particularly attractive. A greater challenge, however, is to transduce the energy input from an external source to a molecular machine into the directed transport of a cargo or the pumping of a substrate against an energy gradient. The organisation and addressing of molecular machines on surfaces and at interfaces is an area where we can expect major advances over the next few years.

In stark contrast to biology, none of mankind's fantastic myriad of present-day technologies (with the exception of liquid crystals) exploit controlled molecular-level motion in any way at all. When we learn how to build synthetic structures that can rectify random dynamic processes and interface their effects directly with other molecular-level substructures and the outside world, it would be very surprising if it did not – at the very least – revolutionalise every

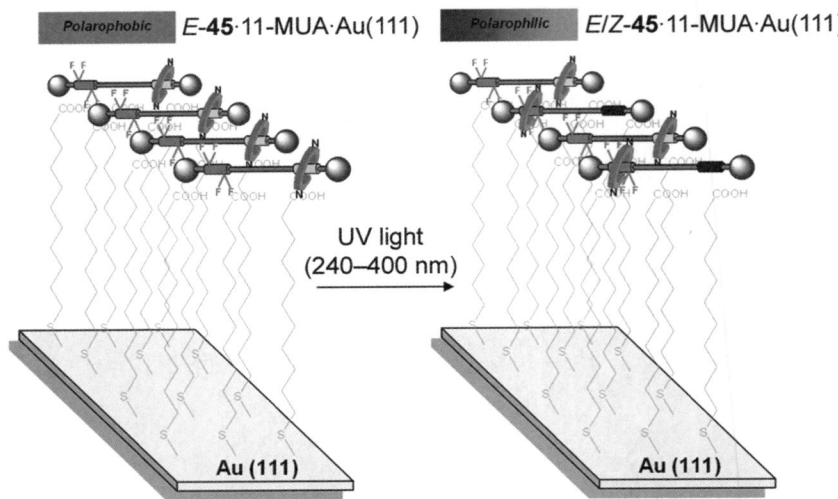

Scheme 1.27 A photoresponsive surface based on switchable fluorinated molecular shuttles. Light-switchable rotaxanes with the fluoroalkane region exposed (*E*-**45**) were physisorbed onto a SAM of 11-MUA on Au(111) deposited onto either glass or mica to create a polarophobic surface, *E*-**45** · 11-MUA · Au(111). Illumination with 240–400 nm light isomerises some of the *E* olefins to *Z* causing a nanometer displacement of the rotaxane threads in the *Z*-shuttles that encapsulates the fluoroalkane units leaving a more polarophilic surface, *E/Z*-**45** · 11-MUA · Au(111). The contact angles of droplets of a wide range of liquids change in response to the isomerisation process.

aspect of functional molecule and materials design. An improved understanding of physics and biology will surely also follow.

References

1. R.P. Feynman, *Eng. Sci.*, 1960, **23**, 22.
2. V. Balzani, A. Credi, F.M. Raymo and J.F. Stoddart, *Angew. Chem. Int. Ed.*, 2000, **39**, 3349.
3. Special Issue on "Molecular Machines", ed. J.F. Stoddart, *Acc. Chem. Res.*, 2001, **34**, 409.
4. Volume on "Molecular Machines and Motors", ed. J.-P. Sauvage, *Struct. Bond.*, 2001, **99**, 1.
5. V. Balzani, M. Venturi and A. Credi, *Molecular Devices and Machines. A Journey into the Nanoworld*, Wiley-VCH, Weinheim, 2003.
6. C.J. Easton, S.F. Lincoln, L. Barr and H. Onagi, *Chem. Eur. J.*, 2004, **10**, 3120.
7. G.S. Kottas, L.I. Clarke, D. Horinek and J. Michl, *Chem. Rev.*, 2005, **105**, 1281.

8. K. Kinbara and T. Aida, *Chem. Rev.*, 2005, **105**, 1377.
9. E.R. Kay and D.A. Leigh, in *Functional Artificial Receptors*, ed. T. Schrader and A.D. Hamilton, Wiley-VCH, Weinheim, 2005.
10. Volume on Molecular Machines, ed. T. R. Kelly, *Top. Curr. Chem.* 2005, **262**, 1.
11. R. Brown, in T*he Miscellaneous Botanical Works of Robert Brown*, ed. J.J. Bennett, Ray Society, London, 1866, **1**, 463.
12. B.J. Ford, *The Microscope*, 1992, **40**, 235.
13. A. Einstein, *Ann. Phys.*, 1905, **17**, 549.
14. J. Perrin, *Atoms*, (English Translation: D.L. Hammick), Constable & Co., London, 1923.
15. M. Haw, *Phys. World*, 2005, **18**(1), 19.
16. P. Hänggi and F. Marchesoni, *Chaos*, 2005, **15**, 026101.
17. J. Renn, *Ann. Phys.*, 2005, **14**, S23.
18. *Molecular Motors*, ed. M. Schliwa, Wiley-VCH, Weinheim, 2003.
19. R.D. Astumian and P. Hänggi, *Phys. Today*, 2002, **55**(11), 33.
20. R.A.L. Jones, *Soft Machines: Nanotechnology and Life*, Oxford University Press, Oxford, 2004.
21. E.M. Purcell, *Am. J. Phys.*, 1977, **45**, 3.
22. T.R. Kelly, *Acc. Chem. Res.*, 2001, **34**, 514.
23. J.P. Sestelo and T.R. Kelly, *Appl. Phys. A*, 2002, **75**, 337.
24. T.R. Kelly, M.C. Bowyer, K.V. Bhaskar, D. Bebbington, A. Garcia, F.R. Lang, M.H. Kim and M.P. Jette, *J. Am. Chem. Soc.*, 1994, **116**, 3657.
25. T.R. Kelly, I. Tellitu and J.P. Sestelo, *Angew. Chem. Int. Ed. Engl.*, 1997, **36**, 1866.
26. T.R. Kelly, J.P. Sestelo and I. Tellitu, *J. Org. Chem.*, 1998, **63**, 3655.
27. A.P. Davis, *Angew. Chem. Int. Ed.*, 1998, **37**, 909.
28. P.W. Atkins, *Physical Chemistry*, Oxford University Press, Oxford, 6th edn, 1998.
29. L. Onsager, *Phys. Rev.*, 1931, **37**, 405.
30. K.L. Sebastian, *Phys. Rev. E*, 2000, **61**, 937.
31. T.R. Kelly, H. De Silva and R.A. Silva, *Nature*, 1999, **401**, 150.
32. T.R. Kelly, R.A. Silva, H. De Silva, S. Jasmin and Y.J. Zhao, *J. Am. Chem. Soc.*, 2000, **122**, 6935.
33. S.P. Fletcher, F. Dumur, M.M. Pollard and B.L. Feringa, *Science*, 2005, **310**, 80.
34. R.K. Bohn and A. Haaland, *J. Organomet. Chem.*, 1966, **5**, 470.
35. X.-B. Wang, B. Dai, H.-K. Woo and L.-S. Wang, *Angew. Chem. Int. Ed.*, 2005, **44**, 6022.
36. J.D. Crowley, B. Bosnich and I.M. Steele, *Chem. Eur. J.*, 2006, **12**, 8935.
37. M.F. Hawthorne and G.B. Dunks, *Science*, 1972, **178**, 462.
38. L.F. Warren and M.F. Hawthorne, *J. Am. Chem. Soc.*, 1970, **92**, 1157.
39. M.R. Churchill and K. Gold, *J. Am. Chem. Soc.*, 1970, **92**, 1180.
40. M.F. Hawthorne, J.I. Zink, J.M. Skelton, M.J. Bayer, C. Liu, E. Livshits, R. Baer and D. Neuhauser, *Science*, 2004, **303**, 1849.

41. *Mechanisms of Protein Folding: Frontiers in Molecular Biology*, Vol. 32, ed. R. H. Pain, Oxford University Press, Oxford, 2nd edn, 2000.
42. F. Ciardelli and O. Pieroni, in *Molecular Switches*, ed. B.L. Feringa, Wiley-VCH, Weinheim, 2001, 399.
43. O. Pieroni, A. Fissi, N. Angelini and F. Lenci, *Acc. Chem. Res.*, 2001, **34**, 9.
44. A. Fersht, *Enzyme Structure and Mechanism*, W. H. Freeman and Company, New York, 2nd edn 1985.
45. J.A. Hardy and J.A. Wells, *Curr. Opin. Struct. Biol.*, 2004, **14**, 706.
46. W.A. Lim, *Curr. Opin. Struct. Biol.*, 2002, **12**, 61.
47. S. Shinkai, M. Ikeda, A. Sugasaki and M. Takeuchi, *Acc. Chem. Res.*, 2001, **34**, 494.
48. L. Kovbasyuk and R. Krämer, *Chem. Rev.*, 2004, **104**, 3161.
49. E.L. Eliel and S.H. Wilen, *Stereochemistry of Organic Compounds*, Wiley-Interscience, New York, 1994.
50. C. Dugave and L. Demange, *Chem. Rev.*, 2003, **103**, 2475.
51. T. Muraoka, K. Kinbara, Y. Kobayashi and T. Aida, *J. Am. Chem. Soc.*, 2003, **125**, 5612.
52. N. Koumura, R.W.J. Zijlstra, R.A. van Delden, N. Harada and B.L. Feringa, *Nature*, 1999, **401**, 152.
53. B.L. Feringa, *Acc. Chem. Res.*, 2001, **34**, 504.
54. B.L. Feringa, N. Koumura, R.A. van Delden and M.K.J. ter Wiel, *Appl. Phys. A*, 2002, **75**, 301.
55. B.L. Feringa, R.A. van Delden and M.K.J. ter Wiel, *Pure Appl. Chem.*, 2003, **75**, 563.
56. M.K.J. ter Wiel, R.A. van Delden, A. Meetsma and B.L. Feringa, *J. Am. Chem. Soc.*, 2005, **127**, 14208.
57. R.W.J. Zijlstra, P.T. van Duijnen, B.L. Feringa, T. Steffen, K. Duppen and D.A. Wiersma, *J. Phys. Chem. A*, 1997, **101**, 9828.
58. N. Koumura, E.M. Geertsema, A. Meetsma and B.L. Feringa, *J. Am. Chem. Soc.*, 2000, **122**, 12005.
59. N. Koumura, E.M. Geertsema, M.B. van Gelder, A. Meetsma and B.L. Feringa, *J. Am. Chem. Soc.*, 2002, **124**, 5037.
60. E.M. Geertsema, N. Koumura, M.K.J. ter Wiel, A. Meetsma and B.L. Feringa, *Chem. Commun.*, 2002, 2962.
61. R.A. van Delden, N. Koumura, A. Schoevaars, A. Meetsma and B.L. Feringa, *Org. Biomol. Chem.*, 2003, **1**, 33.
62. D. Pijper, R.A. van Delden, A. Meetsma and B.L. Feringa, *J. Am. Chem. Soc.*, 2005, **127**, 17612.
63. M.K.J. ter Wiel, R.A. van Delden, A. Meetsma and B.L. Feringa, *J. Am. Chem. Soc.*, 2003, **125**, 15076.
64. T. Fujita, S. Kuwahara and N. Harada, *Eur. J. Org. Chem.*, 2005, 4533.
65. S. Kuwahara, T. Fujita and N. Harada, *Eur. J. Org. Chem.*, 2005, 4544.
66. G. Schill, *Catenanes, Rotaxanes and Knots*, Academic Press, New York, 1971.
67. D.M. Walba, *Tetrahedron*, 1985, **41**, 3161.

68. D.B. Amabilino and J.F. Stoddart, *Chem. Rev.*, 1995, **95**, 2725.
69. G.A. Breault, C.A. Hunter and P.C. Mayers, *Tetrahedron*, 1999, **55**, 5265.
70. *Molecular Catenanes Rotaxanes and Knots*, ed. J.-P. Sauvage and C. O. Dietrich-Buchecker, Wiley-VCH, Weinheim, 1999.
71. P.L. Anelli, N. Spencer and J.F. Stoddart, *J. Am. Chem. Soc.*, 1991, **113**, 5131.
72. A.C. Benniston, *Chem. Soc. Rev.*, 1996, **25**, 427.
73. A.R. Pease, J.O. Jeppesen, J.F. Stoddart, Y. Luo, C.P. Collier and J.R. Heath, *Acc. Chem. Res.*, 2001, **34**, 433.
74. A. Harada, *Acc. Chem. Res.*, 2001, **34**, 456.
75. C.A. Schalley, K. Beizai and F. Vögtle, *Acc. Chem. Res.*, 2001, **34**, 465.
76. J.-P. Collin, C. Dietrich-Buchecker, P. Gaviña, M.C. Jimenez-Molero and J.-P. Sauvage, *Acc. Chem. Res.*, 2001, **34**, 477.
77. F.M. Raymo and J.F. Stoddart, in *Molecular Switches*, ed. B.L. Feringa, Wiley-VCH, Weinheim, 2001, 219.
78. J.-P. Collin, J.-M. Kern, L. Raehm and J.-P. Sauvage, in *Molecular Switches*, ed. B.L. Feringa, Wiley-VCH, Weinheim, 2001, 249.
79. B.X. Colasson, C. Dietrich-Buchecker, M.C. Jimenez-Molero and J.-P. Sauvage, *J. Phys. Org. Chem.*, 2002, **15**, 476.
80. V. Balzani, A. Credi and M. Venturi, *Pure Appl. Chem.*, 2003, **75**, 541.
81. A.M. Brouwer, S.M. Fazio, C. Frochot, F.G. Gatti, D.A. Leigh, J.K.Y. Wong and G.W.H. Wurpel, *Pure Appl. Chem.*, 2003, **75**, 1055.
82. C. Dietrich-Buchecker, M.C. Jimenez-Molero, V. Sartor and J.-P. Sauvage, *Pure Appl. Chem.*, 2003, **75**, 1383.
83. M.C. Jimenez-Molero, C. Dietrich-Buchecker and J.-P. Sauvage, *Chem. Commun.*, 2003, 1613.
84. A.H. Flood, R.J.A. Ramirez, W.Q. Deng, R.P. Muller, W.A. Goddard and J.F. Stoddart, *Aust. J. Chem.*, 2004, **57**, 301.
85. P.M. Mendes, A.H. Flood and J.F. Stoddart, *Appl. Phys. A*, 2005, **80**, 1197.
86. J.-P. Sauvage, *Chem. Commun.*, 2005, 1507.
87. J.-P. Collin and J.-P. Sauvage, *Chem. Lett.*, 2005, **34**, 742.
88. V. Balzani, A. Credi, B. Ferrer, S. Silvi and M. Venturi, *Top. Curr. Chem.*, 2005, **262**, 1.
89. N.N.P. Moonen, A.H. Flood, J.M. Fernández and J.F. Stoddart, *Top. Curr. Chem.*, 2005, **262**, 99.
90. E.R. Kay and D.A. Leigh, *Top. Curr. Chem.*, 2005, **262**, 133.
91. J.-P. Collin, V. Heitz and J.-P. Sauvage, *Top. Curr. Chem.*, 2005, **262**, 29.
92. A.B. Braunschweig, B.H. Northrop and J.F. Stoddart, *J. Mater. Chem.*, 2006, **16**, 32.
93. M.C.T. Fyfe, P.T. Glink, S. Menzer, J.F. Stoddart, A.J.P. White and D.J. Williams, *Angew. Chem. Int. Ed. Engl.*, 1997, **36**, 2068.
94. R.A. Bissell, E. Cordova, A.E. Kaifer and J.F. Stoddart, *Nature*, 1994, **369**, 133.
95. P.T. Glink, C. Schiavo, J.F. Stoddart and D.J. Williams, *Chem. Commun.*, 1996, 1483.

96. P.R. Ashton, P.T. Glink, J.F. Stoddart, P.A. Tasker, A.J.P. White and D.J. Williams, *Chem. Eur. J.*, 1996, **2**, 729.
97. S.J. Cantrill, A.R. Pease and J.F. Stoddart, *J. Chem. Soc. Dalton Trans.*, 2000, 3715.
98. M.-V. Martínez-Díaz, N. Spencer and J.F. Stoddart, *Angew. Chem. Int. Ed. Engl.*, 1997, **36**, 1904.
99. P.R. Ashton, R. Ballardini, V. Balzani, I. Baxter, A. Credi, M.C.T. Fyfe, M.T. Gandolfi, M. Gómez-López, M.-V. Martínez-Díaz, A. Piersanti, N. Spencer, J.F. Stoddart, M. Venturi, A.J.P. White and D.J. Williams, *J. Am. Chem. Soc.*, 1998, **120**, 11932.
100. S. Garaudee, S. Silvi, M. Venturi, A. Credi, A.H. Flood and J.F. Stoddart, *ChemPhysChem*, 2005, **6**, 2145.
101. J.D. Badjic, V. Balzani, A. Credi, S. Silvi and J.F. Stoddart, *Science*, 2004, **303**, 1845.
102. J.D. Badjic, C.M. Ronconi, J.F. Stoddart, V. Balzani, S. Silvi and A. Credi, *J. Am. Chem. Soc.*, 2006, **128**, 1489.
103. C.M. Keaveney and D.A. Leigh, *Angew. Chem. Int. Ed.*, 2004, **43**, 1222.
104. P.R. Ashton, R.A. Bissell, N. Spencer, J.F. Stoddart and M.S. Tolley, *Synlett*, 1992, 914.
105. P.R. Ashton, R.A. Bissell, R. Gorski, D. Philp, N. Spencer, J.F. Stoddart and M.S. Tolley, *Synlett*, 1992, 919.
106. P.R. Ashton, R.A. Bissell, N. Spencer, J.F. Stoddart and M.S. Tolley, *Synlett*, 1992, 923.
107. P.L. Anelli, M. Asakawa, P.R. Ashton, R.A. Bissell, G. Clavier, R. Gorski, A.E. Kaifer, S.J. Langford, G. Mattersteig, S. Menzer, D. Philp, A.M.Z. Slawin, N. Spencer, J.F. Stoddart, M.S. Tolley and D.J. Williams, *Chem. Eur. J.*, 1997, **3**, 1113.
108. D.B. Amabilino, P.R. Ashton, S.E. Boyd, M. Gómez-López, W. Hayes and J.F. Stoddart, *J. Org. Chem.*, 1997, **62**, 3062.
109. H.-R. Tseng, S.A. Vignon and J.F. Stoddart, *Angew. Chem. Int. Ed.*, 2003, **42**, 1491.
110. J.O. Jeppesen, J. Perkins, J. Becher and J.F. Stoddart, *Angew. Chem. Int. Ed.*, 2001, **40**, 1216.
111. J.O. Jeppesen, K.A. Nielsen, J. Perkins, S.A. Vignon, A. Di Fabio, R. Ballardini, M.T. Gandolfi, M. Venturi, V. Balzani, J. Becher and J.F. Stoddart, *Chem. Eur. J.*, 2003, **9**, 2982.
112. H.-R. Tseng, S.A. Vignon, P.C. Celestre, J. Perkins, J.O. Jeppesen, A. Di Fabio, R. Ballardini, M.T. Gandolfi, M. Venturi, V. Balzani and J.F. Stoddart, *Chem. Eur. J.*, 2004, **10**, 155.
113. S.S. Kang, S.A. Vignon, H.-R. Tseng and J.F. Stoddart, *Chem. Eur. J.*, 2004, **10**, 2555.
114. A.H. Flood, A.J. Peters, S.A. Vignon, D.W. Steuerman, H.-R. Tseng, S. Kang, J.R. Heath and J.F. Stoddart, *Chem. Eur. J.*, 2004, **10**, 6558.
115. B.W. Laursen, S. Nygaard, J.O. Jeppesen and J.F. Stoddart, *Org. Lett.*, 2004, **6**, 4167.

116. J.O. Jeppesen, S. Nygaard, S.A. Vignon and J.F. Stoddart, *Eur. J. Org. Chem.*, 2005, 196.
117. J.W. Choi, A.H. Flood, D.W. Steuerman, S. Nygaard, A.B. Braunschweig, N.N.P. Moonen, B.W. Laursen, Y. Luo, E. DeIonno, A.J. Peters, J.O. Jeppesen, K. Xu, J.F. Stoddart and J.R. Heath, *Chem. Eur. J.*, 2006, **12**, 261.
118. A.M. Brouwer, C. Frochot, F.G. Gatti, D.A. Leigh, L. Mottier, F. Paolucci, S. Roffia and G.W.H. Wurpel, *Science*, 2001, **291**, 2124.
119. A. Altieri, F.G. Gatti, E.R. Kay, D.A. Leigh, D. Martel, F. Paolucci, A.M.Z. Slawin and J.K.Y. Wong, *J. Am. Chem. Soc.*, 2003, **125**, 8644.
120. S.A. Vignon, T. Jarrosson, T. Iijima, H.-R. Tseng, J.K.M. Sanders and J.F. Stoddart, *J. Am. Chem. Soc.*, 2004, **126**, 9884.
121. T. Iijima, S.A. Vignon, H.-R. Tseng, T. Jarrosson, J.K.M. Sanders, F. Marchioni, M. Venturi, E. Apostoli, V. Balzani and J.F. Stoddart, *Chem. Eur. J.*, 2004, **10**, 6375.
122. D.S. Marlin, D.G. Cabrera, D.A. Leigh and A.M.Z. Slawin, *Angew. Chem. Int. Ed.*, 2006, **45**, 77.
123. D.S. Marlin, D.G. Cabrera, D.A. Leigh and A.M.Z. Slawin, *Angew. Chem. Int. Ed.*, 2006, **45**, 1385.
124. D.A. Leigh and E.M. Pérez, *Chem. Commun.*, 2004, 2262.
125. A. Altieri, G. Bottari, F. Dehez, D.A. Leigh, J.K.Y. Wong and F. Zerbetto, *Angew. Chem. Int. Ed.*, 2003, **42**, 2296.
126. F.G. Gatti, D.A. Leigh, S.A. Nepogodiev, A.M.Z. Slawin, S.J. Teat and J.K.Y. Wong, *J. Am. Chem. Soc.*, 2001, **123**, 5983.
127. A.C. Benniston, A. Harriman and V.M. Lynch, *Tetrahedron Lett.*, 1994, **35**, 1473.
128. A.C. Benniston, A. Harriman and V.M. Lynch, *J. Am. Chem. Soc.*, 1995, **117**, 5275.
129. V. Balzani, M. Clemente-León, A. Credi, B. Ferrer, M. Venturi, A.H. Flood and J.F. Stoddart, *Proc. Natl. Acad. Sci. USA*, 2006, **103**, 1178.
130. E.R. Kay and D.A. Leigh, *Nature*, 2006, **440**, 286.
131. G. Bottari, F. Dehez, D.A. Leigh, P.J. Nash, E.M. Pérez, J.K.Y. Wong and F. Zerbetto, *Angew. Chem. Int. Ed.*, 2003, **42**, 5886.
132. A.S. Lane, D.A. Leigh and A. Murphy, *J. Am. Chem. Soc.*, 1997, **119**, 11092.
133. T. Da Ros, D.M. Guldi, A.F. Morales, D.A. Leigh, M. Prato and R. Turco, *Org. Lett.*, 2003, **5**, 689.
134. C.G. Gong and H.W. Gibson, *Angew. Chem. Int. Ed. Engl.*, 1997, **36**, 2331.
135. C.G. Gong, T.E. Glass and H.W. Gibson, *Macromolecules*, 1998, **31**, 308.
136. D.A. Leigh, K. Moody, J.P. Smart, K.J. Watson and A.M.Z. Slawin, *Angew. Chem. Int. Ed. Engl.*, 1996, **35**, 306.
137. M. Asakawa, P.R. Ashton, V. Balzani, A. Credi, C. Hamers, G. Mattersteig, M. Montalti, A.N. Shipway, N. Spencer, J.F. Stoddart, M.S. Tolley, M. Venturi, A.J.P. White and D.J. Williams, *Angew. Chem. Int. Ed.*, 1998, **37**, 333.

138. A.M.L. Fuller, D.A. Leigh, P.J. Lusby, A.M.Z. Slawin and D.B. Walker, *J. Am. Chem. Soc.*, 2005, **127**, 12612.
139. D.A. Leigh, P.J. Lusby, A.M.Z. Slawin and D.B. Walker, *Chem. Commun.*, 2005, 4919.
140. D.A. Leigh, J.K.Y. Wong, F. Dehez and F. Zerbetto, *Nature*, 2003, **424**, 174.
141. J.V. Hernández, E.R. Kay and D.A. Leigh, *Science*, 2004, **306**, 1532.
142. E.M. Pérez, D.T.F. Dryden, D.A. Leigh, G. Teobaldi and F. Zerbetto, *J. Am. Chem. Soc.*, 2004, **126**, 12210.
143. M. Asakawa, G. Brancato, M. Fanti, D.A. Leigh, T. Shimizu, A.M.Z. Slawin, J.K.Y. Wong, F. Zerbetto and S.W. Zhang, *J. Am. Chem. Soc.*, 2002, **124**, 2939.
144. G. Bottari, D.A. Leigh and E.M. Pérez, *J. Am. Chem. Soc.*, 2003, **125**, 13360.
145. D.A. Leigh, M.A.F. Morales, E.M. Pérez, J.K.Y. Wong, C.G. Saiz, A.M.Z. Slawin, A.J. Carmichael, D.M. Haddleton, A.M. Brouwer, W.J. Buma, G.W.H. Wurpel, S. León and F. Zerbetto, *Angew. Chem. Int. Ed.*, 2005, **44**, 3062.
146. R.C. Ahuja, P.-L. Caruso, D. Möbius, D. Philp, J.A. Preece, H. Ringsdorf, J.F. Stoddart and G. Wildburg, *Thin Solid Films*, 1996, **285**, 671.
147. C.L. Brown, U. Jonas, J.A. Preece, H. Ringsdorf, M. Seitz and J.F. Stoddart, *Langmuir*, 2000, **16**, 1924.
148. M. Asakawa, M. Higuchi, G. Mattersteig, T. Nakamura, A.R. Pease, F.M. Raymo, T. Shimizu and J.F. Stoddart, *Adv. Mater.*, 2000, **12**, 1099.
149. C.P. Collier, G. Mattersteig, E.W. Wong, Y. Luo, K. Beverly, J. Sampaio, F.M. Raymo, J.F. Stoddart and J.R. Heath, *Science*, 2000, **289**, 1172.
150. C.P. Collier, J.O. Jeppesen, Y. Luo, J. Perkins, E.W. Wong, J.R. Heath and J.F. Stoddart, *J. Am. Chem. Soc.*, 2001, **123**, 12632.
151. Y. Luo, C.P. Collier, J.O. Jeppesen, K.A. Nielsen, E. Delonno, G. Ho, J. Perkins, H.-R. Tseng, T. Yamamoto, J.F. Stoddart and J.R. Heath, *ChemPhysChem*, 2002, **3**, 519.
152. H.B. Yu, Y. Luo, K. Beverly, J.F. Stoddart, H.-R. Tseng and J.R. Heath, *Angew. Chem. Int. Ed.*, 2003, **42**, 5706.
153. Y. Chen, D.A.A. Ohlberg, X.M. Li, D.R. Stewart, R.S. Williams, J.O. Jeppesen, K.A. Nielsen, J.F. Stoddart, D.L. Olynick and E. Anderson, *Appl. Phys. Lett.*, 2003, **82**, 1610.
154. Y. Chen, G.Y. Jung, D.A.A. Ohlberg, X.M. Li, D.R. Stewart, J.O. Jeppesen, K.A. Nielsen, J.F. Stoddart and R.S. Williams, *Nanotechnology*, 2003, **14**, 462.
155. D.R. Stewart, D.A.A. Ohlberg, P.A. Beck, Y. Chen, R.S. Williams, J.O. Jeppesen, K.A. Nielsen and J.F. Stoddart, *Nano Lett.*, 2004, **4**, 133.
156. J.R. Heath and M.A. Ratner, *Phys. Today*, 2003, **56**(5), 43.
157. C.N. Lau, D.R. Stewart, R.S. Williams and M. Bockrath, *Nano Lett.*, 2004, **4**, 569.
158. M.R. Diehl, D.W. Steuerman, H.-R. Tseng, S.A. Vignon, A. Star, P.C. Celestre, J.F. Stoddart and J.R. Heath, *ChemPhysChem*, 2003, **4**, 1335.

159. T.J. Huang, B. Brough, C.-M. Ho, Y. Liu, A.H. Flood, P.A. Bonvallet, H.-R. Tseng, J.F. Stoddart, M. Baller and S. Magonov, *Appl. Phys. Lett.*, 2004, **85**, 5391.
160. Y. Liu, A.H. Flood, P.A. Bonvallet, S.A. Vignon, B.H. Northrop, H.-R. Tseng, J.O. Jeppesen, T.J. Huang, B. Brough, M. Baller, S. Magonov, S.D. Solares, W.A. Goddard, C.-M. Ho and J.F. Stoddart, *J. Am. Chem. Soc.*, 2005, **127**, 9745.
161. J. Berná, D.A. Leigh, M. Lubomska, S.M. Mendoza, E.M. Pérez, P. Rudolf, G. Teobaldi and F. Zerbetto, *Nature Mater.*, 2005, **4**, 704.

CHAPTER 2
Photochemically Controlled Molecular Devices and Machines

VINCENZO BALZANI,* GIACOMO BERGAMINI,
PAOLA CERONI, ALBERTO CREDI AND
MARGHERITA VENTURI

Dipartimento di Chimica "G. Ciamician", Università degli Studi di Bologna, via Selmi 2, 40126 Bologna, Italy

2.1 Introduction

The progress of man's civilisation has always been related to the construction of novel devices and machines. Depending on their use, devices and machines can be very big or very small. The general trend in our "knowledge age" is that of reducing size and weight as much as possible, particularly in the field of information technology. The first electronic computer occupied an entire room, weighed 30 tons, was made of 18 000 valves, and lasted an average of 5.6 h between repairs.[1] Nowadays, a laptop can be placed on our desk, weighs a few kilograms, has more than 40 million transistors and practically does not need any repair. Indeed, the race towards miniaturisation seems to be endless. One can wonder whether, at this stage, we do really need to keep on making things smaller. The answer is that further miniaturisation will not only decrease the size and increase the power of computers, but is also expected to open the way to new technologies,[2] such as nanorobotics.[3]

The miniaturisation of components for the construction of devices and machines is currently pursued by the top-down approach. This approach, which is in the hands of physicists and engineers, consists in manipulating progressively smaller pieces of matter by photolithography and related techniques. The top-down approach, however, is subjected to drastic limitations, including a severe cost escalation, for dimensions smaller than 60 nanometres.[4] But "there is plenty of room at the bottom" for further miniaturisation, as Richard P. Feynman[5] stated in a famous talk to the American Physical Society in 1959. To proceed towards miniaturisation at the nanometre scale, science and technology needs to find new avenues.

A most promising strategy to exploit science and technology at the nanometre scale is offered by the bottom-up approach, which starts from nano- or subnanoscale objects (namely, atoms or molecules) to build up nanostructures. Chemists, being able to manipulate atoms and molecules, are in the ideal position to develop such an approach for the construction of nanoscale devices and machines. In fact, the molecule-by-molecule bottom-up strategy is simply supramolecular chemistry, a discipline that emerged in the late 1970s, consecrated by the award of the Nobel Prize in Chemistry to C.J. Pedersen,[6] D.J. Cram,[7] and J.-M. Lehn[8] in 1987, and extensively developed in recent years.[9]

The (physical) top-down approach has allowed the construction of a variety of solid state microelectronic devices and microelectromechanical systems.[10] It can now be expected that the (chemical) bottom-up approach will move science and technology not only from the micro- to the nanoscale, but also from electronics to photonics and chemionics since light and chemical inputs are convenient ways to power molecular-level devices and to exchange information at the molecular level. Furthermore, the bottom-up approach, taking inspiration from natural nanoscale devices, could displace the interest of scientists from solid state to solution and soft matter.[11]

Up until now scarce attention has been devoted to the possibility of developing photonics at the molecular level because selective excitation of a specific molecule in a supramolecular array is prevented by diffraction-limited spot size considerations. However, progress in near-field optical techniques[12] and electromagnetic energy transport in metal nanoparticle plasmon waveguides[13] seems to overcome this difficulty. Therefore, in the next few years, the bottom-up approach is likely to lead to a wealth of nanodevices in which photonics and chemionics will be integrated to a different extent depending on the function that the device has to perform.

Regardless of the possibility of short-term applications, the development of a set of molecular-level devices and machines based on photochemionics appears to be a worthwhile investment.[14] Systematic research in this field began about 30 years ago with a NATO Workshop on Supramolecular Photochemistry.[15] At that meeting, the first examples of artificial antennae for light harvesting[16] and photoinduced charge-separation reactions mimicking photosynthetic reaction centres[17] were reported. Furthermore, a systematic discussion was presented on the design and construction of functional supramolecular systems that were defined, for the first time, photochemical molecular devices.[18]

Since then, outstanding progress has been made in the design, synthesis and characterisation of photochemical molecular devices and machines. Many recent reviews are available,[19] and an exhaustive monograph[14] has recently been published. In this chapter, some recent achievements in the field of molecular devices and machines operating by light excitation will be illustrated, using almost exclusively examples taken from the work of our laboratory.

2.2 Molecular-level Devices for Processing Light Signals

In supramolecular (multicomponent)[20] systems, light absorption by a specific component can be followed by energy transfer and light emission by another component. The function obtained depends on the specific design of the supramolecular system.

2.2.1 Wires

An important function at the molecular level is photoinduced energy transfer over long distances along predetermined directions. This function can be obtained by linking donor and acceptor components by a rigid spacer, as illustrated in Figure 2.1(a). An example[21] is given by the $[Ru(bpy)_3]^{2+}$–(ph)n–$[Os(bpy)_3]^{2+}$ compounds (bpy = 2,2′–bipyridine; ph = 1,4–phenylene; n = 3, 5, 7) in which excitation of the $[Ru(bpy)_3]^{2+}$ unit is followed by electronic energy transfer to the ground state $[Os(bpy)_3]^{2+}$ unit, as shown by the sensitised emission of the latter. For the compound with $n = 7$ (**1**, Figure 2.1b), the rate constant for energy transfer over the 4.2 nm metal–to–metal distance is $1.3 \times 10^6 \, s^{-1}$.

Spacers with energy levels in between those of the donor and acceptor may help energy transfer (hopping mechanism). Spacers whose energy levels can be manipulated by an external stimulus can play the role of switches for the energy-transfer processes.

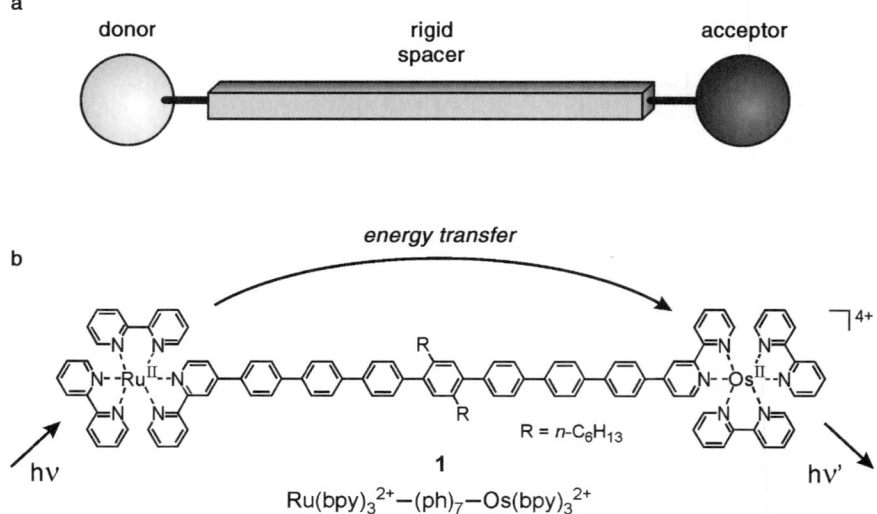

Figure 2.1 Schematic representation of a molecular level wire (a), and an example of photoinduced energy transfer over long distances (b).[21]

2.2.2 Switching Devices

The transfer of electronic energy in a molecular-level wire can be switched on/off by an external stimulus applied to a suitably designed component, incorporated in the wire.

In an appropriately designed compound, the external stimulus can be light. Since, by definition, switching has to be reversible, reversible photochemical reactions have to be used. Photochromic molecules are particularly useful in this regard. An example is given by the D–P–A supramolecular species 2 investigated by Walz et al.[22] (Figure 2.2) in which photoinduced energy transfer from D to A can be switched by photoexcitation of component P. In such a system, the spacer P is a photochromic fulgide molecule that can be transformed by light in a reversible way between a closed P_a and an open P_b configuration. The donor D is an anthryl unit which can be excited at 258 nm, and the acceptor A is a coumarin molecule. When P is in its closed form P_a, its lowest energy level is lower than the energy level of A, so that energy transfer from D to A cannot occur (Figure 2.2) and the sensitised luminescence of the coumarin cannot be observed upon excitation of the anthracene moiety. However, when the P species is isomerised with 520-nm light to yield the P_b isomer, the energy levels are in scale and the sensitised luminescence of the coumarin component at 500 nm can be observed upon excitation of the anthryl component at 258 nm. Since light of three different wavelengths is needed and four different chromophoric units are involved, such a system is not so easy to handle and its behaviour is not really ON/OFF.

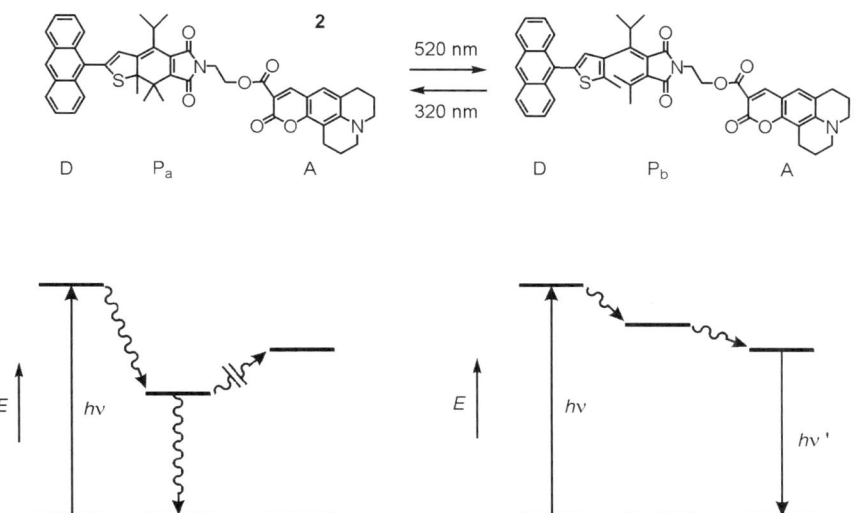

Figure 2.2 Switching of energy transfer from an anthracene moiety to a coumarin moiety by photoisomerisation of a fulgide bridge.[22]

2.2.3 Plug/socket Systems

Supramolecular species whose components are connected by means of non-covalent forces can be disassembled and reassembled[23] by modulating the interactions that keep the components together, thereby allowing switching of energy-transfer processes. Two-component systems of this type are reminiscent of plug/socket electrical devices and, like their macroscopic counterparts, must be characterised by (i) the possibility of connecting/disconnecting the two components in a reversible way, and (ii) the occurrence of an electronic energy flow from the socket to the plug when the two components are connected (Figure 2.3(a)). Hydrogen-bonding interactions between ammonium ions and crown ethers are particularly convenient for constructing molecular-level plug/socket devices since they can be switched ON and OFF quickly and reversibly by means of acid/base inputs.

A plug/socket system that deals with the transfer of electronic energy is illustrated in Figure 2.3(b).[24] The absorption and fluorescence spectra of a CH_2Cl_2 solution containing equal amounts of (±)-binaphthocrown ether **3** and amine **4** indicate the absence of any interaction between the two compounds.

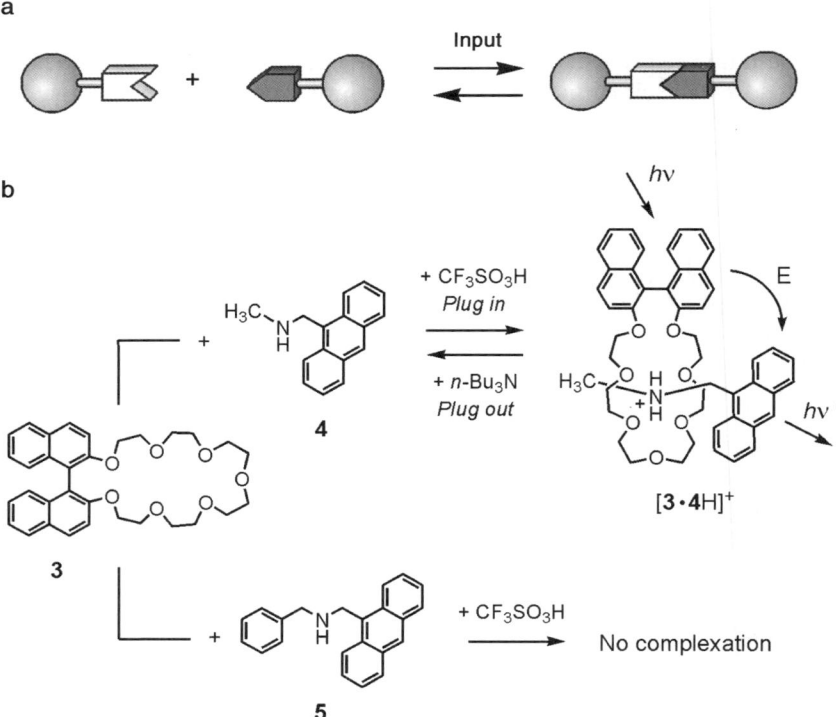

Figure 2.3 Schematic representation of a plug-socket system (a); switching of photo-induced energy transfer by acid/based controlled plug in/plug out of suitable molecular components (b).[24]

Addition of a stoichiometric amount of acid causes profound changes in the fluorescence behaviour of the solution, namely (i) the fluorescence of **3** is quenched, and (ii) the fluorescence of [**4H**]$^+$ is sensitised upon excitation with light absorbed exclusively by the crown ether. These observations are consistent with the formation of a pseudorotaxane-type adduct [**3•4H**]$^+$ wherein very efficient energy transfer takes place from the binaphthyl unit of the crown ether to the anthracenyl group incorporated within the dialkylammonium ion. Such a pseudorotaxane can be disassembled by the subsequent addition of a stoichiometric amount of base, thereby interrupting the photoinduced energy flow, as indicated by the fact that the initial absorption and fluorescence spectra are restored. Interestingly, the plug-in process does not take place when a plug component incompatible with the size of the socket, such as the benzyl-substituted amine **5**, is employed (Figure 2.3(b)).

2.2.4 Molecular Extension Cables

The plug/socket concept can be used to design molecular systems that mimic the function played by a macroscopic electrical extension cable. The function performed by an extension cable is more complex than that played by a plug/socket system since it involves *three* components that must be hold together by *two* connections that have to be controllable *reversibly* and *independently*; in the fully connected system, an electron or energy flow must take place between the remote donor and acceptor units (Figure 2.4(a)).

In the attempt of constructing a molecular-level extension cable for electron transfer, the pseudorotaxane shown in Figure 2.4(b), made of the three components **6**$^{2+}$, [**7H**]$^{3+}$, and **8**, has been synthesised and studied.[25] Component **6**$^{2+}$ consists of two moieties: a [Ru(bpy)$_3$]$^{2+}$ unit, which plays the role of electron donor under light excitation, and a crown ether, which plays the role of a first socket. The ammonium centre of [**7H**]$^{3+}$, driven by hydrogen-bonding interactions, threads as a plug into the first socket, whereas the bipyridinium unit, owing to van der Waals (charge-transfer, CT) interactions, threads as a plug into the third component, **8**, which plays the role of a second socket. In CH$_2$Cl$_2$/CH$_3$CN (98:2 *v/v*) solution, reversible connection/disconnection of the two plug/socket functions can be controlled independently by acid/base and red/ox stimulation, respectively. In the fully connected triad, light excitation of the [Ru(bpy)$_3$]$^{2+}$ unit of component **6**$^{2+}$ is followed by electron transfer to the bipyridinium unit of component [**7H**]$^{3+}$, which is plugged into component **8**. It should be noted that a true elongation cable should contain a plug and a socket at the two ends, instead of two plugs as component [**7H**]$^{3+}$. An improved system of that type has been recently studied.[26] By a suitable choice of the components, molecular elongation cables for energy transfer could also be constructed.

2.2.5 Antenna Systems for Light Harvesting

Dendrimers are well-defined, tree-like macromolecules, with a high degree of order and the possibility of containing selected chemical units in predetermined

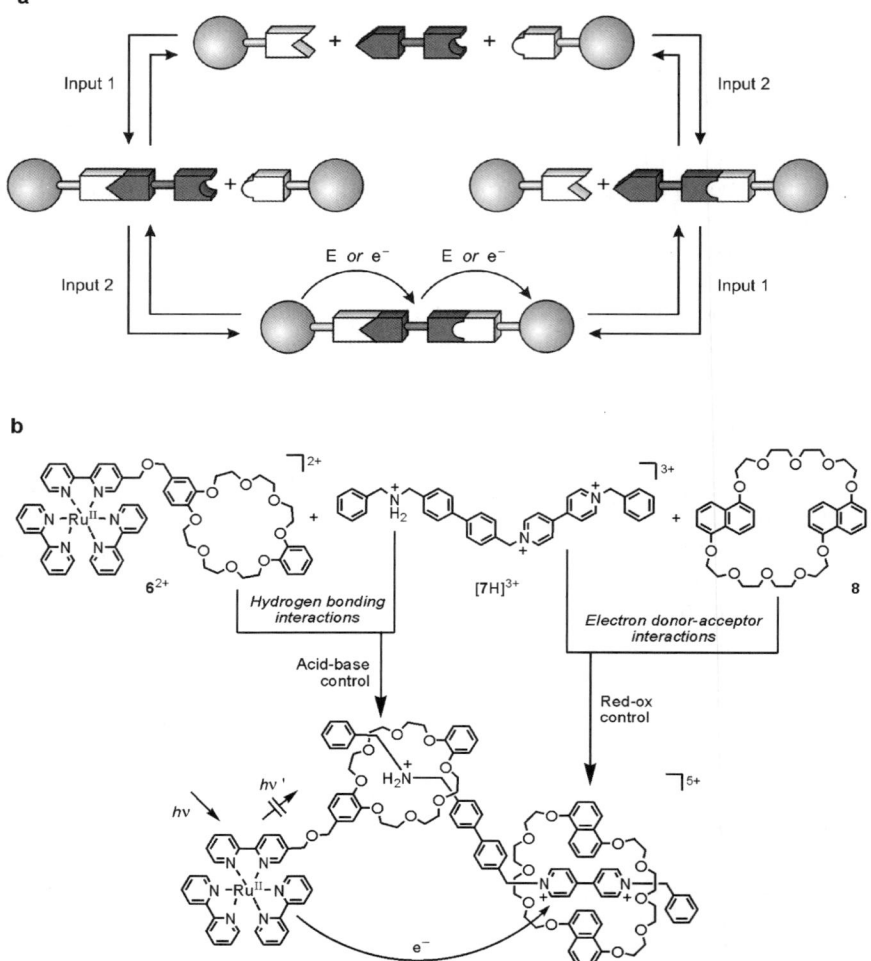

Figure 2.4 Schematic representation of an extension cable (a); a supramolecular system that behaves as a molecular-level extension cable (b).[25]

sites of their structure.[27,28] Particularly interesting dendrimers are those containing photoactive components.[29] Because of the close proximity with other units, a photoactive group of a dendrimer can exhibit different properties compared with those exhibited by the same group when it is isolated. For example, in suitably designed dendritic structures photoexcited units can transfer energy to other components, thereby opening the way towards a number of functions.

In the course of evolution, Nature has succeeded in building up antenna systems that collect an enormous amount of solar energy and redirect it as electronic excitation energy to reaction centres where subsequent conversion

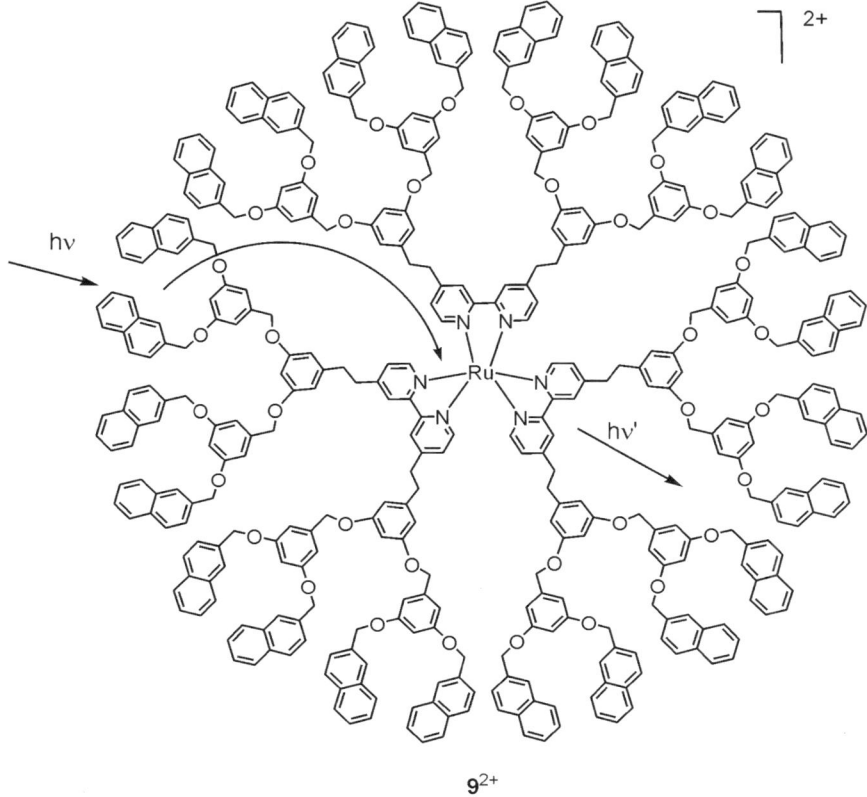

Figure 2.5 Antenna effect in a dendrimer.[33]

into redox chemical energy takes place.[30,31] Suitably designed dendrimers can mimic the light-harvesting function of natural antenna systems since light energy can be channelled by electronic energy transfer towards a specific component of the array.[32]

In the dendritic complex **9^{2+}** shown in Figure 2.5, the 2,2′-bipyridine ligands of the [Ru(bpy)$_3$]$^{2+}$-type core carry branches containing 1,2-dimethoxybenzene- and 2-naphthyl-type chromophoric units.[33] Since such units (as well as the core) are separated by aliphatic connections, the interchromophoric interactions are weak and the absorption spectra of the dendrimer is substantially equal to the sum of the spectra of the chromophoric groups that are present in its structure. The three types of chromophoric groups, namely, [Ru(bpy)$_3$]$^{2+}$, dimethoxybenzene, and naphthalene, are potentially luminescent species. In the dendrimer, however, the fluorescence of the dimethoxybenzene- and naphthyl-type units is almost completely quenched in acetonitrile solution, with concomitant sensitisation of the luminescence of the [Ru(bpy)$_3$]$^{2+}$ core ($\lambda_{max} = 610$ nm). These results show that a very efficient energy-transfer process takes place towards the metal-based dendritic core. It should also be noted that in aerated solution the luminescence intensity of the core is more than twice as intense as that of the $[Ru(bpy)_3]^{2+}$ parent compound because the dendritic branches protect the Ru-based core from dioxygen quenching. Because of the very high absorbance of the naphthyl groups in the UV spectral region, the high energy-transfer efficiency, and the strong emission of the $[Ru(bpy)_3]^{2+}$-type core, dendrimer 9^{2+} exhibits a strong visible emission upon UV excitation even in very dilute (10^{-7} mol L^{-1}) solutions.

2.2.6 Molecular Lenses Capable of Tuning the Colour of Light

An important property of dendrimers is the presence of internal cavities where ions or neutral molecules can be hosted. Energy transfer from the numerous chromophoric units of a suitable dendrimer to an appropriate guest may again result in a light-harvesting antenna system. An advantage shown by such host–guest light-harvesting systems is that the wavelength of the resulting sensitised emission can be changed by using the same dendrimer and different types of guests. Therefore, the light signal can be not only "concentrated",[34] but its colour can also be "tuned".

Dendrimer **10** shown in Figure 2.6 consists of a hexaamine core surrounded by 8 dansyl-, 24 dimethoxybenzene-, and 32 naphthalene-type units.[35] In dichloromethane solution, this dendrimer exhibits the characteristic absorption bands of the component units and a strong dansyl-type fluorescence. Energy transfer from the peripheral dimethoxybenzene and naphthalene units to the fluorescent dansyl units occurs with >90% efficiency. When the dendrimer hosts a molecule of the fluorescent eosin dye (Figure 2.6), the dansyl fluorescence, in its turn, is quenched and sensitisation of the fluorescence of the eosin guest can be observed. Quantitative measurements showed that the encapsulated eosin molecule collects electronic energy from all the 64 chromophoric units of the dendrimer with an efficiency >80% (partial overlapping between dansyl and eosin emissions precludes a better precision). Both intramolecular (*i.e.* within dendrimer) and intermolecular (*i.e.* dendrimer host → eosin guest) energy-transfer processes occur very efficiently by a Förster-type mechanism because of the strong overlap between the emission and absorption spectra of the relevant donor/acceptor units. When eosin is replaced by rose bengal, the characteristic emission of the latter dye is observed. Dendrimers containing amide units can incorporate lanthanide ions like Nd^{3+}, the near-infrared emission (1064 nm) of which can be sensitised by energy transfer from the chromophoric groups of the dendrimer.[36]

2.2.7 Fluorescent Sensors with Signal Amplification

The dendrimers of the poly(propylene amine) family can be easily functionalised in the periphery with luminescent units like dansyl. Each dendrimer nD, where the generation number n goes from 1 to 5, comprises $2^{(n+1)}$ dansyl

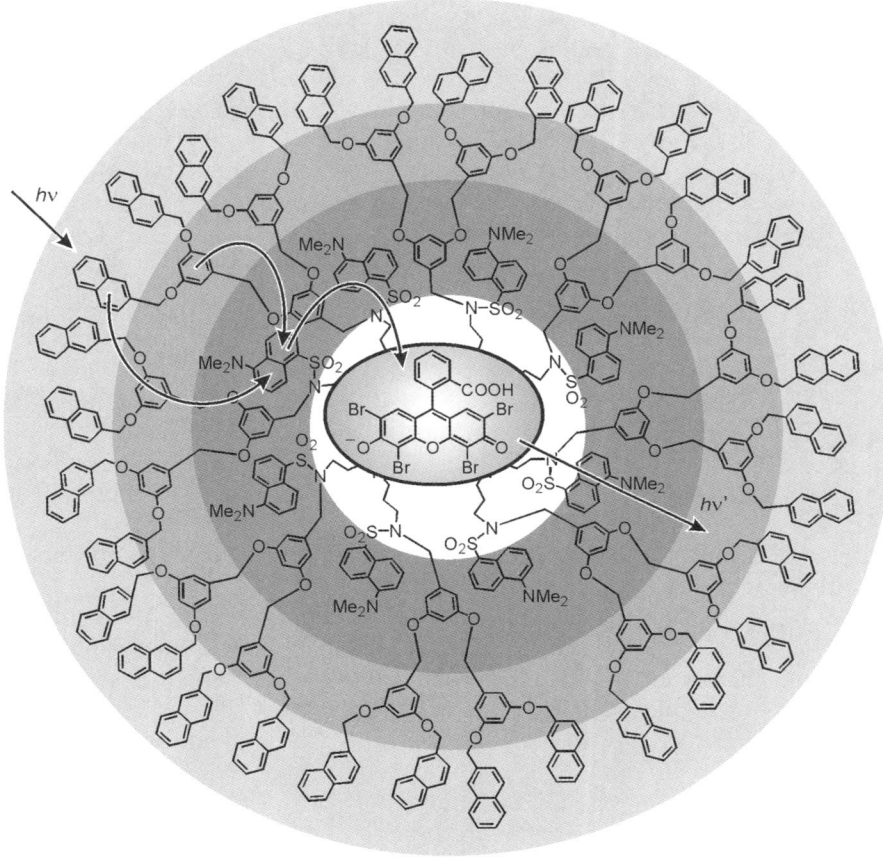

[10⊃eosin]

Figure 2.6 Schematic representation of the energy-transfer processes taking place in a dendrimer that contains three different types of light-harvesting chromophoric units and a hosted eosin molecule.[35]

functions in the periphery and $2^{(n+1)}-2$ tertiary amine units in the interior. Compound **11** (Figure 2.7) represents the fourth generation dendrimer 4D containing 30 tertiary amine units and 32 dansyl functions. The dansyl units behave independently from one another so that the dendrimers display light absorption and emission properties characteristic of dansyl, *i.e.* intense absorption bands in the near UV spectral region ($\lambda_{max} = 252$ and 339 nm; $\varepsilon_{max} \approx 12000$ and 3900 L mol^{-1} cm^{-1}, respectively, for each dansyl unit) and a strong fluorescence band in the visible region ($\lambda_{max} = 500$ nm; $\Phi_{em} = 0.46$, $\tau = 16$ ns in acetonitrile/dichloromethane 5:1 v/v).[37]

Because of the presence of the aliphatic amine units in their interior these dendrimers can play the function of ligands towards transition metal ions.

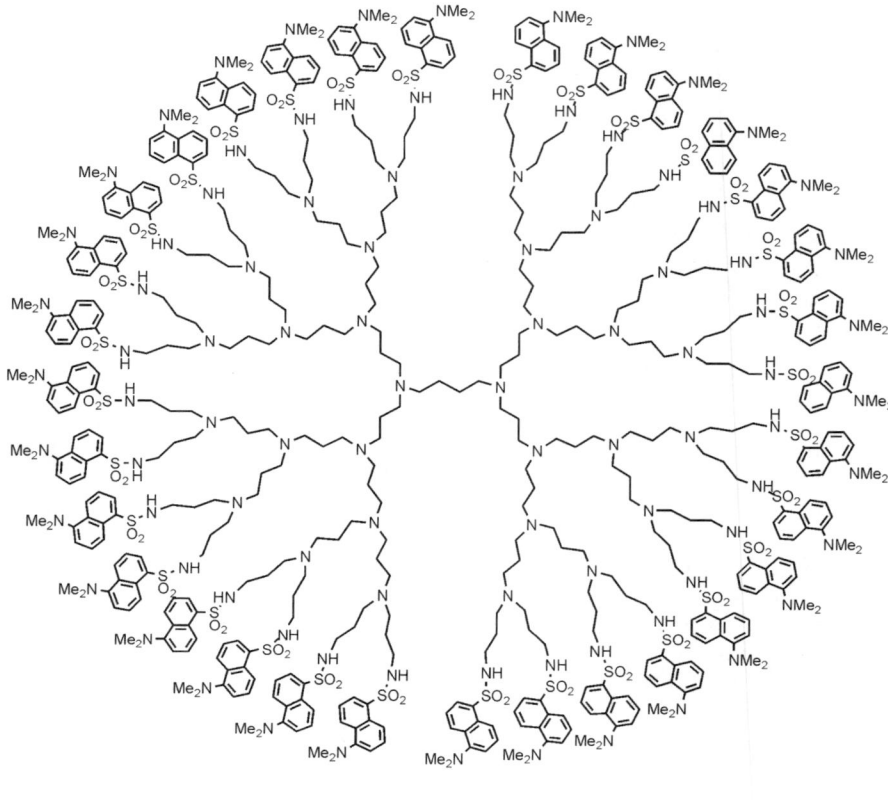

Figure 2.7 Fourth-generation dendrimer of the poly(propylene amine) family functionalised in the periphery with luminescent dansyl units.[37]

Coordination of Co^{2+} ions by **11** has been carefully studied.[38] For comparison purposes, the behaviour of a monodansyl reference compound has also been investigated. The results obtained have shown that: (i) the absorption and fluorescence spectra of a monodansyl reference compound are not affected by addition of Co^{2+} ions; (ii) in the case of the dendrimers, the absorption spectrum is unaffected, but a strong quenching of the fluorescence of the peripheral dansyl units is observed; (iii) the fluorescence quenching takes place by a static mechanism involving coordination of the metal ion, which is a fully reversible process; (iv) along the series of nD dendrimers, a strong amplification of the fluorescence quenching signal is observed with increasing generation. These results show that dendrimers can be profitably used as supramolecular fluorescent sensors for metal ions.

The advantage of a dendrimer for this kind of application is related to the fact that a single analyte can interact with a great number of fluorescent units,

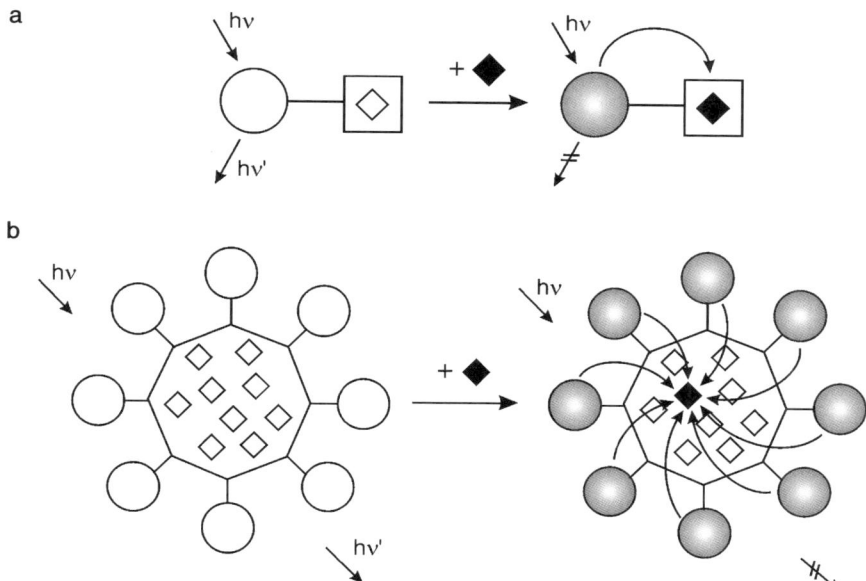

Figure 2.8 Schematic representation of (a) a conventional fluorescent sensor and (b) a fluorescent sensor with signal amplification. Open rhombi indicate coordination sites and black rhombi indicate metal ions. The curved arrows represent quenching processes. In the case of a dendrimer, the absorbed photon excites a single fluorophore component, that is quenched by the metal ion, regardless of its position.

which results in signal amplification. For example, when a Co^{2+} ion enters dendrimer **11**, the fluorescence of all the 32 dansyl units is quenched, with a 32-fold increase in sensitivity with respect to a normal dansyl sensor. This concept is illustrated in Figure 2.8.

2.2.8 Dendrimers for a Multiple Use of Light Signals

Dendrimer **12**, which consists of a benzophenone core and branches that contain six dimethoxybenzene units,[39] can undergo a great variety of photochemical and photophysical processes (see schematic energy-level diagram in Figure 2.9), that include: (i) quenching of the fluorescence of the dimethoxybenzene units by energy transfer to the benzophenone core (antenna effect), (ii) direct and sensitised fluorescence and phosphorescence of the benzophenone core, (iii) hydrogen abstraction of the triplet benzophenone core from solvent molecules, (iv) intramolecular hydrogen abstraction of the triplet benzophenone core from the dendrimer branches, (v) quenching of the phosphorescence of the benzophenone core by energy transfer to Tb^{3+} ions and dioxygen, with sensitised emission in the visible and, respectively, near-infrared spectral region. These results show that suitably designed dendrimers can make multiple use of

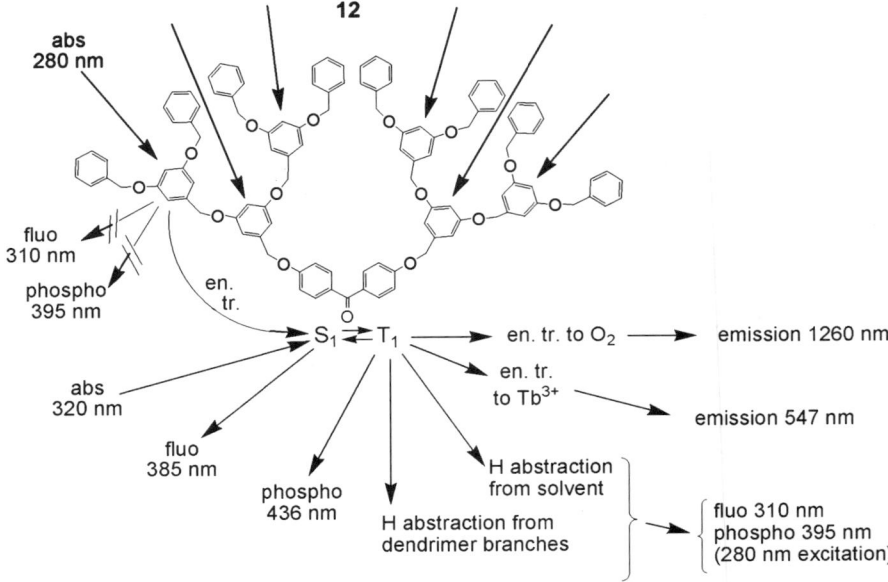

Figure 2.9 Schematic energy-level diagram illustrating the photochemical and photophysical processes that can occur in dendrimer **12**.[39]

light excitation. Indeed, dendrimer **12** can be used to illustrate most of the processes that are discussed in an entire photochemical course.

Figure 2.10 shows a schematic energy-level diagram for dendrimer **13**, consisting of a benzophenone core and branches that contain eight naphthalene units at the periphery.[40] In this dendrimer, excitation of the peripheral naphthalene units is followed by fast singlet–singlet energy transfer to the benzophenone core, but on a longer timescale a back energy transfer process takes place from the triplet state of the benzophenone core to the triplet state of the peripheral naphthalene units. Selective excitation of the benzophenone unit is followed by intersystem crossing and triplet–triplet energy transfer to the peripheral naphthalene units. This sequence of processes, which is made possible by the preorganisation of photoactive units in a dendrimer structure, can be exploited for several purposes. In hydrogen-donating solvents, the benzophenone core is protected from degradation by the presence of the naphthalene units. In solutions containing Tb^{3+}, sensitisation of the green luminescence of such a cation is observed on excitation of both the peripheral naphthalene units and the benzophenone core. Upon excitation of the naphthalene absorption band ($\lambda = 266$ nm) with a laser source, intradendrimer triplet-triplet annihilation of naphthalene excited states leads to delayed naphthalene fluorescence ($\lambda_{max} = 335$ nm), that can also be obtained upon excitation of the benzophenone core at 355 nm. In the latter case, the quite important function called energy upconversion, an example of nonlinear optics,[41] is obtained (excitation at 355 nm; emission at 335 nm).

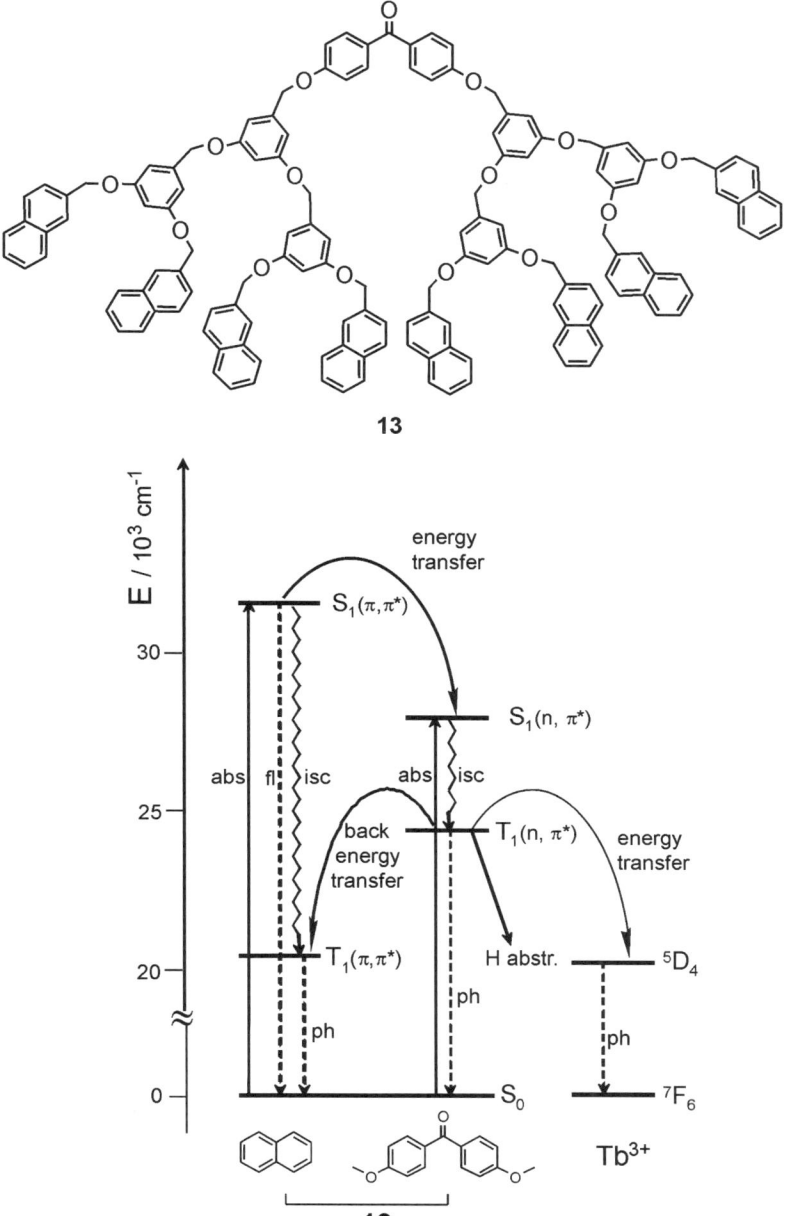

Figure 2.10 Schematic energy-level diagram illustrating the photochemical and photophysical processes that can occur in dendrimer **13**.[40]

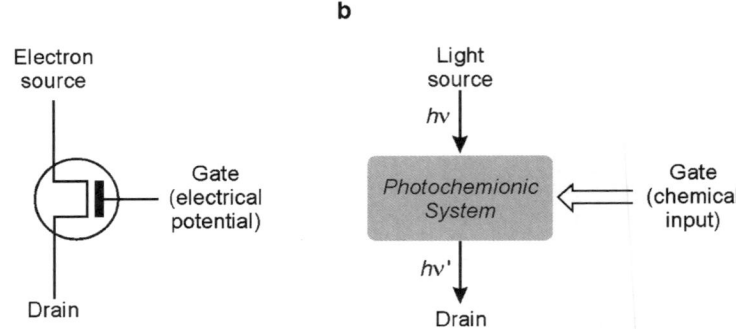

Figure 2.11 Schematic illustration of the similarity between a MOSFET electronic transistor (a) and a photochemionic gate (b).

2.2.9 Logic Gates

In a solid-state transistor the current flowing from a source to a drain can be modulated by a gate potential. It is possible to design molecular-level photochemionic systems that work on a similar principle, except that the source is a light-energy input, the drain is a light-energy output (luminescence), and the gate is a chemical input (Figure 2.11).

For the sake of space, we will only illustrate two examples of light-powered, chemical input(s), optical output logic gates, namely a classical AND gate and a recently studied system that can perform as XOR and XNOR gates. An exhaustive discussion of molecular-level logic gates can be found elsewhere.[42]

2.2.9.1 AND Logic Gate

The AND operator has two inputs and one output (Figure 2.12(a)) and in a simple electrical scheme it can be represented by two switches in series. The best examples of molecular-level AND gates are those based on two chemical inputs and an optical (fluorescence) output, but examples of molecular systems able to process chemical and optical inputs or two optical inputs with AND functions are also known.[42]

Figure 2.12(b) illustrates the case of a system, **14**, consisting of an anthracene, an aliphatic amine, and a crown ether moieties. The fluorescent excited state of the anthracene component is quenched by electron transfer from the amine and the crown ether components, but such a quenching does not occur when the amine is protonated and the crown ether associated with a Na^+ ion, as indicated in the truth table.[43] In methanol, the fluorescence quantum yield in the presence of 10^{-3} mol L^{-1} H^+ and 10^{-2} mol L^{-1} Na^+ is 0.22 (output state 1, fourth line of the truth table, Figure 2.12(c)), whereas none of the three output states 0 has quantum yield higher than 0.009.

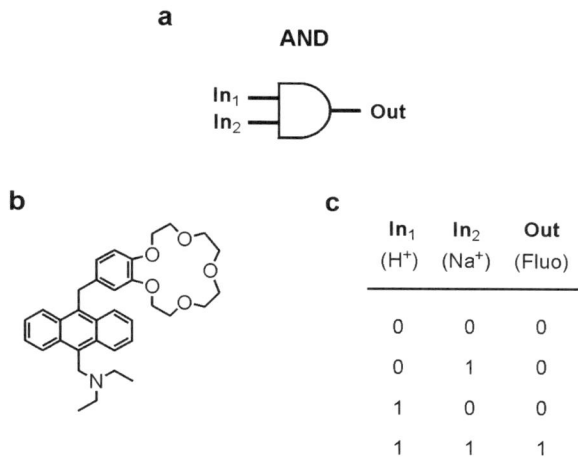

Figure 2.12 Symbolic representation (a), molecular implementation (b), and truth table (c) of an AND logic gate based on a three-component system.[43]

2.2.9.2 XOR and XNOR Logic Gates

The eXclusive OR (XOR) logic gate is particularly important because it can compare the digital state of two signals. If they are different an output 1 is given, whereas if they are the same the output is 0. This logic operation has proven to be difficult to emulate at the molecular scale, but several examples are now available.[42,44,45]

1,4,8,11-Tetraazacyclotetradecane (cyclam) in its protonated forms, can play the role of host towards cyanide metal complexes. In acetonitrile-dichloromethane 1:1 v/v solution acid-driven adducts are formed by [Ru(bpy)(CN)$_4$]$^{2-}$ with a dendrimer **15** consisting of a cyclam core appended with twelve dimethoxybenzene and sixteen naphthyl units (Figure 2.13(a)). Both [Ru(bpy)(CN)$_4$]$^{2-}$ and the dendrimer exhibit characteristic absorption and emission bands, in distinct spectral regions, that are strongly affected by addition of acid. When a solution containing equimolar amounts of [Ru(bpy)(CN)$_4$]$^{2-}$ and **15** is titrated by trifluoroacetic acid, strong spectral changes are observed with isosbestic points maintained up to the addition of two equivalents of acid. The results obtained show that protons promote association of [Ru(bpy)(CN)$_4$]$^{2-}$ and **15** and that after addition of two equivalents of acid a {[Ru(bpy)(CN)$_4$]$^{2-}$•(2H$^+$)•**15**} adduct is formed, in which the two original species share two protons (Figure 2.13(a)). In the adduct, the fluorescence of the naphthyl units is strongly quenched by very efficient energy transfer to the metal complex, as shown by the sensitised luminescence of the latter.[46] The {[Ru(bpy)(CN)$_4$]$^{2-}$•(2H$^+$)•**15**} adduct can be disrupted (i) by addition of a base (1,4 diazabicyclo[2.2.2]octane), yielding the starting species [Ru(bpy)(CN)$_4$]$^{2-}$ and **15**, or (ii) by further addition of triflic acid, with formation of (**15**•2H)$^{2+}$ and protonated forms of [Ru(bpy)(CN)$_4$]$^{2-}$. As a

a

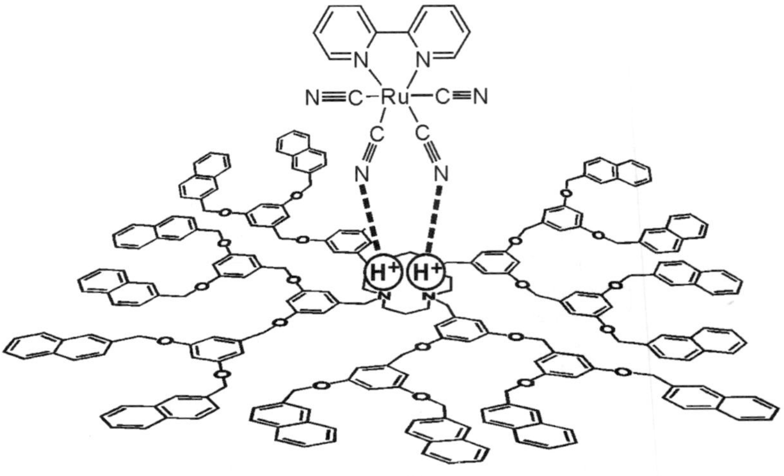

{[Ru(bpy)(CN)$_4$]$^{2-}$•(2H$^+$)•**15**}

b

IN$_1$ (acid)	IN$_2$ (base)	OUT (I$_{335\ nm}$)
0	0	0
0	1	1
1	0	1
1	1	0

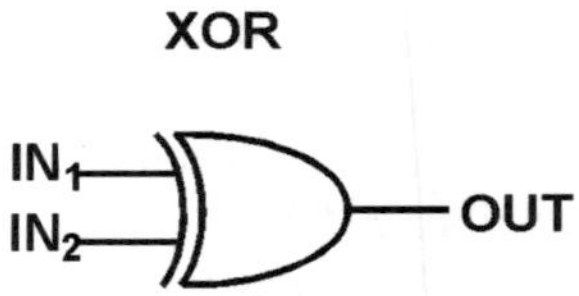

XOR

IN$_1$ (acid)	IN$_2$ (base)	OUT (I$_{680\ nm}$)
0	0	1
0	1	0
1	0	0
1	1	1

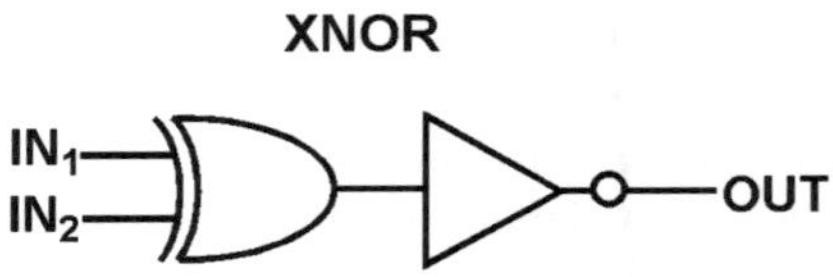

XNOR

Figure 2.13 Proton-driven adduct formation between [Ru(bpy)(CN)$_4$]$^{2-}$ and dendrimer **15** (a). Logic behaviour of the {[Ru(bpy)(CN)$_4$]$^{2-}$•(2H$^+$)•**15**} adduct upon stimulation with acid and base inputs (b).[46]

consequence it has been found that upon stimulation with two chemical inputs (acid and base) {[Ru(bpy)(CN)$_4$]$^{2-}$•(2H$^+$)•**15**} exhibits two distinct optical outputs – a naphthalene-based ($\lambda = 335$ nm) and a Ru-based ($\lambda = 680$ nm) emissions – that behave according to an XOR and an XNOR logic, respectively (Figure 2.13b).

2.3 Light-driven Molecular Machines

In green plants the energy needed to sustain the machinery of life is provided by sunlight. In general, light energy is not used as such to produce mechanical movements, but it is used to produce a chemical fuel, namely ATP, suitable for feeding natural molecular machines.[47] Light energy, however, can directly cause photochemical reactions involving large nuclear movements.[48] A simple example is a photoinduced isomerisation from the lower-energy *trans* to the higher-energy *cis* form of a molecule containing –C=C– or –N=N– double bonds, which is followed by a spontaneous or light-induced back reaction.[48] Such photoisomerisation reactions have indeed been used to design molecular machines driven by light-energy inputs.[49] In supramolecular species, photoinduced electron-transfer reactions can often cause large displacement of molecular components.[14] Indeed, working with suitable systems, an endless sequence of cyclic molecular-level movements can in principle be performed making use of light-energy inputs without generating waste products.[50] Compared to chemical-energy inputs, photochemical-energy inputs offer other advantages, besides the fundamental one of not generating waste products:[19b,i,j] (i) light can be switched on/off easily and rapidly; (ii) lasers provide the opportunity of working in very small space and very short time domains; (iii) photons, besides supplying the energy needed to make a machine work, can also be useful to "read" the state of the system and thus to control and monitor the operation of the machine. In the last few years, a great number of light-driven molecular machines have been developed and the field has been extensively reviewed.[14,19f,g,51,52] We will briefly describe a few examples.

2.3.1 Dethreading/rethreading of Pseudorotaxanes

Dethreading/rethreading of the wire and ring components of a pseudorotaxane resembles the movement of a piston in a cylinder. In order to achieve a light-induced dethreading in such piston/cylinder systems, pseudorotaxane **16**$^{4+}$ has been designed that incorporates a "light-fueled" motor[53] (*i.e.* a photosensitiser) in the ring component (Figure 2.14).[54]

Threading of the wire into the ring is thermodynamically driven because of the electron-acceptor and, respectively, electron-donor properties of the viologen and crown ether units. In deaerated solution excitation of the photosensitiser with visible light in the presence of a sacrificial electron donor (*e.g.*, triethanolamine) causes reduction of the electron–acceptor unit and, as a consequence, dethreading takes place. Rethreading can be obtained by

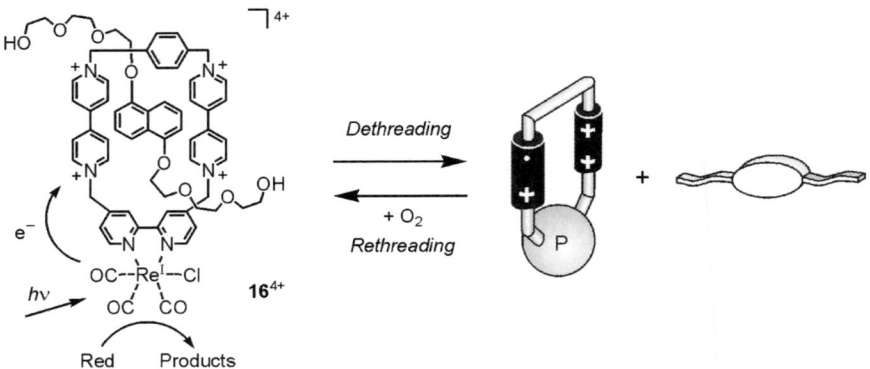

Figure 2.14 Light-driven dethreading of pseudorotaxane 16^{4+} by excitation of a photosensitiser contained in the ring-type component.[54]

allowing oxygen to enter the solution. Through a repeated sequence of deoxygenation and irradiation followed by oxygenation, many dethreading/rethreading cycles can be performed on the same solution without any appreciable loss of signal until most of the reductant scavenger is consumed. It should be pointed out, however, that photochemical systems that rely on such a sensitiser–scavenger strategy produce waste species from the decomposition of the reducing scavenger and from the successive consumption of dioxygen.

A system working on a completely different principle, in which dethreading/rethreading is exclusively governed by light energy without generation of any waste product, is illustrated in Figure 2.15.[55] The thread-like species *trans*-**17**, which contains a π-electron-rich azobiphenoxy unit, and the electron-accepting host 18^{4+} self-assemble very efficiently ($K_{a,trans} = 1.5 \times 10^5 \, \text{L mol}^{-1}$ in MeCN at 298 K) to give a pseudorotaxane. In the pseudorotaxane structure the characteristic fluorescence of free 18^{4+} is completely quenched by charge-transfer interactions. Irradiation with 365-nm light of a solution containing *trans*-**17** and 18^{4+} ($1 \times 10^{-4} \, \text{mol L}^{-1}$, 80% complexed species) causes photoisomerisation of *trans*-**17** to *cis*-**17**. Since the ring interaction with *cis*-**17** ($K_{a,cis} = 1 \times 10^4 \, \text{L mol}^{-1}$) is much weaker than that with *trans*-**17**, photoexcitation causes a dethreading process (Figure 2.15), as indicated by the strong increase in the fluorescence intensity of 18^{4+}. On irradiation at 436 nm or by warming the solution in the dark the *trans* isomer can be reformed and, as a result, it rethreads inside the cyclophane. Since there is no side reaction, the cycle can be repeated for a great number of times.

2.3.2 A Sunlight-powered Nanomotor

The rotaxane 19^{6+} (Figure 2.16) consists of six units suitably chosen and assembled in order to obtain ring shuttling powered by visible light.[56] It

Photochemically Controlled Molecular Devices and Machines 67

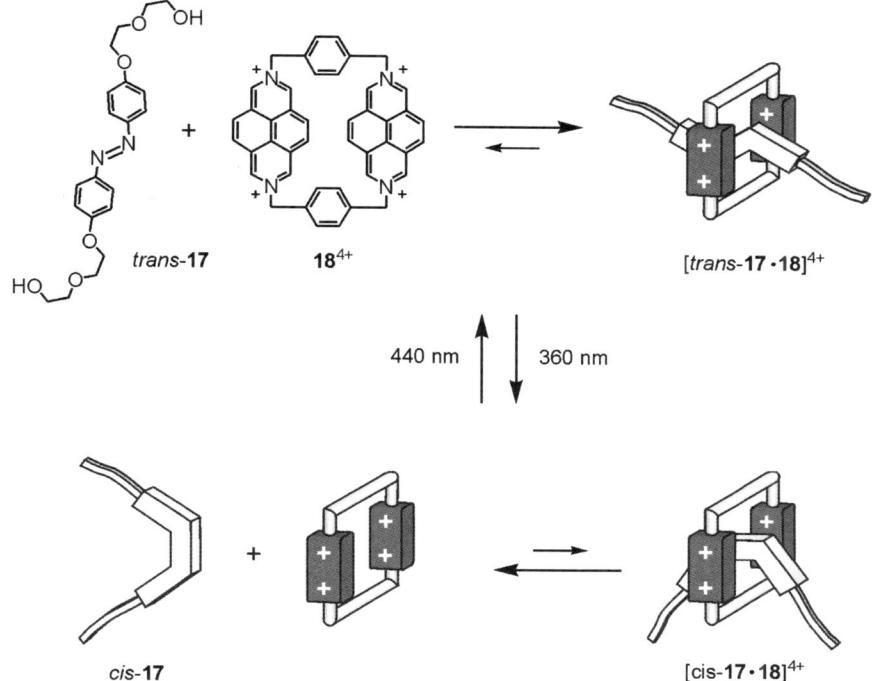

Figure 2.15 Controllable dethreading/rethreading of a pseudorotaxane based on trans–cis photoisomerisation.[55]

comprises a crown ether electron-donor macrocycle **R** (hereafter called the ring), and a dumbbell-shaped component that contains two electron-acceptor recognition sites for the ring, namely a 4,4′-bipyridinium ($\mathbf{A_1}^{2+}$) and a 3,3′-dimethyl-4,4′-bipyridinium ($\mathbf{A_2}^{2+}$) units, that can play the role of "stations" for the ring **R**. Molecular modeling shows that the overall length of $\mathbf{19}^{6+}$ is about 5 nm and the distance between the centres of the two stations, measured along the dumbbell, is about 1.3 nm. Furthermore, the dumbbell-shaped component incorporates a [Ru(bpy)$_3$]$^{2+}$-type (electron-transfer photosensitiser $\mathbf{P}^{2+}$) that also plays the role of a stopper, a *p*-terphenyl-type rigid spacer **S** that has the task of keeping the photosensitiser far from the electron-acceptor units, and finally a tetraarylmethane group **T** as the second stopper.

Electrochemical and nuclear magnetic resonance (NMR) spectroscopic data show that the stable conformation of $\mathbf{19}^{6+}$ is by far the one in which the **R** component is located around the better electron-acceptor station, $\mathbf{A_1}^{2+}$, as represented in Figure 2.16.

The mechanism devised to perform the light-driven shuttling process in the rotaxane $\mathbf{19}^{6+}$ is based on the following four phases (Figure 2.16):

(a) *Destabilisation of the stable conformation*: excitation with visible light of the photoactive unit $\mathbf{P}^{2+}$ (step 1) is followed by the transfer of an

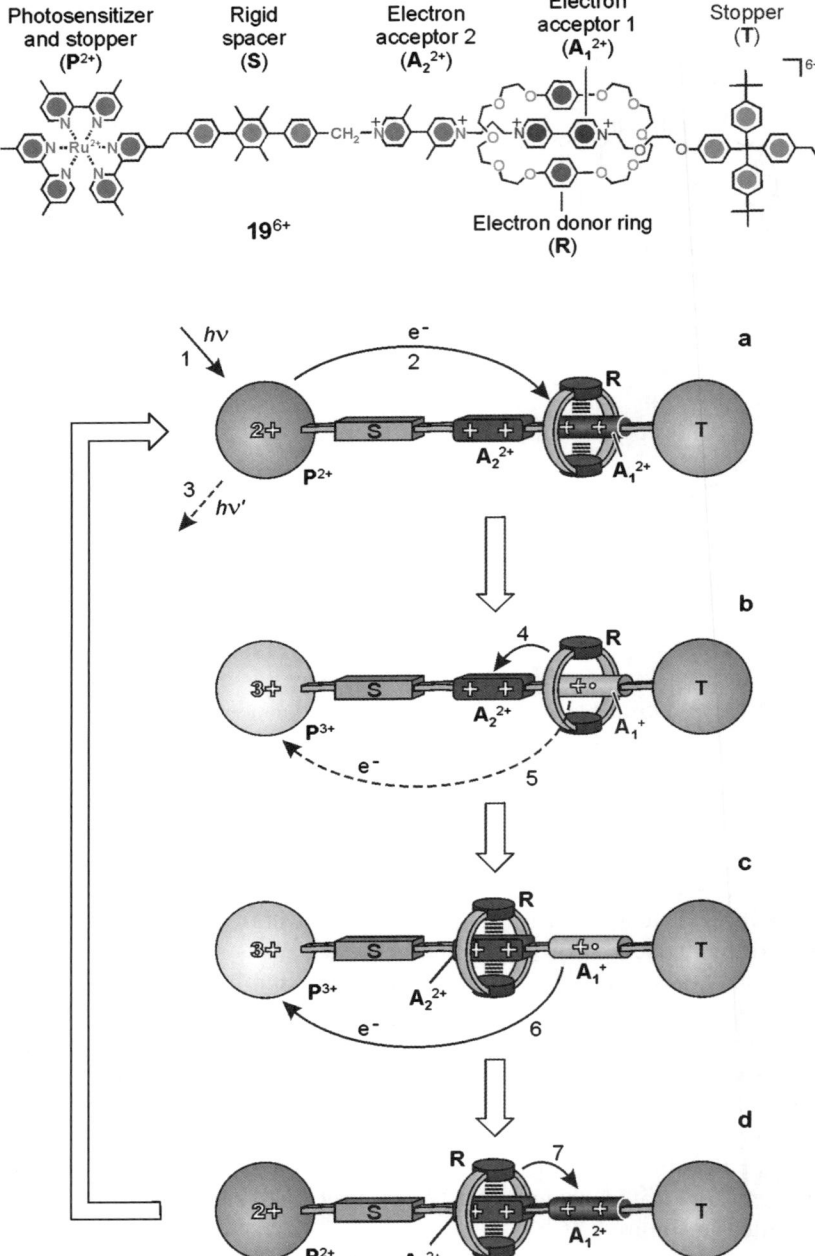

Figure 2.16 Schematic representation of the intramolecular mechanism for the photo-induced shuttling movement in rotaxane 19^{6+}.[56,57]

electron from the $*P^{2+}$ excited state to the A_1^{2+} station, which is encircled by the ring **R** (step 2), with the consequent "deactivation" of this station; such a photoinduced electron-transfer process has to compete with the intrinsic decay of $*P^{2+}$ (step 3).

(b) *Ring displacement*: after reduction ("deactivation") of the A_1^{2+} station to A_1^+, the ring moves by Brownian motion to A_2^{2+} (step 4), a step that has to compete with back electron-transfer from A_1^+ to the oxidised photoactive unit P^{3+} (step 5). This requirement is the most difficult one to meet since step 4 involves only slightly exergonic nuclear motions, whereas step 5 is an exergonic outer-sphere electron transfer process.

(c) *Electronic reset*: a back electron-transfer process from the "free" reduced station A_1^+ to P^{3+} (step 6) restores the electron-acceptor power of the A_1^{2+} station.

(d) *Nuclear reset*: as a consequence of the electronic reset, the ring moves back again by Brownian motion from A_2^{2+} to A_1^{2+} (step 7).

Reversible displacement of the ring between the two stations A_1^{2+} and A_2^{2+} should thus be obtained by light-energy inputs without consumption of chemical fuels and formation of waste products. It has been found[57] that such photoinduced ring displacement does occur, but with very low efficiency (2%). However, efficiency can be considerably improved (about 12%) in the presence of a suitable external electron relay that, without being consumed, slows down the back electron-transfer reaction, thereby leaving more time for ring displacement.

In conclusion, rotaxane **19**$^{6+}$ behaves as an autonomous linear motor powered by visible light. Each phase of the working cycle (Figure 2.16) corresponds, in turn, to the fuel injection and combustion (a), piston displacement (b), exhaust removal (c), and piston replacement (d) of a four-stroke macroscopic engine.

The low efficiency of this nanomotor may seem disappointing, but it should be noted that the fuel (sunlight) is free. Besides being powered by sunlight and operating as an autonomous motor, the investigated system shows other quite interesting properties: it works in mild environmental conditions, it is remarkably stable and it can be driven at high frequency (kHz). In principle, when working by a purely intramolecular mechanism, it is also suitable for operation at the single-molecule level.

2.4 Conclusions

It should be pointed out that the molecular-level devices and machines that have been discussed operate in solution, individually and incoherently. For some applications, they may need to be interfaced with the macroscopic world by ordering them in some way, for example at an interface or on a surface, so that they can behave coherently either in parallel or in series. Research on this topic is developing at a fast growing rate. Furthermore, addressing a single molecular-level device by instruments working at the nanometre level is no longer a dream.[58]

In his famous address to the American Physical Society, R.P. Feynman[5] concluded his reflection on the idea of constructing molecular-level machines as follows: *"What would be the utility of such machines? Who knows? I cannot see exactly what would happen, but I can hardly doubt that when we have some control of the rearrangement of things on a molecular scale we will get an enormously greater range of possible properties that substances can have, and of different things we can do"*. This sentence is still an appropriate comment to the work described in this chapter, but, now that several molecular-level devices and machines have been designed and constructed, something more can be added. Two interesting kinds of non conventional applications of these systems begin to emerge: (i) their logic behaviour can be exploited for information processing at the molecular level and, in the long run, for the construction of molecular-level (chemical) computers;[59] (ii) their mechanical features can be exploited for molecular-level transportation purposes, mechanical gating of molecular-level channels, and nanorobotics.[3b]

Acknowledgements

Financial support from FIRB (Manipolazione molecolare per macchine nanometriche).

References

1. S. McCartney, *ENIAC: The Triumphs and Tragedies of the World's First Computer*, Walker & Company, New York, 1999. See also: www.computerhistory.org.
2. (a) M. Gross, *Travels to the Nanoworld: Miniature Machinery in Nature and Technology*, Plenum, New York, 1999; (b) R. Dagani, *Chem. Eng. News*, 2000, February **28**, 36; (c) A.P. Alivisatos, *Sci. Am.*, 2001, **285**(3), 58; (d) E. Rietman, *Molecular Engineering of Nanosystems*, Springer, New York, 2001; (e) Nanotechnology. Innovation for Tomorrow's World, European Commission, EUR 21 151, 2004, pp. 1–56, available at www.cordis.lu/nanotechnology; (f) S.J. Ainsworth, *Chem. Eng. News*, 2004, April 12, 17; (g) S.R. Morrissey, *Chem. Eng. News*, 2004, April 19, 30; (h) *Nanobiotechnology: Concepts, Applications and Perspectives*, ed. C.M. Niemeyer and C.A. Mirkin, Wiley-VCH, Weinheim, 2004; (i) G. Ozin and A. Arsenault, *Nanochemistry – A Chemical Approach to Nanomaterials*, Royal Society of Chemistry, Cambridge, 2005.
3. (a) K.E. Drexler, *Sci. Am.*, 2001, **285**(3), 66; (b) A.A.G. Requicha, *Proc. IEEE*, 2003, **91**, 1922.

4. R.W. Keyes, *Proc. IEEE*, 2001, **89**, 227; see also the *International Technology Roadmap for Semiconductors (ITRS)*, 2005 Edition, available at http://public.itrs.net.
5. (a) R.P. Feynman, *Eng. Sci.*, 1960, **23**, 22; see also: http://www.feynmanonline.com.
6. C.J. Perdersen, *Angew. Chem. Int. Ed. Engl.*, 1988, **27**, 1021.
7. D.J. Cram, *Angew. Chem. Int. Ed. Engl.*, 1988, **27**, 1009.
8. J.-M. Lehn, *Angew. Chem. Int. Ed. Engl.*, 1988, **27**, 89.
9. (a) *Comprehensive Supramolecular Chemistry*, ed. J.L. Atwood, J.E.D. Davies, D.D. Macnicol and F. Vögtle, Pergamon Press, Oxford, 1996, Vols 1–10; (b) *Physical Supramolecular Chemistry*, ed. L. Echegoyen and A.E. Kaifer, Kluwer, Dordrecht, 1996; (c) *Modular Chemistry*, ed. J. Michl, Kluwer, Dordrecht, 1997; (d) *Transition Metals in Supramolecular Chemistry*, ed. J.-P. Sauvage, Wiley, New York, 1999; (e) *Supramolecular Science: Where It is and Where It is Going*, ed. R. Ungano and E. Dalcanale, Kluwer, Dordrecht, 1999; (f) H.-J. Schneider and A. Yatsimirsky, *Principles and Methods in Supramolecular Chemistry*, Wiley, Chichester, 2000; (g) J.W. Steed and J.L. Atwood, *Supramolecular Chemistry*, Wiley, Chichester, 2000; (h) J.-M. Lehn, *Proc. Natl. Acad. Sci. USA*, 2002, **99**, 4763; (i) *Encyclopedia of Supramolecular Chemistry*, ed. J.W. Steed and J.L. Atwood, Dekker, New York, 2004.
10. (a) I. Amato, *Science*, 1998, **282**, 402; (b) D. Barrow, J. Cefai and S. Taylor, *Chem. Ind.*, 1999, August 2, 591; (c) J.W. Judy, *Smart Mater. Struct.*, 2001, **10**, 1115; (d) J.P. Spatz, *Nature Mater.*, 2005, **4**, 115.
11. R.A.L. Jones, *Soft Machines – Nanotechnology and Life*, Oxford University Press, New York, 2004.
12. (a) D.A. Higgins, D.A. Vanden Bout, J. Kerimo and P.F. Barbara, *J. Phys. Chem.*, 1996, **100**, 13794; (b) M. Irie and K. Matsuda, in *Electron Transfer in Chemistry*, Ed. V. Balzani, Wiley-VCH, Weinheim, 2001, Vol. 5, p. 215; (c) R. Hillenbrand, T. Taubner and F. Keilmann, *Nature*, 2002, **418**, 159; (d) D. Courjon, *Near-field Microscopy and Near-field Optics*, Imperial College Press, London, 2003.
13. (a) J.A. Maier, M.L. Brongersma, P.G. Kik, S. Meltzer, A.A.G. Requicha and H.A. Atwater, *Adv. Mater.*, 2001, **13**, 1501; (b) S.A. Maier, M.D. Friedman, P.E. Barclay and O. Painter, *Appl. Phys. Lett.*, 2005, **86**, 071103.
14. V. Balzani, A. Credi and M. Venturi, *Molecular Devices and Machines – A Journey in the Nano World*, Wiley-VCH, Weinheim, 2003.
15. *Supramolecular Photochemistry*, ed. V. Balzani, Reidel, Dordrecht, 1987.
16. N. Sabbatini, S. Perathoner, V. Balzani, B. Alpha and J.-M. Lehn, in *Supramolecular Photochemistry*, ed. V. Balzani, Reidel, Dordrecht, 1987, p. 187.
17. D. Gust and T.A. Moore, in *Supramolecular Photochemistry*, ed. V. Balzani, Reidel, Dordrecht, 1987, p. 267.

18. V. Balzani, L. Moggi and F. Scandola, in *Supramolecular Photochemistry*, ed. V. Balzani, Reidel, Dordrecht, 1987, p. 1.
19. (a) V. Balzani, A. Credi, F.M. Raymo and J.F. Stoddart, *Angew. Chem. Int. Ed.*, 2000, **39**, 3348; (b) R. Ballardini, V. Balzani, A. Credi, M.T. Gandolfi and M. Venturi, *Acc. Chem. Res.*, 2001, **34**, 445; (c) J.-P. Collin, C.O. Dietrich-Buchecker, P. Gaviña, M.C. Jimenez-Molero and J.-P. Sauvage, *Acc. Chem. Res.*, 2001, **34**, 477; (d) M. Venturi, A. Credi and V. Balzani, in *Handbook on Electron Transfer in Chemistry*, ed. V. Balzani, Wiley-VCH, 2001, **3**, 501; (e) R. Ballardini, M.T. Gandolfi and V. Balzani, in *Handbook on Electron Transfer in Chemistry*, ed. V. Balzani, Wiley-VCH, 2001, **3**, 539; (f) Molecular Switches, ed. B.L. Feringa, Wiley-VCH, Weinheim, 2001; (g) *Struct. Bond.*, 2001, **99**, Special Volume on Molecular Machines and Motors; Guest Editor J.-P.Sauvage; (h) V. Balzani and A. Credi, *Chem. Rec.*, 2001, **1**, 422; (i) R. Ballardini, V. Balzani, A. Credi, M.T. Gandolfi and M. Venturi, *Int. J. Photoen.*, 2001, **3**, 63; (j) V. Balzani, *Photochem. Photobiol. Sci.*, 2003, **2**, 459; (k) V. Balzani, A. Credi, B. Ferrer, S. Silvi and M. Venturi, *Top. Curr. Chem.*, 2005, **262**, 1.
20. V. Balzani, A. Credi and M. Venturi, *Chem. Eur. J.*, 2002, **8**, 5524.
21. B. Schlicke, P. Belser, L. De Cola, E. Sabbioni and V. Balzani, *J. Am. Chem. Soc.*, 1999, **121**, 4207.
22. J. Walz, K. Ulrich, H. Port, H.C. Wolf, J. Wonner and F. Effenberger, *Chem. Phys. Lett.*, 1993, **213**, 321.
23. V. Balzani, A. Credi and M. Venturi, *Proc. Natl. Acad. Sci. USA*, 2002, **99**, 4814.
24. E. Ishow, A. Credi, V. Balzani, F. Spadola and L. Mandolini, *Chem. Eur. J.*, 1999, **5**, 984.
25. P.R. Ashton, R. Ballardini, V. Balzani, E.C. Constable, A. Credi, O. Kocian, S.J. Langford, J.A. Preece, L. Prodi, E.R. Schofield, N. Spencer, J.F. Stoddart and S. Wenger, *Chem. Eur. J.*, 1998, **4**, 2413.
26. B. Ferrer, G. Rogez, A. Credi, R. Ballardini, M.T. Gandolfi, V. Balzani, Y. Liu, H.-R. Tseng and J.F. Stoddart, *Proc. Nat. Acad. Sci.*, 2006, **103**, 18411.
27. (a) C.A. Schalley, F. Vögtle, ed. *Dendrimers V: Functional and Hyperbranched Building Blocks, Photophysical Properties, Applications in Materials and Life Sciences*. In: Top. Curr. Chem., 2003, **228**; (b) G.R. Newkome, C. Moorefield and F. Vögtle, *Dendrimers and Dendrons: Concepts, Syntheses, Perspectives*, VCH, Weinheim, 2001; (c) *Dendrimers and other Dendritic Polymers*, ed. J.M.J. Fréchet and D.A. Tomalia, John Wiley and Sons, Chichester, UK, 2001.
28. For some recent reviews, see: (a) D.A. Tomalia, J.M.J. Fréchet, ed. Special Issue: Dendrimers and Dendritic Polymers. In: *Prog. Polym. Sci.*, 2005, **30** (3–4). (b) R.W.J. Scott, O.M. Wilson and R.M. Crooks, *J. Phys. Chem. B*, 2005, **109**, 692.
29. For some recent reviews, see: (a) P. Ceroni, G. Bergamini, F. Marchioni, V. Balzani, *Prog. Polym. Sci.*, 2005, **30**, 453. (b) F.C. De Schryver,

T. Vosch, M. Cotlet, M. Van der Auweraer, K. Müllen and J. Hofkens, *Acc. Chem. Res.*, 2005, **38**, 514. (c) Goodson, T.G., III. *Acc. Chem. Res.*, 2005, **38**, 99.
30. T. Pullerits and V. Sundström, *Acc. Chem. Res.*, 1996, **29**, 381.
31. T. Ritz, A. Damjanovic and K. Schulten, *ChemPhysChem*, 2002, **3**, 243.
32. V. Balzani, P. Ceroni, M. Maestri and V. Vicinelli, *Curr. Opin. Chem. Biol.*, 2003, **7**, 657.
33. M. Plevoets, F. Vögtle, L. De Cola and V. Balzani, *New. J. Chem.*, 1999, **23**, 63.
34. S. Hecht and J.M.J. Fréchet, *Angew. Chem. Int. Ed.*, 2001, **40**, 74.
35. U. Hahn, M. Gorka, F. Vögtle, V. Vicinelli, P. Ceroni, M. Maestri and V. Balzani, *Angew. Chem. Int. Ed.*, 2002, **41**, 3595.
36. (a) F. Vögtle, M. Gorka, V. Vicinelli, P. Ceroni, M. Maestri and V. Balzani, *ChemPhysChem*, 2001, **2**, 769; (b) V. Vicinelli, P. Ceroni, M. Maestri, V. Balzani, M. Gorka and F. Vögtle, *J. Am. Chem. Soc.*, 2002, **124**, 6461.
37. F. Vögtle, S. Gestermann, C. Kauffmann, P. Ceroni, V. Vicinelli, L. De Cola and V. Balzani, *J. Am. Chem. Soc.*, 1999, **121**, 12161.
38. (a) V. Balzani, P. Ceroni, S. Gestermann, C. Kauffmann, M. Gorka and F. Vögtle, *Chem. Commun.*, 2000, 853; (b) V. Balzani, P. Ceroni, V. Vicinelli, S. Gestermann, M. Gorka, C. Kauffmann and F. Vögtle, *J. Am. Chem. Soc.*, 2000, **122**, 10398.
39. G. Bergamini, P. Ceroni, V. Balzani, F. Vögtle and S.-K. Lee, *ChemPhysChem*, 2004, **5**, 315.
40. G. Bergamini, P. Ceroni, M. Maestri, V. Balzani, S.-K. Lee and F. Vögtle, *Photochem. Photobiol. Sci.*, 2004, **3**, 898.
41. *Handbook of Nonlinear Optics*, ed. R.L. Sutherland, Dekker, New York, NY, 2nd edn, 2003.
42. (a) A.P. de Silva, H.Q.N. Gunaratne, T. Gunnlaugsson, A.J.M. Huxley, C.P. McCoy, J.T. Rademacher and T.E. Rice, *Chem. Rev.*, 1997, **97**, 1515; (b) A.P. de Silva, N.D. McClenaghan and C.P. McCoy, in *Electron Transfer in Chemistry*, ed. V. Balzani, Wiley-VCH, Weinheim, 2001, **5**, 156; (c) A.P. de Silva, N.D. McClenaghan and C.P. McCoy, in *Molecular Switches*, ed. B.L. Feringa, Wiley-VCH, Weinheim, 2001, p. 339; (d) A.R. Pease and J.F. Stoddart, *Struct. Bond.*, 2001, **99**, 189; (e) F. M. Raymo, *Adv. Mater.*, 2002, **14**, 401; (f) V. Balzani, A. Credi and M. Venturi, *ChemPhysChem*, 2003, **4**, 49.
43. A.P. de Silva, H.Q.N. Gunaratne and C.P. McCoy, *J. Am. Chem. Soc.*, 1997, **119**, 7891.
44. A. Credi, V. Balzani, S.J. Langford and J.F. Stoddart, *J. Am. Chem. Soc.*, 1997, **119**, 2679.
45. F. Pina, M.J. Melo, M. Maestri, P. Passaniti and V. Balzani, *J. Am. Chem. Soc.*, 2000, **122**, 4496.
46. G. Bergamini, C. Saudan, P. Ceroni, M. Maestri, V. Balzani, M. Gorka, S.-K. Lee, J. van Heyst and F. Vögtle, *J. Am. Chem. Soc.*, 2004, **126**, 16466.

47. D.-P. Hader and M. Tevini, *General Photobiology*, Pergamon, Oxford, 1987.
48. V. Balzani and F. Scandola, *Supramolecular Photochemistry*, Horwood, Chichester, 1991.
49. See, *e.g.*: (a) S. Shinkai, T. Nakaji, T. Ogawa, K. Shigematsu and O. Manabe, *J. Am. Chem. Soc.*, 1981, **103**, 111; (b) M. Irie and M. Kato, *J. Am. Chem. Soc.*, 1985, **107**, 1024; (c) M. Takeshita and M. Irie, *J. Org. Chem.*, 1998, **63**, 6643; (d) A. Bencini, A. Bernardo, A. Bianchi, M. Ciampolini, V. Fusi, N. Nardi, A.J. Parola, F. Pina and B. Valtancoli, *J. Chem. Soc. Perkin Trans. 2*, 1998, 413; (e) N. Koumura, R.W.J. Zijlstra, R.A. van Delden, N. Harada and B.L. Feringa, *Nature*, 1999, **401**, 152; (f) A. Natanshon and P. Rochon, *Chem. Rev.*, 2002, **102**, 4139; (g) T. Hugel, N.B. Holland, A. Cattani, L. Moroder, M. Seitz and H.E. Gaub, *Science*, 2002, **296**, 1103; (h) N. Komura, E.M. Geertsema, A. Meetsma and B.L. Feringa, *J. Am. Chem. Soc.*, 2002, **124**, 5037; (i) F. Vögtle, M. Gorka, R. Hesse, P. Ceroni, M. Maestri and V. Balzani, *Photochem. Photobiol. Sci.*, 2002, **1**, 45; (j) T. Muraoka, K. Kinbara, Y. Kobayashi and T. Aida, *J. Am. Chem. Soc.*, 2003, **125**, 5612; (k) R.A. van Delden, N. Koumura, A. Schoevaars, A. Meetsma and B.L. Feringa, *Org. Biol. Chem.*, 2003, **1**, 33; (l) L.X. Liao, F. Stellacci and D.V. McGrath, *J. Am. Chem. Soc.*, 2004, **126**, 2181; (m) S. Yagai, T. Karatsu and A. Kitamura, *Chem. Eur. J.*, 2005, **11**, 4054; (n) B. Sapich, A.B.E. Vix, J.P. Rabe and J. Stumpe, *Macromolecules*, 2005, **38**, 10480; (o) M.K.J. ter Wiel, R.A. van Delden, A. Meetsma and B.L. Feringa, *J. Am. Chem. Soc.*, 2005, **127**, 14208; (p) J. Vicario, A. Meetsma and B.L. Feringa, *Chem. Commun.*, 2005, **5910**; (q) R.A. van Delden, M.K.J. ter Wiel, M.M. Pollard, J. Vicario, N. Koumura and B.L. Feringa, *Nature*, 2005, **437**, 1337.
50. See, *e.g.*: A.M. Brouwer, C. Frochot, F.G. Gatti, D.A. Leigh, L. Mottier, F. Paolucci, S. Roffia and G.W.H. Wurpel, *Science*, 2001, **291**, 2124.
51. *Acc. Chem. Res.*, 2001, **34**(6), Special Issue on Molecular Machines; Guest Editor J.F. Stoddart.
52. *Top. Curr. Chem.*, 2005, **262**, Special Volume on Molecular Machines; Guest Editor T.R. Kelly.
53. M. Freemantle, *Chem. Eng. News*, 1998, October 26, 37.
54. P.R. Ashton, V. Balzani, O. Kocian, L. Prodi, N. Spencer and J. F. Stoddart, *J. Am. Chem. Soc.*, 1998, **120**, 11190.
55. V. Balzani, A. Credi, F. Marchioni and J.F. Stoddart, *Chem. Commun.*, 2001, 1860.
56. P.R. Ashton, R. Ballardini, V. Balzani, A. Credi, K.R. Dress, E. Ishow, C.J. Kleverlaan, O. Kocian, J.A. Preece, N. Spencer, J.F. Stoddart, M. Venturi and S. Wenger, *Chem. Eur. J.*, 2000, **6**, 3558.
57. V. Balzani, M. Clemente-León, A. Credi, B. Ferrer, M. Venturi, A.H. Flood and J.F. Stoddart, *Proc. Natl. Acad. Sci.*, 2006, **103**, 1178.
58. (a) S. Weiss, *Science*, 1999, **283**, 1676; (b) *Single Molecule Detection in Solution*, ed. Ch. Zander, J. Enderlein and R.A. Keller, Wiley-VCH,

Berlin, 2002; (c) M.F. Garcia-Parajo, J. Hernando, G.S. Mosteiro, J.P. Hoogenboom, E.M.H.P. van Dijk and N.F. van Hulst, *ChemPhys Chem*, 2005, **6**, 819; (d) R.M. Metzger in Intelligent Materials, ed. M.S. Shahinpoor, H.J. Schneider, 2006, p. XXX.
59. (a) D. Rouvray, *Chem. Brit.*, 2000, **36**(12), 46; (b) P. Ball, *Nature*, 2000, **406**, 118.

CHAPTER 3
Transition-Metal Complex-Based Molecular Machines

BENOÎT CHAMPIN, ULLA LÉTINOIS-HALBES AND
JEAN-PIERRE SAUVAGE*

Laboratoire de Chimie Organo-Minérale, LC 3, UMR 7177 du CNRS, Université Louis Pasteur, Faculté de Chimie, 4 rue Blaise Pascal, 67070, Strasbourg Cedex, France

3.1 Introduction

In the course of the last 15 years, a new field has experienced a spectacular development: the elaboration of dynamic molecular systems for which large-amplitude motions can be induced and controlled from the outside.[1] In this research area, the molecules, often referred to as "molecular machines", display two or several distinct geometries that can be interconverted into one another in a reversible way by various processes.

The rapid growth of this field can be linked to the discovery and a better understanding of numerous dynamic biological systems (motor proteins) whose controlled motions correspond to important biological functions. Such biological motors have been studied in great detail[2] and, for several of them, it has even been possible to visualise the movement while they are in action.[3] Classical examples are ATP synthase, a rotary motor, the actin-myosin complex of the striated muscle, acting as a linear motor, or the kinesins, essential motor proteins able to "walk" on the microtubules and to transport important molecular components of the cell over large distances.

Another important motivation for making artificial molecular machines using the tools of synthetic chemistry is probably more applied, although practical applications of the field are not likely to be discovered in a very near future. The potential of the field in relation to nanodevices as well as to information storage and processing at the molecular level has been abundantly discussed in the recent literature.[4] Artificial "nanomachines" and "nanomotors" are indeed promising, in a long-term perspective, to sort molecules, to transport them across a membrane, to act as valves, able to pilot the delivery of a drug in a given medium, *etc*. Memory storage devices or even molecular computing have also triggered much interest in recent years.

In the present chapter, we will mostly focus on transition-metal-containing molecular machines but purely organic systems are equally important. The contributions of Balzani, Stoddart, Leigh, Harada and others and their coworkers, using catenanes or rotaxanes, represent real breakthroughs.[5] Similarly, the noninterlocking systems proposed by Ferringa, Kelly and others are particularly novel and can also be regarded as pioneering contributions.[6] The photochemical isomerisation of C=C double bonds is a useful process that has been utilised to set molecular fragments in motion within a given molecule.[7]

The use of transition-metal complexes is particularly attractive since metal centres are often electroactive, allowing to induce rearrangements via a metal-localised redox signal, thus circumventing any potential difficulty associated to the generation of organic radicals. In addition, the structure of some transition-metal complexes can be profoundly modified by modifying the pH of the medium or by generating a dissociative excited state, thus allowing to set some parts of the compounds in motion using a chemical signal or a photonic impulse.

It should be noted that the present chapter is by no means exhaustive. We have rather selected a few representative examples in the recent literature but our choice is, of course, arbitrary and reflects the scientific interests of our group.

3.2 Molecular Motions Driven by a Chemical Reaction

3.2.1 Use of a Chemical Reaction to Induce the Contraction/Stretching Process of a Muscle-like Rotaxane Dimer[8]

Linear machines and motors are essential in many biological processes such as, in particular, contraction and stretching of the skeletal muscles. In relation to "artificial muscles", one-dimensional molecular assemblies able to undergo stretching and contraction motions represented thus an exciting target.

A multicomponent system able to contract or stretch under the action of an external chemical signal was designed and made in our group a few years ago. The system is based on a symmetrical doubly threaded topology as represented in Figure 3.1. The motion is easy to visualise: both "strings" (mimicking the myosin-containing thick filament and the actin thin filament of the striated muscle) move along one another but stay together thanks to the rotaxane nature of the system.

The copper-complexed rotaxane dimer 1^{2+} was synthesised (more than 20 steps from commercially available compounds). As shown in Figure 3.2, each "filament" contains both a bidentate chelate (coordinated to copper(I) in compound 1^{2+}) and a tridentate chelate of the terpy type, which is free in the

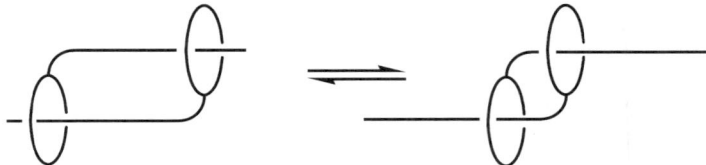

Figure 3.1 Gliding of the filaments in a rotaxane dimer: interconversion of the stretched geometry and the contracted conformation.

copper(I) complex 1^{2+}. The rotaxane dimer was set in motion by exchanging the complexed metal centres. The free ligand, obtained in quantitative yield by reacting the 4-coordinate copper(I) complex 1^{2+} (stretched geometry) with an excess of KCN, was subsequently remetalated with $Zn(NO_3)_2$ affording quantitatively the 5-coordinate Zn^{2+} complex 2^{4+} in the contracted situation (Figure 3.2). The reverse motion, leading back to the extended situation 1^{2+}, could be easily induced upon addition of excess $Cu(CH_3CN)_4^+$. From CPK model estimations, the length of the organic backbone changes from 85 to 65 Å between both situations.

3.2.2 Intramolecular Complexation/Decomplexation Processes as a Means to Make an Intermittent Degenerate Molecular Shuttle

The group of Otera invented a molecular shuttle that can be converted repeatedly between dynamic and static states through alternating intermolecular complexation and decomplexation processes.[9] In principle, a molecular shuttle with two identical "stations" can not be regarded as a molecular machine since the system can not be converted from form A to form B by an external signal. Nevertheless, the present system introduces an interesting concept related to the use of a chemical signal to "freeze" the motion. It is related to the "molecular brake" described some time ago by Kelly et al.[10]

Otera and coworkers designed the [2]rotaxane **3** containing a central bipyridine moiety for effective formation of a chelate and bipyridinium units to act as stations for a crown ether bead (Figure 3.3).

The crown ether moves back and forward between the two equivalent bipyridinium stations. Addition of a solution of $Cu(CH_3CN)_4PF_6$ and subsequent removal of the solvent affords the complex **4**, which is formed by chelation of the central bipyridine ligands to a copper(I) ion. NMR measurements show that the shuttling of the macrocycle is now hampered by the blockade in its path (Figure 3.4).

Treatment of this complex with a suspension of an ion exchange resin leads to the complete decomplexation and the original state of compound **3** is re-established. Yet, this system is not a "real" molecular machine, strictly speaking.

Transition-Metal Complex-Based Molecular Machines

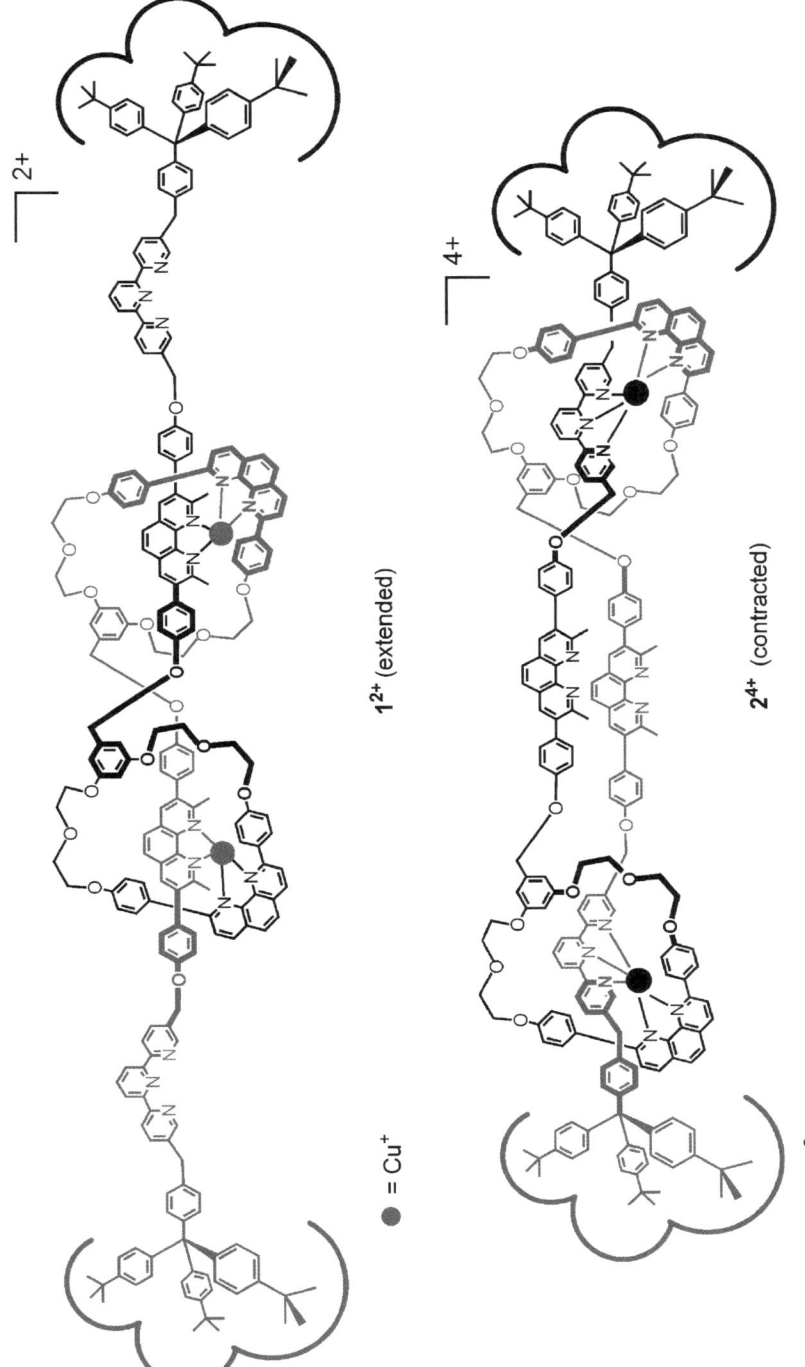

Figure 3.2 The two states of the muscle-like molecule.

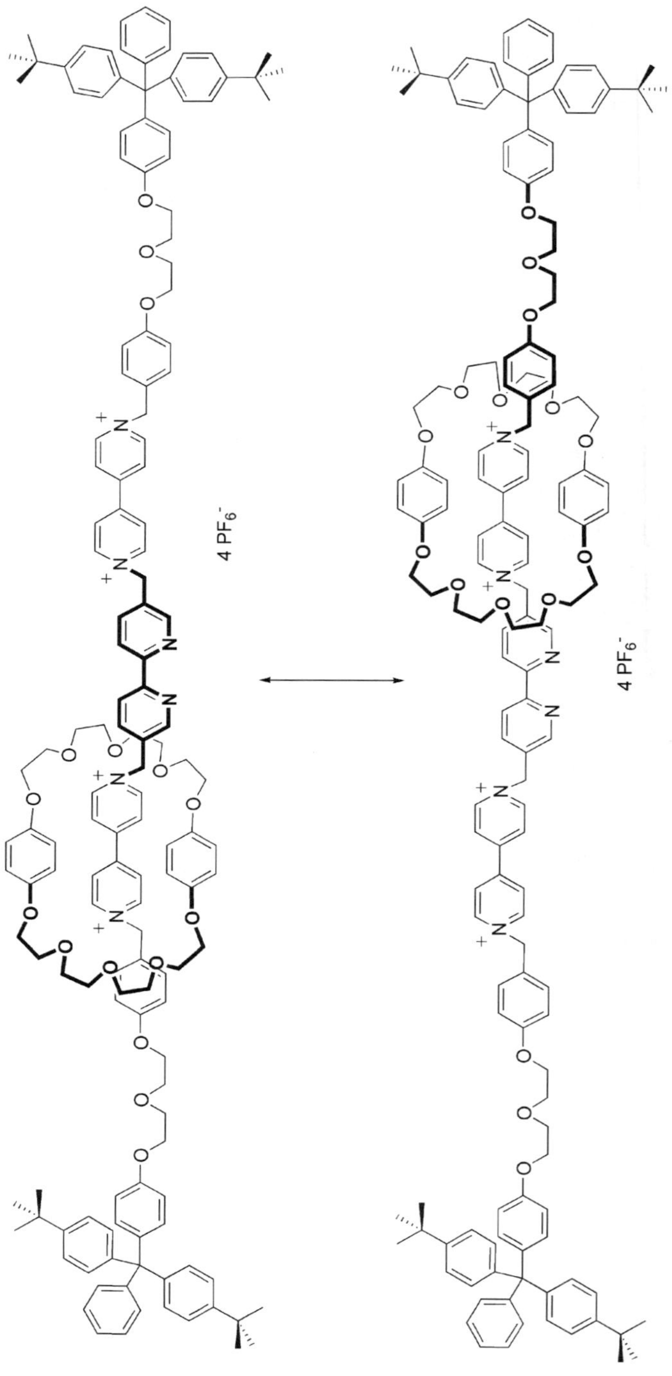

Figure 3.3 Shuttling of the crown-ether along the thread within [2]rotaxane **3**.

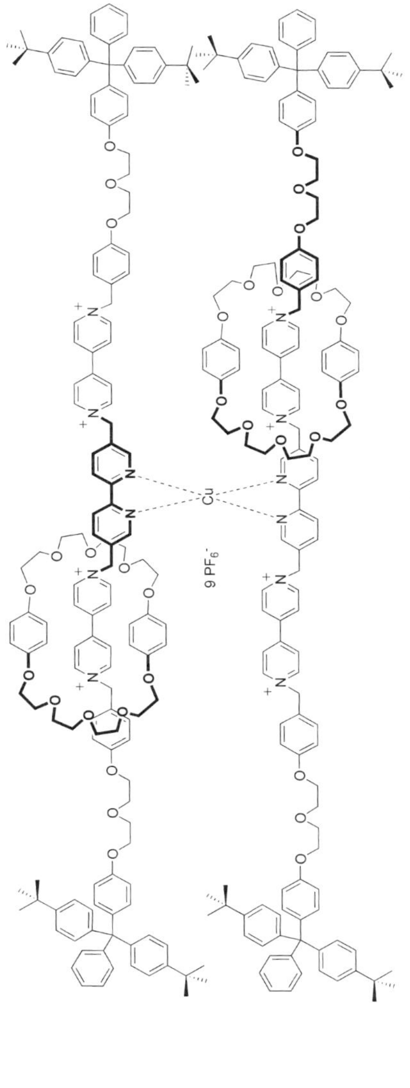

Figure 3.4 Chelation of central bipyridine units to a Cu(I) ion prevents macrocycle from gliding along the thread.

Hence, no redundant reagents are accumulated in the system during the process thus keeping the system clean. Although the switching is not quick at present, the protocol of complexation and decomplexation in combination with an immobilised resin has potential as a new type of molecular switch. Moreover, the construction of a real molecular machine would imply desymmetrisation of the two stations, so that the system could be converted from one form to the other by an external signal.

3.2.3 Molecular Machines Based on Metal-ion Translocation

Molecular machines could also enter the field of molecular recognition by acting as receptors with a useful implemented function: to recognise and bind a substrate only when the proper external stimulus is applied. The molecular machine of Fabbrizzi et al.[11] behaves as such a receptor that merges the advantages of the lock-and-key principle (selectivity) with those of its adaptive behaviour that is activation in the presence of substrates. Moreover, a drastic colour change associated with movement and recognition events turns this system into a very efficient colorimetric sensor.

Macrocycle LH_4 contains two couples of polydentate compartments capable of binding copper(II) ions and comprises two diamide-diamine tetradentate and two pyridine-diamine tridentate binding sets that share the four secondary amino groups.[12] Two distinct colours and types of band are observed as a function of pH.

At low pH (3–6) the amide groups are protonated, thus they are non-coordinating and the two copper(II) ions are coordinated by the two available separated tridentate PDA units. This form is the complex **6**, $Cu_2(LH_4)^{4+}$, in which each copper(II) is tricoordinated by the ligand ($\lambda_{max} = 660$ nm, deep blue), with the other coordination positions occupied by water.[13]

On raising the pH (6–9) two water molecules are deprotonated to give the blue species $Cu_2(LH_4)(OH)_2^{2+}$ and $Cu_2(LH_4)(OH)^{3+}$. But if the pH is raised above 10.5 the four amide protons are released and each copper(II) moves inside one of the two deprotonated diamide-diamine moieties to give the neutral complex **5**, $Cu_2(L)$ (Figure 3.5).[14] This is reflected by a purple-pink solution.

This system was modified into a molecular receptor recognising specifically the substrate imidazole.

The key-and-lock principle could be achieved in this system with imidazole. In a narrower pH range ($10.0 < pH < 10.4$) in the absence of imidazole, the $Cu_2(L)$ species **5** is predominant in its closed form (95%), whereas in the presence of imidazole, it represents 10% of the mixture and the $Cu_2(LH_4)(Im^-)$ species **7** around 90% (Figure 3.6). The species contains a bridging imidazolate anion that forms thanks to the particularly stable $Cu(II)-Im^--Cu(II)$ disposition.[15] Therefore, in this system, it is the substrate itself that makes the cations translocate and causes the system to open, thus allowing binding to take place. This recognition is associated with a neat colour change, thus providing a signal for selective inclusion.

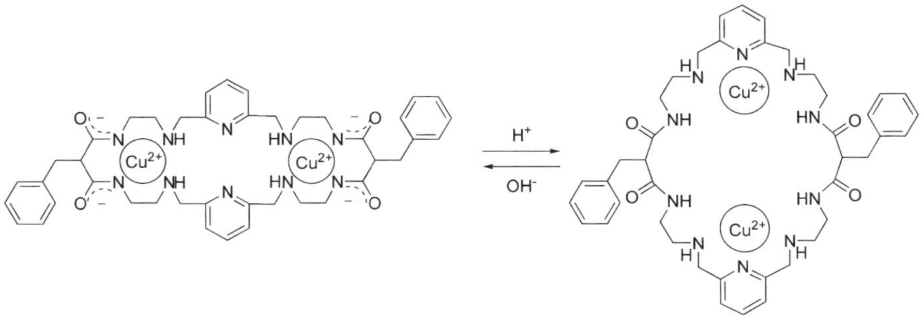

5 : [Cu$_2$(L)]   **6** : [Cu$_2$(LH$_4$)]$^{4+}$

Figure 3.5 pH-dependent intramolecular dislocation of the Cu^{2+} ions.

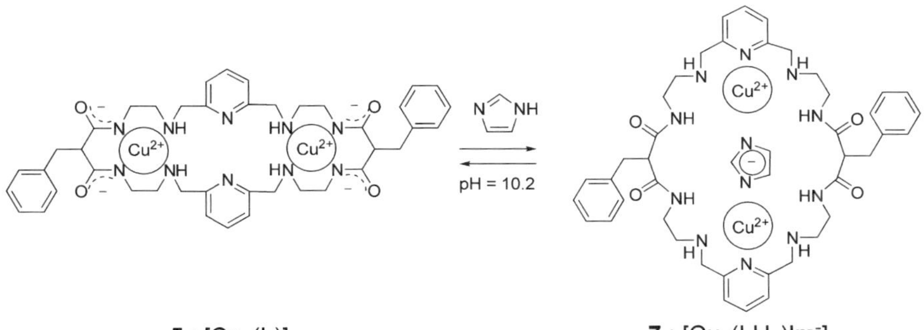

5 : [Cu$_2$(L)]   **7** : [Cu$_2$(LH$_4$)Im$^-$]

Figure 3.6 The imidazole-induced translocation equilibrium at pH 10.2.

It should be noted that many systems not containing transition metals, but where molecular motions are also driven by a chemical reaction, have been reported recently in the literature.[16–18]

3.3 Electrochemically Induced Motions

3.3.1 Transition-metal-complexed Catenanes and Rotaxanes

3.3.1.1 *A Copper-complexed [2]Catenane in Motion with Three Distinct Geometries*

Multistage systems seem to be uncommon, although they are particularly challenging and promising in relation to nanodevices aimed at important electronic functions and, in particular, information storage.[19–22] Among the few examples that have been reported in recent years, three-stage catenanes are particularly significant since they lead to unidirectional rotary motors.[23]

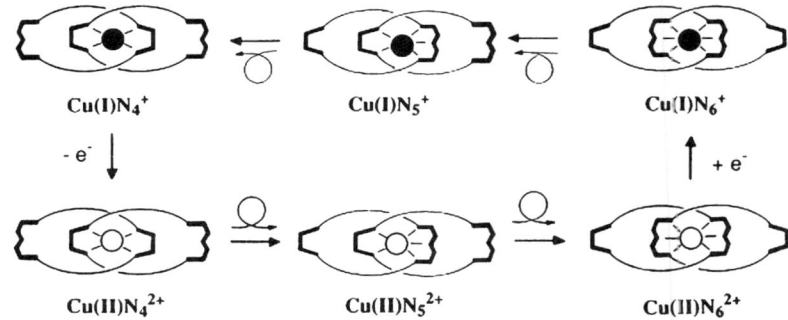

Figure 3.7 A three-configuration Cu(I) catenate whose general molecular shape can be dramatically modified by oxidising the central metal (Cu(I) to Cu(II)) or reducing it back to the monovalent state. Each ring of the [2]-catenate incorporates two different coordinating units: the bidentate dpp unit (dpp) 2,9-diphenyl-1,10-phenanthroline) is symbolised by a U whereas the terpy fragment (2,2':6',2''-terpyridine) is indicated by a stylised W. Starting from the tetracoordinate monovalent Cu complex (Cu(I)N_4^+; top left) and oxidising it to the divalent state (Cu(II)N_4^{2+}), a thermodynamically unstable species is obtained that should first rearrange to the pentacoordinate complex Cu(II)N_5^{2+} by gliding of one ring (left) within the other and, finally, to the hexacoordinate stage Cu(II)N_6^{2+} by rotation of the second cycle (right) within the first one. Cu(II)N_6^{2+} is expected to be the thermodynamically stable divalent complex. The double ring-gliding motion following oxidation of Cu(I)N_4^+ can be inverted by reducing Cu(II)N_6^{2+} to the monovalent state (Cu(I)N_6^+; top right), as represented on the top line of the figure.

In the mid-1990s, our group described a particular Cu-complexed 2-catenane that represents an example of such a multistage compound.[24] The molecule displays three distinct geometries, each stage corresponding to a different coordination number of the central complex (CN = 4, 5, or 6). The principle of the three-stage electrocontrollable catenane is represented in Figure 3.7.

Similarly to the very first and simpler catenane made in our group for which a large-amplitude motion can deliberately be triggered by an external signal,[25] the gliding of the rings in the present system relies on the important differences of stereochemical requirements for coordination of Cu(I) and Cu(II). For the monovalent state the stability sequence is CN = 4 > CN = 5 > CN = 6. On the contrary, divalent Cu is known to form stable hexacoordinate complexes, with pentacoordinate systems being less stable and tetrahedral Cu(II) species being even more strongly disfavoured.

The synthesis of the key catenate Cu(I)N_4^+PF$_6^-$ = 8_4^+ (Figure 3.8(a)) (one should notice that, as usual, the subscripts 4, 5 and 6 indicate the coordination number of the copper centre) derives from the usual three-dimensional template strategy.[26,27]

The visible spectrum of this deep red complex shows a metal-to-ligand charge transfer (MLCT) absorption band (λ_{max} = 439 nm, ε = 2570 mol^{-1} L cm^{-1},

Transition-Metal Complex-Based Molecular Machines 85

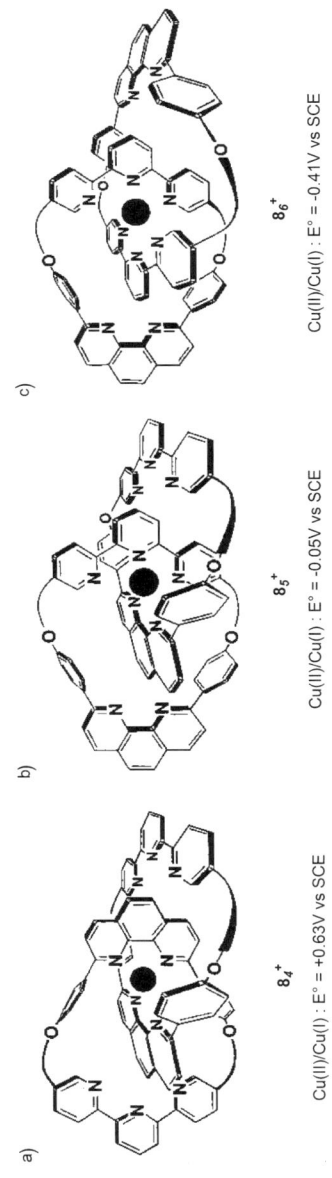

Figure 3.8 The three forms of the copper-complexed catenane, each species being either a monovalent or a divalent complex. (a) 4-coordinate complex (b) 5-coordinate complex (c) 6-coordinate complex.

MeCN). Cyclic voltammetry of a MeCN solution shows a reversible redox process at +0.63 V (vs. SCE). Both the CV data and the UV-vis spectrum are similar to those of other related species.[25,27] The reaction of $\mathbf{8_4}^+$ with KCN afforded the free catenand (not represented), which was subsequently reacted with Cu(BF$_4$)$_2$ to give $\mathbf{8_6}^{2+}$ as a very pale green complex. The hexacoordinate structure of this species was evidenced by UV-vis spectroscopy and electrochemistry. The cyclic voltammogram shows an irreversible reduction at −0.43 V (vs. SCE, MeCN). These data are similar to the ones obtained for the complex Cu(diMe-tpy)$_2$(BF$_4$)$_2$ (diMe-tpy = 5,5″-dimethyl-2,2′:6′,2″-terpyridine).

When a MeCN dark red solution of $\mathbf{8_4}^+$ was oxidised by an excess of NO$^+$BF$_4^-$, a green solution of $\mathbf{8_4}^{2+}$ was obtained. The CV is the same as for the starting complex, and the visible absorption spectrum shows a band at $\lambda_{max} = 670$ nm, $\varepsilon = 810$ mol^{-1} L cm^{-1}, in MeCN, typical of these tetrahedral Cu(II) complexes.[25] A decrease of the intensity of this band was observed when monitoring it as a function of time. This fact is due to the gliding motion of the rings to give the penta- and hexacoordinate Cu(II) complexes, whose extinction coefficients are lower as compared to that for $\mathbf{8_4}^{2+}$ (ca. 125 and 100, respectively). The final product is $\mathbf{8_6}^{2+}$ as indicated by the final spectro- and electrochemical data. A similar behaviour was observed when a solution of $\mathbf{8_4}^+$ was electrochemically oxidised.

When either the $\mathbf{8_6}^{2+}$ solution resulting from this process or a solution prepared from a sample of isolated solid $\mathbf{8_6}^{2+}$(BF$_4^-$)$_2$ were electrochemically reduced at −1 V, the tetracoordinate catenate was quantitatively obtained. The cycle depicted in Figure 3.7 was thus completed. The changeover process for the monovalent species is faster than the rearrangement of the Cu(II) complexes, as previously observed for the previously reported simpler catenate.[25] In fact, the rate is comparable to the CV time scale and three Cu species are detected when a CV of a MeCN solution of $\mathbf{8_6}^{2+}$(BF$_4^-$)$_2$ is performed. The waves at +0.63 V and −0.41 V correspond, respectively, to the tetra- and hexacoordinate complexes mentioned above. By analogy with the value found for the previously reported copper-complexed catenane,[25] the wave at −0.05 V is assigned to the pentacoordinate couple (Figure 3.8(b)).

3.3.1.2 Intramolecular Motion within a Heterodinuclear Bismacrocycle Transition-metal Complex

Wozniak and coworkers described recently the first heterodinuclear bismacrocyclic transition-metal complex $\mathbf{9}^{4+}$ (Figure 3.9) that exhibits potential-driven intramolecular motion of the interlocked crown-ether unit.[28,29] Although the system contains transition metals, the main interaction between the various subunits, which also allowed construction of catenane $\mathbf{9}^{4+}$, is an acceptor–donor interaction of the charge-transfer type.

The reported heterodinuclear catenane should allow a controlled translocation of the crown-ether unit back and forth between two different metal centres

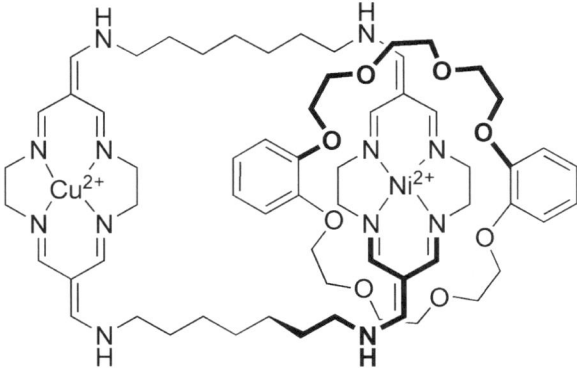

Figure 3.9 Heterodinuclear [2]catenane 9^{4+}.

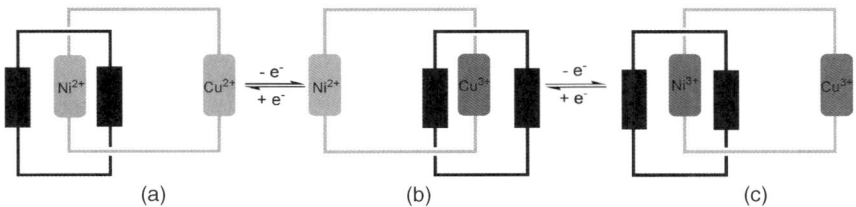

Figure 3.10 Schematic representation of electrochemically controlled molecular motion.

in response to an external stimulus, specifically a potential applied to the electrode (Figure 3.10).

The present system can be set in motion using two consecutive redox signals. The main feature of the machine-like catenane is that the preferred conformation will be that for which the most electrodeficient transition metal macrocyclic complex will lie in between the two aromatic donor fragments of the crown ether.

The bis-macrocyclic ring is positively charged because of the presence of Ni(II) and Cu(II). The crown ether and the bis-azamacrocyclic ring form a sandwich-like structure in such a way that one of the crown-ether aromatic rings is located between the two metal-coordinated macrocyclic rings. The second aromatic ring is located almost parallel to the previous one outside the two linked macrocycles.

Electrochemical characteristics of 9^{4+} show that the Ni(II)/Ni(III) oxidation peak appears at a more positive potential and the reduction Cu(III)/Cu(II) at a more negative potential than in the corresponding bis-macrocycle. These features result from the promoting effect of the crown ether activated by the d^8–d^8 structure. For such short linkers these interactions can be observed. The nickel oxidation signals are split in two. The extent of the splitting is a function of time and temperature. It can be assumed that two different nickel centres, each with a different microenvironment, are the reason for the splitting.

The donor properties of the first group of Ni(II) centres is affected by the vicinity of the electron-rich crown ether (Figure 3.10(a)) while the other one is not (Figure 3.10(b)). At lower frequencies, upon oxidation from copper(II) to copper(III), the crown ether has enough time to relocate from its initial position close to the nickel(II) centre (Figure 3.10(a)) towards the more positively charged copper(III) centre (Figure 3.10(b)). As a result of this relocation, the oxidation to nickel(III) appears at a more positive potential since it is free from the influence of the π-electron-rich crown ether. This "frozen" interconversion within the molecule can be better observed at lower temperature or shorter time scales.[28]

3.3.1.3 A Fast-moving Electrochemically Driven Machine based on a Pirouetting Copper-complexed Rotaxane[30]

The rate of the motion in artificial molecular machines and motors is obviously an important factor. Depending on the nature of the movement, it can range from microseconds, as in the case of organic rotaxanes acting as light-driven molecular shuttles,[31] to seconds, minutes or even hours in other systems involving threading-unthreading reactions[32–33] or metal-centred redox processes based on the Cu(II)/Cu(I) couple.[25]

In order to increase the rate of the motions, a new rotaxane in which the metal centre is as accessible as possible was prepared, the ligand set around the copper centre being thus sterically little hindering compared to previous related systems. Ligand exchange within the coordination sphere of the metal is thus facilitated as much as possible. The two forms of the new bistable rotaxane, **10_4^+** and **10_5^{2+}**, are depicted in Figure 3.11 (as usual, the subscripts *4* and *5* indicate the coordination number of the copper centre). The molecular axis contains a "thin" 2,2'-bipyridine motif, which is less bulky than a 1,10-phenanthroline fragment and thus is expected to spin more readily within the cavity of the ring. In addition, the bipy chelate does not bear substituents in α-position to the nitrogen atoms. **10_4^+** rearranges to the 5-coordinate species **10_5^{2+}** after oxidation and *vice versa*. The electrochemically driven motions were studied by cyclic voltammetry (CV).

A lower limit for the rate constant k of the process can be estimated as >500 s^{-1} (or $\tau < 2$ ms, with $\tau = k^{-1}$).

$$10_5^+ \xrightarrow{k > 500 \text{ s}^{-1}} 10_4^+$$

The rearrangement rate for the 4-coordinate Cu(II) complex is smaller than for the monovalent complex. It is, nevertheless, several orders of magnitude larger than in related catenanes or rotaxanes with more encumbering ligands:

$$10_4^{2+} \xrightarrow{5 \text{ s}^{-1}} 10_5^{2+}$$

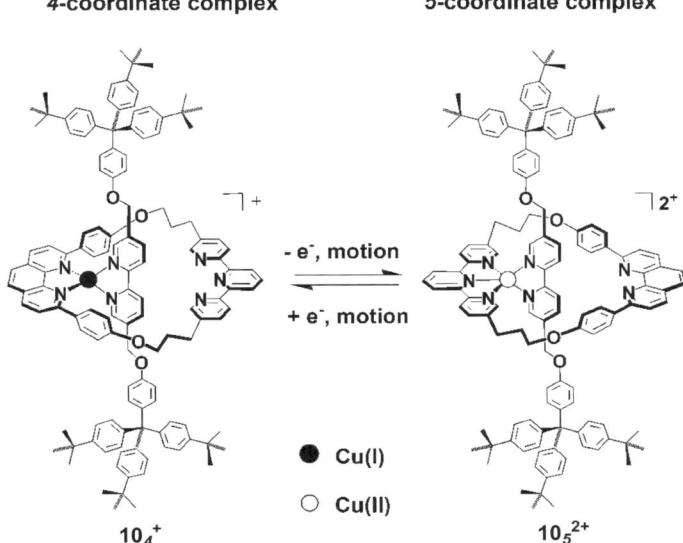

Figure 3.11 Electrochemically induced pirouetting of the ring in rotaxane 10^{n+}; the bidentate chelate and the tridentate fragment are alternatively coordinated to the copper centre.

This example shows that subtle structural factors can have a very significant influence on the general behaviour (rate of movement, in particular) of copper(II/I)-based molecular machines. Further modifications will certainly lead to new systems with even shorter response times.

3.3.2 Other Related Noninterlocking Systems

The first example of redox-driven translocation of a metal centre was based on the couple Fe(III)/Fe(II) and took place in ditopic ligands containing a trishydroxamate compartment,[34,35] suitable for the Fe(III) cation and three bipyridine functions that show a very high affinity towards Fe(II). The translocation was driven through auxiliary redox reactions: reduction of Fe(III) with ascorbic acid and oxidation of Fe(II) with peroxydisulfate. The translocation could be followed both visually and spectrophotometrically (Figure 3.12).

A ditopic ligand was designed in which one compartment displays a selective affinity towards the oxidised metal centre $M^{(n+1)+}$, whereas the other compartment shows a higher affinity towards the reduced cation M^{n+}. On the basis of the assumption that the oxidised cation is hard and the reduced one is soft, the ditopic system should contain a hard compartment (A) and a soft compartment (B). Thus, the hard cation stays in the hard compartment. When the metal centre is reduced to its soft version M^{n+}, it moves to the soft compartment B. Therefore, the metal centre can be translocated at will between A and B in a

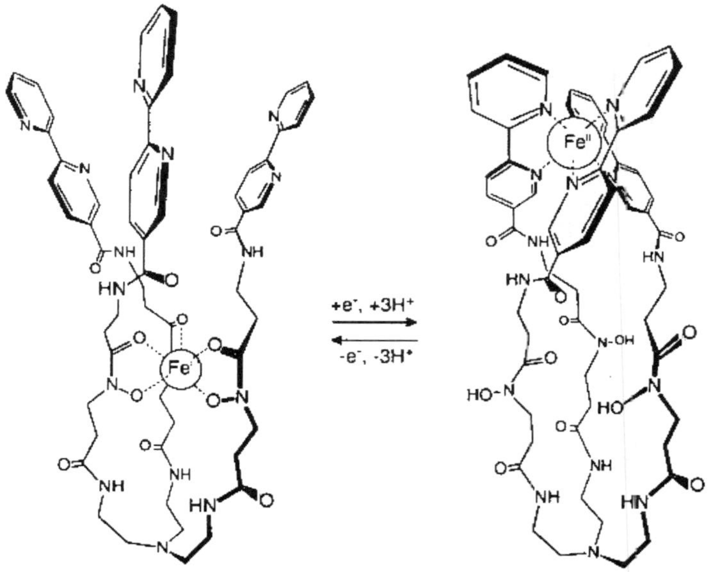

Figure 3.12 Redox-driven translocation of an iron centre within a ditopic system containing a hard and a soft compartment.

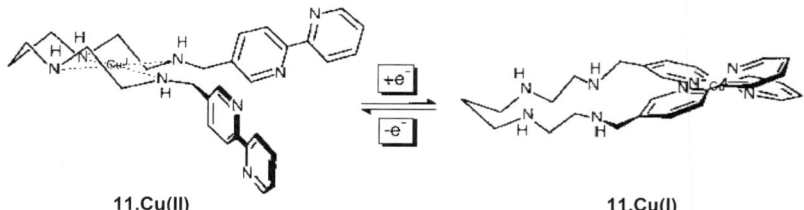

Figure 3.13 Redox-driven translocation of a copper centre based on the Cu^{II}/Cu^{I} change.

repeatable way upon reduction and oxidation of the metal centre in an electrochemical way.

Another example that fits the same mechanistic scheme is provided by the octadentate ligand **11**.[36] The system operates through the Cu(II)/Cu(I) couple. It contains the hard compartment A consisting of four amine groups and the soft compartment B with two 2,2'-bipyridine functions (Figure 3.13).

The translocation process is fast and reversible and can be followed both visually and spectrophotometrically. An MeCN solution containing equimolar amounts of **11** and Cu(II) is blue-violet (d-d absorption band; $\lambda_{max} = 548$ nm, $\varepsilon = 120$ M^{-1}cm^{-1}), which indicates that the oxidised cation resides in the tetramine compartment. On addition of a reducing agent (ascorbic acid), the solution takes the brick-red colour typically observed with the $Cu^{I}(bpy)_2^{+}$

chromophore (MLCT transition, $\lambda_{max} = 430$ nm, $\varepsilon = 1450$ M^{-1} cm^{-1}), indicating that the Cu(II)/Cu(I) reduction process has taken place and that the metal centre has translocated fast to the soft (bpy)$_2$ compartment. On addition of an oxidising reagent, the solution takes again the its original blue-violet colour, indicating that the metal centre (now Cu(II)) has again moved back to the tetramine compartment.

3.4 Light-fuelled Molecular Machines

3.4.1 Photoinduced Decoordination and Thermal Recoordination of a Ring in a Ruthenium(II)-containing [2]Catenane[37]

Our group has recently described multicomponent ruthenium(II) complexes in which one part of the molecule can be set in motion photochemically.[38,39] Among the light-driven molecular machine prototypes that have been described in the course of the last few years, a very distinct family of dynamic molecular systems takes advantage of the dissociative character of ligand-field states in Ru(diimine)$_3^{2+}$ complexes.[40–45] In these compounds, one part of the system is set in motion by photochemically expelling a given chelate, the reverse motion being performed simply by heating the product of the photochemical reaction so as to regenerate the original state. In these systems, the light-driven motions are based on the formation of dissociative excited states. Complexes of the Ru(diimine)$_3^{2+}$ family are particularly well adapted to this approach. If distortion of the coordination octahedron is sufficient to significantly decrease the ligand field, which can be realised by using one or several sterically hindering ligands, the strongly dissociative ligand-field state (^{3}d-d* state) can be efficiently populated from the metal-to-ligand charge transfer (^{3}MLCT) state to result in expulsion of a given ligand. The principle of the whole process is represented in Figure 3.14.

It is thus essential that the ruthenium(II) complexes that are to be used as building blocks of the future machines contain sterically hindering chelates so as to force the coordination sphere of the metal to be distorted from the perfect octahedral geometry.

The [2]catenanes **12^{2+}** and **13^{2+}** were synthesised[46] by using an octahedral ruthenium(II) centre as template. Compound **12^{2+}** consists of a 50-membered ring that incorporates two phen units and a 42-membered ring that contains the bipy chelate. Compound **13^{2+}** contains the same bipy-incorporating ring as **12^{2+}**, but the other ring is a 63-membered ring. Clearly, from CPK model considerations, **13^{2+}** is more adapted than **12^{2+}** to molecular motions in which both constitutive rings would move with respect to one another since the situation is relatively tight for the latter catenane. The light-induced motion and the thermal backreaction carried out with **12^{2+}** or **13^{2+}** are represented in Figure 3.15. They are both quantitative, as shown by UV/Vis measurements and by ^{1}H NMR spectroscopy.

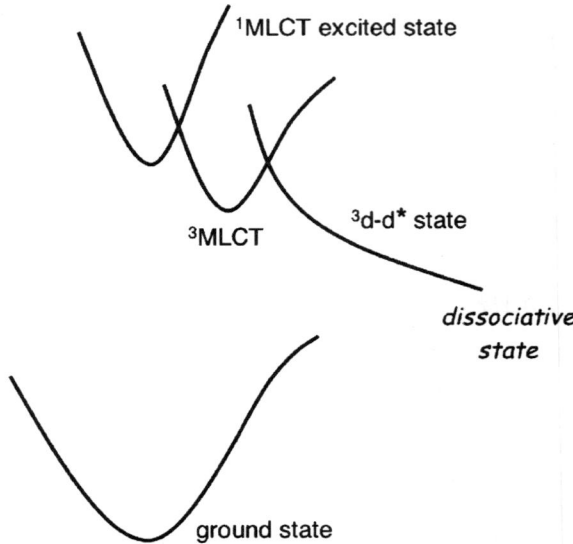

Figure 3.14 The ligand-field state ^{3}d-d* can be populated from the ^{3}MLCT state, provided the energy difference between these two states is not too large: formation of this dissociative state leads to dissociation of a ligand.

The photoproducts, [2]catenanes **12′** and **13′**, contain two disconnected rings since the photochemical reaction leads to decomplexation of the bipy chelate from the ruthenium(II) centre. In a typical reaction, a degassed CH_2Cl_2 solution of **12^{2+}** and $NEt_4^+.Cl^-$ was irradiated with visible light, at room temperature. The colour of the solution rapidly changed from red (**13^{2+}**: $\lambda_{max} = 458$ nm) to purple (**13′**: $\lambda_{max} = 561$ nm) and after a few minutes the reaction was complete. The recoordination reaction **13′** → **13^{2+}** was carried out by heating a solution of **13′**. It is hoped that, in the future, an additional tuneable interaction between the two rings of the present catenanes, **12′** or **13′** will allow better control over the geometry of the whole system. In parallel, two-colour machines will be elaborated, for which both motions will be driven by photonic signals operating at different wavelengths.

3.4.2 A Photochemically Driven Molecular-level Abacus[47,48]

Recently, Credi *et al.*[47,48] reported the design, synthesis and machine-like performance of a [2]rotaxane, in which the ring component can be induced by light excitation to move, that is switch between two different recognition sites or "stations" of the dumbbell-shaped component (Figure 3.17). Such a molecule exhibits an abacus-like geometry and, since it behaves according to binary logic, it could, in principle, be used for information processing[49–53] at the molecular level.

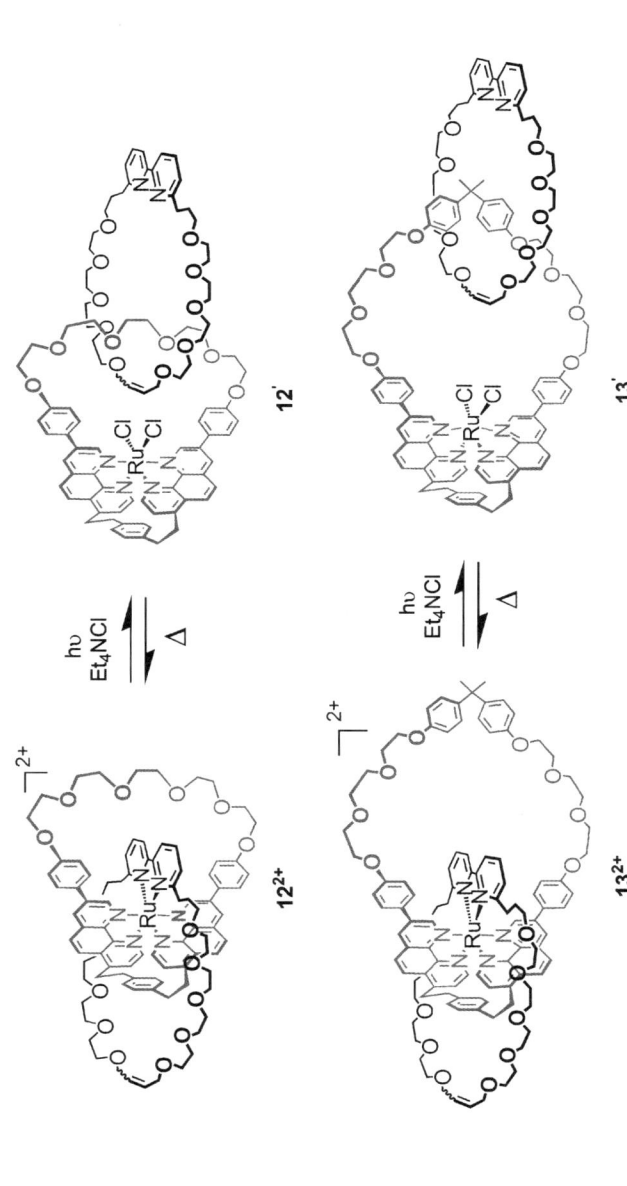

Figure 3.15 Catenanes 12^{2+} or 13^{2+} undergo a complete rearrangement by visible light irradiation: the bipy-containing ring is efficiently decoordinated in the presence of Cl$^-$. By heating the photoproducts 12' or 13', the starting complexes 12^{2+} or 13^{2+} are quantitatively regenerated.

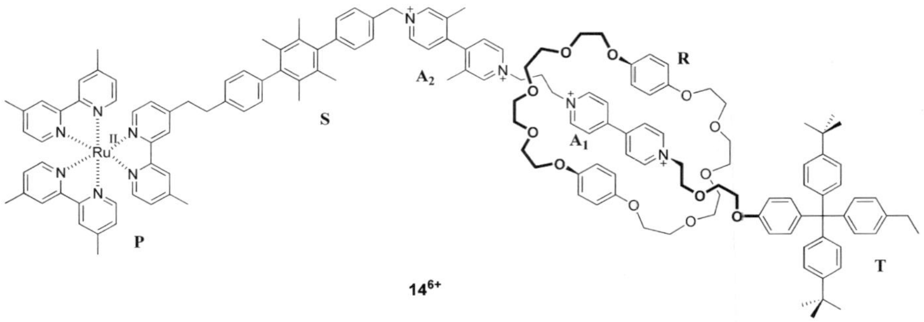

Figure 3.16 Chemical structure of rotaxane **14^{6+}**.

The design principles at the basis of the light-driven molecular machines developed by this group[54] have been employed to obtain the rotaxane **14^{6+}** (Figure 3.16),[47] specifically designed to achieve photoinduced ring shuttling.

This compound is made of a π-electron-donating macrocycle as the ring **R**, and a dumbbell-shaped component that contains (i) a Ru(bpy)$_3^{2+}$-type complex (**P**) as one of its stopper, (ii) a 4,4′-bipyridinium unit (**A$_1$**) and a 3,3′-dimethyl-4,4′-bipyridinium unit (**A$_2$**) as electron-accepting stations, (iii) a p-terphenyl-type ring system as a rigid spacer (**S**), and (iv) a tetraarylmethane group as the second stopper (**T**). The stable translational isomer of rotaxane **14^{6+}** is the one in which the **R** component encircles the **A$_1$** unit, in keeping with the fact that this station is a better electron-acceptor than the other one.

Two working schemes have been devised for the photoinduced switching of **R** between stations **A$_1$** and **A$_2$** – (i) a mechanism based fully on processes that only involves the rotaxane components, that is an intramolecular mechanism, and (ii) a mechanism that requires the help of external reactants, that is a sacrificial mechanism (Figure 3.17).

The results obtained[47] have shown that the photochemically driven switching can be performed successfully by the sacrificial mechanism, which is based on the following operations:

(a) *Destabilisation of the stable translational isomer*: light excitation of the photoactive unit **P** (step 1) is followed by the transfer of an electron from the excited state to the **A$_1$** station, which is encircled by the ring **R** (step 2), with the consequent "deactivation" of this station; such a photoinduced electron-transfer process has to compete with the intrinsic decay of **P*** (step 3).

(b) *Ring displacement after scavenging of the oxidised photoactive unit*: if the solution contains a suitable reductant Red (triethanolamine (TEOA), which is a very good scavenger of oxidised Ru-polypyridine complexes), a fast reduction of Red with **P$^+$** (step 8) competes successfully with the back electron-transfer reaction (step 5). In such a case, the displacement of the ring to **A$_2$** (step 4), even if it is slow, can take place because the originally occupied station remains in its reduced state **A$_1^-$**.

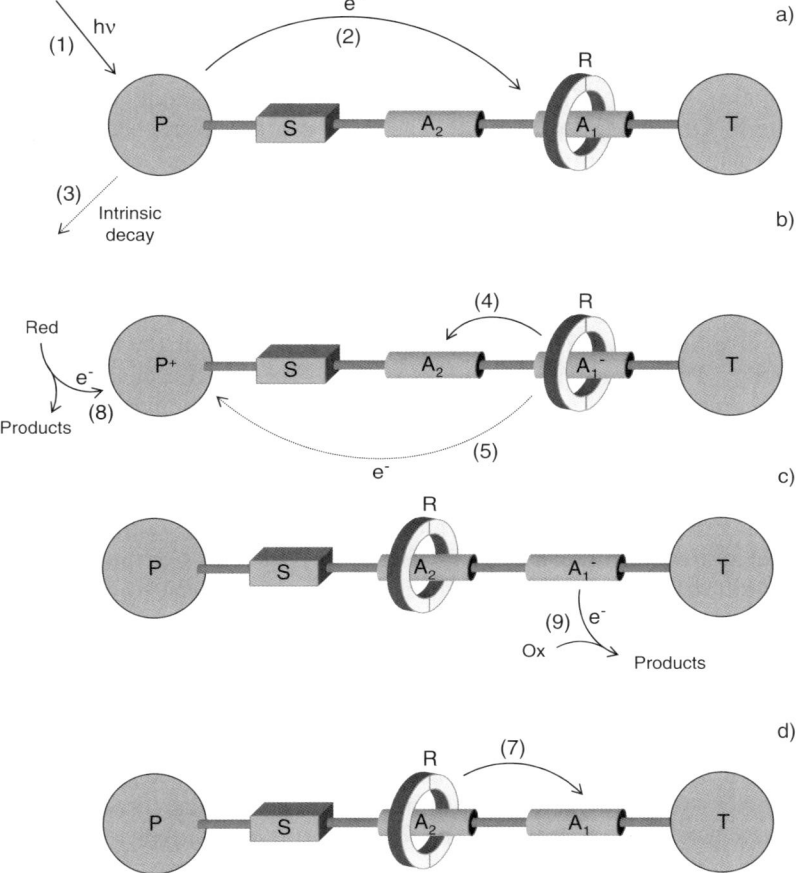

Figure 3.17 Working scheme employing a sacrificial mechanism for the light-driven switching of the ring R between the two stations A_1 and A_2. The dashed lines indicate processes that compete with those needed to make the machine work: steps (3) and (5).

(c) *Electronic reset*: after an appropriate time, restoration of the electron-acceptor power of the A_1 station can be obtained by oxidising A_1^- with a suitable oxidant Ox (dioxygen in this case) (step 9).

(d) *Nuclear reset*: as a consequence of the electronic reset, back movement of the ring from A_2 to A_1 takes place (step 7).

Many investigations are still being carried out in this field. Such mechanical motions with a "power stroke" and a "recovery stroke", which are reminiscent of the workings of a simple piston, have also been reported in other systems not containing a transition metal.[55]

3.5 Conclusion and Prospective

In this chapter, we have discussed a few examples of transition-metal-containing molecular machines, either of the catenane and rotaxane family or not containing interlocking or threading rings. The input can be electrochemical, and it is thus specific of given transition-metal complexes. The systems can be oxidised or reduced at the metal: Cu(II)/Cu(I), Co(III)/Co(II) or Ni(III)/Ni(II) in the examples discussed here. Alternatively, a photonic stimulus can be used to set the system in motion. Here again, the role of the metal centre is essential since the excited state responsible for the absorption of light is a metal-to-ligand charge-transfer (MLCT) excited state but this state is rapidly converted to a dissociative ligand-field (LF) state, which is responsible for the first step of the movement. The third type of stimulus that has been used is chemical. A metal exchange reaction or a recognition reaction allows very significant modification of the shape of a compound: stretching–contraction of a rod-like molecule or reorganisation of a ring-like compound (similar to inflating or deflating the ring). The metal complex can also be used as a "gate" that will prevent the compound from undergoing a large-amplitude motion such as gliding of a ring between two identical stations incorporated in a thread-like fragment. Formation of a complex within the whole molecular system will "freeze" a given situation, whereas dissociation will allow the motion to take place. Strictly speaking, this latter example does not belong to the family of molecular machines since no external stimulus is required to set the system in motion, but it is representative of what can be done to control the rate of molecular motions.

It is still not sure whether the field will lead to applications in a short-term perspective, although spectacular results have been obtained in the course of the last few years in relation to information storage and processing at the molecular level.[4] Apart from this important field, other applications could of course be envisaged: nanomachines for medical applications such as valves and pumps, molecular devices able to sort molecules, active molecular carriers, able to transport selectively given molecules across a membranes, just to cite a few. Nevertheless, it should be kept in mind that practical applications do not usually arise from where one expects them to come.

From a purely scientific viewpoint, the field of molecular machines is particularly challenging and motivating: the fabrication of dynamic molecular systems, with precisely designed dynamic properties, is still in its infancy and will certainly experience a rapid development during the next decades. Such a research area is highly multidisciplinary and requires a high level of competence in synthetic chemistry as well as in physical and materials sciences, which makes it an excellent school for young and ambitious scientists.

Acknowledgments

We would like to thank the CNRS, Région Alsace and the European Commission (BIOMACH) for their financial support.

References

1. (a) V. Balzani, M. Venturi and A. Credi, *Molecular Devices and Machines*, Wiley-VCH, Weinheim, 2003; (b) J.-P. Sauvage, *Molecular Machines and Motors, Structure and Bonding*, Springer, Berlin, Heidelberg; Special issue on Molecular Machines, *Acc. Chem. Res.*, 2001, **34**, 341.
2. M. Schliwa, (ed.), *Molecular Motors*, Wiley-VCH, Weinheim, 2002.
3. (a) R.K. Soong, G.D. Bachand, H.P. Neves, A.G. Olkhovets, H.G. Craighead and C.D. Montemagno, *Science*, 2000, **290**, 1555; (b) H. Noji, R. Yasuda, M. Yoshida and K. Kinosita Jr., *Nature*, 1997, **386**, 299.
4. (a) C.P. Collier, G. Mattersteig, E.W. Wong, Y. Luo, K. Beverly, J. Sampaio, F.M. Raymo, J.F. Stoddart and J.R. Heath, *Science*, 2000, **289**, 1172; (b) A.R. Pease, J.O. Jeppesen, J.F. Stoddart, Y. Luo, C.P. Collier and J.R. Heath, *Acc. Chem. Res.*, 2001, **34**, 433–444.
5. (a) R.A. Bissell, E. Córdova, A.E. Kaifer and J.F. Stoddart, *Nature*, 1994, **369**, 133; (b) J.D. Badjic, V. Balzani, A. Credi, S. Silvi and J.F. Stoddart, *Science*, 2004, **303**, 1845 and references therein; (c) R. Ballardini, V. Balzani, M.T. Gandolfi, L. Prodi, M. Venturi, D. Philp, H.G. Ricketts and J.F. Stoddart, *Angew. Chem. Int. Ed.*, 1993, **32**, 1301; (d) A. Livoreil, J.-P. Sauvage, N. Armaroli, V. Balzani, L. Flamigni and B. Ventura, *J. Am. Chem. Soc.*, 1997, **119**, 12114; (e) N. Armaroli, V. Balzani, J.-P. Collin, P. Gaviña, J.-P. Sauvage and B. Ventura, *J. Am. Chem. Soc.*, 1999, **121**, 4397; (f) H. Murakami, A. Kawabuchi, K. Kotoo, M. Kunitake and N. Nakashima, *J. Am. Chem. Soc.*, 1997, **119**, 7605; (g) V. Balzani, A. Credi, F. Marchioni and J.F. Stoddart, *Chem. Commun.*, 2001, 1860.
6. (a) T.R. Kelly, H. de Silva and R.A. Silva, *Nature*, 1999, **401**, 150; (b) N. Koumura, R.W.J. Zijistra, R.A. van Delden, N. Harada and B.L. Feringa, *Nature*, 1999, **401**, 152; (c) D.A. Leigh, J.K.Y. Wong, F. Dehez and F. Zerbetto, *Nature*, 2003, **424**, 174; (d) E. Katz, O. Lioubashevsky and I. Wilner, *J. Am. Chem. Soc.*, 2004, **126**, 15520; (e) A.N. Shipway and I. Willner, *Acc. Chem. Res.*, 2001, **34**, 421; (f) A. Harada, *Acc. Chem. Res.*, 2001, **34**, 456; (g) S. Zahn and J.W. Canary, *Science*, 2000, **288**, 1404.
7. (a) S. Shinkai, T. Nakaji, T. Ogawa, K. Shigematsu and O. Manabe, *J. Am. Chem. Soc.*, 1981, **103**, 111; (b) S. Shinkai, M. Ishihara, K. Ueda and O. Manabe, *J. Chem. Soc. Perkin Trans. 2*, 1985, 511.
8. (a) M.C. Jiménez, C. Dietrich-Buchecker and J.-P. Sauvage, *Angew. Chem. Int. Ed. Engl.*, 2000, **39**, 3284; (b) M.C. Jiménez, C. Dietrich-Buchecker and J.-P. Sauvage, *Chem. Eur. J*, 2002, **8**, 1456.
9. L. Jiang, J. Okano, A. Orita and J. Otera, *Angew. Chem. Int. Ed.*, 2004, **43**, 2121–2124.
10. T.R. Kelly, M.C. Bowyer, K. Vijaya Bhaskar, D. Beddington, A. Garcia, F. Lang, M.H. Kim and M.P. Jette, *J. Am. Chem. Soc.*, 1994, **116**, 3657.
11. L. Fabrizzi, F. Foti, S. Patroni, P. Pallavicini and A. Taglietti, *Angew. Chem. Int. Ed.*, 2004, **43**, 5073–5077.

12. V. Amendola, L. Fabrizzi, C. Mangano, H. Miller, P. Pallavicini, A. Perotti and A. Taglietti, *Angew. Chem. Int. Ed.*, 2002, **41**, 2553–2556.
13. K.D. Karlin, J.C. Hayes, S. Juen, J.P. Hutchinson and J. Zubieta, *Inorg. Chem.*, 1982, **21**, 4108–4109.
14. V. Amendola, C. Brusoni, L. Fabrizzi, H. Miller, P. Pallavicini, A. Perotti and A. Taglietti, *J. Chem. Soc. Dalton Trans.*, 2001, 3528.
15. P.K. Coughlin, S.J. Lippard, A.E. Martin and J.E. Bulkowski, *J. Am. Chem. Soc.*, 1980, **102**, 7616–7617.
16. J.D. Badjic, V. Balzani, A. Credi, S. Silvi and J.F. Stoddart, *Science*, 2004, **303**, 1845.
17. S.A. Vignon, T. Jarrosson, T. Iijima, H.-R. Tseng, J.K.M. Sanders and J.F. Stoddart, *J. Am. Chem. Soc.*, 2004, **126**, 9884–9885.
18. D.A. Leigh and E.M. Pérez, *Chem. Commun.*, 2004, 2262–2263.
19. D.A. Parthenopoulos and P.M. Rentzepis, *Science*, 1989, **245**, 843.
20. I. Willner, R. Blonder and A. Dagan, *J. Am. Chem. Soc.*, 1994, **116**, 3121.
21. M. Irie, O. Miyatake and K. Uchida, *J. Am. Chem. Soc.*, 1992, **114**, 8715.
22. S.L. Gilat, S.H. Kawai and J.-M. Lehn, *Chem. Commun.*, 1993, 1439.
23. D.A. Leigh, J.K.Y. Wong, F. Dehez and F. Zerbetto, *Nature*, 2003, **424**, 174.
24. D. Cardenas, A. Livoreil and J.-P. Sauvage, *J. Am. Chem. Soc.*, 1996, **118**, 11980.
25. A. Livoreil, C.O. Dietrich-Buchecker and J.-P. Sauvage, *J. Am. Chem. Soc.*, 1994, **116**, 9399.
26. C.O. Dietrich-Buchecker, J.-P. Sauvage and J.-P. Kintzinger, *Tetrahedron Lett.*, 1983, **24**, 5095.
27. C.O. Dietrich-Buchecker and J.-P. Sauvage, *Tetrahedron*, 1990, **46**, 503.
28. B. Korybut-Daszkiewicz, A. Więckowska, R. Bilewicz, S. Domagata and K. Wozniak, *Angew. Chem. Int. Ed.*, 2004, **43**, 1668–1672.
29. B. Korybut-Daszkiewicz, A. Więckowska, R. Bilewicz, S. Domagata and K. Wozniak, *J. Am. Chem. Soc.*, 2001, **123**, 9356–9366.
30. I. Poleschak, J.-M. Kern and J.-P. Sauvage, *Chem. Commun.*, 2004, 474.
31. A.M. Brouwer, C. Frochot, F.G. Gatti, D.A. Leigh, L. Mottier, F. Paolucci, S. Roffia and G.W.H. Wurpel, *Science*, 2001, **291**, 2124.
32. P.R. Ashton, R. Ballardini, V. Balzani, I. Baxter, A. Credi, M.C.T. Fyfe, M.T. Gandolfi, M. Gómez-López, M.-V. Martínez-Díaz, A. Piersanti, N. Spencer, J.F. Stoddart, M. Venturi, A.J.P. White and D.J. Williams, *J. Am. Chem. Soc.*, 1998, **120**, 11932.
33. V. Balzani, A. Credi, G. Mattersteig, O.A. Mattews, F.M. Raymo, J.F. Stoddart, M. Venturi, A.J.P. White and D.J. Williams, *J. Org. Chem.*, 2000, **65**, 1924.
34. C. Canevet, J. Libman and A. Shanzer, *Angew. Chem. Int. Ed.*, 1996, **35**, 2657–2660.
35. L. Zeeikovich, J. Libman and A. Shanzer, *Nature*, 1995, **374**, 790–792.
36. V. Amendola, L. Fabrizzi, C. Mangano and P. Pallavicini, *Acc. Chem. Res.*, 2001, **34**, 488–493.

37. P. Mobian, J.-M. Kern and J.-P. Sauvage, *Angew. Chem. Int. Ed. Eng.*, 2004, **43**, 2392.
38. J.-P. Collin, A.-C. Laemmel and J.-P. Sauvage, *New J. Chem.*, 2001, **25**, 22.
39. A.-C. Laemmel, PhD thesis, University of Strasbourg.
40. M. Adelt, M. Devenney, T.J. Meyer, D.W. Thompson and J.A. Treadway, *Inorg. Chem.*, 1998, **37**, 2616.
41. J. Van Houten and J. Watts, *Inorg. Chem.*, 1978, **17**, 3381.
42. H.F. Suen, S.W. Wilson, M. Pomerantz and J.L. Walsch, *Inorg. Chem.*, 1989, **28**, 786.
43. D.V. Pinnick and B. Durham, *Inorg. Chem.*, 1984, **23**, 1440.
44. B. Durham, J.V. Caspar, J.K. Nagle and T.J. Meyer, *J. Am. Chem. Soc.*, 1982, **104**, 4803.
45. S. Tachiyashiki and K. Mizumachi, *Coord. Chem. Rev.*, 1994, **132**, 113.
46. P. Mobian, J.-M. Kern and J.-P. Sauvage, *Helv. Chim. Acta.*, 2003, **86**, 4195.
47. P. R. Ashton, R. Ballardini, V. Balzani, A. Credi, K. R. Dress, E. Ishow, C. J. Kleverlaan, O. Kocian, J. A. Preece, N. Spencer, J. F. Stoddart, M. Venturi and S. Wenger, *Chem. Eur. J.*, 2000, **6**, 19, 3558.
48. A. Credi and B. Ferrer, *Pure Appl. Chem.*, 2005, **77**, 6, 1051.
49. A.P. de Silva, H.Q.N. Gunaratne and C.P. McCoy, *Nature*, 1993, **364**, 42–44.
50. A.P. de Silva, H.Q.N. Gunaratne, T. Gunnlaugsson, A.J.M. Huxley, C.P. McCoy, J.T. Rademacher and T.E. Rice, *Chem. Rev.*, 1997, **97**, 1515–1566.
51. A. Credi, V. Balzani, S.J. Langford and J.F. Stoddart, *J. Am. Chem. Soc.*, 1997, **119**, 2679–2681.
52. F. Pina, A. Roque, M.J. Melo, M. Maestri, L. Belladelli and V. Balzani, *Chem. Eur. J.*, 1998, **4**, 1184–1191.
53. A.P. de Silva, H.Q.N. Gunaratne and C.P. McCoy, *J. Am. Chem. Soc.*, 1999, **121**, 1393–1394.
54. P.R. Ashton, V. Balzani, O. Kocian, L. Prodi, N. Spencer and J.F. Stoddart, *J. Am. Chem. Soc.*, 1998, **120**, 11190.
55. A.M. Brouwer, C. Frochot, F.G. Gatti, D.A. Leigh, L. Mottier, F. Paolucci, S. Roffia and G.W.H. Wurpel, *Science*, 2001, **291**, 2124.

CHAPTER 4
Chemomechanical Polymers

HANS-JÖRG SCHNEIDER* AND KAZUAKI KATO

FR Organische Chemie, Universität des Saarlandes, D 66041 Saarbrücken/ Germany

4.1 Introduction

Chemomechanical polymers are intelligent materials that respond to external chemical stimuli by changing their mechanical properties, in particular their shape. In most cases these materials are polymeric gels, in which exposure to chemical compounds leads to volume changes. Chemomechanical polymers thus provide an intriguing way to new actuator and also sensor systems, which can operate without any additional measuring devices, including transducers or transmitters, and also without external power supply.[1] An important aspect for future uses is the possible miniaturisation of such devices, which can go down to the nanoscale, taking advantage then also of high speed and energy-saving property of such systems. Most promising applications involve self-sustained systems for drug delivery,[2] for artificial muscles,[3] or for machines driven by chemical interactions. In polyelectrolyte gels swelling and contraction can be induced by pH gradients; these can also be produced electrically (as in electrodialysis), thus allowing the design of both pH-operated and microelectromachines.[4] Several chemomechanical elements in particular for pH sensing have been already described.[5] Although science must go a long way until the efficiency of biological actuators such as the myosin-actin complex can be approximated with completely artificial systems, the first steps in this direction have been taken. Other chapters in this monograph illustrate the development of, e.g., molecular machines on a molecular level, until now confined mostly to observation in solution, but also recent combinations of natural and artificial polymers for ATP-driven macroscopic movements.

The presence of suitable binding sites in smart synthetic polymers opens the way to highly selective molecular recognition also for endogenic substances, such as specific metal ions, amino acids, peptides, nucleotides, carbohydrates including glucose. This allows construction of, e.g., drug-release devices that

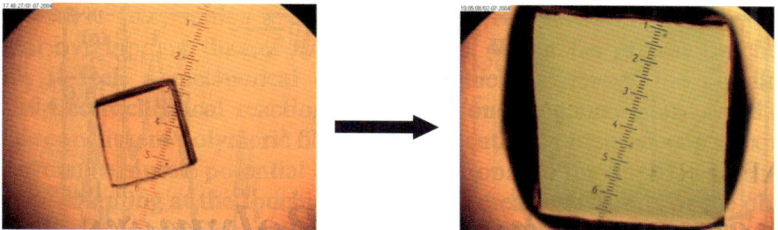

Figure 4.1 Size increase of a water-saturated chemomechanical hydrogel film piece (1.1 × 0.9 × 0.4 mm) induced by a 0.25 mM solution of a dipeptide (gly-gly in the presence of Cu(II) ions, pH 4.5; see Section 4.8).

are triggered by the level of such effector compounds in the body; selective uptake of, *e.g.*, toxic compounds could also be operated this way. Supramolecular chemistry has in the last years developed hundreds of synthetic hosts for all kind of desired ligands.[6] Until now these receptors have been mostly used in solution, but in principle can be implemented in functional materials such as chemomechanical polymers. Due to the necessarily statistical nature of synthetic polymers with sufficient flexibility structural insights are more difficult, but are also of lesser importance than characterisation of their performance as smart materials. The present chapter aims to highlight principles and recent development in the field of chemomechanical polymers, with an emphasis on the *function* of these intelligent materials, on their performance in the presence of different chemical stimuli, and to a lesser degree on the underlying structural and mechanical properties. Figure 4.1 visualises with one example how the interaction of a peptide in less than 0.25 mM concentration can trigger an about 20-fold volume increase of the already water-saturated chemomechanical hydrogel. It will be discussed below how the sensitivity of chemomechanical polymers can be further enhanced by miniaturisation of the polymer particles and by increasing the affinity of effectors to the host units. Of course solvents, in particular water alone, lead in the first place to sizeable swelling of gels, reaching, *e.g.*, water content of up to 99%. We will always characterise the chemomechanical changes on materials, which have been exposed already to the medium under the same condition, but without the chosen chemical effector.

4.2 Chemomechanical Polymers Triggered by pH

Until recently most of the known chemomechanical materials are those that change their mechanical property, in particular their shape, by changes of the acidity of the surrounding medium. Generally, all polymers bearing ionisable groups show expansion upon ionisation, *e.g.*, the many gels containing aminogroups will swell at lower pH. The historical starting point of this can be seen in a contribution by Katchalsky, Kuhn and others[7], who described – albeit in

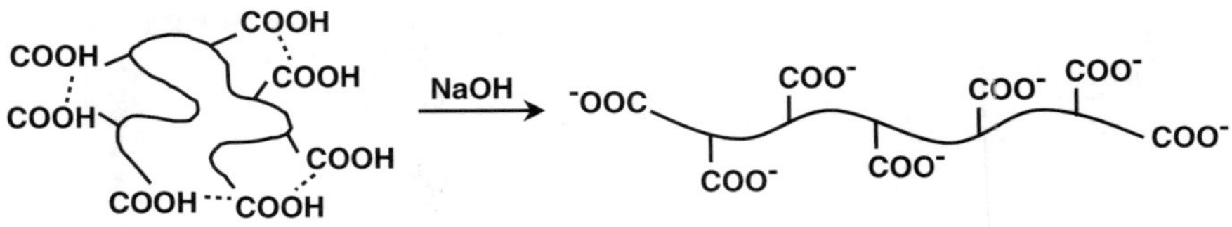

Scheme 4.1 Model for the viscosity increase by deprotonation of polyacrylic acid.

solution – already in 1950 the reversible viscosity increase of polyacrylic acid (PA) under basic condition. This can be ascribed primarily to a chain unfolding produced by loss of hydrogen bonds that can materialise with protonated carboxylic groups, and by a chain-elongating repulsion between the carboxylate anions at higher pH (Scheme 4.1).

Almost countless polymer gels have since then been described that exhibit shape changes at different pH values,[8] often described as phase transitions,[9] which may involve a multitude of states.[9,10] The theory of pH-induced polymer swelling was developed already in the late 1980s.[11] Particularly gels containing randomly distributed both positively and negatively charged groups can exhibit more than two, either collapsed or swollen, phases; they may exibit maximum swelling at two different pH values, which depend on the relative pK values.[12] For example, chitosan-polyacrylic acid gels exhibit three maximum expansions at pH 3, pH 6 and pH 8.[13] A heuristic description has been proposed[9] between a loose swollen state, which is stable at lower temperatures, and a shrunken, collapsed state, which results, e.g., from temperature increase. In the swollen state a target molecule – i.e. a drug – is believed to adsorb preferentially by single contacts with loose receptor sites of the polymer network, whereas in the shrunken state multipoint interactions can lead to higher affinity. Therefore, the chemomechanical effect is a function of the applied temperature.

Gels have also been used with imprinting techniques, in which a target molecule serves for generation of complementary binding cavities.[14,15] After imprinting, polymerisation and the following substrate release in the swollen state a collapsed state can retain sufficient memory to bind a ligand with increased selectivity, although the effects until now were rather modest.[10,16] Thus, a thermosensitive gel showed concentration-dependent swelling induced only by the compound that was used for imprinting gel such as adrenaline or ephedrine.[17]

Hydrogels from phosphorylcholine and polymethacrylic acid and esters also allow pH-dependent drug release; they form networks that, in line with the measured expansions, with the modulus and with FT-IR spectra, show swelling only at higher pH, ascribed as in Scheme 4.1 to the formation of the intramolecular crosslinks by hydrogen bonds with carboxyl groups.[18] In some gels undergoing phase transitions within a narrow range of temperature and also of pH values there is no stable intermediate between a swollen and a collapsed state.[19,20]

With microgels derived from poly(methacrylic acid /nitrophenyl acrylate) the pH range of the swelling response is proportional to the solution pK(a)s of their

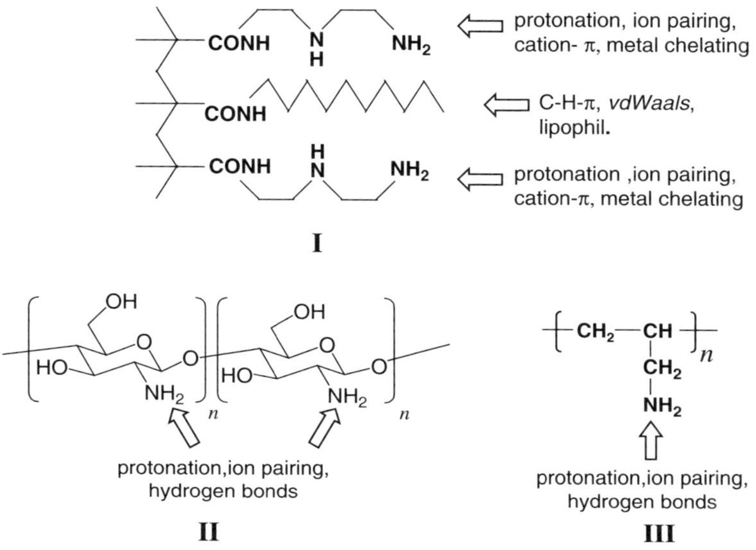

Scheme 4.2 Structures of some chemomechanical polymers (**I**: poylmethacrylic acid with supramolecular binding functions, **II**: chitosan, **III**: polyallylamine).

functional groups.[21] Bicomponent fibrous hydrogel membranes from poly(vinyl alcohol)/poly(acrylic acid) exhibit pH-dependent swelling in two directions of the material.[22] Chitosan **II** (Scheme 4.2) itself forms a gel by using simple inorganic or organic acids.[23] Although such gels slowly dissolve under acidic conditions[24] in the absence of a crosslinker,[25] the chitosan gels themselves, are under close to neutral conditions, stable enough for many applications. Semi-interpenetrating hydrogel networks of chitosan and polyacrylamide show increased swelling below pH = 7.[26] Poly(ethylene oxide) grafted methacrylic acid and acrylic acid hydrogels were studied as a drug carrier for the protection of insulin from the acidic environment of the stomach.[27] Copolymers containing and galactosamine glutamate forming fiber hydrogel show under acidic conditions dense packing and shrinking.[28] Similarly, semi-interpenetrating networks of crosslinked copolymer acrylamide/acrylic acid polyallylamine can serve as an effective pH-stimulated drug-delivery system, as investigated by equilibrium and oscillatory swelling techniques.[29] Two-component cross-linked chitosan derivatives show swelling at pH 7.4 and above due to ionisation of the carboxylic acid groups in the gel.[30] Strong temperature dependence of pH-triggered hydrogels composed of poly(2-ethyl-2-oxazoline) and chitosan was reported.[31] Microgels from poly(methacrylic acid-co-nitrophenyl acrylate) were selectively derivatised with carboxylic acid, glutamic acid, hydroxamic acid, sulfonic acid, and ethanol functional groups, yielding pH- and NaCl-induced swelling response and drug loading, proportional to the solution pK(a)s of their functional groups.[21] Polyallylamine (PAH, **III**, Scheme 4.2), after crosslinking, exhibits a sharp swelling at pH 7, although its pK value of

9.7[32] would suggest a jump at higher pH; this behaviour was explained by counteracting ion pairing with the special hydroxyphenylsulfonate used in this study.[33] In contrast to most other ionic hydrogels the pH-stimulated PAH polymers **III** show a very slow reverse contraction after swelling.[34]

The hydrogel **I** derived from reaction of poly(methyl)methacrylate (PMMA), diethylenetriamine and dodecylamine (Scheme 4.2) provides both pH-sensitive and lipophilic binding groups; it seems to be the only known gel that displays a symmetric pH profile above and below pH 7 (Figure 4.2).[35] Xanthan-chitosan gels were reported to be sensitive not only to high pH with a maximum swelling at pH 10, but to some degree also at pH 0, where, however, the gel dissolves in the course of swelling.[36]

The chemomechanical response to varying pH depends strongly on the ionic strength of the medium (Figure 4.2(b)).[37] Independent measurements indicate that similar expansions as, e.g., at pH 2 and 12 in the presence of salts (Figure 4.2(a)) occur with e.g., related sodium chloride concentrations alone at neutral pH; thus, the lower and higher pH effects are also due to increased stronger ionic strength by the necessary acid or base concentrations at such pH values.[35b] In accord with this, one found that polysaccharide-derived gels

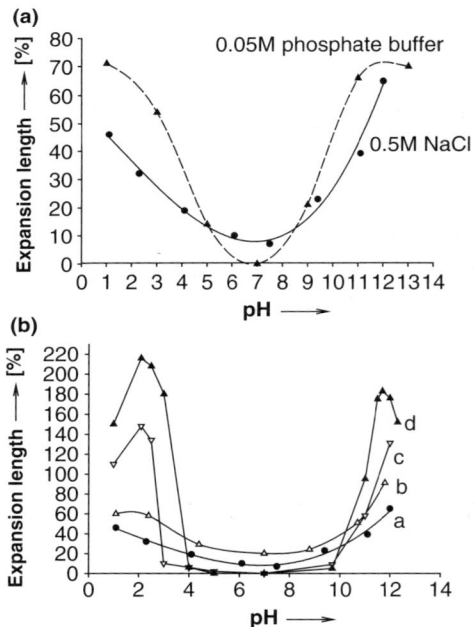

Figure 4.2 (a) Size changes of a PMMA-derived hydrogel (I, Scheme 4.1) as function of pH; calculated maximum volume change 390%; in 0.05 M phosphate buffer (circles), and in 0.5 M sodium chloride (triangles).; (b) pH expansion profiles at different salt concentrations; in 0.5 M (●–a), 0.05 M (△–b), and 0.025 M (▽–c) sodium chloride solution, respectively, and in water with very dilute HCl or NaOH (▲–d).[35]

decrease the swelling with an increase in the ionic strength of the salt solutions, due to a charge-screening effect for monovalent cations, as well as due to ionic crosslinking by multivalent cations.[38]

Unspecific *salt effects* are often due to changes of the ionic strength,[39] as shown, *e.g.*, with beta-hairpin peptides that self-assemble into hydrogel nanostructures consisting of semiflexible fibrillar assemblies.[40] Circular dichroism spectroscopy indicates in absence of salt unfolded peptides; an increased ionic strength screens electrostatic interactions between charged amino acids within the peptide with subsequent beta-hairpin formation. Adsorption affinities for various substrates decrease by orders of magnitude with increasing salt concentration.[41] A moderate selectivity with respect to different alkali salts is seen with the hydrogel **I**, as illustrated in Scheme 4.3.[35b] The partially reversed effects in presence of, *e.g.*, sodium phosphate will be discussed below in the context of cooperativity effects. It should be noted that reverse contraction after expansion of ionic polymers always needs treatment with another salt or buffer as replacement of the gegenion neutralising the charge of the backbone.

Anion effects can to some degree be selective also without introduction of selective binding groups into gels. An alternative used not for chemomechanical properties but for sensing of different phosphate derivatives rests on fluorescent artificial receptors that upon guest binding can dynamically change the location between aqueous cavity and hydrophobic fibers provided under semiwet conditions of a supramolecular hydrogel, as observed by confocal laser scanning microscopy.[42] As found recently, swelling of chitosan **II** is not only a function of pH, but also of the acids used. Expansion stimulated by acetic acid starts already at pH 6.2, which is close to the pKa of chitsoan. In contrast, acids such as hydrochloric acid or phosphoric acid expand at lower pH (Figure 4.3).[45] The particular effect of the polyvalent phosphoric acid can be tentatively ascribed to the ionic crosslinking counteracting the expansion.[25] Ionic crosslinking with concomitant gel contraction exists only in the presence of anions, while free acids, formed depending on their pK values, allow expansion. Monocarboxylic

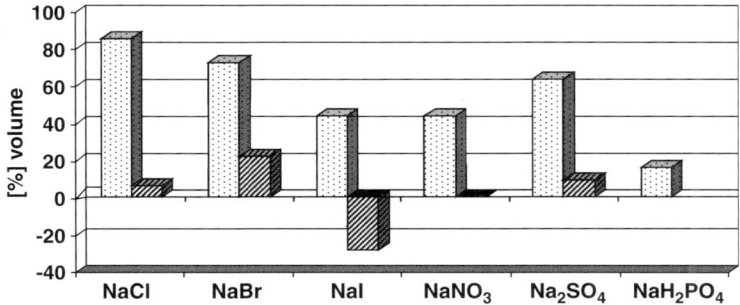

Scheme 4.3 Volume changes ([%]) induced by different anions on the PAA-derived polymer **I**; [NaX] = 0.10 M; left bars (always expansion): in pure water, pH 7.3; right bars (partially contraction): in presence of 0.02 M phosphate buffer, pH 7.0.[35b]

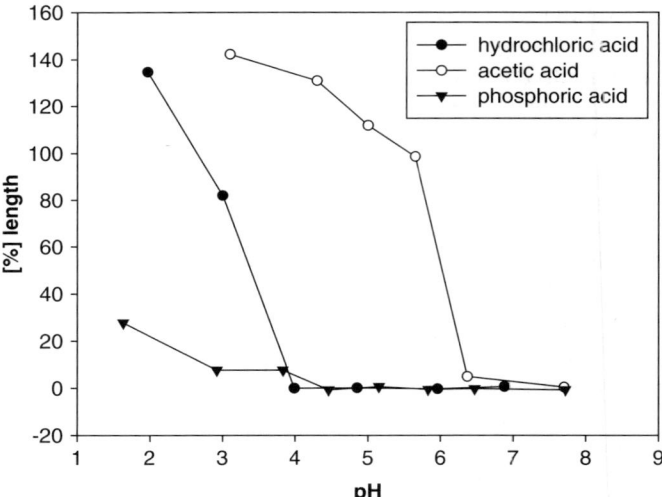

Figure 4.3 pH profiles of expansion of chitosan film in the presence of different acids; (●) 0.1 M hydrochloric acid, (○) 0.1 M acetic acid and (▼) 0.1 M phosphoric acid; starting with gel at pH 7.

Table 4.1 Volume expansions of chitosan gel **II** triggered by different acids at pH 5.[*1]

Acid	pKa	[%] length	Acid	pKa	[%] length
Hydrochloric acid (0.1 M)	–	0 ± 1.5	Oxalic acid (10 mM)	1.23, 4.19	3 ± 1.5
Acetic acid (0.1 M)	4.75	850 ± 7.4	Malonic acid (10 mM)	2.83, 5.69	40 ± 3.0
Phosphoric acid (0.1 M)	2.12, 7.21, 12.67	3 ± 0.6	Succinic acid (10 mM)	4.16, 5.16	760 ± 3.0
Sulfuric acid (0.1 M)	1.92	23 ± 1.5	Glutaric acid (10 mM)	4.34, 5.41	790 ± 3.0
Benzoic acid (20 mM)	4.19	640 ± 9.0	Tartaric acid (10 mM)	2.98, 4.34	26 ± 4.5
Cyclohexanoic acid (50 mM)	–	1020 ± 3.0			

[*1] All at pH 7 ± 0.2; deviations [%] from triplicate measurements; 30 mM phosphate buffer was used only in case of sulfuric acid.

acids such as benzoic or cyclohexanoic acid behave like acetic acid, whereas dicarboxylic acids exhibit a dependence on the chain length separating the carboxylic groups: glutaric and succinic acid, having similar pK values as the monoacids, behave like these, whereas oxalic, tartaric and malonic acid with their low pK values, due to the proximity of the carboxylgroups, lead to reduced chemomechanical effects, again ascribed to counteracting crosslinking with the anions, in analogy to phosphate (Table 4.1).

Chitosan obtained by ionic crosslinking with either triphosphate TPP or polyphosphate PP was found to swell at degrees that depend on the mode of preparation. The swelling at high pH was ascribed to deprotonation of the chitosan amine groups with disruption of the crosslinking salt bridges. However, if the gels with TPP or PP were prepared at pH 6.8 the pH-induced swelling was small, whereas gels prepared at pH 1.2 showed large swelling above pH 8.[25] In PAH III gels, crosslinked with glutardialdehyde, iodide induces larger contraction compared to chloride. It was concluded that the higher polarisability of iodide ions result in enhanced ion-pair formation and thereby decreased osmotic pressure with a collapse of the gel.[34]

4.3 Particle-size Effects and Kinetics

Swelling kinetics are mostly controlled by effector diffusion, described by Fick's laws. Expansion of xanthan-chitosan-derived gels upon pH change seems to be mainly controlled by the diffusion of mobile ions, except that at pH values below 10 the degree of ionisation during swelling also may affect the swelling rate.[43,44] Figure 4.4 illustrates that with the PA-derived polymer **I** the kinetics of both expansion and the fully reversible contractions follow pseudo-first–order equations, the same holds for the corresponding ab- and desorption of the effectors.[35]

The size and shape of the polymer particles used have a distinct influence on the expansion and contraction of the chemomechanical materials. The rate of volume changes depend on the effector concentration gradients, the speed of effector diffusion into the particles, and therefore also on the surface to volume ratio. Thus, film pieces of chitosan gel **II** show, with histidine and acetic acid as effector rates, which increase with the surface to volume ratio S/V (Figure 4.5).[45]

Another, until now often overlooked, influence of the polymer particle size on the volume change is due to the effect, that smaller particles may need a smaller amount of effector for saturation.[46] The possible sensitivity increase

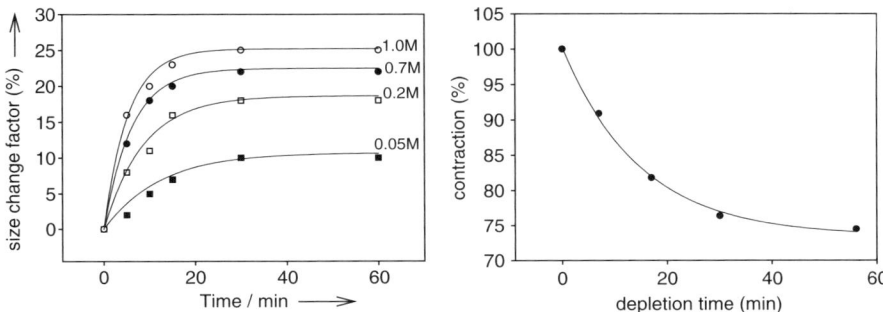

Figure 4.4 Kinetics of expansion and desorption (action of AMP on polymer **I**, least square fit to first-order equation.[35b]

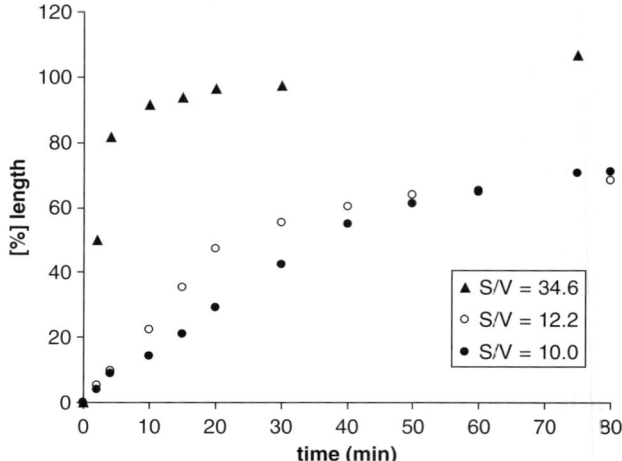

Figure 4.5 Expansion kinetics, and surface to volume S/V effects with chitosan; elongation induced by 50 mM L-histidine and 0.1 M acetic acid, in 30 mM phosphate buffer at pH 5.0. Approximate "half-life" time $t_{1/2}$ for 50% of the maximum expansion: $t_{1/2} = 42$, 32 and 3 min for S/V = 10.0, 12.2 and 34.6, respectively.

with the decrease of a particle volume has been described in the context of sensor miniaturisation.[47] The sensitivity increase, however, holds only as long as the affinity of an effector towards the sensing material is so high that all, or nearly all, effector molecules are absorbed, independent of the external effector concentration. As long as the affinity is high enough one can indeed observe that the effector concentration that is required to reach a certain volume change is a function of the used chemomechanical particle size (Figure 4.6).[46]

4.4 Water Uptake and Release

Dry hydrogels take up much water (up to 99%, e.g., in the case of polyallylamine[45]); conversion from neutral to ionic condition invariably will lead to the uptake of more water needed for solvation of both the cationic and anionic sites (Scheme 4.4(a)). Exposure of the swollen polymer to different effectors in aqueous solution can also lead to significant additional weight and volume increase (Scheme 4.4(b)) due to the necessary solvation of the effector. Alternatively, effector absorption can lead to water release with concomitant contraction as consequence of, e.g., counteracting ionic crosslinking (Scheme 4.4(c)). It is known that water transport through gels is significantly affected by the pH of the environment.[48] Chitosan hydrogels grafted with lactic or glycolic acid were characterised by FT-IR and differential scanning calorimetry (DSC), and the presence of three different water types was proposed.[49] Carboxymethylcellulose gels showed water uptake as a function of the used crosslinking; after sulfating the gels are pH sensitive. The water uptake of such

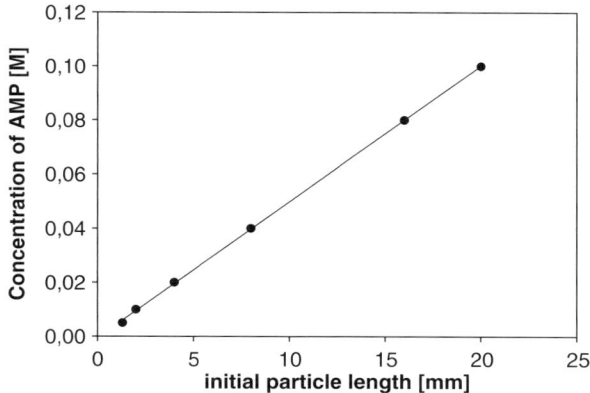

Figure 4.6 Effector AMP concentration needed for a certain expansion (here 35% in volume) as a function of the polymer particle size (variable length, with constant width and thickness; in 0.02 M phosphate buffer, pH = 7.0). Volume expansions of up to 300% are possible under the working conditions with the smaller particle, but not with the larger ones.[46]

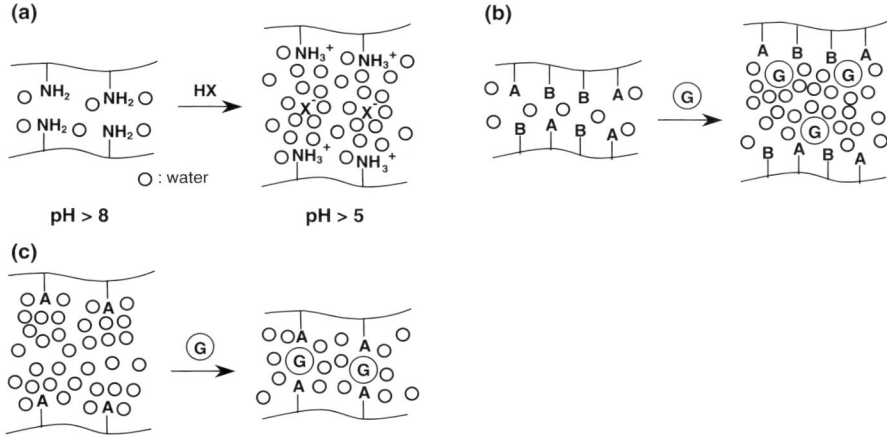

Scheme 4.4 (a) Water uptake as result of pH change with concomitant charge increase, (b) water uptake as result of complexation with guest **G**, and (c) water release as result of, *e.g.*, ionic cross-linking with guest **G**.

gels was studied by FT-IR spectra; hydrogen-bond formation between the chains was proposed to explain the correlation found between water uptake properties and the chemical composition of the gels.[50] Albumin gels showed at the isoelectric point a minimum of swelling, and expansion above and below this.[51] In crosslinked chitosan/polyether hydrogels the total water content and the amount of bound water is related to the pH values of the environment; the diffusion coefficient from the kinetics is correlated with the water state of the

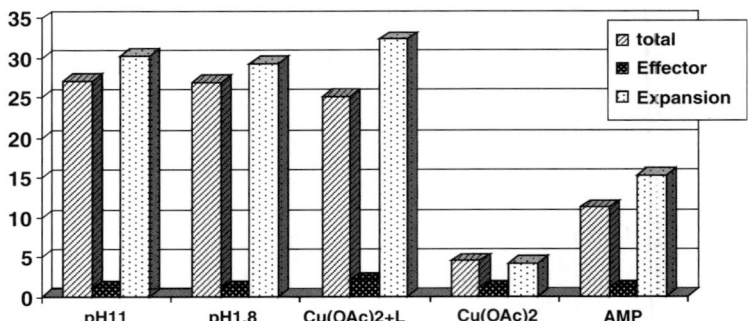

Scheme 4.5 Weight increase compared to expansion. Values scaled per mg of wet polymer, in mg of total weight increase due to water and effector, and in mg increase due to effector alone (upper limit as estimated from absorption measurements and complexometry); volume expansion Δv (from 1 mm^3 wet), average from length and width increase. Weight increase is within the error due to water-content increase.[35b]

swollen hydrogel, and likely is related to the relative composition of bound and nonbound water.[52] Hydrogels from fluorenyl-ala-ala dipeptides release 40% water upon addition of the known strong binding ligand vancomycin.[53]

Scheme 4.5 shows the weight increase which accompanies the effector absorption with the PA polymer **I**; independent measurements of water content before and after expansion establish that the weight increase is almost entirely due to water uptake. At the same time the scheme illustrates the abovementioned water uptake by protonation and deprotonation at low and high pH; in comparison to pH 7 there is a water content increase by a factor of $f = 26 \pm 1$. The swelling induced by other effectors reflects the need of these molecules inside the gel for the effector solvation, leading to a expansion that goes far beyond the weight and volume increase needed by the effector alone; in line with this, the swelling increases with the size of structurally related effectors.[35b] It has long been known that in supramolecular complex formation, in particular with ion pairing, solvation differences before and after complex formation play a major role, leading also to distinct dynamic differences between water inside and outside supramolecular cavities.[54] With chemomechanical hydrogels the water uptake accompanying solvation becomes directly measurable.

4.5 Concentration Profiles

Unless abrupt phase transitions[9,10,19] occur one may expect saturation-type of profiles as a function of the effector concentration until all the binding sites of a chemomechanical polymer are occupied. Figure 4.7 illustrates with polymer **I** that such isotherm-like curves can be observed, unless the cooperativity effects discussed below play a role. The curves present only approximately normal

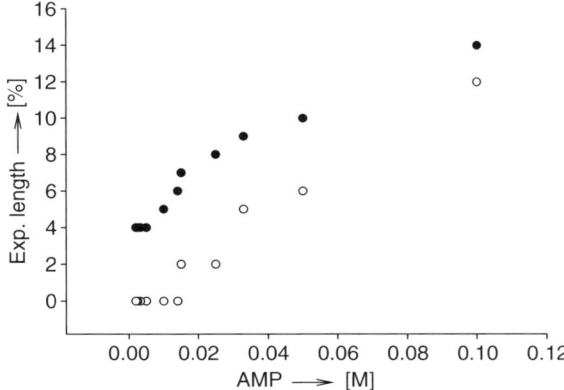

Figure 4.7 Expansion (length) as function of AMP concentration; polymer **A**, film size $5.0 \times 2.0 \times 0.4$ mm, pH 7.0. Lower trace (○), in the absence of buffer; upper trace (●), in the presence of 0.02 M NaH_2PO_4 buffer.[35b]

saturation isotherms, but allow an affinity towards the polymer to be inferred that resembles apparent association constants of related host–guest complexes in aqueous solution. Thus, the apparent binding constant K, *e.g.*, for AMP under the conditions in Figure 4.7 amounts to about 20 M^{-1}, roughly comparable to K values reported in homogeneous solution for the interaction of AMP and ethylenediamine-type host compounds.[55] From the first part of the particularly discontinuous concentration profiles with transition-metal ions[56] and the maximum expansions reached there one can estimate affinities that are, as expected, much higher, *e.g.*, for Cu(II) ions with an apparent K value around 10^5 to 10^6 M^{-1}. However, the profiles not only depend on the presence of additional salts or buffers, but also show a lag period before swelling begins (Figure 4.7). In contrast, UV-vis measurements show, as expected, absorption already at low concentrations, which are too low for an expansion. The expansion starts at a concentration at which presumably the effector starts to move inside the gel after first saturating the surface.[35b]

4.6 Cooperativity and Logical Gate Functions

In solution supramolecular complexes have been already designed in such a way that two different chemical entities are required in order to give a certain response.[57] Here, *e.g.*, a fluorescence signal is only emitted (or quenched) from a host–guest complex if the pH has a certain value, or if a second compound is present. Such systems are either based on cooperativity between different guest molecules, or on allosteric effects between two or more distinct binding and conformationally coupled sites. As mentioned above, pH-induced swelling of a chemomechanical polymer often depends also on, *e.g.*, salt concentrations or rather ionic strength, which may be considered as simple, although rather unselective cooperativity. Only recently has it been shown that with a chemomechanical polymer bearing several recognition functions, such as **I**, can selective response be materialised between several effectors[58] (Scheme 4.6). The cooperativity can be so strong that a volume change occurs with, *e.g.*, peptides only if a certain metal cation is present (see Section 4.8). Such gel-based logical AND gate systems require no spectroscopic detection and transducers, but communicate directly to the outside world.

The expansion stimulated on the functional polymer **I** by different effectors depends markedly on the pH, thus representing an efficient macroscopic AND gate system. Thus, the expansion magnitude induced by the nucleotides UMP and AMP is reversed in going from pH 7 to pH 11 (Scheme 4.7).[35]

Logical AND gates in the sense of positive cooperativity are not only seen by the influence of pH but also between different effectors. For example, with polymer **I** the expansion induced by AMP reaches a distinct maximum only at a certain concentration of phosphate anions as the second effector (Figure 4.8).[58] With other effectors negative cooperativity has also been observed; *e.g.*, the presence of sodium salts distinctly lowers the expansion induced in polymer **I** by zinc acetate (Figure 4.9).[56] The ethylenediamine units in **I** allows chelation and selective volume changes with transition-metal ions;[59] as expected, low pH values lead to a reversal of the expansions, since the ethylenediamine units become protonated – another manifestation of cooperativity with pH. The

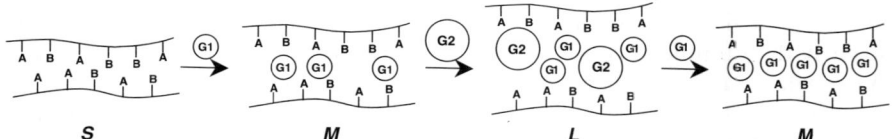

Scheme 4.6 Cooperativity effects with two guest molecules **G1** and **G2**; *S*: small/tight; *M*: medium/loose; *L*: large/loose network.

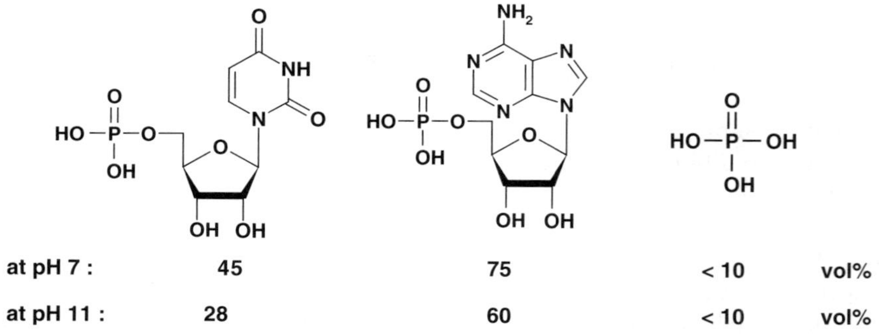

at pH 7 :	45	75	< 10	vol%
at pH 11 :	28	60	< 10	vol%

Scheme 4.7 Expansions (in % volume) triggered by nucleotides UMP and AMP and phosphate (0.1 M) at different pH values; chemomechanical polymer **I**.[35]

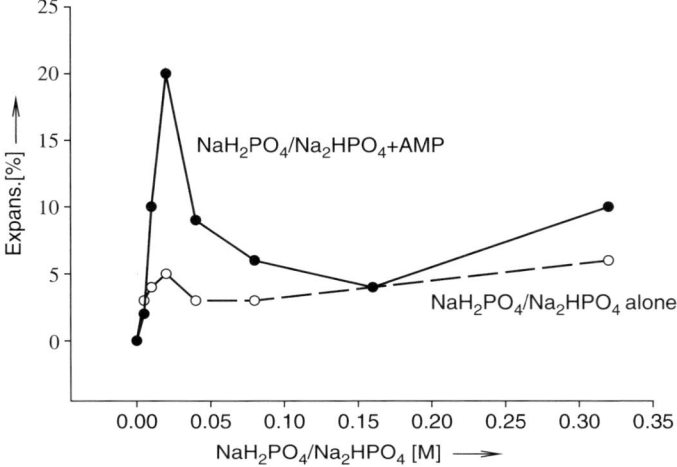

Figure 4.8 Cooperativity between AMP and phosphate in the expansion with polymer **I**. (Expansion given in one dimension).[58]

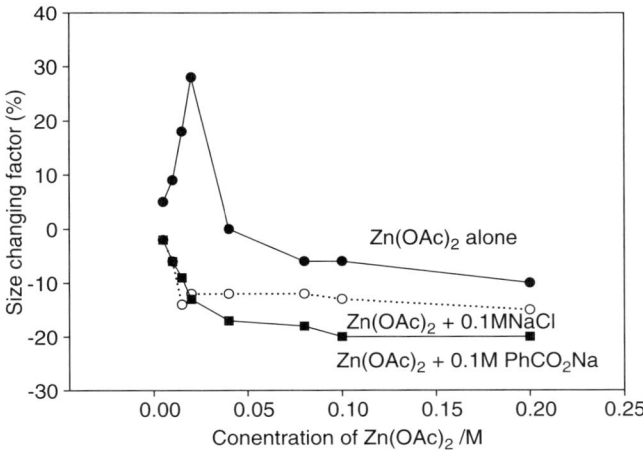

Figure 4.9 Expansions with the chelating polymer **I** induced by zinc(II)acetate with and without the presence of sodium choride or sodium benzoate as the second effector.[35b,56]

critical effector concentration at which the expansion reaches a maximum has been shown to correlate with the size of the polymer particle used, for the reasons discussed above in the context of the relation between size and sensitivity.[46]

4.7 Selectivity with Organic Effector Molecules

The introduction of even simple receptor groups such as ethylenediamine ("ene") units and lipophilic alkyl chains in polymer **I** already leads to relatively large differences in swelling by various organic compounds. They can be understood and planned on the basis of interaction mechanisms known from supramolecular chemistry. As indicated in Scheme 4.2 the ene units allow not only metal chelation, and pH-dependent protonation but also hydrogen bonding as well as N^+ cation-π effects. That the latter play an essential role is obvious from the large chemomechanical effects as long there are aryl units in the effector molecule: the saturated cyclohexane-carboxylate has, within the error, no effect in contrast to the almost isosteric benzoic acid (Scheme 4.8).[35] That electrostatic attraction, or ion pairing with the cationic N^- sites in **I** is an additional prerequisite for any size changes of the gel is clear from the inactivity of electroneutral compounds. Finally, the alkyl groups introduced into polymer **I** provide for lipophilic interaction, which varies between the different organic moieties of the effectors. In consequence, not only nucleotides (see Scheme 4.7), but even structural isomers such as different benzene dicarboxylicacids can be distinguished (Scheme 4.8). Lipophilic interactions are also responsible for the unique contraction of polymer **I** upon exposure to ammonium compounds bearing longer alkyl chains; Scheme 4.9 illustrates the systematic change from the usual expansion to shrinking.[35] It should be borne in mind that small expansion differences can be sufficient to switch, *e.g.*, release of a drug from a suitable container, on or off.

Another binding mechanism, often used in supramolecular complexes as well as in nature, are stacking interactions between aromatic units.[60] Introduction of anthrhyl groups into chitosan **II** provided for large interaction sites with aromatic effectors, although the substitution degree was low. With various amino acids one found expansions that as expected increase with the lipophilicity of the residues (Scheme 4.10).

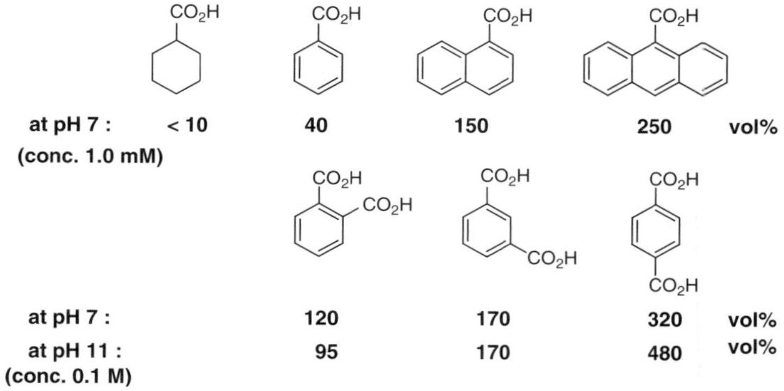

at pH 7 : (conc. 1.0 mM)	< 10	40	150	250 vol%
at pH 7 :		120	170	320 vol%
at pH 11 : (conc. 0.1 M)		95	170	480 vol%

Scheme 4.8 Selectivity between different organic effector compounds. (0.02–0.05 M phosphate buffer; the effects at pH 11 are corrected for difference between pH 7 and pH 11 alone (390 vol%)).[35a]

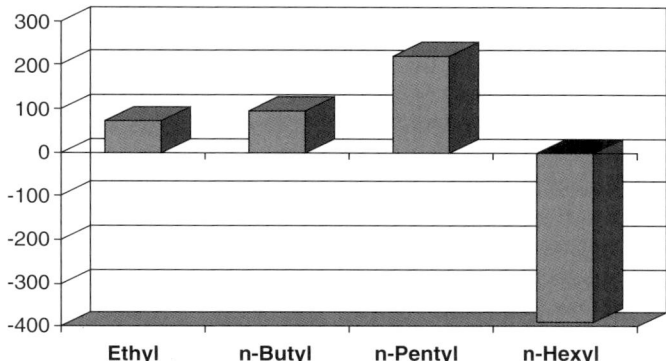

Scheme 4.9 Volume changes changes [%]) of polymer **I** with peralkylammonium hydroxides R$_4$NOH (after correction for pH-induced change, pH 12.4 to 12.7).[35b]

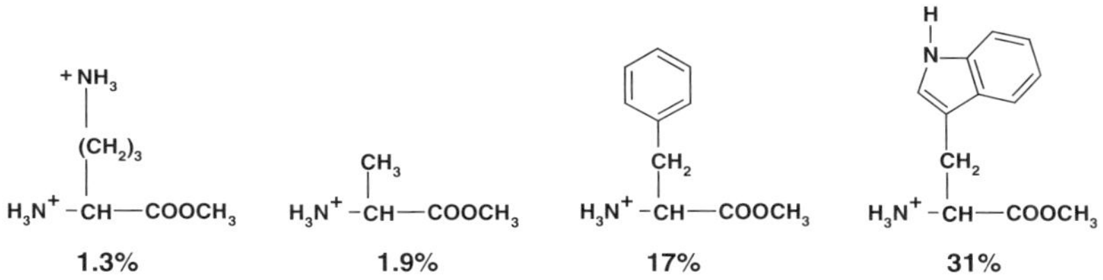

$^+$NH$_3$ CH$_3$ CH$_2$—C$_6$H$_5$ CH$_2$—indole
(CH$_2$)$_3$
H$_3$N$^+$—CH—COOCH$_3$ H$_3$N$^+$—CH—COOCH$_3$ H$_3$N$^+$—CH—COOCH$_3$ H$_3$N$^+$—CH—COOCH$_3$
1.3%   1.9%   17%   31%

Scheme 4.10 Volume expansions [%] on chitosan-anthrhyl polymer with different amino acid esters; pH and salt effects deducted; (pH 5.0; 0.1 M effector).[64]

Even without introduction of special recognition groups, some chemomechanical polymers can exhibit a moderate selectivity against special effector compounds. Chitosan gel **II** for example shows expansion differences not only between few carboxylic acids (see Table 4.1), but also between basic and nonbasic amino acids, which for this purpose need not to be protected but can be used in their native form. The largest expansions are observed with histidine (Table 4.2); due to the chiral chitosan backbone one could expect also some chiral discrimination, which is materialised with the most strongly interacting histidine to a moderate, yet promising degree. The difference in swelling depends on the acid anion with which it is competing, showing a preference for the D-enantiomer over the L-form that amounts to 145% *vs*. 120% in the presence of acetate; to 134% *vs*. 123% in the presence of benzoate, whereas no difference was observed in the presence of chloride.

Even the simple polyallylamine **III**, crosslinked for gel formation, exhibits some interesting selectivity. Using small-angle X-ray scattering (SAXS, Figure 4.10) it has been shown that incorporation of strongly bound salts such as arylsulfonates leads to ordered structures, with distances between the polymer

Table 4.2 Volume expansions of chitosan gel **II** triggered by basic amino acids in presence of different acids.[*1]

Acid	L-histidine	L-lysine	L-arginine	control[*2]
0.07 M chloride	120 ± 9.5	9 ± 2	9 ± 1	44 ± 1
0.10 M acetate	86 ± 4.5	69 ± 3	56 ± 8	586 ± 5
0.10 M phosphate	-3 ± 2	-6 ± 1	-6 ± 1	0.9 ± 0.3
0.10 M surfate	-6 ± 0.3	-3 ± 1.5	-9 ± 2.5	23 ± 1.5
0.08 M benzoate	310 ± 5	218 ± 8.5	192 ± 13	325 ± 8

[*1]All measured in 30 mM phosphate buffer at pH 5.0; deviations [%] from triplicate measurements; expansion with other aminoacids such as ala, phe, trp: < 10%.
[*2]Expansion volume without amino acids.

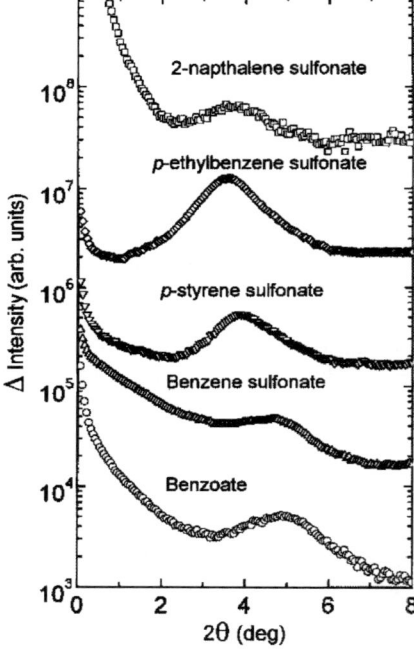

Figure 4.10 SAXS patterns of collapsed gels derived from polyallylamine **III** in various aromatic sodium salt solutions (0.07 M); taken with permission from Ref. 34.

backbone that increase with the size of the guest molecules.[34] Such effectors lead to weight decrease of, *e.g.*, 95%; re-swelling is possible by treatment with an excess of weaker effectors such as NaCl that cause only 40% weight loss. Recent studies reveal that the PAH polymer gel **III** allows even some isomeric compounds to be distinguished (Scheme 4.11).[45] Generally, ionic crosslinking leads to a volume decrease with all kind of carboxylic acids, with significant

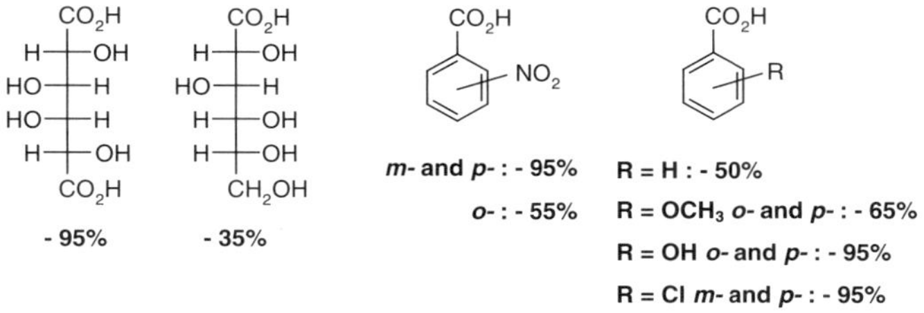

Scheme 4.11 Volume change of PAH gel **III** in the presence of different acids effectors (10 mM) at pH = 7.[45]

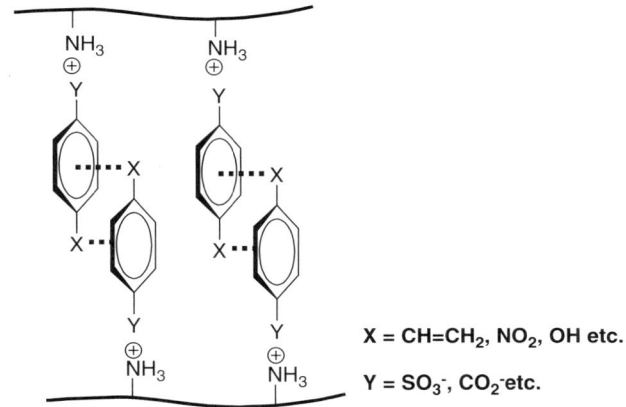

Figure 4.11 Model of a polyallylamine **III**-derived gel with substituted aromatic acids as counterion spacers.

differences between natural glucoronic and glucaric acids. Surprisingly, benzoic acids with suitable substituents also lead to large contractions of similar size as the dicarboxylic acids, presumably as a result of stacking interactions between the aryl rings in analogy to Figure 4.11. Nitrosubstituents have been shown to interact strongly with other aromatic moieties by dispersive forces,[61] which would explain why the nitrobenzoic acids lead to particular sizeable contraction as long as the nitrogroup is not too close to the anionic carboxylate center (*cf*. Scheme 4.11, R = *o*-NO$_2$).

4.8 Ternary Complex Formation for Amino Acids and Peptides as Effectors

The metal chelating groups in polymer **I** allow binding, in addition, also of effectors that otherwise would not interact at all with the chemomechanical polymer. It is known from solution chemistry that, *e.g.*, Cu(II) or Zn(II) cations

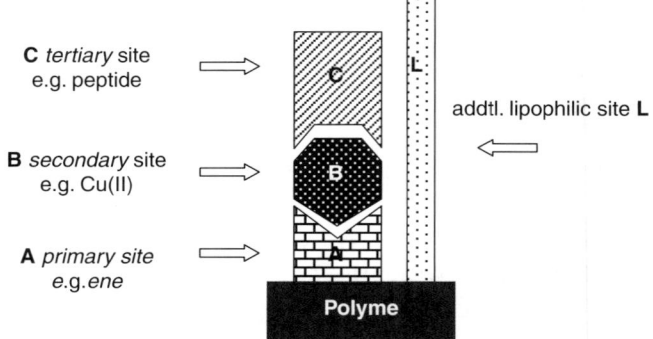

Scheme 4.12 Ternary complex formation.[64]

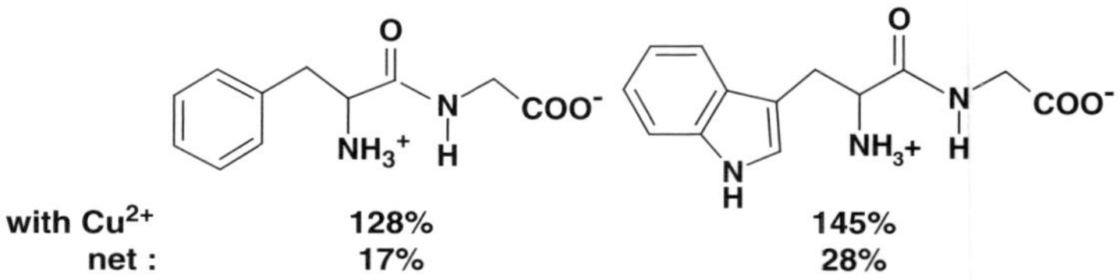

| with Cu^{2+} | 128% | 145% |
| net: | 17% | 28% |

Scheme 4.13 Examples of expansion (in one dimension) in ternary complexes; Cu(II) and Effector 0.25 mM polymer **I** pH 4.5. Net: effect of Cu^{2+} alone deducted.[64]

form ternary mixed complexes with a number of chelating agents including amino acids and peptides.[62] Such cocomplexation has been used successfully in supramolecular associations, *e.g.*, for sensor application.[63] The additional interaction groups of polymer **I** (Scheme 4.12)[64] provide for additional discrimination with the amino acids side groups, which is visible in swelling differences between different peptides (Scheme 4.13). With some special chelating agents the expansions reach a record 475% on top of the swelling produced by the metal ion itself. Removal of the metal cations by decomplexing agents leads to a reversible contraction of the swollen gel as a function of the decomplexing agent chelating strength.

4.9 Selectivity by Covalent Interactions/ Glucose-triggered Size Changes

Several substrates of biological importance such as carbohydrates exert notoriously weak intermolecular interactions, particular in water. The most practical way to overcome this problem, *e.g.*, for the glucose-responsive supramolecular systems consists in the formation of boronic esters, for which NMR studies in solution[65] suggest crosslinking with two boronic residues

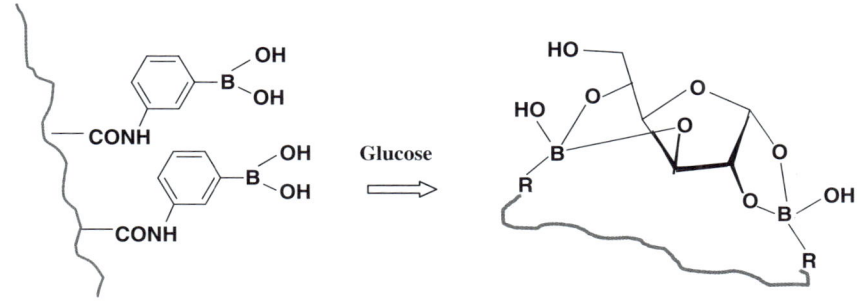

Scheme 4.14 Boronic ester formation with glucose.

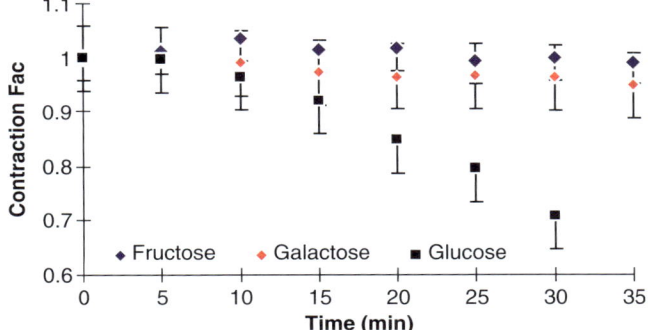

Figure 4.12 Selective response of a boronic-ester equipped polymer to glucose over fructose and galactose. Measurement under blood plasma condition with all sugars at the 0.005 M concentration.[69]

(Scheme 4.14). The use of such esters for the determination of the configuration of carbohydrates has long been known,[66] but has been implemented into sophisticated supramolecular sensing systems only during recent years.[67,68] It has been shown that this principle can be transferred to chemomechanical polymers.[69,70] Thus, reaction of poly(methyl methacrylate) with 3-aminophenylboronic acid, simultaneously processed with diethylenetriamine, dodecylamine, and butylamine, yields a hydrogel, which after swelling in water exhibits sizeable contraction with glucose, presumably due to the disappearance of the negatively charged boronate anions (Figure 4.12). The response is selective for glucose under blood plasma conditions with only 5 mM sugar concentration, holding obvious promise for self-sustained delivery devices, *e.g.*, for insulin release.[71]

4.10 Conclusions

On the basis of the now vast experience of supramolecular chemistry it will be possible to develop polymers that are selectively stimulated by all kinds of possible molecules in the environment. Such selective polymers allow engineering of actuator devices that can drive a micromachine or, *e.g.*, deliver a drug, entirely self-controlled. Asymmetric bilayers made from films of different chemical nature and thereby different response to the environment can be used for an increased macroscopic movements by bending of the film stripes.[72] As discussed in Section 4.3, particle miniaturisation can enhance both the speed and the sensitivity of response of a chemomechanical polymer. Cooperative effects between several different effectors can lead to high selectivity under defined conditions, and allow stimulation by otherwise inactive compounds. Biomacromolecules including proteins can also be used both as the basis for chemomechanical polymers with active and selective recognition sites, as well as effectors. Materials responsive to glucose have been already developed on the basis of lectin, specifically of concanavalin,[73] and also using enzymes such as glucose oxidase.[74] Grafting an antigen and the corresponding antibody into a polymer network the binding of the free antigen can trigger a gel-volume change.[75] The biomimetic translation of selective molecular recognition into the outside world with completely self-sustained and fully automatic devices will be of particular significance for future biomedical applications.

References

1. (a) *Polymer Sensors and Actuators*, ed. O. Osada and D.E. Rossi, Springer, Berlin, 1999; (b) L. Dai, *Intelligent Macromolecules for Smart Devices*; Springer, London, 2004.
2. For leading references on environment-sensitive hydrogels for drug delivery see: *Advances in Controlled Drug Delivery: Science, Technology and Products*, ed. S.M. Dinh and P. Liu, Washington, DC, 2003, ACS symposium series 846, Washington DC; *Advances in Biomaterials and Drug Delivery Systems*, G.-H. Hsiue, T. Okano, U.Y. Kim, H-W. Sung, N. Yui and K.D. Park, Princeton International Publishing, 2002; J. Kopecek, *Eur. J. Pharm. Sci.*, 2003, **20**, 1–16; J. Kopecek, *Nature*, 2002, **417**, 388; N.A. Peppas and Y. Huang, *Pharm. Res.*, 2002, **19**, 578–587; E. Oral and N.A. Peppas, *J. Biomed. Mater. Res. Part A*, 2004, **68A**, 439; M.E. Byrne, K. Park and N.A. Peppas, *Adv. Drug Deliv. Rev.*, 2002, **54**, 149; D. Kaneko, J.P. Gong and Y. Osada, *J. Mater. Chem.*, 2002, **12**, 2169; P. Gupta, K. Vermani and S. Garg, *Drug Discov. Today*, 2002, **7**, 569; S.V. Vinogradov, T.K. Bronich and A.V. Kabanov, *Adv. Drug Deliv. Rev.*, 2002, **54**, 135; J.Z. Hilt and M.E. Byrne, *Adv. Drug Deliv. Rev.*, 2004, **56**, 1599; S. Murdan, *J. Control. Release*, 2003, **92**, 1; T. Miyata, T. Uragami and K. Nakamae, *Adv. Drug Deliv. Rev.*, 2002, **54**, 79; Y. Qiu and K. Park, *Adv. Drug Deliv. Rev.*, 2001, **53**, 321; M. Kumar and N. Kumar, *Drug Devel. Indus. Pharm.*, 2001, **27**, 1; D.T. Eddington and D.J. Beebe, *Adv. Drug Deliv. Rev.*, 2004, **56**, 199.

3. Y. Osada, *Adv. Mater.*, 1991, **3**, 107; D. Kaneko, J.P. Gong and Y. Osada, *J. Mater. Chem.*, 2002, **12**, 2169; M. Shahinpoor, *Proc. SPIE-Intern. Soc. Opt. Eng.*, 1994, **2189**, 134.
4. M. Shahinpoor, *Proc. SPIE-Intern. Soc. Opt. Eng.*, 1994, **2189**, 265, see also M. Shahinpoor, *Electrochimica Acta*, 2003, **48**, 2343.
5. For leading references on chemomechanical polymers as pH-sensitive sensors see: L. Zhang and W.R. Seitz, *Anal. Bioanal. Chem.*, 2002, **373**, 555; M.T.V. Rooney and W.R. Seitz, *Anal. Comm.*, 1999, **36**, 267; S. Ikeda, H. Kumagai, T. Sakiyama, C.H. Chu and K. Nakamura, *Biosci. Biotechnol. Biochem.*, 1995, **59**, 1422; T. Schalkhammer, C. Lobmaier, F. Pittner, A. Leitner, H. Brunner and F.R. Aussenegg, *Sens. Actuators B-Chem.*, 1995, **24**, 166; M.T.A. Ende, C.L. Bell, N.A. Peppas, G. Massimo and P. Colombo, *Int. J. Pharm.*, 1995, **120**, 33; W. C. Michie, B. Culshaw, I. McKenzie, M. Konstantakis, N. B. Graham, C. Moran, F. Santos, E. Bergqvist and B. Carlstrom, *Opt. Lett.*, 1995, **20**, 103.
6. See, *e.g.*, *Comprehensive Supramolecular Chemistry*, Vols 1–11; J.M. Lehn, J.L. Atwood, J.E.D. Dvies, D.D. MacNicol, D.D. MacNicol and F. Vögtle, Series editors, Pergamon/Elsevier, Oxford etc, 1996; J.-M. Lehn, *Supramolecular Chemistry. Concepts and Perspectives*, Weinheim, Wiley VCH 1995; F. Vögtle, *Supramolecular Chemistry: An Introduction*, Weinheim, Wiley VCH 1993; *Encyclopedia of Supramolecular Chemistry*, ed. J.L. Atwood and J.W. Steed, Marcel Dekker, Inc., New York, 2003; H.-J. Schneider and A. Yatsimirsky. *Principles and Methods in Supramolecular Chemistry*, Wiley, Chichester, 2000; J.W. Steed and J.L. Atwood, *Supramolecular Chemistry*, Wiley, New York, 2000.
7. W. Kuhn, B. Hargitay, A. Katchalsky and H. Eisenberg, *Nature*, 1950, **165**, 514.
8. For leading references on pH sensitive gels see, *e.g.*, Reviews: R. Yoshida, *Curr. Org. Chem.* 2005, **9**, 1617; D.T. Eddington, D.J. Beebe, *Adv. Drug. Deliv. Rev.* 2004, 199; D.N. Robinson and N.A. Peppas, *Macromolecules*, 2002, **35**, 3668; D.N. Robinson, and N.A. Peppas, *Macromolecules*, 2002, **35**, 3668; E. Roux, M. Francis, F.M. Winnik and J.C. Leroux: *Int. J. Pharm.*, 2002, **242**, 25; E. Roux, M. Lafleur, E. Lataste, P. Moreau and J.C. Leroux, *Biomacromolecules*, 2003, **4**, 240; T. Watanabe, K. Ito, C. Alvarez-Lorenzo, A.Y. Grosberg and T. Tanaka, *J. Chem. Phys.*, 2001, **115**, 1596; K. Suzuki, T. Yumura, Y. Tanaka and M. Akashi, *J. Bioact. Compat. Polym.*, 2001, **16**, 409; K.S. Soppimath, A.R. Kulkarni and T.M. Aminabhavi, *J. Control. Release*, 2001, **75**, 331; K.F. Arndt, D. Kuckling and A. Richter, *Polym. Adv. Techn*, 2000, **11**, 496; G. Gerlach, M. Guenther, J. Sorber, G. Suchaneck, K.F. Arndt and A. Richter, *Sens. Actuators B-Chem.*, 2005, **111**, 555; D. Kuckling, H.J.P. Adler, K.F. Arndt, L. Ling and W.D. Habicher, *Macromol. Symp.*, 1999, **145**, 65; D. Kuckling, A. Richter and K.F. Arndt, *Macromol. Mater. Eng.*, 2003, **288**, 144; Q.X. Wang, H. Li and K.Y. Lam, *J. Polym. Sci., Part B: Polym. Phys.*, 2006, **44**, 326; M. El-Sherbiny, R.J. Lins, E.M. Abdel-Bary and D.R.K. Harding, *Eur. Polym. J.*, 2005, **41**, 2584; S.L. Chen, M.Z. Liu, S.P. Jin and

Y. Chen, *J. Appl. Polym. Sci.*, 2005, **98**, 1720; T. Caykara and I. Aycicek, *J. Polym. Sci. A Polym. Chem.*, 2005, **43**, 2819; F. Weiss and H. Finkelmann, *Macromolecules*, 2004, **37**, 6587; K.N. Plunkett and J.S. Moore, *Langmuir*, 2004, **20**, 6535; T. Caykara, M. Dogmus and O. Kantoglu, *J. Polym. Sci. A Polym. Chem.*, 2004, **42**, 2586; B.D. Johnson, D.J. Beebe and W. Crone, *Mater. Sci. Eng. C- Biomim. Supramol. Systems*, 2004, **24**, 575; Z.M. Shakhsher, I. Odeh, S. Jabr and W.R. Seitz, *Microchim. Acta*, 2004, **144**, 147; M.J. Molina, M.R. Gomez-Anton and I.F. Pierola, *Macromol. Chem. Phys.*, 2002, **203**, 2075, S.L. Zhou, S. Matsumoto, H.D. Tian, H. Yamane, A. Ojida, S. Kiyonaka and I. Hamachi, *Chem. Eur. J.*, 2005, **11**, 1130; M. Yazdani-Pedram, J. Retuert and R. Quijada, *Macromol. Chem. Phys.*, 2000, **201**, 923; M. Sauer; D. Streich and W.Meier, *Adv. Mater.* 2001, **13**, 1649.
9. K. Ito, J. Chuang, C. Alvarez-Lorenzo, T. Watanabe, N. Ando and A.Y. Grosberg, *Prog. Polym. Sci.*, 2003, **28**, 1489.
10. M. Annaka and T. Tanaka, *Physica A*, 1994, **204**, 40.
11. See L. Brannonpeppas and N.A. Peppas, *J. Control. Rel.*, 1989, **8**, 267; L. Brannonpeppas and N.A. Peppas, *Chem. Eng. Sci.*, 1991, **46**, 715; R.A. Siegel and B.A. Firestone, *Macromolecules*, 1988, **21**, 3254 and references cited therein.
12. M. Annaka and T. Tanaka, *Nature*, 1992, **355**, 430.
13. G.R. Mahdavina, M.J. Zohuriaan-Mehr and A. Pourjavadi, *Polym. Adv. Technol.*, 2004, **15**, 173.
14. G. Wulff, *Chem. Rev.*, 2002, **102**, 1; K. Mosbach, *Trends Biochem. Sci.*, 1994, **19**, 9; C. Alexander, L. Davidson and W. Hayes, *Tetrahedron*, 2003, **59**, 2025.
15. M.E. Byrne, E. Oral, J.Z. Hilt and N.A. Peppas, *Polym. Adv. Technol.*, 2002, **13**, 798; H. Asanuma, T. Hishiya and M. Komiyama, *Adv. Mater.*, 2000, **12**, 1019; R.A. Bartsch and M. Maeda, ed., *Molecular and Ionic Recognition with Imprinted Polymers, ACS Symp. Ser.*, **703**, 1998, Washington DC.
16. C. Alvarez-Lorenzo, O. Guney, T. Oya, Y. Sakai, M. Kobayashi, T. Enoki, Y. Takeoka, T. Ishibashi, K. Kuroda, K. Tanaka, G.Q. Wang, A.Y. Grosberg, S. Masamune and T. Tanaka, *Macromolecules*, 2000, **33**, 8693; T. Moritani and C. Alvarez-Lorenzo, *ibid.*, 2001, **34**, 7796.
17. M. Watanabe, T. Akahoshi, Y. Tabata and D. Nakayama, *J. Am. Chem. Soc.*, 1998, **120**, 5577.
18. K. Nam, J. Watanabe and K. Ishihara, *Polymer*, 2005, **46**, 4704.
19. See eg T. Hirotsu, Y. Hirokawa and T. Tanaka, *J. Chem. Phys.*, 1987, **87**, 1392.
20. K. Podual and N. A. Peppas, *Polym. Int.*, 2005, **54**, 581.
21. G.M. Eichenbaum, P.F. Kiser, D. Shah, S.A. Simon and D. Needham, *Macromolecules*, 1999, **32**, 8996.
22. X. Jin and Y.L. Hsieh, *Polymer*, 2005, **46**, 5149.
23. C. Iversen, A.L. Kjoniksen, B. Nystrom, T. Nakken, O. Palmgren and T. Tande, *Polym. Bull.*, 1997, **39**, 747; M. Hamdine, M.C. Heuzey and

A. Bégin, *Int. J. Biol. Macromol.*, 2005, **37**, 134; A. Montembault, C. Viton and A. Domard, *Biomacromolecules*, 2005, **6**, 653.
24. H.-M. Kam, E. Khor and L.-Y. Lim, J. Biomed. Mater. Res., *Part B*, 1999, **48**, 881.
25. F.-L. Mi, S.-S. Shyu, T.-B. Wong, S.-F. Jang, S.-T. Lee and K.-T. Lu, *J. Appl. Polym. Sci.*, 1999, **74**, 1093.
26. S.J. Kim, S.R. Shin, N.G. Kim and S.I. Kim, *J. Macromol. Sci. -Pure Appl. Chem.*, 2005, **A42**, 1073.
27. Y.M. Lim, Y.M. Lee and Y.C. Nho, *Macromolec. Res.*, 2005, **13**, 327.
28. S.L. Zhou, S. Matsumoto, H.D. Tian, H. Yamane, A. Ojida, S. Kiyonaka and I. Hamachi, *Chem. Eur. J.*, 2005, **11**, 1130.
29. Y.X. Zhang, F.P. Wu, M.Z. Li and E.J. Wang, *Polymer*, 2005, **46**, 7695.
30. F.L. Mi, H.F. Liang, Y.C. Wu, Y.S. Lin, T.F. Yang and H.W. Sung, *J. Biomater. Sci. –Polym. Edn*, 2005, **16**, 1333.
31. S.J. Kim, K.J. Lee, I.Y. Kim, D.I. Shin and S.I. Kim, *J. Appl. Polym. Sci.*, 2006, **99**, 1100.
32. S. Kobayashi, M. Tokunoh, T. Saegusa and F. Mashio, *Macromolecules*, 1985, **18**, 2357.
33. E.D. Oliveira, S.G. Hirsch R.J. Spontak and S.H. Gehrke, *Macromolecules*, 2003, **36**, 6189 and references cited therein.
34. G.V. Rama Rao, T. Konishi and N. Ise, *Macromolecules*, 1999, **32**, 7582.
35. (a) H.-J. Schneider, T. Liu and N. Lomadze, *Angew. Chem. Int. Ed. Engl.*, 2003, **42**, 3544; (b) H.-J. Schneider, T. Liu and N. Lomadze, *Eur. J. Org. Chem.*, 2006, 677.
36. C.H. Chu, T. Sakiyama and T. Yano, *Biosci. Biotechnol. Biochem.*, 1995, **59**, 717.
37. A. Pourjavadi, H. Hosseinzadeh and R. Mazidi, *J. Appl. Polym. Sci.*, 2005, **98**, 255.
38. A. Pourjavadi, M. Sadeghi and H. Hosseinzadeh, *Polym. Adv. Technol.*, 2004, **15**, 645.
39. K.W. Seo, D.J. Kim and K.N. Park, *J. Ind. Eng. Chem.*, 2004, **10**, 794.
40. J.P. Schneider, D.J. Pochan, B. Ozbas, K. Rajagopal, L. Pakstis and J. Kretsinger, *J. Am. Chem. Soc.*, 2002, **124**, 15030; B. Ozbas, J. Kretsinger, K. Rajagopal, J.P. Schneider and D.J. Pochan, *Macromolecules*, 2004, **37**, 7331.
41. T. Watanabe, K. Ito, C. Alvarez-Lorenzo, A.Y. Grosberg and T. Tanaka, *J. Chem. Phys.*, 2001, **115**, 1596.
42. S. Yamaguchi, L. Yoshimura, T. Kohira, S. Tamaru and I. Hamachi, *J. Am. Chem. Soc.*, 2005, **127**, 11835.
43. C.H. Chu, H. Kumagai, T. Sakiyama, S. Ikeda and K. Nakamura, *Biosci. Biotechn. Biochem.*, 1996, **60**, 1627.
44. Y. Chu, P.P. Varanasi, M. J. McGlade and S. Varanasi, *J. Appl. Polym. Sci.*, 1995, **58**, 2161.
45. K. Kato and H.-J. Schneider, unpublished results.
46. H.-J. Schneider, L. Tianjun and N. Lomadze, *Chem. Commun.*, 2004, 2436.

47. R. Kopelman and S. Dourado, *Proc. SPIE-Intern. Soc. Opt. Eng.*, 1996, **2836**, 2; H.A. Clark, R. Kopelman, R. Tjalkens and M.A. Philbert, *Anal. Chem.*, 1999, **71**, 4837.
48. B. Kim, K. La Flamme and N.A. Peppas, *J. Appl. Polym. Sci.*, 2003, **89**, 1606.
49. X. Qu, A. Wirsen and A. C. Albertsson, *Polymer*, 2000, **41**, 4589, see also X. Qu, A. Wirsen and A.C. Albertsson, *J. Appl. Polym. Sci.*, 1999, **74**, 3186.
50. R. Barbucci, A. Magnani and M. Consumi, *Macromolecules*, 2000, **33**, 7475.
51. H.Y. Park, I.H. Song, J.H. Kim and W.S. Kim, *Int. J. Pharm.*, 1998, **175**, 231.
52. Y.L. Guan, L. Shao, J. Liu and K. DeYao, *J. Appl. Polym. Sci.*, 1996, **62**, 1253.
53. Y. Zhang, H. Gu, Z. Yang and B. Xu, *J. Am. Chem. Soc.*, 2003, **125**, 13680.
54. K. Bhattacharyya, *Acc. Chem. Res.*, 2003, **36**, 95, see also R.U. Lemieux, *Acc. Chem. Res.*, 1996, **29**, 373 and references cited therein.
55. See J. Aguilar, P. Diaz, F. Escarti, E. Garcia-Espana E, L. Gil, C. Soriano and B. Verdejo, *Inorg. Chim. Acta*, 2002, **339**, 307 and references cited therein.
56. H.-J. Schneider, Liu Tianjun, *Chem. Commun.*, 2004, 100.
57. Reviews: A.P. de Silva, D.B. Fox, A.J.M. Huxley and T.S. Moody, *Coord. Chem. Rev.*, 2000, **205**, 41; A.R. Pease and J.F. Stoddart, *Struct. Bond.*, 2001, **99**, 189; F.M. Raymo, *Adv. Mater.*, 2002, 401; V. Balzani, A. Credi and M. Venturi, *ChemPhysChem*, 2003, **4**, 49.
58. H.-J. Schneider, T. Liu, N. Lomadze and B. Palm, *Adv. Mater.*, 2004, **16**, 613.
59. For other metal ion-responsive gels see, *e.g.*, J. Ricka and T. Tanaka, *Macromolecules*, 1985, **18**, 83; J.H. Holtz and S.A. Asher, *Nature*, 1997, **389**, 829.
60. E.A. Meyer, R.K. Castellano and F. Diederich, *Angew. Chem.-Int. Ed.*, 2003, **42**, 1210.
61. H.-J. Schneider and T. Liu, *Angew. Chem. Int. Ed. Engl.*, 2002, **41**, 1368.
62. See, *e.g.*, H. Sigel and R.B. Martin, *Chem. Rev.*, 1982, **82**, 385; E. Farkas and I. Sovago, in: *Amino Acids, Peptides and Proteins*, 2002, **33**, 295; S. Aoki and E. Kimura, *Chem. Rev.*, 2004, **104**, 769; G. Licini and P. Scrimin, *Angew. Chem. Int. Ed. Engl.*, 2003, **42**, 4572.
63. See, *e.g.*, (a) J.W. Canary and B.C. Gibb, *Progr. Inorg. Chem.*, 1997, **45**, 1; (b) F.M. Raymo and J.F. Stoddart, *Chem. Ber.*, 1996, **129**, 981; (c) J.W. Steed, *Coord. Chem. Rev.*, 2001, **215**, 171; (d) L. Fabbrizzi, M. Licchelli, F. Mancin, M. Pizzeghello, G. Rabaioli, A. Taglietti, P. Tecilla and U. Tonellato, *Chem. Eur. J.*, 2002, **8**, 94.
64. N. Lomadze and H.-J. Schneider, *Tetrahedron Lett.*, 2005, **46**, 751.
65. See, *e.g.*, M. Bielecki, H. Eggert, J.C. Norrild, *J. Chem. Soc., Perkin Trans. 2*, 1999, 449 and references cited therein.
66. J. Boeseken, *Adv. Carbohydrate Chem.*, 1949, **4**, 189.
67. J. Yoon and A.W. Czarnik, *J. Am. Chem. Soc.*, 1992, **114**, 5874.

68. S. Shinkai, M. Takeuchi and A. Ikeda, in Ref (a), p. 183; T.D. James, K.R.A.S. Sandanayake and S. Shinkai, *Angew. Chem. Int. Ed. Engl.*, 1996, **35**, 1910; W. H. Wang, O. Rusin, X. Y. Xu, K.K. Kim, J.O. Escobedo, S.O. Fakayode, K.A. Fletcher, M. Lowry, C.M. Schowalter, C.M. Lawrence, F.R. Fronczek, I.M. Warner and R.M. Strongin, *J. Am. Chem. Soc.*, 2005, **127**, 15949; and references cited therein.
69. S. Kabilan, J. Blyth, M.C. Lee, A.J. Marshall, A. Hussain, X.P. Yang and C.R. Lowe, *J. Mol. Recog.*, 2004, **17**, 162; A. Matsumoto, T. Kurata, D. Shiino and K. Kataoka, *Macromolecules*, 2004, **37**, 1502; K. Kataoka, H. Miyazaki, M. Bunya, T. Okano and Y. Sakurai, *J. Am. Chem. Soc.*, 1998, **120**, 12694; I. Hisamitsu, K. Kataoka, T. Okano and Y. Sakurai, *Pharmac. Res.*, 1997, **14**, 289; A. Kikuchi, K. Suzuki, O. Okabayashi, H. Hoshino, K. Kataoka, Y. Sakurai and T. Okano, *Anal. Chem.*, 1996, **68**, 823.
70. G. Samoei, W. Wang, J.O. Escobedo, X. Xu, H.-J. Schneider and R.M. Strongin, *Angew. Chem. Int. Ed. Engl.*, 2007, **46**, 2694; for related papers see, *e.g.*, A. Matsumoto, S. Ikeda, A. Harada and K. Kataoka, *Biomacromolecules*, 2003, **4**, 1410; S.A. Asher, V.A. Alexeev, A.C. Sharma, A.V. Goponenko, S. Das, I.K. Lednev, C.S. Wilcox, D.N. Finegold, *J. Am. Chem. Soc.* **2003**, *125*, 3322.
71. For a gel with glucose imprinting see E. Oral and N.A. Peppas, *J. Biomed. Mater. Res. Part A*, 2004, **68A**, 439.
72. Z. Hu, X. Zhang and Y. Li, *Science*, 1995, **269**, 525; F.-M. Li, S.-J. Chen, F.-S. Du, Z.-Q. Wu and Z.-C. Li, in *Field ResponsivePolymers*; ed. I.M. Kahn and J.S. Harrison, *ACS Symp. Ser.*, **726**, Washington, DC, 1999.
73. T. Miyata, A. Jikihara, K. Nakamae and A.S. Hoffman, *J. Biomater. Sci.-Polym. Ed.*, 2004, **15**, 1085; J.J. Kim and K. Park, *J. Control. Rel.*, 2001, **77**, 39; A.A. Obaidat and K. Park, *Biomaterials*, 1997, **18**, 801; S.J. Lee and K. Park, *J. Mol. Recog.*, 1996, **9**, 549; A.A. Obaidat and K. Park, *Pharmac. Res.*, 1996, **13**, 989; T. Miyata, A. Jikihara, K. Nakamae and A.S. Hoffman, *Macromol. Chem. Phys.*, 1996, **197**, 1135; E. Kokufata, Y.Q. Zhang and T. Tanaka, *Nature*, 1991, **351**, 302.
74. H. Suzuki, A. Kumagai, K. Ogawa and E. Kokufuta, *Biomacromolecules*, 2004, **5**, 486; S.H. Yuk, S.H. Cho and S.H. Lee, *Macromolecules*, 1997, **30**, 6856; K. Matsuda, H. Orii, M. Hirata and E. Kokufuta, *Polym. Gels Networks*, 1994, **2**, 299.
75. T. Miyata, N. Asami and T. Uragami, *Nature*, 1999, **399**, 766.

CHAPTER 5

Ionic Polymer Metal Nanocomposites as Intelligent Materials and Artificial Muscles

MOHSEN SHAHINPOOR*

Department of Mechanical Engineering and Biomedical Engineering Center, College of Engineering, University of Maine, Orono, Maine 04469, USA

5.1 Summary

Basic recent results, properties and characteristics of ionic polymer conductor nanocomposites (IPCNC) and ionic polymer metal nanocomposites (IPMNC) as biomimetic multifunctional distributed nanosensors, nanoactuators, nano-transducers and artificial muscles are briefly discussed in this chapter. In particular, the chapter starts with some fundamental considerations on biomimetic distributed nanosensing and nanoactuation and then expands its coverage to some recent advances in manufacturing techniques, force optimisation, 3D fabrication of IPMNCs, recent modeling and simulations, sensing and transduction and product development. The chapter also covers some recent industrial and medical applications including multifingered grippers (macro-, micro-, nano-), biomimetic robotic fish and caudal-fin actuators, diaphragm micropump, multistring musical instruments, linear actuators made with IPMNCs, IPMNC-based data glove and attire, IPMNC-based heart compression/assist devices and systems, a wing-flapping flying system made with IPMNCs and a host of others.

---

*Formerly, Artificial Muscle Research Institute (AMRI), Division of Neurological Surgery, School of Medicine, University of New Mexico, Albuquerque, New Mexico 87131, USA

5.2 Introduction

Recent findings on ionic polymer conductor nanocomposites (IPCNCs) and ionic polymer metal nanocomposites (IPMNCs) as biomimetic distributed nanosensors, nanoactuators and artificial muscles and electrically controllable polymeric network structures have been presented recently.[1-5] Furthermore, in Ref. [1], methods of fabrication of several electrically and chemically active ionic polymeric gel muscles such as polyacrylonitrile (PAN), poly(2-acrylamido-2-methyl-1-propane sulfonic) acid (PAMPS), and polyacrylic-acid-bis-acrylamide (PAAM) as well as a new class of electrically active composite muscle such as ionic polymeric conductor composites (IPCCs) or ionic polymer metal composites (IPMCs) made with perfluorinated sulfonic or carboxylic ionic membranes are introduced and investigated that have resulted in six US patents regarding their fabrication and application capabilities as distributed biomimetic nanoactuators, nanotransducers and nanosensors. Several apparatuses for modeling and testing of the various IPMNC artificial muscles are described to show the viability of the application of both chemoactive and electroactive muscles. Furthermore, fabrication methods of PAN fibre muscles in different configurations such as spring-loaded fibre bundles, biceps, triceps, ribbon-type muscles, and segmented fibre bundles to make a variety of biomimetic sensors and actuators are also reported in Ref. [1].

Theories and numerical simulations associated with ionic polymer gels electrodynamics and chemodynamics are also discussed, analysed and modeled for the manufactured material.

In this chapter we concentrate on perfluorinated sulfonic ionic multifunctional materials, as potentially powerful ionic polymers for biomimetic distributed nanosensing, nanoactuation, nanorobotics, nanotransducers for power conversion and harvesting, as well as artificial muscles for medical and industrial applications.

It must be noted that widespread electrochemical processes and devices utilise poly(perfluorosulfonic acid) ionic polymers. These materials exhibit,[1-5] good chemical stability, remarkable mechanical strength, good thermal stability, and high electrical conductivity when sufficiently hydrated and made into a nanocomposite with a conductive phase such as metals, conductive polymers or graphite. A number of physical models have been developed to understand the mechanisms of water and ion transport in ionic polymers and membranes.[5] Morphological features influence transport of ions in ionic polymers. These features have been studied by a host of experimental techniques including: small- and wide-angle X-ray scattering, dielectric relaxation, and a number of microscopic and spectroscopic studies.

The emerging picture of the morphology of ionic polymers is that of a two-phase system made up of a polar fluid (water)-containing ion cluster network surrounded by a hydrophobic polytetrafluoroethylene (PTFE) medium. The integrity and structural stability of the membrane is provided by the PTFE backbones and the hydrophilic clusters facilitate the transport of ions and water in the ionic polymer. These nanoclusters have been conceptually described as containing an interfacial region of hydrated, sulfonate-terminated perfluoroether side chains surrounding a central region of polar fluids. Counterions such as Na^+ or Li^+ are to be found in the vicinity of the sulfonates. It must be noted that the length of the side chains has a direct bearing on the separation between ionic domains, where the majority of the polar fluids resides, and the nonpolar domains.

High-resolution NMR of some perfluoroionomer shows an unusual combination of a nonpolar, Teflon-like backbone, with polar and ionic side branches. Liu and Schmidt-Rohr[6] obtained the first high-resolution NMR spectra of solid perfluorinated polymers by combining 28 kHz magic-angle spinning (MAS) with rotation-synchronised 19F pulses. Their NMR studies enable more detailed structural investigations of the nanometre-scale structure and dynamics of PTFE based ionomers. It has also been well established[1-6] that anions are tethered to the polymer backbone and cations (H^+, Na^+, Li^+) are mobile and solvated by polar or ionic liquids within the nanoclusters of size 3–5 nm.

5.3 Three-dimensional Fabrication of IPMNCs

It is well understood that all commercially available (as-received) perfluorinated ion-exchange polymers are in the form of hydrolysed polymers and are semicrystalline and may contain ionic clusters. The membrane form of these polymers has a typical thickness in the range of approximately 100–300 μme. Such a small thickness of commercially available membranes permits fast mass transfer for use in various chemical processes. Knowing that such as-received semicrystalline membranes are not melt processable, they are not suitable for the fabrication of three-dimensional electroactive materials or other composite forms. Of course the raw resin form of perfluorinated ionic membranes is available and is melt processable to various 3D shapes. However, the subsequent 3D body will have to be chemically treated (hydrolysed) to become ionic and this process of hydrolysation is quite complex.

In a previous study[10] a newly developed fabrication method was reported that can scale-up or down the IPMC artificial muscles in a strip size of μm–cm thickness, using a liquid form of perfluorinated ionic polymers. By meticulously evaporating the solvent (isopropyl alcohol) out of the solution, recast ionic polymer can be obtained.[10] A number of these samples are shown in Figure 5.1.

5.4 Manfacturing Methodologies

Manufacturing an IPMNC begins with selection of an appropriate ionic polymeric material. Often, ionic polymeric materials are manufactured from

Figure 5.1 (a) An eight-finger synthetic muscle. It has a thickness of approximately 2 mm: (b) A coil-type synthetic muscle. This coil type muscle creates a linear actuation motion.

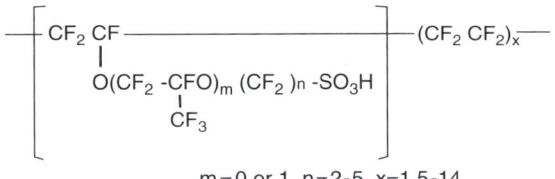

Figure 5.2 Perfluorinated acid polymers.

polymers that consist of fixed covalent ionic groups. The currently available ionic polymeric materials that are convenient to be used as IPMNCs are:

1. Perfluorinated polyalkenes with short side chains terminated by ionic groups typically sulfonate or carboxylate (SO_3^- or COO^-) for cation exchange or ammonium cations for anion exchange (see Figure 5.2). The large polymer backbones determine their mechanical strength. Short side chains provide ionic groups that interact with water and the passage of appropriate ions.

2. Styrene/divinylbenzene-based polymers in which the ionic groups have been substituted from the phenyl rings where the nitrogen atom is fixed to an ionic group (Figure 5.3).

5.5 Manufacturing Steps

The essential steps in manufacturing of IPMNCs are:

Roughen the material surface where it will serve as an effective electrode. These steps include sandblasting, glass-bead blasting or sandpapering the surface of the polymer in order to increase the surface-area density where platinum-salt penetration and reduction occurs as well as ultrasonic cleaning and chemical

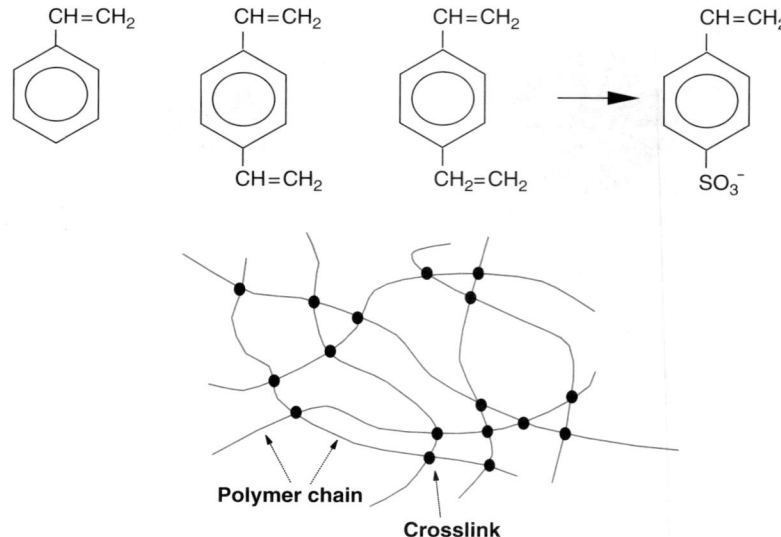

Figure 5.3 Styrene/divinylbenzene-based ion-exchange materials (top) and its structural form (bottom).

cleaning by acid boiling (HCl or HNO_3 low concentrates). The second step is to incorporate the ion-exchanging process using a metal-complex solution such as tetra-amine platinum chloride hydrate as an aqueous platinum complex ($Pt(NH_3)_4Cl_2$ or $Pt(NH_3)_6Cl_4$) solution. Although the equilibrium condition depends on the types of charge of the metal complex, such complexes were found to provide good electrodes.

The third step (initial making of platinum ionic polymer composites) is to reduce the platinum complex cations to the metallic state in the form of nanoparticles by using effective reducing agents such as an aqueous solution of sodium or lithium borohydride (5%) at favourable temperature (*i.e.* 60°C). Platinum black-like layers deposit near the surface of the material.

The final step (surface-electrode-placement process) is intended to effectively grow Pt (or other novel metals, a few micrometre thickness) on top of the initial Pt surface to reduce the surface resistivity. This entire process is also called REDOX process in chemistry.

5.6 Electrically Induced Robotic Actuation

In perfluorinated sulfonic acid polymers there are relatively few fixed ionic groups. They are located at the end of side chains so as to position themselves in their preferred orientation to some extent. Therefore, they can create hydrophilic nanochannels, so called *cluster networks*.[7,8]

Such configurations are drastically different in other polymers such as styrene/divinylbenzene families that limit, primarily by crosslinking, the ability of the ionic polymers to expand due to their hydrophilic nature. Basically, the cations attract water molecules and thus they separate from the polymer backbone charged pendant groups and gather around them a number of water molecules, thus expanding or swelling the network.

Once an electric field is imposed on such a network, the conjugated and hydrated cations rearrange to accommodate the local electric field and thus the network deforms, which in the simplest of cases such as in thin membrane sheets, spectacular bending is observed (Figures 5.4 and 5.5) under small electric fields such as tens of volts per millimetre.

Typical experimental deflection curves are depicted in Figures 5.6, 5.7 and 5.8.

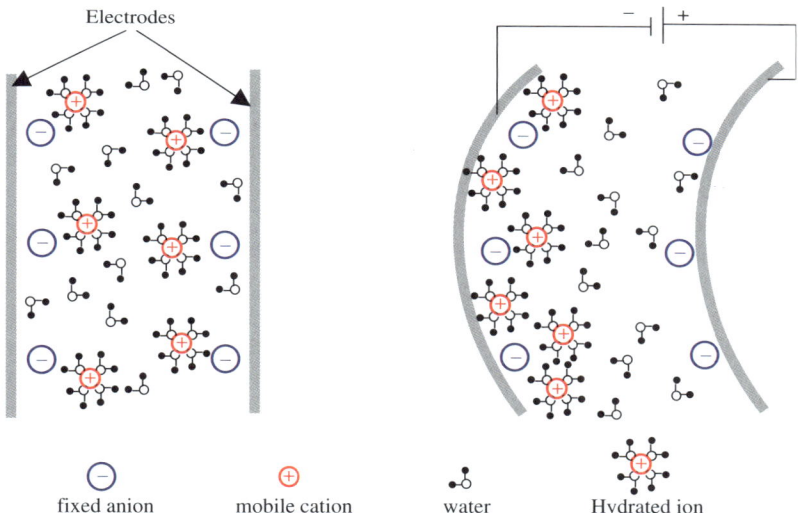

Figure 5.4 Simple bending in an electric field.

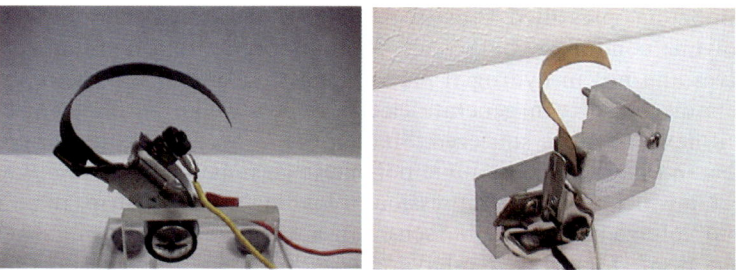

Figure 5.5 Typical deformation of strips (10 mm × 80 mm × 0.34 mm) of ionic polymers under a step voltage of 4 V.

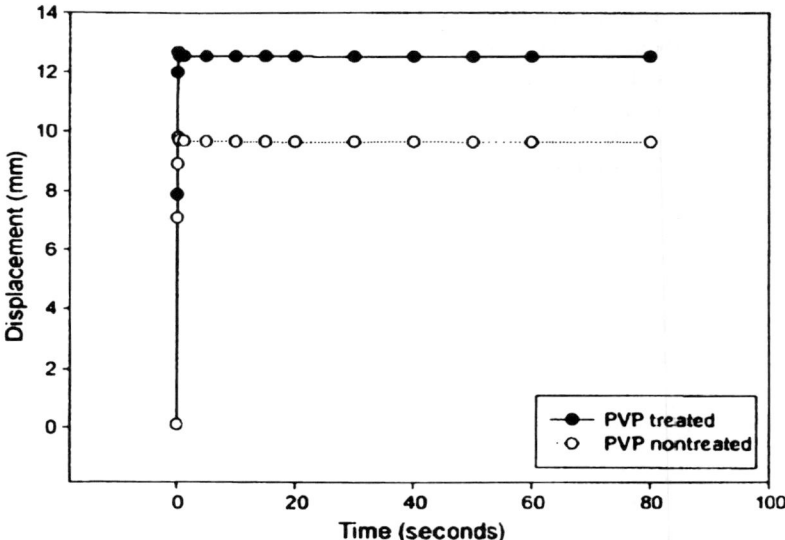

Figure 5.6 Step response displacement characteristics of IPMNC samples (δ: arc length, L_o: effective beam length) and $L_o = 1.5$ inch (bottom).

Typical frequency dependent dynamic deformation characteristics of IPMCs are depicted in Figure 5.7.

Once an electric field is imposed on an IPMNC cantilever, in the cantilever polymeric network the hydrated cations migrate to accommodate the local electric field. This creates a pressure gradient across the thickness of the beam and thus the beam undergoes bending deformation (Figure 5.5) under small electric fields such as tens of volts per millimetre. Figure 5.8 depicts typical force and deflection characteristics of cantilever samples of IPMNC artificial muscles.

5.7 Distributed Nanosensing and Transduction

Shahinpoor[9] has presented a review on sensing and transduction properties of ionic polymer conductor composites. Shahinpoor,[10,11] reported that IPMCs by themselves and not in hydrogen pressure electrochemical cells as reported by Sadeghipour et al.[12] can generate electrical power like an electromechanical battery if flexed, bent or squeezed. Shahinpoor[10,11] reported the discovery of a new effect in ionic polymeric ionic polymeric gels, namely the *ionic flexogelectric* effect in which flexing, compression or loading of IPMC strips in air created an output voltage like a dynamic sensor or a transducer converting mechanical energy to electrical energy. Keshavarzi et al.[13] applied the transduction capability of IPMC to the measurement of blood pressure, pulse rate, and rhythm measurement using thin sheets of IPMCs. Motivated by the idea of measuring

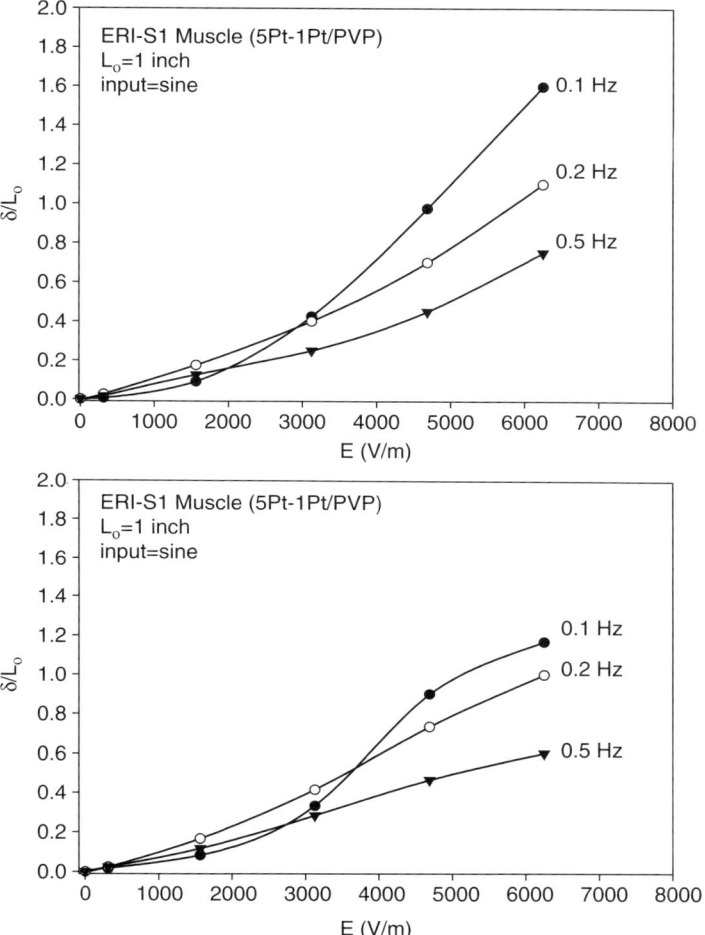

Figure 5.7 Displacement characteristics of an IPMC, ERI-S1 (δ: arc length, L_o: effective cantilever beam length) $L_o = 1.0$ inch.

pressure in the human spine, Ferrara et al.[14] applied pressure across the thickness of an IPMC strip while measuring the output voltage. Typically, flexing of such material in a cantilever form sets them into a damped vibration mode that can generate a similar damped signal in the form of electrical power (voltage or current) as shown in Figure 5.9.

The experimental results for mechanoelectrical voltage generation of IPMCs in a flexing mode are shown in Figure 5.10. Figure 5.11 depicts the power outputs for a sample of thin sheets of IPMCs.

The experimental results depicted in Figures 5.10 and 5.11 showed that almost a linear relationship exists between the voltage output and the imposed displacement of the tip of the IMPNC sensor (Figure 5.10).

134   *Chapter 5*

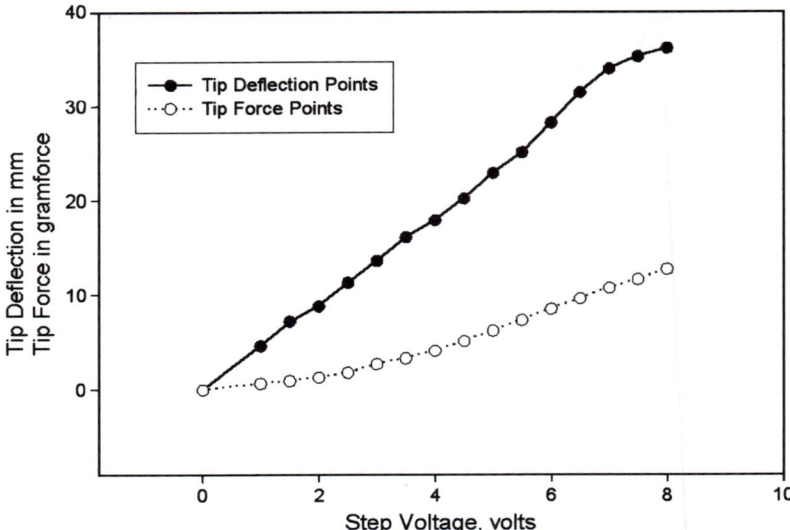

Figure 5.8 Variation of tip-blocking force and the associated deflection if allowed to move versus the applied step voltage for a 1 cm × 5 cm × 0.3 mm IPMNC Pt-Pd sample in a cantilever configuration.

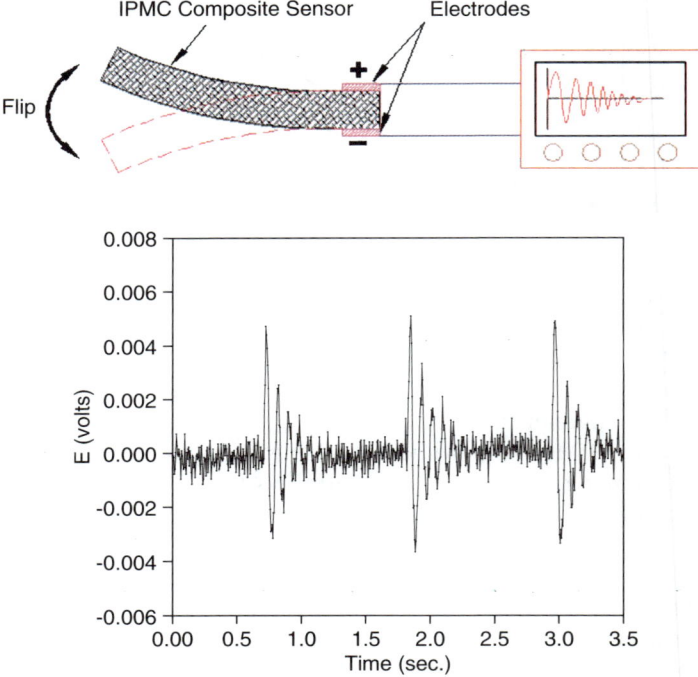

Figure 5.9 A typical voltage response of an IPMC strip (1 cm × 4 cm × 0.2 mm) under oscillatory mechanical excitations.

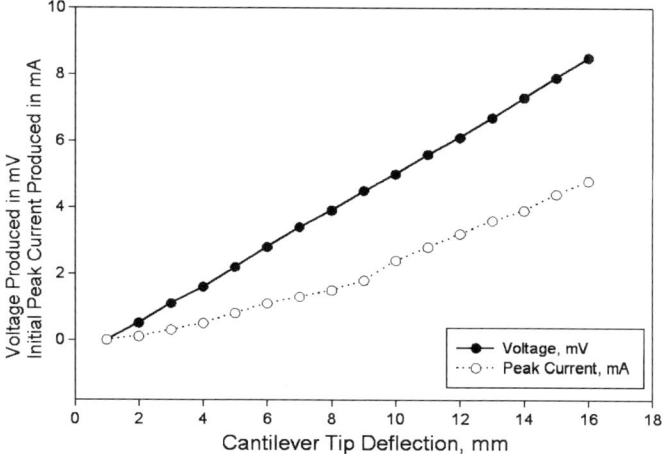

Figure 5.10 Typical voltage–current output of IPMC samples.

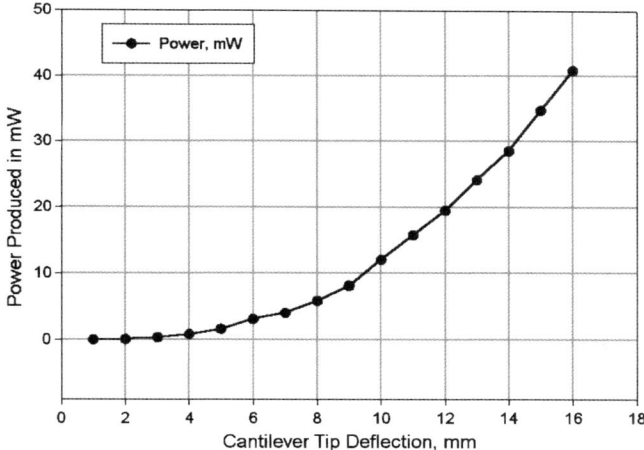

Figure 5.11 Typical power output of IPMC samples.

IPMNC sheets can also generate power under normal pressure. Thin sheets of IPMNC were stacked and subjected to normal pressure and normal impacts and were observed to generate large output voltage. Endoionic motion within IPMNC thin sheet batteries produced an induced voltage across the thickness of these sheets when a normal or shear load was applied. A material testing system (MTS) was used to apply consistent pure compressive loads of 200 N and 350 N across the surface of an IPMNC 2×2 cm sheet. The output pressure response for the 200 N load (73 psi) was 80 mV in amplitude and for the 350 N load (127 psi) it was 108 mV. This type of power generation may be useful in the

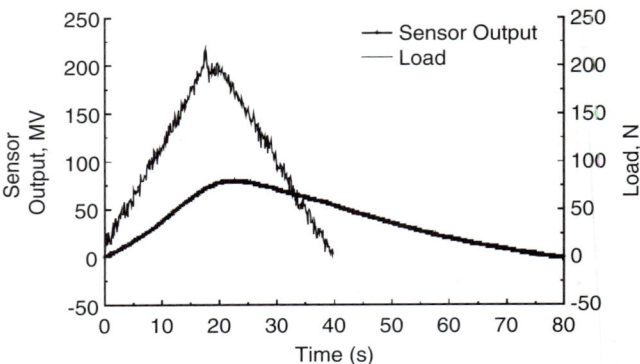

Figure 5.12 Output voltage due to normal 90-degree impact of 200 N load on a 2 cm × 2 cm × 0.2 mm IPMC sample.

Figure 5.13 Cantilever and load-cell configuration for measuring the tip-blocking force of IPMNC samples.

heels of boots and shoes or places where there is a lot of foot or car traffic. Figure 5.12 depicts the output voltage of the thin sheet IPMNC batteries under 200 N normal load. The output voltage is generally about 2 mV/cm length of the IPMNC sheet.

As far as force generation is concerned, IPMNCs generally have a very high force density as depicted in Figure 5.8. Figure 5.13 displays the cantilever and load-cell configuration for measuring the tip blocking force of typical samples of IPMNCs.

5.8 Modeling and Simulation

As recent as 2000, Nobel Laureate Pierre de Gennes et al.[15] presented the first phenomenological theory for sensing and actuation in ionic polymer metal composites. Asaka and Oguro[16] discussed the bending of polyelectrolyte membrane–platinum composites by electric stimuli and presented a theory on actuation mechanisms in IPMC by considering the electro-osmotic drag term in transport equations. Nemat-Nasser and Li[17] discussed a modeling on the electromechanical response of ionic polymer–metal composites based on electrostatic attraction/repulsion forces in IPMCs. Later, Nemat-Nasser[18] proposed a revised version of an earlier paper and stressed the role of hydrated cation transport and mobility within the clusters and polymeric networks in IPMCs. Nemat-Nasser and Wu[19] have proposed a discussion on the role of backbone ionic polymer and in particular sulfonic versus carboxylic ionic polymers, as well as the effect of different cations such as K^+, Na^+, Li^+, Cs^+ and some organometallic cations on the actuation and sensing performance of IPMCs. Tadokoro,[20] and Tadokoro et al.[21,22] have presented an actuator model of IPMC for robotic applications on the basis of physicochemical phenomena. A recent comprehensive review by Shahinpoor and Kim[4] on modeling and simulation of ionic polymeric artificial muscles discusses the various modeling approaches for the understanding of the mechanisms of sensing and actuation of ionic polymers and the notion of ion mobility.

Let us now summarise the underlying principle of the IPMCs actuation and sensing capabilities, which can be described by the standard Onsager formulation using linear irreversible thermodynamics. When *static conditions* are imposed, a simple description of the *mechanoelectric effect* is possible based upon two forms of transport: *ion transport* (with a current density, J, normal to the material) and *solvent transport* (with a flux, Q, we can assume that this term is the water flux). The conjugate forces include the electric field, E, and the pressure gradient, $-\nabla p$. The resulting equation has the concise form of,

$$J(x,y,z,t) = \sigma E(x,y,z,t) - L_{12} \nabla p(x,y,z,t) \quad (5.1)$$

$$Q(x,y,zt) = L_{21} E(x,y,z,t) - K \nabla p(x,y,z,t) \quad (5.2)$$

where σ and K are the material electric conductance and the Darcy permeability, respectively. A cross coefficient is usually $L = L_{12} = L_{21}$. The simplicity of the above equations provides a compact view of the underlying principles of both actuation, transduction and sensing of the IPMNCs as also shown in Figures 5.7–5.10.

When we measure the *direct* effect (actuation mode, Figure 5.8), we work (ideally) with electrodes that are impermeable to ion species flux, and thus we have $Q = 0$. This gives:

$$\nabla p(x,y,z,t) = \frac{L}{K} E(x,y,z,t) \quad (5.3)$$

This $\nabla p(x,y,z,t)$ will, in turn, induce a curvature κ proportional to $\nabla p(x,y,z,t)$. The relationships between the curvature κ and pressure gradient $\nabla p(x,y,z,t)$ are fully derived and described in de Gennes et al.[15] Let us just mention that $(1/\rho_c) = \mathbf{M}(\mathbf{E})/YI$, where $\mathbf{M}(\mathbf{E})$ is the local induced bending moment and is a function of the imposed electric field $\mathbf{E}$, Y is the Young's modulus (elastic stiffness) of the strip that is a function of the hydration H of the IPMC and I is the moment of inertia of the strip. Note that locally $\mathbf{M}(\mathbf{E})$ is related to the pressure gradient such that in a simplified scalar format:

$$\nabla p(x,y,z,t) = (\mathbf{M}/I) = Y/\rho_c = Y\kappa_E \quad (5.4)$$

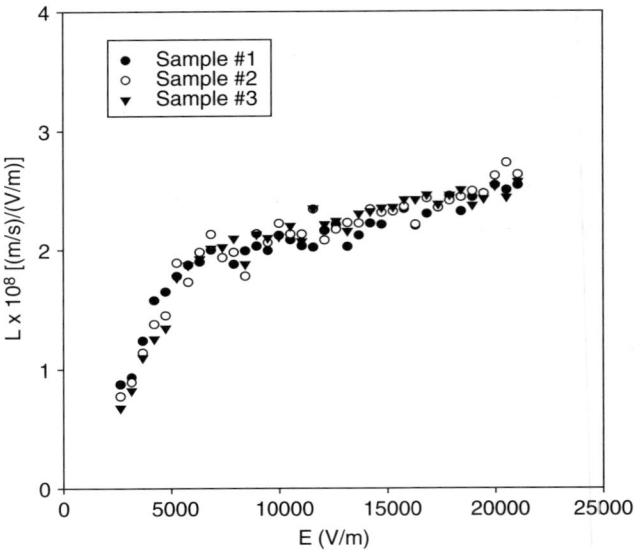

Figure 5.14 Experimental determination of Onsager coefficient L using three different samples.

Now from eqn (5.4) it is clear that the vectorial form of curvature $\underset{\sim}{\kappa}_E$ is related to the imposed electric field **E** by:

$$\underset{\sim}{\kappa}_E = (L/KY)\,\underset{\sim}{E} \qquad (5.5)$$

Based on this simplified model the tip bending deflection δ_{max} of an IPMC strip of length l_g should be almost linearly related to the imposed electric field due to the fact that:

$$\underset{\sim}{\kappa}_E \cong [2\underset{\sim}{\delta}_{max}/(l_g^2 + \underset{\sim}{\delta}_{max}^2)] \cong 2\underset{\sim}{\delta}_{max}/l_g^2 \cong (L/KY)\,\underset{\sim}{E} \qquad (5.6)$$

The experimental deformation characteristics depicted in Figure 5.14 is clearly consistent with the above predictions obtained by the above linear irreversible thermodynamics formulation that is also consistent with eqns (5.5) and (5.6) in the steady-state conditions and has been used to estimate the value of the Onsager coefficient L to be of the order of $10^{-8}\,m^2/V\,s$. Here, we have used a low-frequency electric field in order to minimise the effect of loose water back diffusion under a step voltage or a dc electric field. Other parameters have been experimentally measured to be $K \sim 10^{-18}\,m^2/CP$, $\sigma \sim 1\,A/mV$ or S/m. Figure 5.14 depicts a more detailed set of data pertaining to Onsager coefficient L.

5.9 Smart-Product Development

Environmental Robots Incorporated (www.environmental-robots.com) is currently the world's leader in developing medical and industrial products based on ionic polymer metal nanocomposites (IPMNCs) and ionic polymer conductor nanocomposites (IPCNCs) as distributed biomimetic nanosensors, nanoactuators, nanotransducers and artificial muscles. These products include two artificial muscle science kits for both contractile fibre bundles of PAN muscles as well as bending flexing and sensing silver- and platinum-based IPMCs, an assortment of ionic polymeric artificial muscles, an assortment of polymeric micromuscles, sensor and actuators packages, packages of ionic polymeric muscles, dynamic signal generators to drive the ionic polymeric muscles and a host of other ionic polymeric muscle products.

5.10 Medical, Engineering and Industrial Applications

Mechanical grippers, biomimetic noiseless swimming robotic fish and fin actuators, diaphragm pump designs, linear actuators metering valves, musical instruments in which IPMNC fibres are used as strings and generate a musical note if picked or played, flat key boards, data attire and Braille alphabet,

Figure 5.15 Some IPMC-based products by ERI: (a) biomimetic robotic fish and caudal-fin actuators, (b) Solid-state wing-flapping plane.

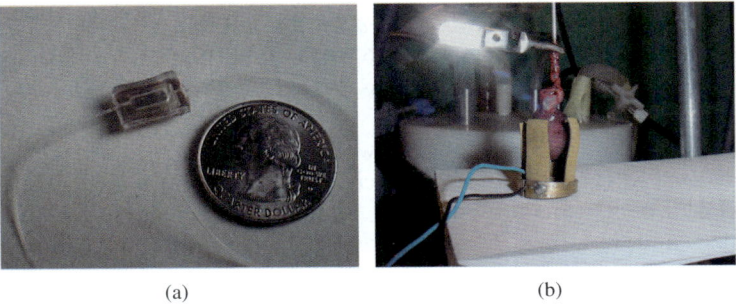

Figure 5.16 Some IPMC-based products by ERI: (a) IPMC diaphragm micropump, (b) IPMC miniheart compression and assist device.

artificial ventricular or cardiac-assist muscles, surgical tool, peristaltic pumps, artificial smooth muscle actuators, composite wing-flap, resonant flying machine and a host of others, as discussed in Refs 1 and 5. Figures 5.15 and 5.16 depict some of these products.

This concludes our coverage of IPMNCs as multifunctional intelligent materials with distributed nanosensing, nanoactuation and nanotransduction capabilities.

References

1. M. Shahinpoor, K.J. Kim and Mehran Mojarrad, *Ionic Polymeric Artificial Muscles*, ERI/AMRI Press, Albuquerque, New Mexico, 2004.
2. M. Shahinpoor and K.J. Kim, *Smart Mater. Struct. Int. J.*, 2001, **10**, 819–833.
3. K.J. Kim and M. Shahinpoor, *Smart Mater. Struct. (SMS)*, 2003, **12**, No. 1, 65–79.
4. M. Shahinpoor and K.J. Kim, *Smart Mater. Struct. Int. J.*, 2004, **13**, No. 4, 1362–1388.
5. M. Shahinpoor and K.J. Kim, *Smart Mater. Struct. Int. J.*, 2005, **14**, No. 1, 197–214.
6. S.-F. Liu and K. Schmidt-Rohr, *Macromolecules*, 2001, **34**, 8416–8418.
7. T.D. Gierke and W.Y. Hsu, in: A. Eisenberg, H.L. Yeager, ed., *Perfluorinated Ionomer Membranes*, ACS, Washington, DC, 1982, pp. 283–307.
8. T.D. Gierke, G.E. Munn and F.C. Wilson, *ACS Symp. Ser.*, 1982, **180**, 195–216.
9. M. Shahinpoor, IMECE2004-60954, *Proceedings of ASME-IMECE2004, 2004 ASME International Mechanical Engineering Congress and RD&D Exposition*, November 13–19, 2004, Anaheim Convention Center/Hilton, Anaheim, California, 2004.
10. M. Shahinpoor, 1995, *Proc. SPIE 1995 North American Conference on Smart Structures and Materials*, February 28–March 2, 1995, San Diego, California, 2441, paper no. 05, pp. 42–53.
11. M. Shahinpoor, *ICIM'96 and Third European Conference on Smart Structures and Materials*, 1996, pp. 1006–1011, June 1996, Lyon, France.
12. K. Sadeghipour, R. Salomon and S. Neogi, *Smart Mater. Struct. J.*, 1992, **1**, 172–179.
13. A. Keshavarzi, M. Shahinpoor, Kwang J. Kim and J. Lantz, *Elactroactive Polymers*, SPIE publication number 3669-36, 1999, pp. 369–376.
14. L. Ferrara, M. Shahinpoor, K.J. Kim, B. Schreyer, A. Keshavarzi, E. Benzel and J. Lantz, *Elactroactive Polymers*, SPIE publication number 3669-45,1999, pp. 394–401.
15. P.G. de Gennes, K. Okumura, M. Shahinpoor and K.J. Kim, 2000, *Europhys. Lett.*, 2003, **504**, pp. 513–518.
16. K. Asaka and K. Oguro, *J. Electroanal. Chem.*, 2000, **480**, pp. 186–198.

17. S. Nemat-Nasser and J.Y. Li, *J. Appl. Phys.*, 2000, **87, No. 7**, 3321–3331.
18. S. Nemat-Nasser, *J. Appl. Phys*, 2002, **92, No. 5**, 2899–2915.
19. S. Nemat-Nasser and Y. Wu, *J. Appl. Phys.*, 2003, **93**, 5255–5267.
20. S. Tadokoro, *Proc. IEEE, ICRA*, 2000, 1340–1346.
21. S. Tadokoro, S. Yamagami, T. Takamori and K. Oguro, *Proc. SPIE*, 2000, **3987**, 92–102.
22. S. Tadokoro, T. Takamori and K. Oguro, *Proc. SPIE*, 2001, **4329**, 28–42.

CHAPTER 6

Artificial Muscles, Sensing and Multifunctionality

TORIBIO FERNÁNDEZ OTERO

Universidad Politécnica de Cartagena, ETSII. Centre for Electrochemistry and Intelligent Materials (CEMI), Paseo de Alfonso XIII, Aulario II, 30203 Cartagena, Spain

6.1 Introduction

Most of the new technological advances developed by humans are inspired on natural systems, organs, functions or devices present in living creatures and developed through millions of years of biological evolution. Natural systems and devices combine efficiency and intellectual, scientific and technological elegance. Moreover, the constituted basic resources: organic molecules, water and mineral salts, are present in all countries around the world. By contrast, current technology is based on strategic resources, very concentrated in only a few countries.

The biological functions developed by organs involve: electric pulses, aqueous media, and organic molecules bearing specific properties, ionic interchanges and chemical reactions. In order to optimise the evolutionary effort most of the developed molecules and biological polymers, *i.e.* collagen, are multifunctional, bearing simultaneously different properties: mechanical, optical, *etc*.

The human efforts concerning technological developments try to mimic most of the biological functions as well as the efficiency of the biological organs. Among artificial organic materials conducting polymers (CP), discovered at the end of the 1970s, and in particular the electrochemical behaviour of CPs bear a good deal of properties and functions that, as we will try to show here, are much more similar to the biological ones than any other originated by any man-made material.

6.2 Materials

Seven different families of materials constitute conducting polymers:[1]

- *basic conducting polymers* (polypyrrole, polyaniline, polythiophene, polyphenylvinilene, *etc.*), each of then able to originate, by chemical or electrochemical oxidation, hundreds of oxidised materials (polymer/counterion/solvent) depending on the salt and solvent used,
- *substituted polymers* (one or several hydrogen atoms from the basic monomers can be substituted inducing new properties in both the monomer and the new polymer),
- *self-doped polymers* (one of the substituents is an ionic group),
- *copolymers* (the basic monomer is a dimer, a trimer, or a tetramer constituted by two or more basic monomers),
- *polymer/macroion blends*: the synthesised oxidised material contains macroions that are not interchanged with the solution during redox processes;
  - in *hybrid materials* the macroion is inorganic meanwhile in
  - *polymeric or organic blends* the macroion is a poly anion or an organic macroion,
- *composites* with different organic or inorganic materials.

Each of those families is constituted by different subfamilies of conducting polymer/balancing counterion and solvent. During electrochemical oxidation, or reduction, positive or negative charges are stored along the polymeric chains and balancing counterions are forced to penetrate from the solution. Depending on the salt used to balance the charge during oxidation/reduction processes, we obtain a different material that can produce, at least from a theoretical point of view, a different family of electrochemical devices that we will see below. This is an open window to the future: the development and optimisation of those different device families and their technological applications will require a lot of theoretical and experimental work.

6.3 Electrochemical Behaviour of Conducting Polymers in Aqueous Solution

All the above-described families of conducting polymers are suitable for oxidation processes *transforming the neutral molecules into oxidised material* storing positive charges along the chains.[2-5] The transformation from a neutral material to an oxidised material occurs through two different kinds of electrochemical reactions:

a. prevailing anion interchange during redox processes giving an increase of volume during oxidation (Figure 6.1) and a decrease of volume during

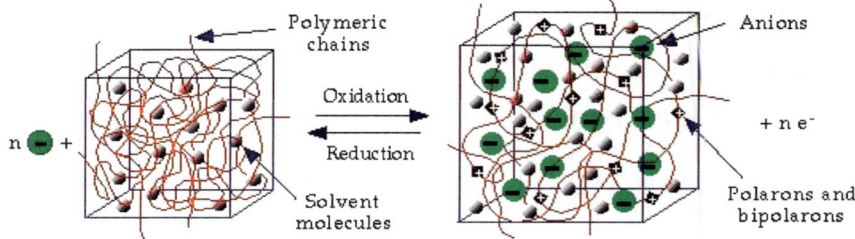

Figure 6.1 Schematic representation of the reversible volume change associated with the electrochemical reactions of polypyrrole in electrolytes. (Reproduced from: T.F. Otero, Y. Osada and D.E. De Rossi, ed., *Polymer Sensors and Actuators*, Springer-Verlag, Berlin, 2000, p. 304).

reduction (accepted for families 1, 2, 4, 5 and 7):

[1]

$$(pPy^0)_s + n(Cl^-)_{aq} + mH_2O \leftrightarrow [(pPy^{n+})_s(Cl^-)_n (H_2O)_m]_{gel} + (ne^-)_{metal}$$

neutral chains   oxidised chains

b. prevailing cation interchange (Figure 6.2), shrinking the material during the oxidation processes and swelling during reduction (accepted for families 3 and 6):

[2]

$$[(pPy^0)[n(MA^-)n(C^+)]_{(s)} \xrightarrow[(Red)]{(oxid)} [(pPy^{n+})(MA^-)_n]_s + n(C^+)_{aq} + (ne^-)_{metal}$$

neutral chains   oxidised chains

Where the different subindexes mean: s, solid and aq, aqueous solution. MA^- represents any macroscopic anion trapped inside the CP during polymerisation, pPy represent the polypyrrole (or any other CP) chains and C^+ represent a cation. The role of the aqueous molecules in this case is not so clear due to the presence of ionic species in the material, whatever their oxidation state.

Reactions [1] and [2] produce positive charges on the polymeric chains and the process, from a physical point of view is named *p doping*.

Moreover some conducting polymers have an electronic affinity high enough to suffer *transitions from the neutral state to a reduced state*, storing negative charges (*n doping*) on the chains at high cathodic potential (a very stable solvent, or stable ambient is required for this reaction):

[3]

$$(pPy^0)_s + n(C^+)_{aq} + (ne^-)_{metal} \leftrightarrow [(pPy^{n-})_s(C^+)_n (H_2O)_m]_{gel}$$

neutral chains   reduced chains

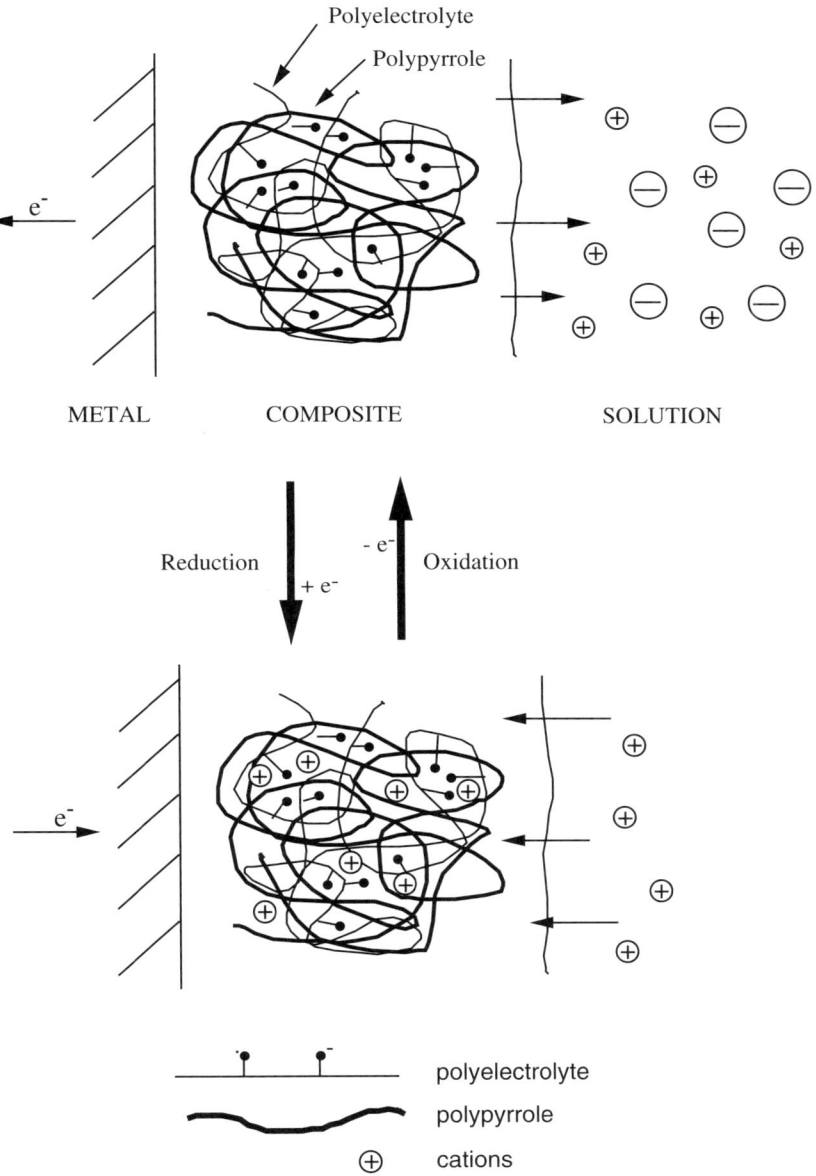

Figure 6.2 Metal/polymeric blend/solution system. The polymeric blend is constituted by entangled structures of a CP (*i.e.* polypyrrole) and a polyelectrolyte (*i.e.* polyvinyl sulfonate) obtained by electrogeneration from a solution containing the monomer and the solved polyelectrolyte. During oxidation/reduction, cations are expulsed or inserted form the solution and the polymer shrinks/swells, respectively (Reproduced from Ref. 18).

So, two basic redox (reverse oxidation/reduction) processes can be considered from neutral molecules of CP:

reduced chains ↔ neutral chains ↔ oxidised chains
n doping (a)   (b) p doping

All of the above-described reactions are simplified expressions, an initial approach to real processes. Any film of a conducting polymer acts as a polymeric membrane. As for any other membrane (biological or commodity polymers) equilibrium is stated between solvent, anions and cations in the polymer matrix and the electrolyte. So, during oxidation or reduction processes the charge balance is stated by simultaneous interchange of anions, cations and solvent across the polymer/solvent interface. From the different families of CPs one of the reactions, [1] or [2], use to prevail.

We will now focus our attention on basic CPs but, after consideration of the specific prevailing ionic interchange, similar reasoning can be extended to any other family of CPs. Any oxidised film of a basic electrogenerated conducting polymer is constituted by entangled, crosslinked and partially degraded polymeric chains bearing positive charges, plus those counterions required to compensate the charges, and water. Being more precise, the polymer is a mixed material (conjugated chains, crosslinking points and degraded monomeric units), the exact composition of which is a function of the conditions of synthesis and of the previous history of the material that induce new crosslinking and degradation points. The injection of electrons to the oxidised chains promotes the compensation of the positive charges along the chains, the relocation of the double bonds, the expulsion of counterions and water towards the solution. The polymer shrinks, helped by the increase of Van de Waals attractive forces between neutral polymeric chains. During electrochemical oxidation reverse processes occur with the introduction of counterions and water. *The process is equivalent to traditional doping of inorganic semiconductors by ionic implantation* processes using ionic bombardment. *Now, extremely soft conditions are used*: low external energy (mV), atmospheric pressure and normal laboratory conditions, resulting in a reverse infinitesimal process under perfect electrochemical control.

Most of the CP based on the polythiophene families also can be reduced and reoxidised from the neutral state, with generation or elimination, respectively, of negative charges along the chains and entrance or expulsion of cations from or to the solution.

So, any electrochemical process involving conducting polymers includes: *flow of electrons* across the polymer/metal interface, *flow of ions* and *water* molecules across the polymer/solution interface, reorganisation *(breaking and formation) of double bonds* along the chains with *conformational movements* and *shrinking* or *swelling* processes. Those changes of volume are induced by the electrochemical reactions that generate strong changes on the molecular interactions (polymer–polymer, polymer–solvent, polymer–ions) along the reaction time.

6.4 Nonstoichiometric, Soft, and Wet Materials

During the oxidation process of a polymeric film we can extract from each polymeric chain: one, two, ..., n electrons, requiring penetration from the solution of one, two, ..., n counterions to balance the positive charges generated along the polymeric chains inside the polymer matrix. The electrochemical reaction is reversible, so it can be stopped at any point of the chain-oxidation process, it can be reversed to any previous oxidation state or advanced up to any subsequent point. Any intermediate point between the neutral chain and the overall oxidised chain:

$$CP, (CP^+)A^-, (CP^{2+})A^-_2, (CP^{3+})A^-_3, (CP^{4+})A^-_4, (CP^{5+})A^-_5, (CP^{6+})A^-_7,..., (CP^{n+})A^-_n$$

(where A^- represents the counterion balancing charge) is stable, and n can go from one to millions, as a function of the chain length. So, these are nonstoichiometric materials that counterion composition can change in a continuous, infinitesimal and reverse way from cero (reduced material) up to a 40 to 60% (depending on the counterion molecular weight). The nonstoichiometric nature was demonstrated and visualised using electrochromic films that oxidise under nucleation processes: when the growth of the oxidised nuclei is interrupted those nuclei are partially reduced oxidising the reduced regions (Figure 6.3) until a uniform final state.[6,7]

During oxidation of a very thin polymeric film the electronic conductivity of the material shifts along 7 to 12 orders of magnitude, from a nonconductor material to conductivities similar to those of metals. Never before, during the history of human development, has any property of a natural or artificial material changed over such a large range. The discovery of conducting polymers, related to this unprecedented property, by *MacDiarmid, Heeger and Shirakawa* were awarded with the 2000 Nobel Prize for Chemistry. The technological possibilities of those materials as flexible semiconductors have attracted the attention of more than 90% of the scientists involved in this fascinating field, trying to reproduce all the microelectronics using now flexible polymeric devices and products without any restriction related to the material support (avoiding the silicon servitude) for the electronic components.

There exists another unprecedented and unparalleled possibility that, to the present moment, has only been recognised and has attracted the interest of a small percentage of those scientist involved in this area: the nonstoichiometric nature of the material allowing a continuous, infinitesimal and reversible change on the counterion material composition, from cero to 50% w/w. In fact, the variation range of the real conducting polymers decreases very fast when the thickness of the film grows. Thick material films shrink during reduction and closes (the average pore diameter between chains becomes less than the counterion diameter) entrapping a significant amount of counterions inside the polymer matrix and the concomitant positive charges on the chains, keeping such a high conductivity that uniform metal deposits (Figure 6.4) can be obtained by flow of high currents. The reduction can only be completed and

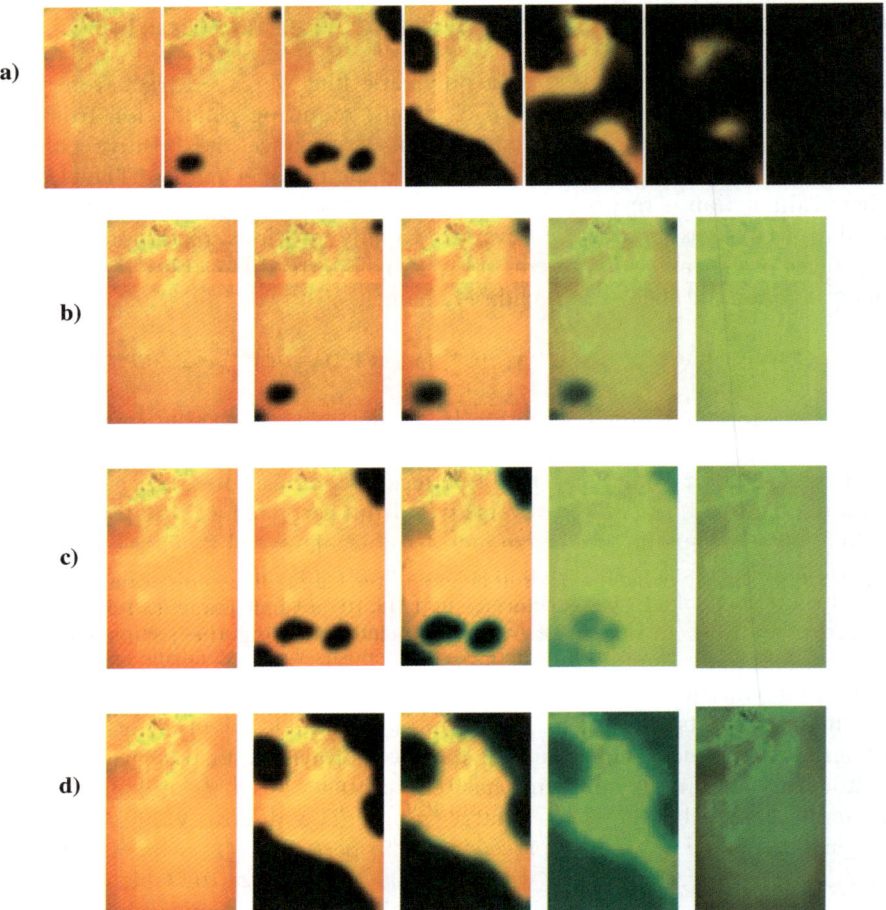

Figure 6.3 Polypyrrole films compacted by cathodic polarisation at −1200 mV, maintained for 60 s, and oxidised by a potential step to 100 mV at constant temperature (25°C) in 0.1 M LiClO$_4$ acetonitrile solution, showing the formation and growth of nuclei (dark blue) on the compacted and reduced film (yellow) (a) The film was oxidised at 100 mV for 0, 3, 5, 8, 12, 15 and until the end of the nucleation/coalescence and oxidation processes. The procedure was repeated switching off the oxidation/nucleation after: (b) 3 s, (c) 5 s and (d) 8 s. After switching off the electrical contacts new pictures were taken at intervals of 3 s to follow the evolution of the system. (Reproduced from Ref. 6.)

the counterions expelled from the matrix under application of high cathodic overpotentials for long periods of time. So, for films thicker than 1 µm, the composition can be changed from 0.001 up to 50%. Anyway, never before has a material having so large a variation in the material composition been available for chemists, physicists or engineers. This surprisingly ignored fact implies that

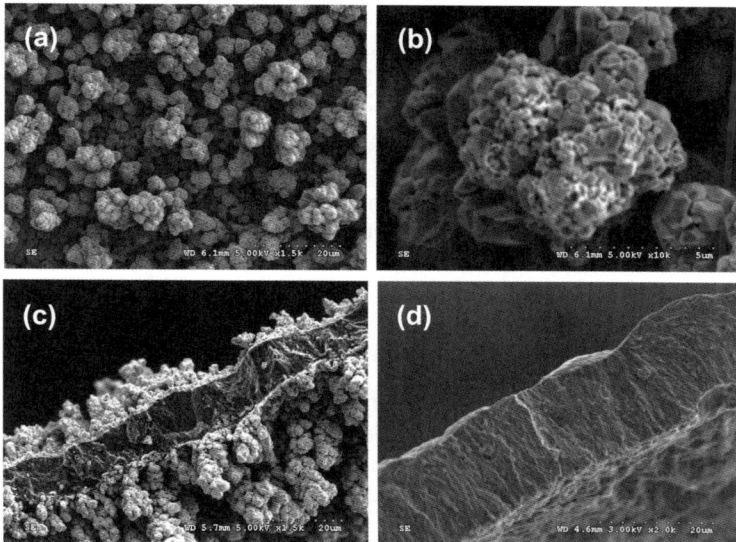

Figure 6.4 SEM secondary electron images of the Cu-modified polypyrrole electrode ($E_{Cu} = -2.1$ V). (a) Surface at 1.5×10^3 magnifications ($\times 1.5 \times 10^3$), (b) details of Cu crystallites at 10^4 magnification, (c) cross section of the Cu-modified electrode where copper deposits with a dendritic structure and uniform distribution on the electrode are observed in both faces of the film ($\times 1.5 \times 10^3$), (d) cross section of a nonmodified pPy film ($\times 2.0 \times 10^3$). The Cu was electrodeposited at -1.5 V in 0.5 M LiClO$_4$ + 0.1 M CuSO$_4$ aqueous solution on a self-sustained polypyrrole electrode that previously was deeply reduced at -2.0 V for 30 min, in 0.1 M LiClO$_4$ acetonitrile solutions. (Reproduced from Ref. 111.)

any property related to the chemical composition must also change in a continuous, infinitesimal and reverse way.

6.5 Electrochemical Properties

One explanation for the low interest of the scientific community in this fact can be linked to the need for a liquid electrolyte and the simultaneous entrance and expulsion of solvent molecules. So, the materials, moreover nonstoichiometric, are soft and wet. By tradition soft and wet materials only were used by humans from their natural origin and in most cases as food. Only a few cases of technological applications of artificial gels and polymers can be found, most of them related to the food industry or pharmacology, almost never were classical technological and industrial applications involved. Scientists and engineers associate the presence of water close to industrial materials with corrosion, deterioration and degradation processes. Despite these prejudgments, we will show here that the electrochemistry of soft and wet conducting polymers presents unparalleled multifunctional and biomimetic properties, the magnitude of which is changed in a reversible and infinitesimal way over a large range, as do the counterion compositions, under perfect electrochemical control.

Those properties changing as a function of the material composition and attracting more scientist attention are: electrochemomechanical (change of volume), electrochromic, charge storage, chemical storage and electroporosity.

6.5.1 Electrochemomechanical Properties

The entrance and expulsion of counterions and water driven by the electrochemical reaction promotes reversible changes on the material volume. These changes are under control of the electrochemical reaction and can be applied to generate a macroscopic movement and mechanical energy: this is an electrochemomechanical property[8,9] recently reviewed in Ref. 10 and 11.

6.5.2 Electrochromic Properties

The reversible reorganisation of the double bonds along the polymeric chains generates and destroys chromophores (polarons and bipolarons) adsorbing light in the UV-vis and near-IR regions of the spectra. The colour of the material can be changed, as the concentration of chromophores change, by the electrochemical reaction in a continuous, reverse and infinitesimal way. This is an electrochromic property, similar to that shown during electrochemical processes using oxides of transition metals[12–14] and recently reviewed.[15]

6.5.3 Charge Storage

Transition from neutral to oxidised polymers implies the storage of positive charges along the polymeric chains. Moreover, transition from a neutral to a reduced polymer implies the storage of negative charges along the polymeric chains. So CPs are suitable materials for the storage of both positive or negative charges along chains.[12–14,16]

6.5.4 Porosity

A film of a basic conducting polymer in the neutral estate is a compacted structure with pores across the structure having a small diameter. During oxidation coulombic repulsions appear among the emerging positive charges in neighbouring chains with simultaneous reorganisation of the double-bond distribution forcing conformational changes with progressive increase of the average pore diameters when the oxidation progresses. Under a transversal electric field, increasing currents (flow of counterions) are observed for increasing oxidation states of the polymer. The average pore diameter, and the transversal current, decreases under progressive reduction.[17,18]

6.5.5 Electron/Chemical Transduction

During reverse electrochemical processes any electronic interchange between the CP and the metallic wire supporting the electronic flow from or towards the potentiostat, is linked to the simultaneous interchange of chemical ions between the CP and the solution.[19–21] This must be a unique relationships, each injected electron forces the interchange of a one valence chemical ion in order to maintain charge balance. This is a unique electron/chemical transduction suitable for the reversible storage and release under control of chemical and pharmacological compounds in the polymer matrix. Moreover, the equilibrium potential of each polymer/chemical compound could define new simultaneous sensing properties.

6.5.6 Unparalleled Simultaneous Sensing Possibilities

The reverse electrochemical processes, [1] [2] or [3], suppose that any intermediate composition of the material (from neutral to overall oxidation, or from neutral to overall reduction) is linked to an electrochemical equilibrium between polymeric species, counterions and water inside the polymer, ions and water in solution, and electronic energy in the metal. This means that any physical (mechanical, optical, electrical, magnetic, gravity, *etc*.) or chemical (ionic strength, counterion concentration, ionic force, temperature, *etc*.) acting on the chemical equilibrium, must promote a simultaneous change in the energy level of the electrons in the metal wire from the potentiostat to the CP. Consequently, for any device based on an electrochemical property and their change with the material composition along the electrochemical reaction we can control, using only the two connecting wires, the change of the property (imposing a current flow) and we can sense (following the device potential) the influence of any physical or chemical ambiental change.[22–24] Any of these devices will work simultaneously as an actuator and a sensor, as do the natural organs, overcoming any unprecedented device that requires separate actuating and sensing systems.

6.6 Multifunctional and Biomimicking Properties

All the above-described properties are linked to any of the electrochemical reactions: [1] [2] or [3], taking place in a soft and wet material, the composition of which resembles that of the animal organs: water, ions and polymeric (organic) molecules. Every property is linked to one or several functions that also seem inspired by biological organs:[3,25]

Property	Action	Inspired organ
Electrochemomechanical	Change of volume	Muscles
Electrochromic	Change of colour	Mimetic skins
Charge storage	Current generation	Electric organs
Electroporosity	Transversal ionic flow	Membrane
Chemical or pharmacological storage	Chemical modulation	Glands
Electron/ion transduction	ΔV (chem/phys. properties)	Biosensors
Electron/neurotransmitter	Channel V action	Nervous interface

6.7 Natural Muscles

Muscles are efficient devices (molecular motors) working at constant temperature to transform chemical energy, form glucose, into mechanical energy and heat. At the moment, among those machines produce by human technology to transform stored chemical energy into mechanical energy, only fuel cells work at constant temperature. Fuel cells are able to transform fuels (hydrogen or organic compounds) into CO_2, H_2O, electrical energy (intermediate between chemical and mechanical energies) and heat. So, only this type of device is expected to be able to compete in the future with muscle efficiencies. The fact of working at constant temperature, far away from the servitude imposed by Carnot's principle, makes natural muscles much more efficient that any man-made machine able to transform chemical energy from wood, coal, petrol-derivative fuels or natural fuels into mechanical energy by combustion: internal combustion devices, steam engines, turboreactors. Moreover, any machine containing these motors produces quite rudimentary movements and a lot of noise (acoustic and electromagnetic) and ambient deterioration, even though very useful for human development.

The actuation of any natural muscle is based on molecular motors constituted by polymeric chains of actin and myosin. The basic structure for the actuation is the sarcomere, where actin and myosin chains are organised as quasilongitudinal fibres perpendicular to the sarcomere walls. The nervous (electric) impulse transporting from the brain the actuation order promotes the liberation of calcium ions inside the sarcomere. This concentration increase induces, through chemical transformation of ATP into ADP, conformational movements of the actuating polymeric chains and the contraction of the muscle. So, the actuation of a natural muscle involves: (a) aqueous media, (b) an electric pulse arriving from the brain (the pulse generator) to the muscle through the nervous system, (c) liberation of calcium ions inside the sarcomere, (d) chemical reactions, (e) conformational changes along natural polymeric chains (actin and myosin) with change of the sarcomere volume and (d) water interchange. This natural motor is able to generate quite elegant and gentle movements. Moreover, the actuation involves simultaneous sensing processes allowing a perfect consciousness of both the mechanical movements and the physical interactions between the muscle and its environment and the muscles can self-repair: they are intelligent devices.

6.8 Devices based on the Electrochemical Properties of Conducting Polymers

6.8.1 Artificial Muscles

So, if we try to mimic natural muscles or natural molecular motors we shall attempt to include, at least, electric pulses and polymeric chains. Some pioneering devices were constructed in the 1950s using films of polymeric gels immersed in aqueous solutions.[26–29] Interaction between electric fields, or electric currents, and swelled polymeric gels only attracted the interest of a few scientists up to the end of the 1980s and the beginning of the 1990s.[30–33] Then, a fast development of the field took place stimulated by the interest to reproduce commercial piezoelectric or electrostrictive (electromechanical) devices developed with inorganic materials, but using now similar properties, the control of dimensional changes in polymeric materials by electric fields. The beginning of this explosive interest overlapped the discovery of intrinsically conductive polymers as well as that of the reverse variation of their volume when submitted to reverse electrochemical oxidation/reduction processes. This controlled volume variation envisages the construction of new electrochemomechanical devices, bearing new and unexpected actuating and sensing possibilities.

6.8.1.1 Polymeric Electrochemomecanical and Electromechanical Muscles

Any of the developed devices based on the interaction between electric fields, or electric currents, and polymers was named artificial muscle. We can summarise the present state-of-the-art of the named artificial muscles, actuation of which involves polymers, electric fields and electric currents, by means of their classification in two main areas:

- *Artificial muscles responding mainly to electric fields, E (electromechanical actuators)*, with the dimensions of the variation of the electroactive polymer proportional to:

    E^2 : Electrostrictive actuators
    E : Piezoelectric actuators
           Ferroelectric actuators
           Electrostatic actuators
           Electrokinetic actuators (electro-osmotic)

- *Artificial muscles responding* mainly to electric charges, with the dimensions of the variation under control of the electrochemical reaction, and proportional to:

    Q : *Electrochemomechanical actuators*

Usually, artificial muscles based on electrostrictive, piezoelectric, electrostatic or ferroelectric materials have been manufactured as thin films of dry polymer (named electroactive material), both sides coated with a thin metallic film required to apply the electric field. Electrokinetic artificial muscles are constituted by films of polymeric gel (polymer, solvent and salt) and two electrodes, located as close as possible to the material, or coating both material sides, and required to apply the electric field, which drives the electro-osmotic process. Any of the actuators described in this paragraph has a triple layer structure: metal/polymer/metal (Figure 6.5). The presence of water and ions in electrostrictive, piezoelectric or ferroelectric polymers produces an overlapping of both actuation processes: electromechanical and electrokinetics.

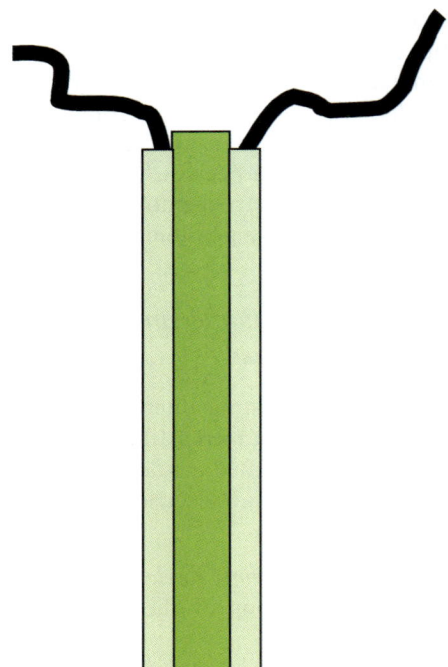

Figure 6.5 Triple-layer structure of electromechanical (EM) actuators and one kind of electrochemomechanical (ECM) actuator. In EM actuators the electroactive polymer constitutes the internal layer, the two external ones are sputtered metals. In ECM the CPs constitute the external layers supported by an internal polymeric or ionic conducting film.

Whatever the electromechanical device the high applied potentials (from several volts to thousand of volts) in aqueous media induce water hydrolysis. Oxygen and hydrogen evolve at the anode/polymer and at the cathode/polymer interfaces, respectively, producing large pH changes. The electrochemical reactions consume charge, decreasing the efficiency of the electromechanical process. The chemical variations induce polymeric degradation, destroying the device.

In the absence of water other chemical reaction is required for the actuation of these devices. The applied electric field forces fast conformational movements on the polymeric chains and concomitant macroscopic changes of volume, which relax in the absence of the electric field. Well-established electrostatic and mechanical (elasticity) models applied to polymeric materials are used to model the responses. A great theoretical structure has been developed for the description of inorganic actuators from the Curie's time, providing a strong background for the development of polymeric electromechanical muscles. These electromechanical actuators, or artificial muscles, are treated in detail in other chapters of this book.

Here, we will focus our attention on those artificial muscles based on the volume changes generated in CP by electrochemical reactions in electrolytes: electro-chemo-mechanical actuators. The presence of chemical reactions, solvents and salts gives these actuators a great similitude to natural muscles. The interaction between the different chemical and physical properties acting on both, equilibrium and rate of the driving electrochemical reaction provides the electrochemomechanical actuators with unique simultaneous sensing and actuating possibilities that will be introduced below.

6.8.1.2 Electrochemomechanical Muscles or Actuators: Volume Variation

The two electrochemical reactions [1] and [2] are accepted as the origin of the volume variations taking place in films of CP under current flow in the presence of an electrolyte. When the interchange of anions prevails (accepted for families 1, 2, 4, 5 and 7) during the redox processes, the volume of the film swells during oxidation and shrinks during electrochemical reduction. Under prevailing cation interchange (accepted for families 3 and 6), the material shrinks during oxidation and swells during reduction. The important role played by the water interchange in both oxidised and reduced materials, is an unsolved point, in particular when prevailing cation interchanges are present.

Considering the nonstoichiometry of the oxidised material, and the electrochemical control of the material composition, we will expect that the volume variation will be under control of the consumed charge.[34–36] So, acceleration, stopping or reversing the movement must be related to rising direct currents, current-flow interruption or reversing the direction of the direct current flow.

Dimensional changes, or volume variations in films of CPs have been followed by different methodologies, or estimated from experimental densities and weights of dried oxidised and reduced films, in some of the pioneering studies of the electrochemistry of conducting polymers.[37–49] These changes were confirmed at microscopic level by "*in-situ*" AFM, ellipsometry, conductivity measurements, *in-situ* electrogravimetry and other techniques.[55–56]

6.8.1.3 Electrochemical Basic Molecular Motors

We can imagine an ideal and lineal polymeric chain connected to a metallic electrode and immersed in an electrolyte (Figure 6.6). The strong intramolecular interactions originating a coil compact structure of the chain.[1] Under oxidation consecutive electrons (from 1 to n) are extracted from the chain, positive charges are generated and the above-described processes now progressively produce the transformation of the coil-like structure to a rod-like structure. The process can be stopped at any intermediate position, or reversing from any intermediate position by stopping or reversing the direction of the

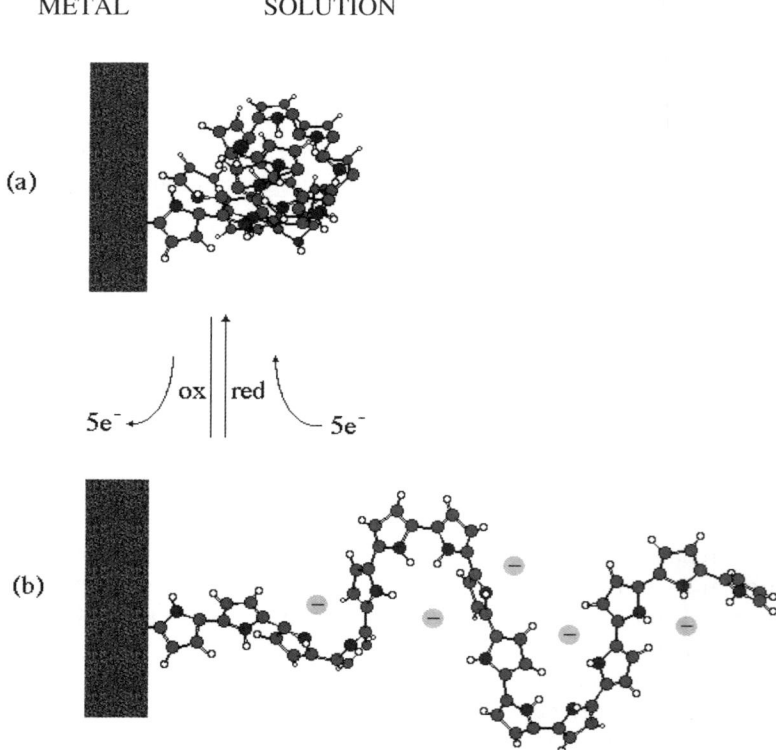

Figure 6.6 Molecular motor: reverse conformational changes (mechanical energy) stimulated by oxidation or reduction of the polymeric chain in an electrolyte. (a) Reduced chain, (b) oxidised chain. (Figure from: *Modern Aspect of Electrochemistry*, number 33, R.E. White *et al.*, ed., Kluwer Academic/Plenum Publishers, New York, 1999, p. 344.)

current flow (the direction of the electrochemical reaction): this is the basic molecular motor working under electrochemical stimulation of the conformational movements with electrochemical control of the intramolecular interactions. This basic molecular actuator includes: electric pulses, ions and water interchanges between the polymer and the solution, chemical reactions, stimulation of the conformational movements along polymeric chains and changes in the inter- and intramolecular interactions. These processes occurring in soft and wet materials mimic, at the molecular level, the consecutive events involved in the actuation of a natural anisotropic muscle.[57,58]

6.8.1.4 Structures and Designs of Artificial Muscles

At the moment, our control of the polymeric synthesis does not allow the obtainment of linear and parallel chains linking two nanoscopic films of a rigid and conducting material. This device will mimic the behaviour of the sarcomere. But only three-dimensional films of CP given three-dimensional changes of volume are available.

In order to translate these three-dimensional microscopic changes of volume into a uniform macroscopic movement, devices having different structures are being designed: very thin bending monolayers, bilayers or triple layers, or as longitudinal movements of monolayers, tubes, combination of bending double layers or triple layers, *etc.*

6.8.1.5 Bending Structures

Asymmetrical Monolayers. We can design monolayers of the same CP, but having an internal asymmetry able to produce asymmetric swelling or shrinking processes across the film under the same electrochemical process (oxidation or reduction), or a half of the film swells meanwhile the second half shrinks.[59–75] The last films can be electrogenerated thanks to the different materials available for the same CP. So, a film of a CP having prevalent anion interchange can be electrogenerated on a metal electrode. Then, the coated electrode is translated to a new monomeric solution where the electrogeneration goes on forming a cation-prevalent interchange film of a different, or of the same, CP. The film peeled from the electrode and used as an anode in an electrolyte shows an asymmetric change of volume: swelling the fraction of the film where the anion interchange with the solution prevails; meanwhile the fraction where the interchange of cations prevails, shrinks. The film requires the presence of a metallic counterelectrode to allow the current flow.

Some other possibilities are being explored to produce asymmetric monolayers: by physical means growing the CP on adsorbed and porous materials, or by electrochemical means generating a film of CP with a concentration gradient of counterions, by generating a gradient crosslinking network, or by generating a bilayer of the conducting polymer with a macroanion (shrinks by oxidation) and then of the same conducting polymer with a small anion (swells by oxidation), or even placing a metal sheet between both films. These asymmetric films also produce a bending movement and also need an electrolyte and a counterelectrode in order to allow the flow of the current and to produce the bending movement. On the counterelectrode the current flow will generate new chemicals and pH variations, promoting the progressive deterioration of the actuating film and spending most of the consumed electrical energy.

6.8.1.6 Bilayers

A film of a conducting polymer (with prevalent anion interchange, or cation interchange) electrogenerated on a metallic electrode is stuck to a polymeric tape. The bilayer CP/tape is removed from the electrode. The new bilayer is used as a new working electrode in an electrolyte. The mechanical stress gradient generated across the bilayer interface by the swelling/shrinking processes, induced by the electrochemical reactions [1] [2] or [3], depending on the family of the studied CP film, originates the bending movement of the bilayer free end.[8-9,75] When the film of CP has a prevailing interchange of anions, the oxidation originates the swelling of the CP, pushing the bilayer and staying at the convex side of the bended device (Figure 6.7). CPs having cation-prevailing interchange suffer shrinking processes during oxidation trailing the device and staying at the concave side of the bended device. To allow current flow, a metallic counterelectrode is required. A good fraction of the consumed electrical energy is wasted to produce the electrochemical reactions required to allow the current flow through the counterelectrode.

6.8.1.7 Triple Layers

Using a two-sided polymeric tape and sticking one film of a conducting polymer by the side, avoiding any contact between the two films of CP we produce a triple layer. One of the CP films is connected to the working electrode output from a potentiostat. The second CP film is connected to the counter-electrode (CE) output and short-circuited with the reference electrode (RE). By immersion of the triple layer in an electrolyte we complete the structure of the electrochemical cell allowing the current to flow through them (Figure 6.8). Now, the same current flows through the two polymeric films: the anode swells (preferential anionic interchange) and pushes the device, the cathode shrinks and trails the device.[6] By means of the reference electrode we follow, when a current flow was imposed to the device from the potentiostat/galvanostat, the evolution of the potential difference between the CP film acting as WE and the CP film acting as CE: this is the muscle potential.

6.8.1.8 Structures giving Lineal Movements

Fibres and Films. Different methodologies are being used to obtain fibres of conducting polymers.[76,77] A film of a conducting polymer removed from the

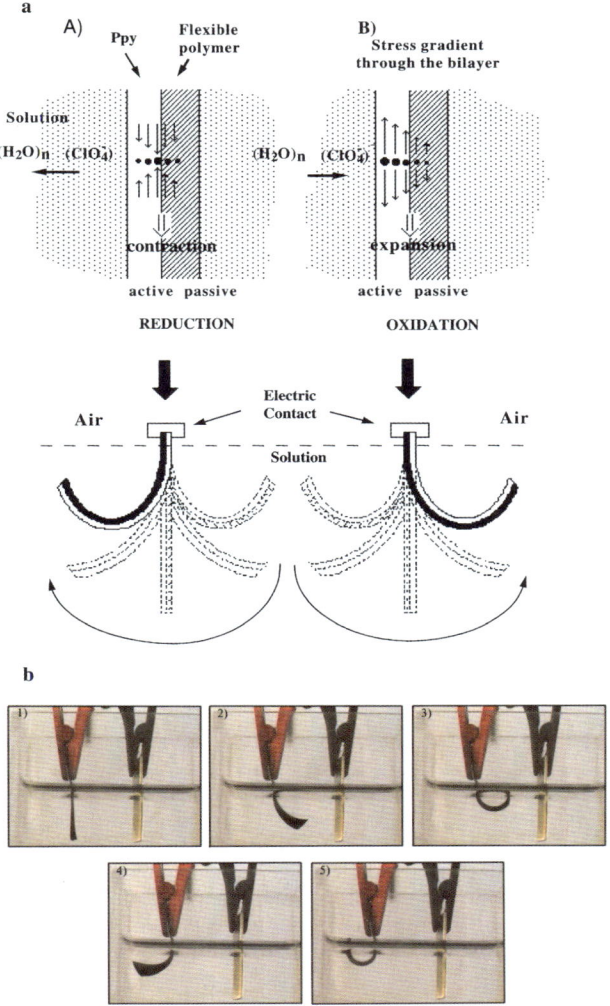

Figure 6.7 (a) Scheme of a bilayer CP (active) adherent, flexible and nonconducting polymer tape (passive). Scheme of the ionic interchanges between the CP and the solution during the electrochemical reactions produced by the current and of the stress gradients between both layers induced by the ionic interchanges and concomitant volume changes in the active layer of CP (Reproduced from TF Otero in Ref. 18). (b) Macroscopic movements in solution (bottom) during anodic or cathodic (oxidation or reduction) current flow through a CP/tape device. Different positions of a bilayer muscle showing the metallic counterelectrode and the solution: 1 to 3 flow of an anodic current of 15 mA, 3 to 5 flow of a cathodic current of -15 mA. (Reproduced from Ref. 18)

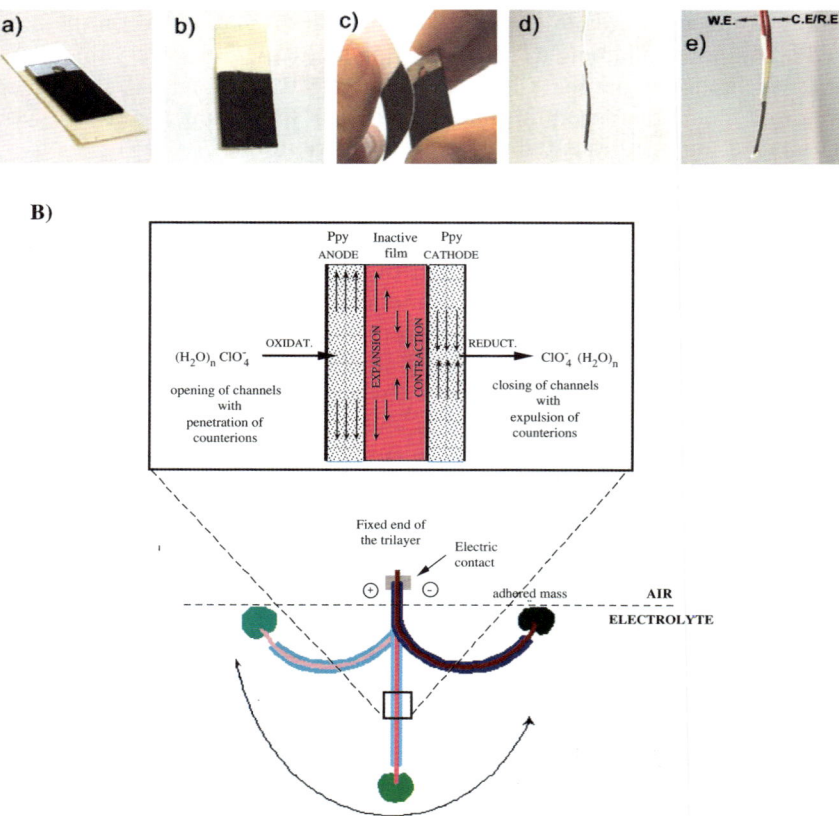

Figure 6.8 (a) Consecutive steps for the construction of a triple-layer muscle (CP/tape/CP) starting from electrogenerated films of a CP on metallic electrodes. (b) Scheme of: ionic interchanges, induced stress gradients and generated angular movements during current flow. (Reproduced from Ref. 1)

metal electrode where it was electrogenerated can be used in a universal mechanical test equipment, provided with an electrochemical cell in order to follow lineal displacements and mechanical energies produced under different chemical, electrical or physical conditions.[78–104] Bundles of these films or fibres are checked to produce vertical displacements of weights. These individual films or bundles are used as working electrodes, therefore a metal counterelectrode is required to allow the current flow, generating similar problems to those described above for bending asymmetric films.

6.8.1.9 Tubes and Films with Metal Supports

A second approach consists of the electropolymerisation of a conducting polymer on helical metallic wires up to the generation of a tube, or on zigzag metal wires to generate films, *etc.*[105–110] Those efforts are based on the named conducting/nonconducting transition during the reduction of very thin films of CPs, which should originate a drop on the electrical potential along a self-supported film and a nonuniform actuation. The supporting metal wire should guarantee a uniform potential and current distribution.

Most of the described experimental conditions involve devices working under partial oxidation of the material, never attaining deep reduction states, so having a high electronic conductivity. For such thick films (several μm) as are usually employed for the construction of macroscopic actuators the experimental results indicate, and the ESCR model predicts, a very difficult reduction completion, *i.e.* time consuming and at very high cathodic overpotentials.[111,112] The polymeric entanglement is closed under partial reduction, entrapping counterions and keeping a conductivity always higher than 10^{-2} S cm^{-1}. Under most of the experimental conditions this gives a uniform actuation, not requiring embedded metallic wires to guarantee a uniform conductivity along the device and a uniform bending along the device, as proved by digital image analysis.[113,114] These structures also need a metallic counterelectrode to allow the current flow.

6.8.1.10 Combination of Bending Structures

The present state of the technology presents an open problem, common to the electrochemical and electromechanical devices and related to the structure: how to design a basic element (macroscopic or microscopic) giving a longitudinal movement and including either: actuating electrode, actuating counterelectrode and reference electrode that repeated by *n* times through the space can give three-dimensional devices having any shape and any volume (Figure 6.9), the properties of which are *n* times the property of the basic element. The design has to guarantee the electric contacts and the uniform distribution of the electric current by every constitutive element across the three-dimensional muscle. Different combinations of triple layers[115] and bilayers[116] are being constructed and checked: the key point is to keep simultaneous sensing and actuating capabilities despite the complexity of the device. (A parallel problem for electromechanical actuators requires the uniform distribution of the electric field on the different elements of the three-dimensional device.)

6.8.1.11 Microdevices and Microtools

Any material, such as metals, that can be electrodeposited or electrogenerated can be applied during microtechnological processes to produce micrometric structures. The electrochemical synthesis of the films of conducting polymers, and the electrochemical actuation, are suitable for the construction of elegant and imaginative microdevices and microtools constituted by bilayers by using microelectronic technologies.[117–132]

6.8.1.12 Establishing a New Paradigm

Any of the described structures for actuators based on CP include a new paradigm related to traditional inorganic actuators, or to the new polymeric actuators (piezoelectric, electrostrictive, ferroelectric, electrokinetics, *etc.*): the

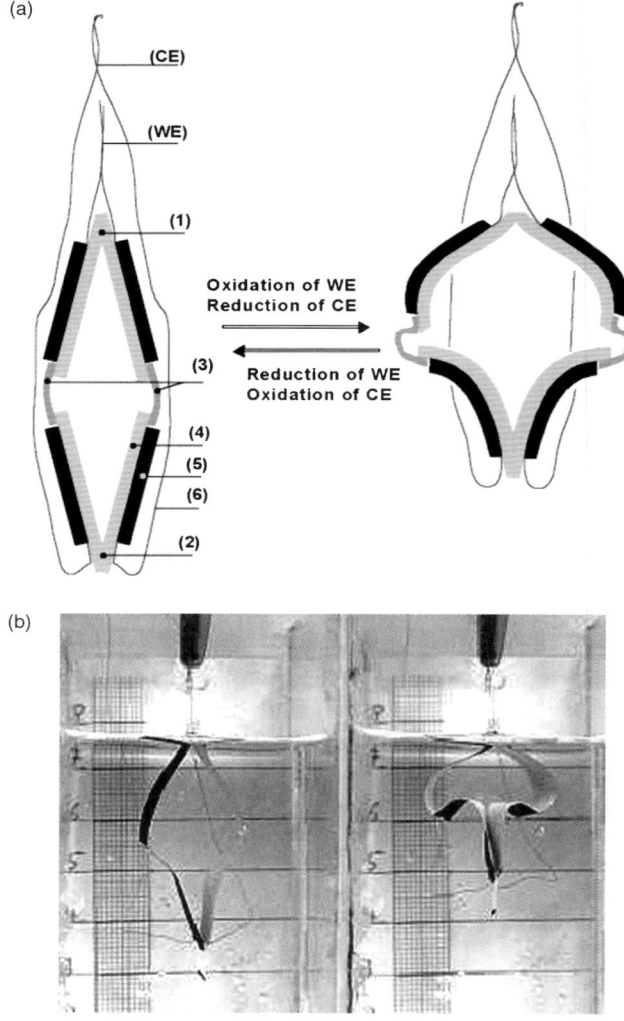

Figure 6.9 (a) Scheme of a complex element constituted by 4 bilayers CP/tape, the two on the top acting as working electrode and the two on the bottom, as counterelectrode short-circuited with the reference electrode and two lateral plastic hinges. (b) Pictures of the extended and contracted positions (21% of the original length) of the element in 1 M LiClO$_4$ aqueous solution by flow of 2.5 mA during 90 s. (c) Proposed combination of lineal elements. (Reproduced from Ref. 116.)

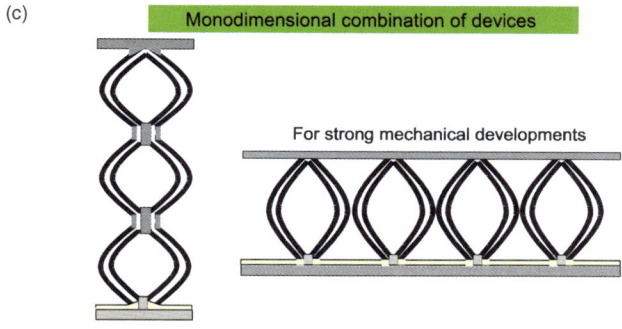

Figure 6.9 (Continued).

electrode material is, at the same time, the active material and the support for the current flow, and for the electrochemical reaction the origin of the volume variations. Moreover, actuation requires the presence of an electrolyte as the counterion supplier. All the inorganic actuators (and also all the new polymeric ones mimicking them) use two inert metallic electrodes in order to produce and to apply an electric field on the actuating material. That means that three layers constitute all the electromechanical devices: metal-electrode/actuating-polymeric-material/metal-electrode. So, it is not possible to imagine an electromechanical actuator constituted by an isolated film of a piezoelectric or electrostrictive material acting, at the same time as the electrode to support the electric field.

6.8.1.13 Sensing Muscles

The electrochemical equilibrium stated by a muscle under a stationary state between the partial oxidised polymer, ions and water in solution and electrons on the connecting metal wire implies that any mechanical, electrical, optical, thermal, magnetic or chemical variable acting on the equilibrium must influence the potential of the device. Working under constant current that means that the evolution of the device potential will shift towards lower or higher values, depending on the way that every variable acts on the chemical equilibrium. The increase of these variables shifting the chemical equilibrium towards the formation of the oxidised material (electrolyte concentration or temperature) will promote lower potentials (Figure 6.10). Those variables favouring a shift towards the reduced material (mechanical stress) will promote an increase on the electric potential.[22–24,115–116,133–141] When a device is constituted by a triple layer, or by combination of double or triple layers the muscle potential is the potential difference between those films acting as anode and those acting as cathode.

As expected, the muscle potential and the consumed electrical energy change linearly as a function of the different variables. The muscle potential and the overall consumed electrical energy required to describe the same movement

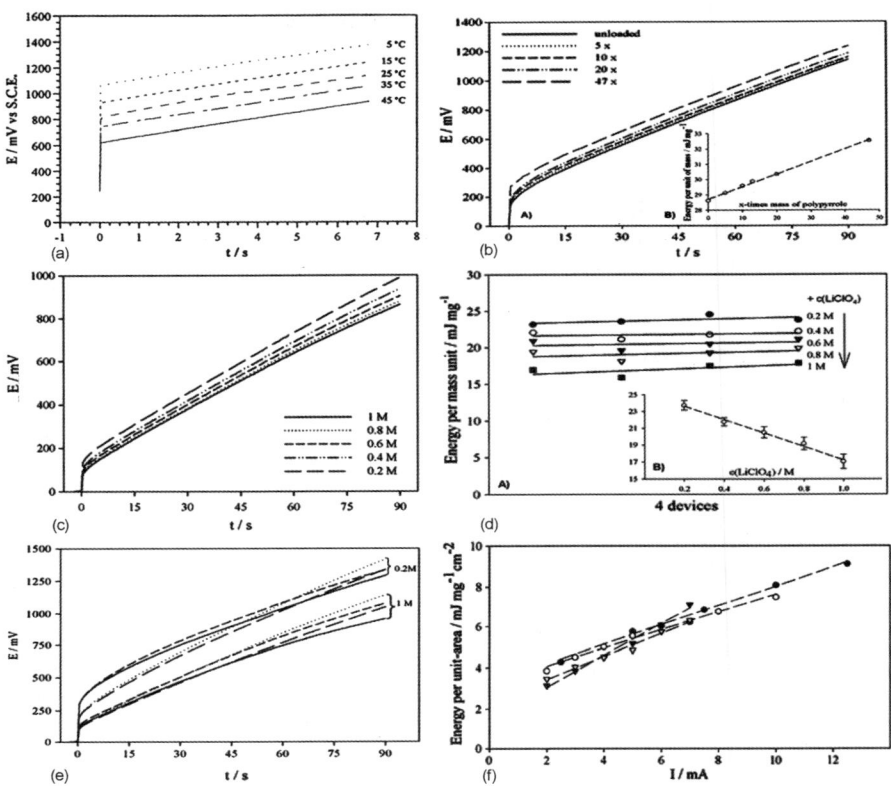

Figure 6.10 Evolution of the muscle potential of: (a) a bilayer, followed vs. a reference electrode Ag/AgCl, during an angular movement of 90°, at a constant current, under different temperatures; (b) of a triple layer shifting different weights, the number indicates times the weight of conducting polymer in the device, adhered to their bottom; (c) of the devices from Figure 6.9 in different concentrations of the electrolyte; (d) lineal evolution of the consumed electrical energy in (c) as a function of the concentration indicating the sensing abilities of the device, (e) and (f) reproducibility for different complex elements. (Reproduced from Refs. 133, 141 and 116.)

increase when mechanical stress (increasing pieces of metal trailed by the muscle adhered to their bottom), or rising currents were applied.[115–116,133–141] The muscle potential and the consumed electrical energy decrease when the electrolyte concentration, or the temperature, increase.

6.8.1.14 *Tactile Muscles and Primitive Conscious Systems*

The mechanical sensing characteristics (increase of the muscle potential when the weight of the trailed weight rises) predict the possibility of developing a tactile sensor. If an obstacle is located in the middle of the muscle path, the muscle moves free, under a constant current, the evolution of the muscle potential overlapping that of the free muscle. When the muscle touches the obstacle it feels a mechanical resistance, which influences the anodic and cathodic electrochemical reactions occurring on the concomitant anodic and cathodic constituent polymeric films. Consequently, a potential step is expected to occur on the muscle potential being proportional to the opposed mechanical resistance emerging at the touching moment: increasing weights of the obstacle will produce increasing potential steps, as those found experimentally (Figure 6.11). Both the potential step and the consumed electrical energy after the touching moment follow a linear evolution as a function of the obstacle weight.[142,143] When the muscle is not able to shift the obstacle the potential steps very fast by several volts, and the muscle degrades if we do not switch off the current.

These results allow the construction of a conscious device: a current source connected to the electrochemomechanical artificial muscle, a single computer with very simple software able to follow the potential evolution, detecting the potential jump and quantifying the potential step. These results are immediately transformed into useful information: the touching moment, the weight of the obstacle and if the machine is able to shift the obstacle or not and in this case to switch off the current. Moreover, the sensing abilities of the device can provide, simultaneously, information about the ambient conditions: working temperature or electrolyte concentration. All that information also could be obtained through the only two connecting wires carrying actuating and sensing signals. The described ensemble: sensing muscle, generator and software constitutes a quite primitive conscious device.

When these mechanical sensing abilities are tested using a film of CP in the electrolyte under different preloads, a linear relationship is found whatever the studied variable, but the consumed electrical energy decreases, at the beginning for increasing preloads, goes through a critical minimum preload (Figure 6.12) and then increases linearly with increasing mechanical stresses.[144,145] Madden's group only went through the decreasing-potentials region. So, apparently when we studied the abilities of bilayer or trilayer muscles for sensing trailed weights or the mechanical resistances opposed by obstacles, always we are located over this second part, but it should be expected that some opposite results (decrease of the consumed energy for increasing weights) can be found when actuators constructed using very thin bilayers or self-supported asymmetric films should be used as mechanical sensors for small weights or mechanical resistances.

6.8.1.15 Muscles Working in Air

Despite the unprecedented simultaneous actuating and sensing properties, the human technology is based on dry materials and dry devices. In this way different attempts are being performed to develop all solid-state actuators.[8,146–154] The most common structure is: *conducting polymer/ionic conductive tape/ conducting polymer*. Also *encapsulation* should be convenient for this kind of

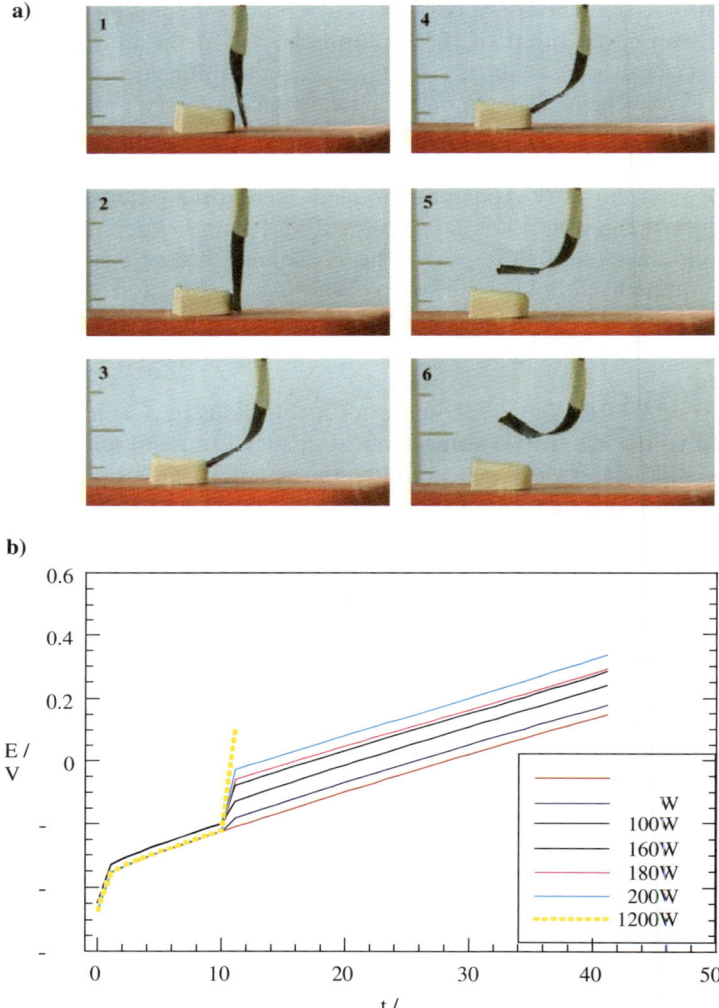

Figure 6.11 (a) Movement of 90° of the free end of an artificial muscle under flow of 5 mA at 25°C in 1 M LiClO$_4$ aqueous solution, touching an obstacle of 3000 mg after 10 s of movement starts and sliding it later. The dimensions of the device were: 2 × 1 cm, and each polypyrrole film weighs 6 mg, being 13 μm thick. (b) Evolution of the working potential of a triple layer, which meets and slides different obstacles after 10 s of current flow. The weight of the obstacles range between 50 to 1200 times the active polypyrrole weight. (c) Black: Evolution of the working potential of the free muscle (without any obstacle). Red: the muscle touches and slides an obstacle weighing 3600 mg after 10 s of current flow. The shaded area shows the electric energy consumed by the muscle to push and slide the obstacle. (d) Evolution of the working potential (or muscle potential) increment after touching the obstacle as a function of the obstacle weight/polypyrrole weight. Each polypyrrole film weights 6 mg, and device area was 2.5 cm^2. (Reproduced from Refs. 142 and 143.)

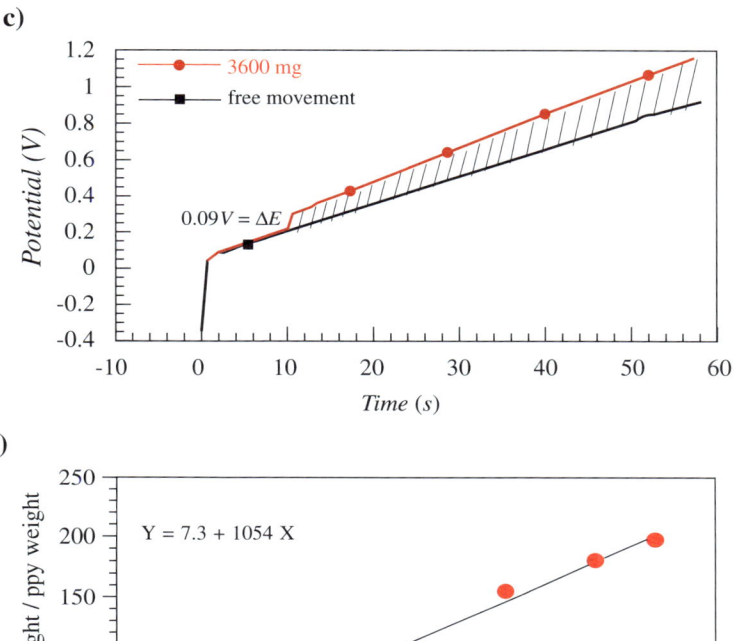

Figure 6.11 (Continued).

device. Different approaches produce absorption mechanical sensors, based on the reverse bending by absorption/desorption of water, organic solvents or different compounds.

6.8.1.16 Difficulties

Considering that one of the aims should be the production of three-dimensional muscles (microscopic and macroscopic) for the construction of any kind of tools and robots, the present state of this emerging area present some difficulties to be overcome.

For very fast actuation rates <0.1 s and bending angles greater than 90° conducting limitations appear: electronic and, mainly, ionic. The fast electronic difficulties along a film of CP can be overcome with intermediate metal deposition: this includes extra stiffness and peeling processes. To overcome a possible low ionic conductivity of the electrolyte ionic liquids[155–166] are being

a)

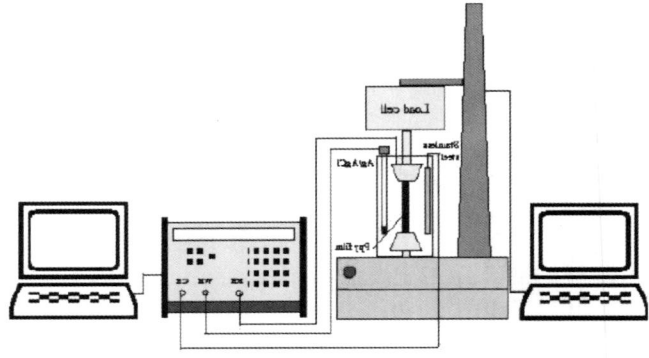

b)

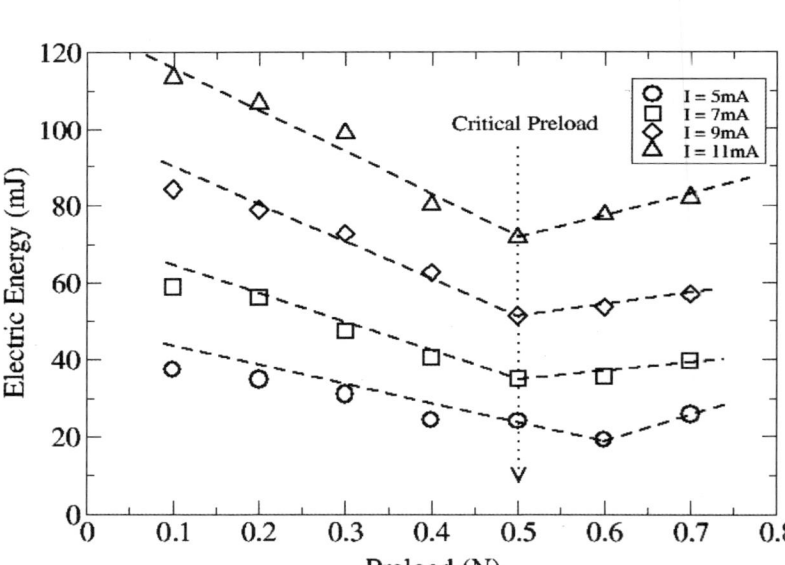

Figure 6.12 (a) Experimental equipment for "*in-situ*" mechanical characterisation under electrochemical reactions. (b) Electric energy consumed during oxidation process of a free-standing polypyrrole film by flow of 5, 7, 9 and 11 mA under different preloads ranging between 0.1 and 0.7 N. (Reproduced from Ref. 144.)

checked: results point to extended lifetimes without increasing rates, probably related to a dried ambient and diffusion/relaxation difficulties inside the material, respectively. Here, the problem is to identify the rate-limiting step: ionic conduction in the electrolyte or ionic diffusion inside the polymer.

Most of the devices developed to produce directly linear movements require a metallic counterelectrode (Figure 6.7(b)) losing, at least, a half of the consumed energy to produce chemicals that contribute to degrade the conducting polymer. When we try to put together several, n, of these elements in order to obtain n times the mechanical energy produced by one of those devices, an irregular electric-field distribution promotes nonuniform movements. Tactile properties and, for still-increasing resistances, sensing abilities are lost.

The use of bending devices including actuating anode and cathode (short-circuiting the reference and counterelectrodes) requires the design of complex structures as basic elements by combination of bilayers or triple layers, able to translate these bending movements into linear ones. There, new experimental difficulties appear. Non uniform resistances on the electrical connections between the copper wires and the films of conducting polymers produce irregular movements. Mechanical rigidities emerge by a combination of several basic elements, decreasing the amplitude of the linear movement and losing some of the simultaneous sensing properties being the subtlest, the tactile sense, the most evanescent.

6.8.1.17 Mechanical Characterisation of Materials and Devices

Whatever the structure of the polymeric actuator or working with single films, or fibres a mechanical characterisation of the materials and devices is required. On the other hand, meanwhile new structures keeping sensing and tactile properties and able to produce large movements and mechanical energy are designed, most of the efforts can be concentrated on the generation of new materials that improve the dimension variations and the mechanical characteristics during oxidation/reduction processes.[76-111] This requires the synthesis of new materials from the families stated above and their mechanical characterisation determining the evolution of their length under a constant mechanical tension, or following the evolution of the mechanical tension, at a constant length of the film, in the presence of the electrolyte under different chemical, electrical, mechanical, optical, magnetic or thermal conditions (Figure 6.12). At the moment, the claimed evolution of the material length during the first voltammetric cycle has been improved from a few per cent to longer that 30%. This means it surpasses the natural muscles by almost one order of magnitude. Nevertheless, a faster decrease is observed on the consecutive cycles on this percentage of variation length. The present state of the attained mechanical characteristics and efficiencies were recently reviewed.[10,167,168]

6.8.1.18 Actuators as Products

Despite these difficulties, the evolution of those new devices, is so fast that many new companies are being created trying to develop new products and applications: EAMEX in Japan, Quantum Technology Pty. Ltd in Australia (development of Braille screens), Micromuscles AB in Sweden, Molecular

Mechanisms in Massachusetts, are some of the pioneers and new companies developing different medical tools or devices for medical applications.

The great technological interest is stimulated by the fact that developing new polymeric actuators is a priority area for receiving priority economical support from DARPA (USA) and ESA (EU) agencies. This implies a strong competition to get funds for developing new devices and to get new products at the expense, in most of the cases, of basic developments (polymerisation and degradation kinetics, or models) so fundamental in this emerging area of electrochemomechanical devices. Now, after 13 years of academic research the time for the first important stage of technological developments has arrived. A new world of soft and wet, sensing and tactile (conscious) actuating machines is awaiting for the intellectual energy and engineering ability of young scientists.

6.8.2 Other Electrochemically based Properties and Devices: Electrochromic Devices

As stated above, the extraction of consecutive electrons from the chains of a conducting polymer promotes, among the described processes, disappearing of double bonds and formation of new double bonds, now on the C–C between two monomeric units, with creation of conjugated structures denominated polarons and bipolarons. Most of the neutral polymers adsorb high-energy photons from the UV region, corresponding to the distance between the HOMO levels and the LUMO levels. The new polaronic and bipolaronic, bonding and antibonding, empty levels are energetically located between these HOMO and LUMO levels: lower-energy photons from the visible and IR are required to promote electrons to them from the HOMO level and they adsorb photons from the visible and IR.[15,169–173] These polarons and bipolarons behave as chromophores.

The concentration of these new chromophores is under the control of the electrochemical reaction, which is reversible, and under the control of infinitesimal charges. So, the adsorption of visible light can be changed in a reverse way by infinitesimal steps, as the driving electrochemical reaction and concentration of chromophores do (Figure 6.13).

So we can produce any device: a smart mirror, a smart window or glasses, *etc*. All of them will contain an electrochemical cell. The basic electrodes are: mirror-polished metal, a glass coated with a sputtered metal film, a glass or a transparent plastic coated with a transparent film of a conducting material – the most usual is indium tin oxide (ITO electrode) – or any of the previous electrodes coated with a uniform film of an electrochromic conducting polymer: metal, glass/metal or ITO/glass). The construction of the devices requires a thin transparent membrane film, allowing the ionic flow between the two constituent electrodes and avoiding any electronic contact, which will short-circuit and destroy the device. Two main device structures are possible: electrode (metal or ITO)/membrane/CP electrode, or CP electrode/membrane/CP

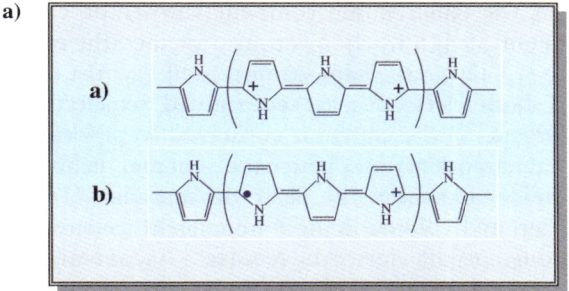

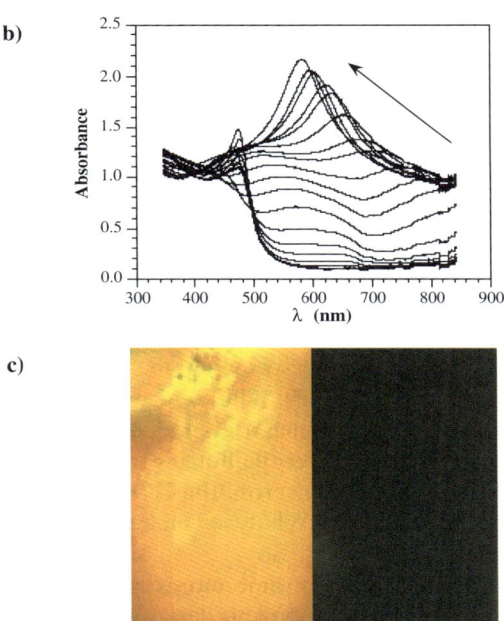

Figure 6.13 (a) Covalent-bond structure in polarons (here radical cations) and bipolarons (here dications) with conjugated distribution of electrons and charges. (b) Continuous increase of the absorbance and shift on the polaronic (600 nm to 490 nm) and bipolaronic (>850 to 590 nm) during a potential sweep from −900, to 500 mV for polypyrrole films from 0.1 M LiClO$_4$ propylene carbonate solution. (c) colour presented by the reduced (at −900 mV) and oxidised (at 500 mV) film. (Reproduced from TF Otero et al., Langmuir, 1999, **15**, 1323.)

electrode. The most suitable for long-lifetime devices is the second one. The structure must guarantee a perfect and uniform adherence among the constituent layers in order to avoid nonuniform colour changes.

The device works as any electrochemical cell: any current flow originates the transformation of the same number of chemical equivalents on both electrodes and the flow of the same number of equivalents of charge through the membrane (per equivalent a fraction is transported by the positive ions present in the system, t_+, the rest being transported by the negative ions, t_-: [$t_+ + t_- = 1$]. If both electrodes are constituted by electroactive materials, conducting polymers, all the charge is consumed to produce the concomitant oxidation and reduction reactions, with expulsion and inclusion of ions and the concomitant volume changes of the electrodic materials.

If these reactions and volume changes were not considered during the device design, the lifetime of the device is reduced. Asymmetric volume changes produce a progressive peeling of the different constituent adhered layers, or asymmetric mechanical stress distributions and nonuniform colours. When one of the electrodes does not have any material electroactive, *i.e.* metal, metal-coated glass, ITO; new chemicals are formed during current flow, thus promoting the degradation of the polymeric membrane, arriving at and degrading the electrochromic conducting polymer. The lifetime of the device is shortened.

We will focus our attention now on the electrochromic window constituted by two complementary electrochromic materials. This means that the absorbance must increase at the same time in both films: one of the material films, the one acting as an anode, oxidises; at the same time the second material film, the one acting as a cathode, reduces. The absorbance is reduced when the current flows in the opposite direction in both electrodic materials. Among researchers there exists an idea that this experimental arrange avoids degradation processes. Nevertheless, the lifetime of the devices can be very short if, during reverse current flow and cyclic colour changes, the potential of one or both constituent materials move outside the potential range for reverse electrochemical behaviour established for that material under the device experimental conditions. Outside this potential range the flowing current must meet another substance, ions, solvent, CP degradation reactions, the transformation of which supports the remaining current flow. Whatever the reaction the final result is a degradation of the electroactive CP. So a good fitting of the real cycling potential to the potential range of reverse electrochemical reactions for the studied material is a very important point to design long-lifetime devices. The scientific community has paid not much attention to this point.

6.8.3 Batteries

The electrochemical reaction of a CP describes storage of positive charges when the CP oxidises from the neutral state and storage of negative charges when the CP reduces from the neutral state: Ref. 3. Vol 4, Chapter 11.[174] The reduction state is only attained by some of the CP as most of the thiophene families. The availability of thin films of ionic-conducting membranes allows the construction of both, hybrid batteries (ELECTRODE/membrane/CP), where the material of the electrode is nonpolymeric (metal, alloy, oxide, intercalation compound, *etc.*) or all polymeric batteries (CP/membrane/CP). This means that a plethora of different kind of batteries where at least one of the electrodes is a conducting polymer can be designed. We will focus our attention here on those all-polymeric batteries: conducting polymers constitute both electrodes. Still now many different kinds of designs can be stated, which we will attempt to classify according to the structure and according to the cell potential.

Two basic physical structures can be considered: *salt storage* – all polymeric batteries are those where the concentration of electrolyte in the membrane changes during charge and discharge processes, and *rocking chair* – all polymeric batteries, where the salt concentration in the membrane remains constant but with a continuous flow entrance of one of the ions from a polymeric electrode and the same simultaneous exit amount towards the second electrode (Figure 6.14). The sense of the ionic flow is opposite under charge or discharge. The rocking-chair batteries attract most of the scientific interest. Under ideal conditions only one of the ions participates in the ionic transportation between electrodes, being the transport number $t_+ = 1$, or $t_- = 1$. This means that the two constituent polymeric electrodes interchange anions during redox processes (basic polymers, substituted polymers, copolymers), or that the interchange of anions prevails (with the anion's transport number in the membrane close to unity) even in liquid electrolytes. The second general possibility is that both electrodes interchange cations (self-doped polymers, polymeric blends, or hybrid materials).

Related to the cell potential, conducting polymers offer two main new possibilities for the batteries design related to the more classical materials. The large potential window for the oxidation/reduction process of some CPs allows the design of low-potential cells: *neutral CP/membrane/oxidised CP*, the potential ranging *from a few mV to 1V*. With this design we can even *use the same polymer for the two constituent electrodes* obtaining a great device equilibrium from either structural, chemical or volume-balance points of view. After discharge, both electrodes have the same oxidation state. On recharge,

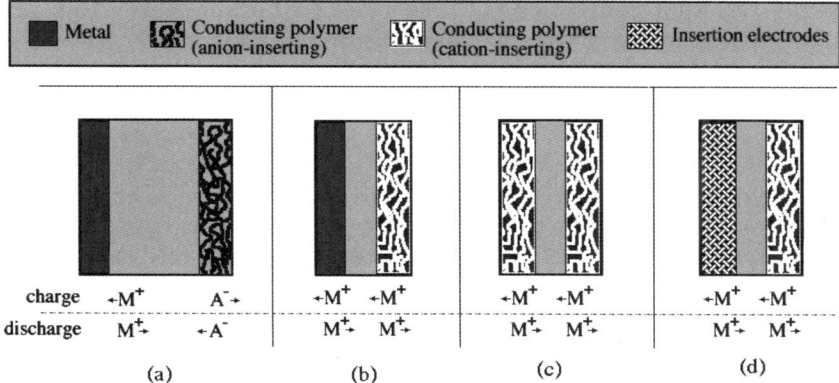

Figure 6.14 Different structures for batteries including conducting polymers. The grey central structure represents an electrolyte for the accumulation of salt (a), or a very thin ionic conductor membrane (polymeric).

whatever the sense of the recharge current, the initial structure is attained, one of the electrodes in a neutral state and the second in an oxidised state. Some of these structures allow very fast charge and discharge processes under pure diffusion kinetic control: *they were named supercapacitors*. The second structure provides great cell potential: *reduced polymer/membrane/oxidised polymer*. The cell potential ranges between *4 V and 1.5 V*. This structure is very sensitive to moisture or contamination: the high potential promotes their decomposition with generation of new radials and degradation of the polymeric materials.

A brilliant future is expected for those all polymeric batteries, with a new and unparalleled possibility – *green* batteries: new CP suffering biodeterioration. Nevertheless, a lot of basic effort is required to understand how to balance the processes there involved: material potential ranges, volume variations, limiting kinetic processes and overpotentials, stabilisation and degradation, equilibrium/sensing/kinetics relationships.

6.8.4 Membranes and Electron/Ion (or Electron/Chemical) Transducers

When a basic CP is oxidised the emerging positive charges along the chains promote coulombic repulsions between them that together with the stimulated conformational movements generate free volume between the chains. The average diameter of the intermolecular pores is expected to increase.

Membranes are constructed using films of CPs. Under polarisation at different anodic potentials different oxidation states are obtained (Figure 6.15). By applying a potential sweep between two metallic electrodes located at each side of the membrane, the ionic current flowing across it increases when the oxidation degree increases. The process is reversible and the current, at a constant potential across the membrane decreases when the membrane is partially reduced. For the same oxidation state of the membrane the transversal ionic current decreases when the diameter of the anion increases. So porosity, volume changes and chemical storage and release look interlinked.[37,39,175–190]

These results, if quite spectacular, produce very low ionic currents for large anions showing how limited our control of the polymer synthesis is. From a theoretical point of view no limitation exists to the synthesis of a polymer film that by oxidation develops pores thicker enough to allow the flow of large anions. Looking for the best conditions of polymerisation to produce these films (tailored process) is one of the challenges for the coming years.

6.9 Theoretical Models

These electrochemical reactions and devices constitute, for any modeling attempt, a great asset: they include crossing interactions between polymeric chains, ions and solvent. Moreover, covalent bonds are reversibly broken and

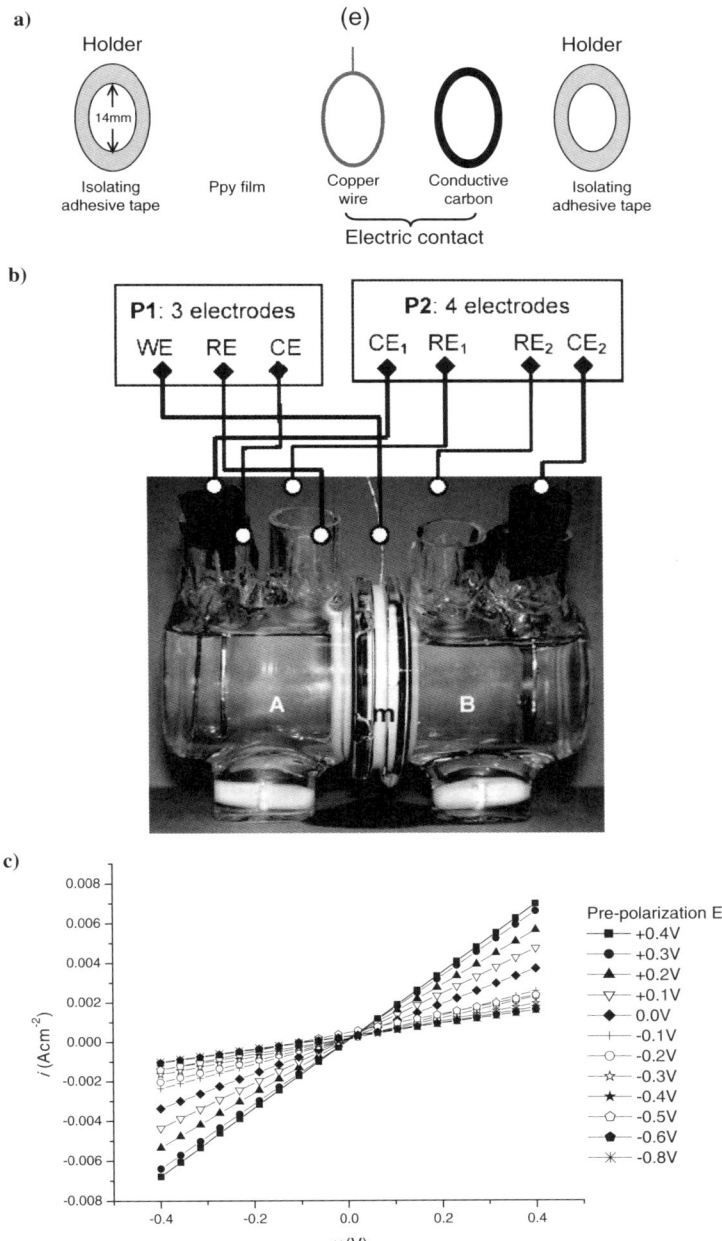

Figure 6.15 (a) Scheme for the construction of a membrane; (b) Experimental system for the control of the membrane, located between compartments A and B; (c) experimental results of the ionic current flowing across the membrane oxidised at steady state by polarisation at different potentials: the current and the porosity increase for increasing oxidation states, decreasing for decreasing oxidation states. (Reproduced from Ref. 213.)

reconstructed, charges are stored along the polymeric chains and the molecular interactions also are changed in a reversible way. The theoretical description and prediction of properties and devices requires the quantification of the chemical reactions produced by electrical currents (electrochemistry) involving polymers (polymer science) and ions to generate changes of volume and mechanical energy (thermodynamics and mechanics). This is a complex task, but very stimulating because most of the unmodeled biological functions also involve electric currents, polymeric molecules and chemical reactions. In this context, one of the weakest points to improve electrochemomechanical muscles and any other electrochemical device based on CPs is the lack of a solid theoretical background.

We can focus our attention on artificial muscles and on changes of volume produced by an electrochemical reaction. In physics the *elasticity* is in charge of the theoretical description and quantification of both the energies and dimensional changes involved in materials submitted to *external* mechanical stresses. So, an initial approach to the theoretical description of involved energies and dimensional changes observed in conducting polymers could be through well-established elastic models, despite the fact that here changes of volume come from internal processes. Elastic models were developed from empirical studies of materials having a constant composition and well-defined Young's and Poisson's constants. Here, the composition of a conducting polymer changes during the actuation process and we need to define how the elastic constants "change" as a function of the composition. The electrochemical kinetics, a macroscopic part of the physical chemistry, describes the composition change and the equilibrium potential for different compositions, so important for the quantification of physical and chemical sensing, is described by *thermodynamics*. Considering that we are working with polymers the theoretical description of any physical or chemical aspect of the material must include the *polymer science*.

The electrochemical kinetics includes charge transfer, diffusion or migration processes as possible limiting steps for the electrochemical reaction responsible for the change of volume and for the actuation. Nevertheless, the presence of slow migration processes inside the polymer matrix only could be linked to slow free-volume generation/destruction by slow conformational changes of the chains stimulated by the electrochemical restructuring of the double bonds along those chains. Polymer science describes these conformational changes, and the different molecular forces playing a key role during the actuation: polymer–polymer, polymer–solvent, polymer–anions and polymer–cations interactions. These forces change along the actuation: from strong polymer–polymer attractions in a reduced polymer, to strong polymer–polymer repulsions and strong charged polymer–anion and strong charged polymer–water attraction in fully oxidised polymer.

This complex system is very stimulating because of their great similitude with most of the processes, molecular interactions and changes on those molecular interactions occurring to develop functions by organs in living creatures. Up to now, only some aspects of the problem have been treated.

6.9.1 Elastic Models

An initial approach to a theoretical description of artificial muscles is by ignoring all the electrochemical reactions taking place during the actuation, as well as the simultaneous change in the material composition. In a separate approach we can follow the evolution of the electrochemical potential and current defining the consumed electrochemical energy. Simultaneously, through a mechanical test, we follow the evolution of the material length under constant stress, or the evolution of the produce force by the material for a constant length.[191–193] From the consumed electrical energy and from the produced mechanical work the efficiency of the process can be defined. The inclusion of the charge is attempted through diffusion processes required to compensate the charge of a supposed double layer.[167] These are just empirical approaches very suitable to describe some mechanical results if the elastic constants of both oxidised and neutral materials are quite close. Models to treat overlaps with polymeric or electrochemical models, theoretical predictions of new results or improvements in the efficiency of the devices do not yet exist.

6.9.2 Electrochemical Models

Most of the efforts devoted to describe the electrochemical responses (voltammetric, chronoamperometric, coulometric, impedances, quartz crystal microbalance) of the conducting polymers ignore the need to measure accurately the change of volume induced by the electrochemical reaction. This fact can reveal the unfitness of the traditional electrochemical models to treat three-dimensional electrodes at the molecular level. These entire models are founded on the hypotheses that whatever the structure of the electrode, even those extremely porous, presents a well-defined liquid/solid two-dimensional interface. In this context, all the electrochemical responses including structural information about shrinking, compaction, relaxation and swelling processes, never treated by the traditional electrochemistry, were considered by the literature as "anomalous" or "memory" effects. The term anomalous was used because maxima were found in chronoamperograms where after a very fast increase of the current when the potential was stepped, an exponential decrease was expected, as described for a Cottrell process, or because an important potential shift, related to that expected for the polymer oxidation, is observed on the voltammetric maxima. The term memory effect was used because they appear after polarisation, for constant times, at high cathodic overpotentials (working with polymers where the interchange of anions prevails during redox processes); or at anodic overpotentials for those polymers having prevailing cation interchange. Moreover, the evolution of the involved charge presents a clear asymmetry between oxidation and reduction processes. In this context, the electrochemistry of conducting polymers is modeled as traditional porous electrodes,[194–198] taking into account the viscoelasticity[199–202] or by considering a percolation model of conducting/non conducting regions,[203–208] or pure osmotic processes.[209] The charge asymmetry is explained by the presence of two different chemical subsystems at the oxidation state[111,210–213] or establishing three different regimes for the interchange of ions between the solution and the electrolyte.[214] When the anomalous electrochemical responses are modeled several (more than four in some cases) adjustable parameter are included, without any physical meaning and changing their values from one experimental result to the next.

6.9.3 Relaxation Models

Our group is developing a different approach: a relaxation model. This possibility looks suitable to overlap those macroscopic (electrochemistry, thermodynamics and elasticity) and microscopic (polymer science) theories required for a complete description of the electrochemistry of conducting polymers and their correlated properties. Similar problems emerged during the attempts to describe polarisation currents in commodity polymers, or magnetic and mechanical properties of materials. The most suitable solution always came from a relaxation model.

The involved electrochemical reaction has a kinetic constant defined, under suitable (nonanomalous) conditions, by the Buttler–Volmer equation. As every kinetic constant it includes a temporal dimension (s^{-1}), linked here to the relaxation time (τ). This relaxation time includes the structural molecular changes inducing volume variations: it is the time required to relax one mol of polymeric segments (having an average length λ) during the electrochemical reaction, generating the suitable free volume to lodge counterions coming from the solution. This is a statistical (Arrhenius) process: $\tau = \tau_0\, e^{(-\Delta H/RT)}$, occurring faster in those chains having a greater energy.

Here, ΔH includes all the energetic components (thermodynamics, electrochemistry, polymer science and elasticity). Our task (that of scientists) is to develop from this equation an explicit expression able to describe, or predict, those experimental facts related to each of the models (physical or chemical) there involved. The answer to the question of how far those descriptions and predictions can arrive is, from our point of view, that most of the living processes involving electric currents, chemical reactions, conformational movements in biopolymers, aqueous and ionic interchanges can be there included: most of the functions developed by living creatures.

Since 1994 some of my students have been focused on the development of relaxation equations able to describe and predict, without adjustable parameters, the anomalous electrochemical responses for chronoamperometric, voltammetric and coulometric results, as a function of the experimental variables: polarisation (compaction) overpotential, polarisation (compaction) time, relaxation and swelling (oxidation or reduction, according to the studied polymeric family) potential, electrolyte concentration, relaxation temperature, compaction temperature or solvent.[215–227] Also, a preliminary approach was started to the treatment of artificial muscles.[217] The attained results show the potentialities of the relaxation model to describe, without adjustable parameters, both, electrochemical and mechanical results.

Moreover, if the polymer science is there included, our relaxation model must describe the same facts that the free-volume theory does for amorphous polymers. The obtained equations predict how the relaxation processes and the free volume must change as a function of the polymer–solvent interactions, obtaining through cheap and easy electrochemical experiments, the Flory χ constant, the origin of all polymer science.[228] This fact opens a solid route to make explicit obtaining polymeric magnitudes through cheap electrochemical measurements and, in parallel, great possibilities to describe technological and biological interactions between electrochemistry and polymer science.

6.9.4 Molecular Dynamics Treatment

All the previous approaches can be considered as up to bottom procedures. Starting from macroscopic experiments we try to generate theoretical models able to explain at a molecular level, if possible, these results and, moreover, suitable to predict the results for new experiments. This is compatible with a different bottom to up approach by accounting all the molecular interactions taking place in a system where hundreds of oligomeric chains and counterions and thousands of solvent molecules are allowed to evolve following the dynamic laws to describe the molecular movements of a stationary, reduced or partially oxidised films, their density, the solvent content and distribution, *etc*. The molecular dynamic treatment can only be extended to stationary states of those systems constituted by a few hundred short oligomers (8 to 15 monomeric units, neutral or under different oxidation states), up to (the number is a function of the oxidation deep) a similar number of counterions and a few thousand solvent molecules. This is a good complementary approach to visualise, at the molecular level, the distribution of the polymer chains (following their conformational movements), solvent and counterions (Figure 6.16) for different stationary states of oxidation[229–231] allowing the theoretical calculation of densities, volume changes under oxidation, diffusion coefficients or evolution of the water (solvent) content in oxidised polymers. These are improved steps related to those previous molecular dynamics approaches in solid conducting polymers.[232–236] Improving computing capabilities and by establishing new algorithms the possibility of treating more complex systems (longer chains, branched or crosslinked chains containing some degraded; sp^3, carbons, counterion–coion couples) is open.

6.10 Final Remarks

Any of the above-considered devices are constituted by an electrochemical cell. They include: anode, cathode and electrolyte. This means that a triple-layer muscle or a polymeric smart window work, at the same time, as a battery. A

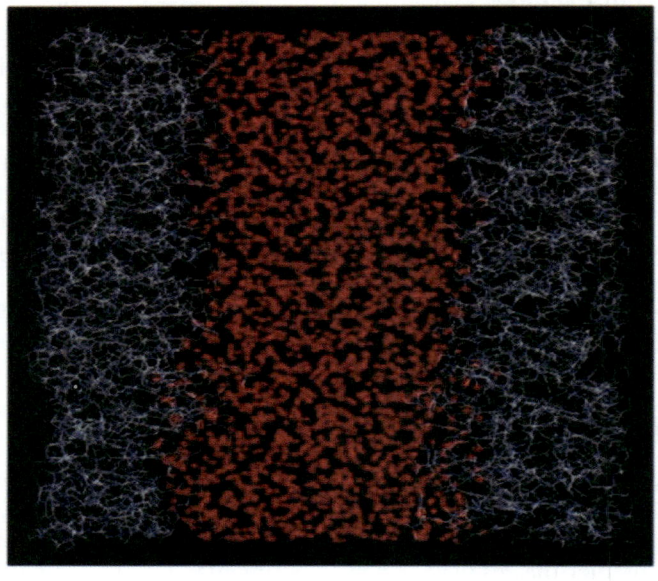

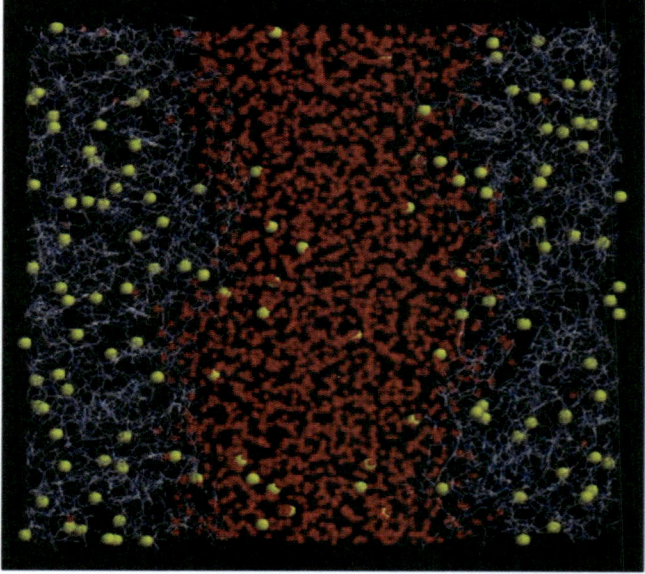

Figure 6.16 Snapshots obtained by molecular dynamics simulation of: (a) 128 reduced polypyrrole molecules (decamers), 64 on every of the lateral layers, and 2677 water molecules, the red beads correspond to the oxygen of the water molecules, and (b) 128 oxidised molecules bearing one positive charge per decamer, 128 chloride ions (yellow beads) and 2549 water molecules. Every chloride replaced a water molecule. (Reproduced from Ref. 229.)

good design of the devices must allow a partial recovering of the consumed charge. Moreover, for some specific applications it should be possible to design movable batteries and polymeric batteries with a final electrochromic film; both of them should show the state of charge.

Those researchers working on the development of all polymeric devices, where both electrodes are constituted by CPs bearing some electrochemical property (triple-layer muscles, polymeric electrochromic windows, polymeric batteries) are submitted to a great stress to produce solid devices. When the solvent is avoided the kinetics (movement rate of a muscle, rate of the absorbance change, or times required to charge or discharge) move from a diffusion-limiting step to a much slower conformational relaxation limiting step: just against the goal stated by most of the projects. Moreover, suitable solvents and electrolytes act as plastifiers favouring conformational movements and decreasing stiffness.

The development of new devices requires a solid theoretical background that does not yet exist. Those researchers working with batteries use models based on porous electrodes, ignoring the existence of changes of volume: swelling, shrinking, compaction and relaxation, which will control either adherence between electroactive electrodes, membrane and container, charge and discharge rates and lifetime of the device. For most engineers working with artificial muscles and actuators the key point is the mechanical behaviour and the properties of the material treated as a pure material and including the charge as a new parameter without any major consideration of the electrochemistry and polymer science. For electrochromic windows the chromophore concentration change is linked to a charge flowing through a material, closer to the electroluminescent processes than to any interchange of ions, swelling, shrinking or molecular compaction process. Thus, the development of new devices bearing such unparalleled properties as those above-described is outside the scope of scientists and engineers. It is necessary for the young scientist to undertake the tasks of putting together electrochemistry, polymer science, mechanics and optics developing a model able to describe any electrochemical property of these materials.

New materials and a greater control of the synthesis will be required to produce tailored materials: the best material does not exist, but the best for a well-defined application. The best material to construct a muscle will provide a great volume variation per unit of charge, so it is worse to construct a solid and rigid battery peeling apart the anode, cathode and membrane during the first discharge. The best test to check our ability to produce suitable materials should be the construction of membranes. If we are able to control the average pore diameter both under swelled or under compacted states, we should be able to produce smart membranes for any ion diameter, artificial glands for any application in medicine, agriculture, food, *etc.*, nervous interfaces, faster volumetric artificial muscles, smarter windows, and batteries for any kind of application. The integrated theoretical model will allow prediction of the behaviour of most of those devices, so close to that of life functions.

References

1. T.F. Otero, *Modern Aspects of Electrochemistry*, Vol. 33. (ed.) J. O-M. Bockris, R.E. White and B.E. Conway, Plenum, New York, 1999, pp. 307–434.
2. *Handbook of Conducting Polymers*, ed. T. Stotheim, R. Elsenhaumer and J. Reynolds. Marcel Dekker Inc., New York, 1998.
3. *Handbook of Organic Conductive Molecules and Polymers*, ed. Hari Singh Nalwa, John Wiley & Sons, Chichester, 1997.
4. P. Chandrasekhar, *Conducting Polymers, Fundamental and Applications*, Kluwer Academic Publishers, Boston, 1999.
5. G.G. Wallace, G.M. Spinks, L.A.P. Kane-Maguire and P.R. Teasdale, *Conductive Electroactive Polymers. Intelligent Materials*, CRC Press, Boca Ratón, 2003.
6. T.F. Otero and I. Boyano, *Chem Phys Chem*, 2003, **4**, 868.
7. T.F. Otero, I. Boyano, M.T. Cortés and G. Vázquez, *Electrochim. Acta*, 2004, **49**, 3719.
8. T.F. Otero, E. Angulo, J. Rodríguez and C. Santamaría. ES 2048086; T.F. Otero, J. Rodríguez and C. Santamaría ES 2062930, 1992.
9. T.F. Otero, E. Angulo, J. Rodríguez and C. Santamaría, *J. Electroanal. Chem.*, 1992, **341**, 369.
10. E. Smela, *Adv. Mater.*, 2003, **15**, 481.
11. M.T. Cortes and J.C. Moreno, *e-Polymers*, 2003, Art 41.
12. A.F.J. Diaz, *Electroanal. Chem.*, 1980, **111**, 111.
13. F.B. Kaufman, Polymer-modified Electrodes, *J. Electroanal. Chem.*, 1980, **36**, 422.
14. A.F. Díaz, *J. Electroanal. Chem.*, 1981, **115**, 129.
15. R.J. Mortimer, A.L. Dyer and J.R. Reynolds, *Displays*, 2006, **27**, 2.
16. L.W. Shaklette, N.S. Murthy and R.H. Baughman, *Mol. Cryst. Liq. Cryst.*, 2001, **121**, 1985.
17. P. Burgmayer and R.W. Murray, *J. Am. Chem. Soc.*, 1982, **104**, 6139.
18. *Intrinsically Conducting Polymers: An Emerging Technology*, ed. M. Aldissi, Kluwer Academic Publishers, NATO ASI Series, **246**, 1993.
19. C. Weidlich, K.M. Mangold and K. Juttner, *Electrochim. Acta.*, 2005, **50**, 1547.
20. C. Dusemund and G. Schwitzgebel, *Ber. Bunsen-Ges.-Phys. Chem. Chem. Phys.*, 1991, **95**, 1543.
21. R.M. Torresi and S.L.D. Maranhao, *J. Electrochem. Soc.*, 1999, **146**, 4179.
22. J.M. Sansiñena, Doctoral Thesis. University of the Basque Country (UPV/EHU). 1998.
23. T.F. Otero and J.M. Sansiñena, *J. Bioelectrochem. Bioenerg.*, 1995, **38**, 411.
24. T.F. Otero and M.T. Cortés, *Sens. Actuators B-Chem.*, 2003, **96**, 152.
25. T.F. Otero, *Electrochemistry and Conducting Polymers: An Emerging and Accesible Technological Revolution*, in *New Organic Materials*, ed. N. Martin and C. Seoane, University Complutense, Madrid, 1994, pp. 205–237.

26. A. Katchalsky and M. Zwick, *J. Polym. Sci.*, 1955, **16**, 221.
27. A. Katchalsky and H. Eisenberg, *Nature*, 1950, **166**, 267.
28. W. Kuhn, B. Hargitay, A. Katchalsky and H. Eisenberg, *Nature*, 1950, **165**, 514.
29. A. Katchalsky, *Experientia.*, 1949, **5**, 319.
30. Y. Osada and Y. Saito, *Makromol. Chem.-Macromol. Chem. Phys.*, 1975, **176**, 2761.
31. T. Tanaka and D.J. Fillmore, *J. Chem. Phys.*, 1979, **70**, 1214.
32. T. Tanaka, *Science*, 1982, **218**, 467.
33. Y. Osada and M. Hasebe, *Chem. Lett.*, 1985, 1285.
34. *Handbook of Conducting Polymers*, ed. T. Stotheim, R. Elsenhaumer and J. Reynolds, Marcel Dekker Inc., New York, 1998, pp. 1015–1028.
35. *Polymer Sensors and Actuators*, ed. D. de Rossi and Y. Osada, Springer-Verlag, Berlin, 2000, pp. 295–323.
36. *Structural Biological Materials, Design and Structure-Properties Relationships*, ed. M. Elices and R.W. Cahn, Pergamon Materials Series, Amsterdam, 2000, pp. 187–220.
37. P. Burgmayer and R.W. Murray, *J. Electrochem. Soc.*, 1983, **130**, C117.
38. R.H. Baughman, L.W. Shacklette, R.L. Elsembaumer, E. Plitcha and C. Becht, *Conducting Polymer Electromechanical Actuator*, In *Conjugated Polymer Materials. Opportunities in Electronics, Optoelectronics and Molecular Electronics*, ed. J.L. Brédas and R.R. Chance, Kluwer Academic Publishers, Netherlands, 1990.
39. R.H. Baughman and L.W. Shacklette, *Science and Application of Conducting Polymers*, ed. W.R. Salaneck, D.T. Clark and E.J. Samuelson, Adam Hilger, Bristol, 1991, p. 47.
40. Q.B. Pei and O. Inganas, *J. Phys. Chem.*, 1992, **96**, 10507.
41. P. Burgmayer and R.W. Murray, *J. Phys. Chem.*, 1984, **88**, 2515.
42. N. Mermilliod, L. Zuppiroli and B. Francois, *Mol. Crys. Liq. Crys.*, 1982, **86**, 1971.
43. B. Francois, N. Mermilliod and L. Zuppiroli, *Synth. Met.*, 1981, **4**, 131.
44. K. Okabayashi, F. Goto, K. Abe and T. Yoshida, *Synth. Met.*, 1987, **18**, 365.
45. K. Okabayashi, F. Goto, K. Abe and T. Yoshida, *J. Electrochem. Soc.*, 1989, **136**, 1986.
46. M. Slama and J. Tanguy, *Synth. Met.*, 1989, **28**, C171.
47. Q.B. Pei and O.J. Inganas, *Phys. Chem.*, 1993, **97**, 6034.
48. Q.B. Pei and O.J. Inganas, *Phys. Chem.*, 1992, **96**, 10507.
49. S. Shimoda and E. Smela, *Electrochim. Acta.*, 1998, **44**, 219.
50. E. Smela and N. Gadegaard, *J. Phys. Chem. B*, 2001, **105**, 9395.
51. E. Smela and N. Gadegaard, *Adv. Mater.*, 1999, **11**, 953.
52. L. Lizarraga, E.M. Andrade and F.V. Molina, *J. Electroanal. Chem.*, 2004, **561**, 127.
53. E.M. Andrade, F.V. Molina, M.I. Florit and D. Posadas, *Electrochem. Solid State*, 2000, **3**, 504.
54. X.W. Chen and O. Inganas, *Synth. Met.*, 1995, **74**, 159.

55. C. Barbero and R. Kotz, *J. Electrochem. Soc.*, 1994, **141**, 859.
56. M.F. Suárez and R.G. Compton, *J. Electroanal. Chem.*, 1999, **462**, 211.
57. D.W. Hurry, in *Biomolecular Materials by Design*, ed. M. Alper, H. Bayley, D. Kaplan and M. Navia, Materials Research Society, Vol. 330, Pittsburgh, 1994.
58. L. Stryer, *Bioquímica*, Reverté, Barcelona, 1988.
59. M. Onoda, T. Okamoto, K. Tada and H. Nakayama, *Jpn. J. Appl. Phys. Part 2*, 1999, **38**, L1070.
60. S. Shakuda, S. Morita, T. Kawai and K. Yoshino, *Jpn. J. Appl. Phys. Part 1*, 1993, **32**, 5143.
61. M. Onoda, K. Tada and H. Nakayama, *Synth. Met.*, 1999, **102**, 1321.
62. T. Okamoto, Y. Kato, K. Tada and M. Onoda, *Thin Solid Films*, 2001, **393**, 383.
63. H.L. Wang, J.B. Gao, J.M. Sansiñena and P. McCarthy, *Chem. Mater.*, 2002, **14**, 2546.
64. J.M. Sansiñena, J.B. Gao and H.L. Wang, *Adv. Funct. Mater.*, 2003, **13**, 703.
65. M. Onada, H. Shonaka and K. Tada, *Curr. Appl. Phys.*, 2005, **5**, 194.
66. T. Okamoto, K. Tada and M. Onada, *Jpn. J. Appl. Phys.* Part 1, 2000, 39, 2854; Part 2, 1999, **38**, L1070.
67. M. Onada, Y. Kato, H. Shonaka and K. Tada, *Elect. Eng. Jpn.*, 2004, **149**, 7.
68. M. Onoda and K. Tada, *IEICE Trans. Elect.*, 2004, **E87C**, 128.
69. G.Y. Han and G.O. Shi, *J. Electroanal. Chem.*, 2004, **569**, 169.
70. G.Y. Han and G.O. Shi, *Sens. Actuators B-Chem.*, 2004, **99**, 525.
71. W. Takashima, S.S. Pandey and K. Kaneto, *Sens. Actuators B-Chem.*, 2003, **89**, 48.
72. W. Takashima, S.S. Pandey and K. Kaneto, *Synth. Met.*, 2003, **135**, 61.
73. W.G. Li, C.L. Johnson and H.L. Wang, *Polymer*, 2004, **45**, 4769.
74. H. Okuzaki and T. Hattori, *Synth. Met.*, 2003, **135**, 45.
75. Q.B. Pei and O. Inganas, *Adv. Mater.*, 1992, **4**, 277.
76. J. Madden, I.W. Hunter and R.J. Gilbert, *Gastroenterology*, 2002, **122**, S1111.
77. A. Mazzoldi, C. Degl'Innocenti, M. Michelucci and D. De Rossi, *Mater. Sci. Eng. C-Bio. S.*, 1998, **6**, 65.
78. A. Mazzoldi, F. Carpi and D. DeRossi, *Ann. Chim. Sci. Des Mater.*, 2004, **29**, 55.
79. A. DellaSanta, D. DeRossi and A. Mazzoldi, *Smart Mater. Struct.*, 1997, **6**, 23.
80. A. DellaSanta, D. DeRossi and A. Mazzoldi, *Synth. Met.*, 1997, **90**, 93.
81. A. Della Santa, A. Mazzoldi, C. Tonci and D. De Rossi, *Mater. Sci. Eng. C-Bio. S.*, 1997, **5**, 101.
82. K. Kaneto, Y. Sonoda and W. Takashima, *Jpn. J. Appl. Phys. Part 1*, 2000, **39**, 5918.
83. D.J. Irvin, S.H. Goods and L.L. Whinnery, *Chem. Mater.*, 2001, **13**, 1143.

84. G.M. Spinks, L. Liu, G.G. Wallace and D.Z. Zhou, *Adv. Funct. Mater.*, 2002, **12**, 437.
85. W. Lu, A.G. Fadeev, B.H. Qi, E. Smela, B.R. Mattes, J. Ding, G.M. Spinks, J. Mazurkiewicz, D.Z. Zhou, G.G. Wallace, D.R. MacFarlane, S.A. Forsyth and M. Forsyth, *Science*, 2002, **297**, 983.
86. L. Bay, K. West, P. Sommer-Larsen, S. Skaarup and M. Benslimane, *Adv. Mater.*, 2003, **15**, 310.
87. S. Hara, T. Zama, S. Sewa, W. Takashima and K. Kaneto, *Chem. Lett.*, 2003, **32**, 576.
88. S.S. Pandey, W. Takashima and K. Kaneto, *Thin Solid Films*, 2003, **438**, 206.
89. G.M. Spinks, D.Z. Zhou, L. Liu and G.G. Wallace, *Smart Mater. Struct.*, 2003, **12**, 468.
90. S. Hara, T. Zama, W. Takashima and K. Kaneto, *Polymer. J.*, 2004, **36**, 151.
91. S. Hara, T. Zama, W. Takashima and K. Kaneto, *J. Mater. Chem.*, 2004, **14**, 1516–1517.
92. S.S. Pandey, W. Takashima and K. Kaneto, *Sens. Actuators B-Chem*, 2004, **102**, 142.
93. W. Lu, E. Smela, P. Adams, G. Zuccarello and B.R. Mattes, *Chem. Mater.*, 2004, **16**, 1615.
94. B.H. Qi, W. Lu and B.R. Mattes, *J. Phys. Chem. B*, 2004, **108**, 6222.
95. S. Hara, T. Zama, W. Takashima and K. Kaneto, *Polymer. J.*, 2004, **36**, 933.
96. A. Tanaka, W. Takashima and K. Kaneto, *Chem. Lett.*, 2004, **33**, 1470.
97. T. Zama, S. Hara, W. Takashima and K. Kaneto, *B. Chem. Soc. Jpn.*, 2004, **77**, 1425.
98. G.M. Spinks, B.B. Xi, D.Z. Zhou, V.T. Truong and G.G. Wallace, *Synth. Met.*, 2004, **140**, 273.
99. E.A. Moschou, S.F. Peteu, L.G. Bachas, M.J. Madou and S. Daunert, *Chem. Mater.*, 2004, **16**, 2499.
100. W. Lu, A.G. Fadeev, B.M. Qi, E. Smala, B.R. Mattes, J. Ding, G.M. Spinks, J. Mazurkiewics, D.Z. Zhou, G.G. Wallace, D.R. MacFarlane, S.A. Forsyth and M. Forsyth, *Science*, 2002, **297**, 983.
101. K.J. Lin, S.J. Fu, C.Y. Cheng, W.H. Chen and H.M. Kao, *Angew. Chem. Int. Ed. Int.*, 2004, **43**, 4186.
102. J. Joo and M. Pyo, *Electrochem. Solid State*, 2003, **6**, E27.
103. M. Pyo, C.C. Bohn, E. Smela, J.R. Reynolds and A.B. Brennan, *Chem. Mater.*, 2003, **15**, 916.
104. J.Y. Quyang and Y.F. Li, *Polymer*, 1997, **38**, 3997.
105. J. Ding, L. Liu, G.M. Spinks, D.Z. Zhou, G.G. Wallace and J. Gillespie, *Synth. Met.*, 2003, **138**, 391.
106. J. Ding, D.Z. Zhou, G. Spinks, G. Wallace, S. Forsyth, M. Forsyth and D. MacFarlane, *Chem. Mater.*, 2003, **15**, 2392.
107. S. Hara, T. Zama, S. Sewa, W. Takashima and K. Kaneto, *Chem. Lett.*, 2003, **32**, 800.

108. S. Hara, T. Zama, W. Takashima and K. Kaneto, *Synth. Met.*, 2004, **146**, 47.
109. S. Hara, T. Zama, A. Ametani, W. Takashima and K. Kaneto, *J. Mater. Chem.*, 2004, **14**, 2724.
110. P.G.A. Madden, J.D.W. Madden, P.A. Anquetil, N.A. Vandesteeg and I.W. Hunter, *J Oceanic Eng.*, 2004, **29**, 696.
111. T.F. Otero and M.J. Ariza, *J. Phys. Chem. B*, 2003, **107**, 13954.
112. T.F. Otero, O. Sara Costa, M.J. Ariza and M. Marquez, *J. Mater. Chem.*, 2005, **15**, 1662.
113. R. Verdú, J. Morales, A. Fernández-Romero, M.T. Cortés, T.F. Otero and L. Weruaga, Mechanical Characterization of Artificial Muscles With Computer Vision, In *Electroactive Polymer Actuators and Devices*, Vol. 4695, ed. Y. Bar-Cohen, SPIE, 2002, p. 253.
114. R. Verdú, R. Berenguer, J. Morales, G. Vázquez, T.F. Otero and L. Weruaga, 3D Mechanical characterization of artificial muscles with stereoscopic computer vision and active contours, *Proceedings of International Conference on Image Processing*, Barcelona, Spain, 2003.
115. T.F. Otero, M.T. Cortés and I. Boyano, Macroscopic devices and complex movements developed with artificial muscles, In *Electroactive Polymer Actuators and Devices*, SPIE, Vol. 4695, ed. Y. Bar-Cohen, 2002, pp. 395–402.
116. T.F. Otero and M.J. Broschart, *Appl. Electrochem.*, 2006, **36**, 205–214.
117. E. Smela, O. Inganas, Q. Pei and I. Lundstrom, *Adv. Mater.*, 1993, **5**, 630.
118. E. Smela, O. Inganas and I. Lundstrom, *Science*, 1995, **268**, 1735.
119. D. Pede, E. Smela, T. Johansson, M. Johansson and O. Inganas, *Adv. Mater.*, 1998, **10**, 233.
120. E.W.H. Jager, E. Smela, O. Inganas and I. Lundstrom, *Synth. Met.*, 1999, **102**, 1309.
121. E.W.H. Jager, E. Smela and O. Inganas, *Sens. Actuators B-Chem.*, 1999, **56**, 73.
122. E. Smela, M. Kallenbach and J. Holdenried, *J. Microelectromech. S.*, 1999, **8**, 373.
123. E. Smela, *J. Micromech. Microeng.*, 1999, **9**, 1.
124. E. Smela, *Adv. Mater.*, 1999, **11**, 1343.
125. M. Roemer, T. Kurzenknabe, E. Oesterschulze and N. Nicoloso, *Anal. Bioanal. Chem.*, 2002, **373**, 754.
126. L.M. Low, S. Seetharaman, K.Q. He and M.J. Madou, *Sens. Actuators B-Chem.*, 2000, **67**, 149.
127. E.W.H. Jager, *Science*, 2000, **290**, 1540.
128. E.W.H. Jager, *Science*, 2000, **288**, 2335.
129. E.W.H. Jager, O. Inganas and I. Lundstrom, *Science*, 2000, **288**, 2335.
130. E.W.H. Jager, E. Smela and O. Inganas, *Science*, 2000, **290**, 1540.
131. E.W.H. Jager, O. Inganas and I. Lundstrom, *Adv. Mater.*, 2001, **13**, 76.
132. C. Immerstrand, K. Holmgren-Peterson, K.E. Magnussen, E. Jager, M. Krogh, M. Skoglund, A. Selbing and O. Inganas, *Mater. Res. Bull.*, 2002, **27**, 461.

133. T.F. Otero and J.M. Sansiñena, *Bioelectrochem. Bienerg.*, 1997, **42**, 117.
134. T.F. Otero and M.T. Cortes, *Chem. Commun*, 2004, 284.
135. T.F. Otero and J.M. Sansiñena, *Adv. Mater.*, 1998, **10**, 491.
136. T.F. Otero, *Electrochemomechanical Devices based on Conducting Polymers*, in *Polymer Sensors and Actuators*, ed. D. de Rossi and Y. Osada, Springer-Verlag, Oxford, 2000, pp. 295–323.
137. T.F. Otero and M.T. Cortés, *Sens. Actuators B-Chem.*, 2003, **96**, 152.
138. T.F. Otero, J.M. Sansiñena, H. Grande and J. Rodriguez, *Portugalæ Electrochim. Acta*, 1995, **13**, 499.
139. M.T. Cortés, T.F. Otero, A. Vázquez and I. Boyano, *Electrochim. Acta*, 2001, **19**, 263.
140. T.F. Otero, S. Villanueva, M.T. Cortés, S.A. Cheng, A. Vázquez, I. Boyano, D. Alonso and R. Camargo, *Synth. Met.*, 2001, **119**, 419.
141. T.F. Otero and M.T. Cortés *Electrochemical Characterization and Control of Triple-layer Muscles*, in *Smart Structures and Materials 2000: Electroactive Polymer Actuators and Devices, Proc. SPIE*, ed. Y. Bar-Cohen, 2000, **3987**, 252–260; and 2001, **4329**, 93–100.
142. T.F. Otero and M.T. Cortés, *Adv. Mater.*, 2003, **15**, 279.
143. T.F. Otero, M.T. Cortés, I. Boyano and G. Vázquez, *Non-stoichiometry and Tactile Muscles with Conducting Polymers*, in *Smart Structures and Materials 2000: Electroactive Polymer Actuators and Devices, Proc. SPIE*, ed. Y. Bar-Cohen, 2004, **5385**, 425–432.
144. T.F. Otero, J.J. López Cascales and G. Vázquez Arenas, Mechanical Characterization of Free-Standing Polypyrrole Film, *Mater. Sci. Eng. C*, 2007, **290**, 241–249.
145. J.D.W. Madden, P.G.A. Madden, P.A. Anquetil and I.W. Hunter, *Mater. Res. Soc. Symp. Proc.*, 2002, 698.
146. K. Kaneto, M. Kaneko, Y. Min and A.G. MacDiarmid, *Synth. Met.*, 1995, **71**, 2211.
147. J.M. Sansiñena, V. Olazabal, T.F. Otero, C.N. Polo da Fonseca and Macro-A. De Paoli, *Chem. Commun.*, 1997, 2217–2218.
148. H. Okuzaki, *Kobunshi Ronbunshu*, 2005, **62**, 362.
149. H. Yan, K. Tomizawz, H. Ohno and N. Toshima, *Macromol. Mater. Eng.*, 2003, **288**, 578.
150. D.Z. Zhou, G.M. Spinks, G.G. Wallace, C. Tiyapiboonchaiya, D.R. MacFarlane, M. Forsyth and J.Z. Sun, *Electrochim. Acta.*, 2003, **48**, 2355.
151. J.D. Madden, R.A. Cush, T.S. Kanigan, C.J. Brenan and I.W. Hunter, *Synth. Met.*, 1999, **105**, 61.
152. H. Okuzaki, *J. Intell. Mater. Syst. Struct.*, 1999, **10**, 465.
153. T.W. Lewis, L.A.P. Kane-Maguire, A.S. Hutchison, G.M. Spinks and G.G. Wallace, *Synth. Met.*, 1999, **102**, 1317.
154. F. Vidal, C. Plesse, D. Teyssie and C. Chevrot, *Synth. Met.*, 2004, **142**, 287.
155. W. Lu, A.G. Fadeev, B.H. Qi, E. Smela, B.R. Mattes, J. Ding, G.M. Spinks, J. Mazurkiewicz, D.Z. Zhou, G.G. Wallace, D.R. MacFarlane, S.A. Forsyth and M. Forsyth, *Science*, 2002, **297**, 983.

156. G.G. Wallace, *Abst. Am. Chem. Soc.*, 2003, **226**, U640.
157. J.M. Pringle, M. Forsyth, D.R. MacFarlane, K. Wagner, S.B. Hall and D.L. Officer, *Polymer*, 2005, **46**, 2047.
158. H. Randriamahazaka, C. Plesse, D. Teyssie and C. Chevrot, *Electrochim. Acta.*, 2005, **50**, 1515.
159. Y.K. Koo, B.H. Kim, D.H. Park and J. Joo, *Mol. Cryst. Liq. Cryst.*, 2004, **425**, 333.
160. P. Danielsson, J. Bobacka and A. Ivaska, *J. Solid State Electron.*, 2004, **8**, 809.
161. M.D. Bennett and D.J. Leo, *Sens. Actuators A-Phys.*, 2004, **115**, 79.
162. F. Vidal, C. Plesse, D. Teyssie and C. Chevrot, *Synth. Met.*, 2004, **142**, 287.
163. G.M. Spinks, B.B. Xi, D.Z. Zhou, V.T. Truong and G.G. Wallace, *Synth. Met.*, 2004, **140**, 273.
164. D.Z. Zhou, G.M. Spinks, G.G. Wallace, C. Tiyapiboonchaiya, D.R. MacFarlane, M. Forsyth and J.Z. Sun, *Electrochim. Acta.*, 2003, **48**, 2355.
165. J. Ding, D.Z. Zhou, G. Spinks, G. Wallace, S. Forsyth, M. Forsyth and D. MacFarlane, *Chem. Mater.*, 2003, **15**, 2392.
166. W. Lu, A.G. Fadeev, B. Qi and B.R. Mattes, *Synth. Met.*, 2003, **135**, 139.
167. P.G.A. Madden, J.D.W. Madden, P.A. Anquetil, N.A. Vandesteeg and I.W. Hunter, *IEEE J. Oceanic Eng.*, 2004, **29**, 696.
168. J.D.W. Madden, N.A. Vandesteeg, P.A. Anquetil, P.G.A. Madden, A. Takshi, R.Z. Pytel, S.R. Lafontaine, P.A. Wieringa and I.W. Hunter, *IEEE J. Oceanic Eng.*, 2004, **29**, 706.
169. S.A. Jenekhe and D.I. Kiserow, *ACS Symp. Ser.*, 2005, **888**, 2–15.
170. A.A. Argun, P.H. Aubert, B.C. Thompson, I. Schwendeman, C.L. Gaupp, J. Hwang, N.J. Pinto, D.B. Tanner, A.G. MacDiarmid and J.R. Reynolds, *Chem. Mater.*, 2004, **16**, 4401–4412.
171. L. Groenendaal, G. Zotti, P.H. Aubert, S.M. Waybright and J.R. Reynolds, *Adv. Mater.*, 2003, **15**, 855–879.
172. D.R. Rosseinsky and R.J. Mortimer, *Adv. Mater.*, 2001, **13**, 783.
173. P. Chandrasekhar, B.J. Zay, D. Ross, T. McQueeney, G.C. Birur, T. Swanson, L. Kauder and D. Douglas, *ACS Symp. Ser.*, 2005, **888**, 66–79.
174. P. Chandrasekhar, *Conducting Polymers, Fundamental and Applications*, Kluwer Academic Publishers, Boston, 1999.
175. C. Ehrenbeck and K. Juttner, *Electrochim. Acta.*, 1996, **41**, 1815.
176. K. Juttner and C. Ehrenbeck, *J. Solid State Electron.*, 1998, **2**, 60.
177. P.N. Bartlett, P.R. Birkin, M.A. Ghanem and C.S. Toh, *J. Mater. Chem.*, 2001, **11**, 849.
178. X.W. Chen, K.Z. Xing and O. Inganas, *Chem. Mater.*, 1996, **8**, 2439.
179. X.W. Chen and O. Inganas, *Synth. Met.*, 1995, **74**, 159.
180. V.M. Schmidt, D. Tegtmeyer and J. Heitbaum, *J. Electroanal. Chem.*, 1995, **385**, 149.
181. M. Ryan, E. Bowden and J. Chambers, *Anal. Chem.*, 1994, **66**, R360.

182. H. Yang and J. Kwak, *J. Phys. Chem. B*, 1997, **101**, 774.
183. B. Piro, T.A. Nguyen, J. Tanguy and M.C. Pham, *J. Electroanal. Chem.*, 2001, **499**, 103.
184. M. Kendig and M. Hon, *Corrosion*, 2004, **60**, 1024–1030.
185. Y.L. Li, K.G. Neoh and E.T. Kang, *J. Biomed. Mater. Res. Part A*, 2005, **73**, 171.
186. B. Massoumi and A. Entezami, *J. Bioact. Compat. Polym.*, 2002, **17**, 51.
187. B. Massoumi and A. Entezami, *Eur. Polym. J.*, 2001, **37**, 1015.
188. M. Pyo and J.R. Reynolds, *Chem. Mater.*, 1996, **8**, 128.
189. M. Pyo, G. Maeder, R.T. Kennedy and J.R. Reynolds, *J. Electroanal. Chem.*, 1994, **368**, 329.
190. H. Shinohara, M. Aizawa and H. Shirakawa, *Chem. Lett.*, 1985, 179–182.
191. F. Carpi and D. De Rossi, *Mater. Sci. Eng. C-Biomimetic Supramol. Syst.*, 2004, **24**, 555.
192. G.M. Spinks, T.E. Campbell and G.G. Wallace, *Smart Mater. Struct.*, 2005, **14**, 406.
193. A. Della Santa, A. Mazzoldi, C. Tonci and D. De Rossi, *Mater. Sci. Eng. C-Biomimetic Mater. Sens. Syst.*, 1997, **5**, 101.
194. J. Bisquert, G. Garcia-Belmonte, F. Fabregat-Santiago and P.R. Bueno, *J. Electroanal. Chem.*, 1999, **475**, 152.
195. J. Bisquert, G. Garcia-Belmonte and A. Pitarch, *ChemPhysChem.*, 2003, **4**, 287.
196. G. Garcia-Belmonte and J. Bisquert, *Electrochim. Acta.*, 2002, **47**, 4263.
197. G. Garcia-Belmonte, J. Bisquert, E.C. Pereira and F. Fabregat-Santiago, *J. Electroanal. Chem.*, 2001, **508**, 48.
198. J. Bisquert, G.G. Belmonte, F.F. Santiago, N.S. Ferriols, M. Yamashita and E.C. Pereira, *Electrochem. Commun.*, 2000, **2**, 601.
199. A.R. Hillman, I. Efimov and M. Skompska, *Faraday Discuss.*, 2002, **121**, 423.
200. A.R. Hillman, I. Efimov, M.J. Swann and S. Bruckenstein, *J. Phys. Chem.*, 1991, **95**, 3271.
201. M. Skompska, A. Jackson and A.R. Hillman, *Phys. Chem. Chem. Phys.*, 2000, **2**, 4748.
202. H.L. Bandey, A.R. Hillman, M.J. Brown and S.J. Martin, *Faraday Discuss.*, 1997, **107**, 105.
203. K. Aoki and M. Kawase, *J. Electroanal. Chem.*, 1994, **377**, 125.
204. K. Aoki, *J. Electroanal. Chem.*, 1994, **373**, 67.
205. M. Pyo, G. Maeder, R.T. Kennedy and J.R. Reynolds, *J. Electroanal. Chem.*, 1994, **368**, 329.
206. K. Aoki, *J. Electroanal. Chem.*, 1993, **348**, 273–282.
207. K. Aoki, *J. Electroanal. Chem.*, 1992, **334**, 279–290.
208. K. Aoki, *J. Electroanal. Chem.*, 1991, **310**, 1–12.
209. L. Bay, T. Jacobsen. S. Skaarup and K. West, *J. Phys. Chem. B*, 2001, **105**, 8492–8497.
210. M.A. Vorotyntsev and J. Heinze, *Electrochim. Acta.*, 2001, **46**, 3309–3324.

211. M.A. Vorotyntsev, E. Vieil and J. Heinze, *J. Electroanal. Chem.*, 1998, **450**, 121–141.
212. M.A. Vorotyntsev, E. Vieil and J. Heinze, *Electrochim. Acta.*, 1996, **41**, 1913–1920.
213. T.F. Otero and M.J. Ariza, *Colloid Surf. A*, 2005, **270–271**, 226.
214. M.A. Vorotyntsev, E. Vieil and J. Heinze, *J. Electrochem.*, 1995, **31**, 1027–1035.
215. T.F. Otero, H. Grande and J. Rodríguez, *J. Electroanal. Chem.*, 1995, **39b4**, 211.
216. T.F. Otero, H. Grande and J. Rodríguez, *Electrochim. Acta*, 1996, **41**, 1863–1869.
217. T.F. Otero, H. Grande and J. Rodríguez, *J. Phys. Org. Chem.*, 1996, **9**, 381.
218. T.F. Otero and H. Grande, *J. Electroanal. Chem.*, 1996, **414**, 171.
219. T.F. Otero, H. Grande and J. Rodríguez, *Synth. Met.*, 1996, **83**, 205.
220. T.F. Otero, H. Grande and J. Rodríguez, *J. Phys. Chem.*, 1997, **101**, 3688.
221. T.F. Otero, H. Grande and J. Rodríguez, *J. Phys. Chem. B*, 1997, **101**, 8525.
222. H. Grande and T.F. Otero, *J. Phys. Chem. B*, 1998, **102**, 7535.
223. T.F. Otero, I. Cantero and H. Grande, *Electrochim. Acta.* 1999, **44**, 2053.
224. T.F. Otero and I. Boyano, *J. Phys. Chem. B.*, 2003, **107**, 6730.
225. T.F. Otero and I. Boyano, *J. Phys. Chem. B*, 2003, **107**, 4269.
226. T.F. Otero, M.T. Cortés and I. Boyano, *J. Electroanal. Chem.*, 2004, **562**, 135.
227. T.F. Otero and J. Padilla, *J. Electroanal. Chem.*, 2004, **561**, 167.
228. H. Grande, T.F. Otero and I. Cantero, *J. Non-Cryst. Solids*, 1998, **235–237**, 619–622.
229. J.J. Lopez Cascales, A.J. Fernández and T.F. Otero, *J. Phys. Chem. B*, 2003, **107**, 9339–9343.
230. J.J. Lopez Cascales and T.F. Otero, *J. Chem. Phys.*, 2004, **120**, 1951–1057.
231. J.J. López Cascales and T.F. Otero, *Macromol. Theory Simul.*, 2005, **14**, 40–48.
232. A.C. Kolbert, N.S. Sariciftci, K.U. Gaudl, P. Bauerle and M. Mehring, *J. Am. Chem. Soc.*, 1991, **113**, 8243–8246.
233. M. Bee, D. Djurado, J. Combet, M. Telling, P. Rannou, A. Pron and J.P. Travers, *Phys. B*, 2001, **301**, 49–53.
234. A.J. Dianoux, G.R. Kneller, J.L. Sauvajol and J.C. Smith, *J. Chem. Phys.*, 1994, **101**, 634–644.
235. G.M. Lombardo, G. Piccitto and R. Pucci, *Philos. Mag. B*, 1994, **69**, 925–930.
236. J. Corish and D.A. Morton-Blake, *Eff. Defects Solids*, 2002, **157**, 805–821.

CHAPTER 7
Electrochemically Controllable Polyacrylonitrile-Derived Artificial Muscle as an Intelligent Material

KWANG J. KIM* AND KIYOUNG CHOE

Active Materials and Processing Laboratory, Mechanical Engineering Department, University of Nevada, Reno, NV 89557, USA

7.1 Polyacrylonitrile in General

Polyacrylonitrile (PAN) fibres in an active form (PAN or PAN gel modified by annealing/crosslinking and partial hydrolysis) are known to elongate and contract when immersed in caustic and acidic solutions, respectively.[1–12] As can be seen in Figure 7.1, the change in length for these pH-activated fibres is typically greater than 100% contraction/expansion of PAN. As an artificial muscle, PAN has many advantages, such as high force-to-weight ratio, silent operation, and muscular behaviour motion. In order for the PAN to be used for any further actuation applications, decisive parameters of PAN gel fibre need to be documented. These include the mechanical properties and force-generation characteristics for both chemically driven and electrically driven systems.

The basic unit of commercially available PAN fibre (Mitsubishi Rayon Co., Japan) starts with a "single strand" that could be handled properly from the engineering standpoint of interest. One PAN strand consisted of approximately two thousand filaments. The typical diameter of each filament is approximately 10 μm in the raw state and 30 μm in the fully elongated state (gel).

The advantage of PAN fibre among other electrolyte gels is that PAN fibre gel has good mechanical properties, which can compare to those of biological muscles. The large volume change of PAN fibre gel also allows the reduction in

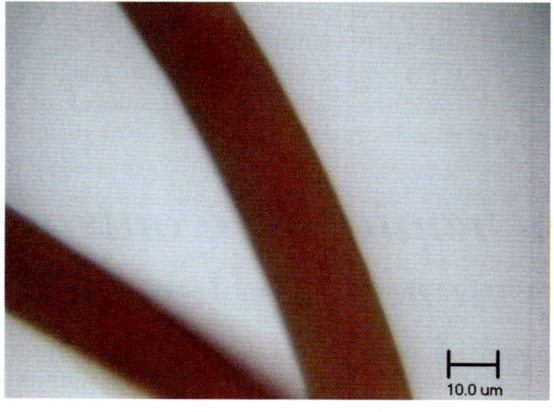

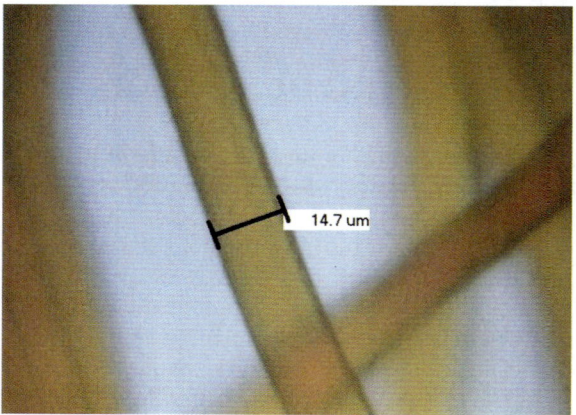

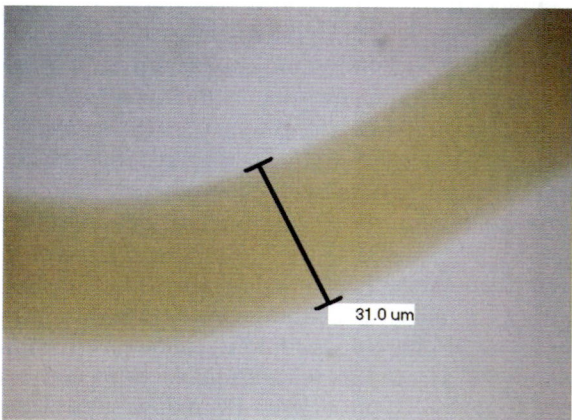

Figure 7.1 PAN fibres in different states. *Top*: oxidised PANs (prior to activation); *Middle*: at low pH contracted PAN (1 N HCl); *Bottom*: at high pH expanded PAN (1 N LiOH).

size of the gel, which is an important factor in determining response time. PAN fibre is also readily available from industrial sources. This results in the need to intensify the study of PAN fibre and make the material quite promising as high-performance artificial muscle. For most cases the shrinking and swelling characteristics of polymer gels are dependent on radical groups in the molecular structure. For example, pH-sensitive poly(2-hydroxyethyl methacrylate-co-acrylic acid) gel swells in high pH solution and shrinks in low pH solution and has a carboxyl group. On the other hand, polymer gels including basic groups show opposite pH-sensitive behaviour. Activated PAN fibres have crosslinked nitrogen-containing ring structures and carboxylic acid groups. The crosslinked structures give PAN mechanical properties that are much stronger than other types of ionic gels, while the carboxyl groups make *oxy-*PAN fibres shrink in acidic solutions and swell in basic solutions. It can be seen that PAN fibres are able to convert chemical energy directly into mechanical motion. Following the early work by Katchalsky and Zwick[13] and Hickey and Peppas,[14] the possible explanation about contraction and expansion of modified PAN fibre is illustrated in Figure 7.2. In an acidic solution, the carboxyl groups bond with hydrogen ions since the PAN fibre gels have a compact structure exuding water out of the polymer network. The response time of the swelling/shrinking process typically depends on the diffusion process of such ions and the related solvent water. Based on ion diffusion theory, the response time of swelling may be predicted as proportional to the square of the gel-fibre diameter. The surface/volume ratio also affects the response time. Note that

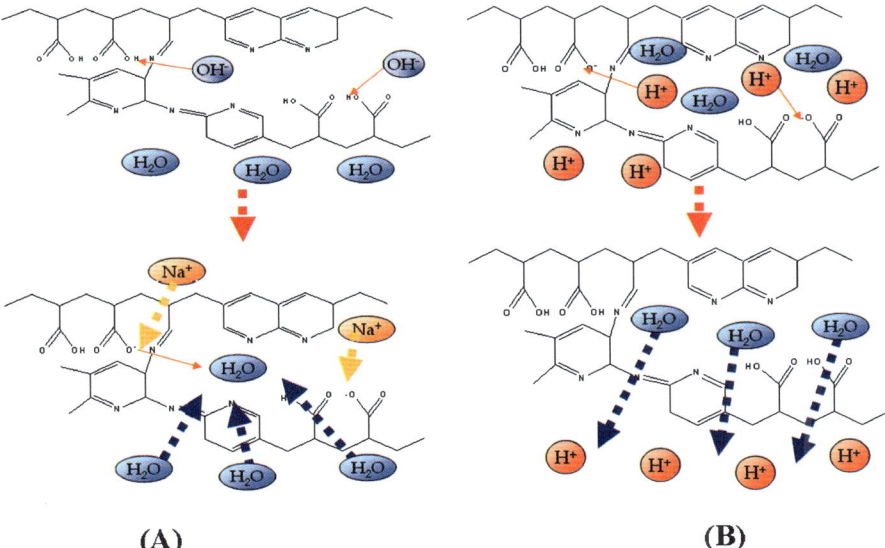

Figure 7.2 Elongation and contraction of PAN fibre in basic (A) and acidic (B) solution.[15]

such pH-induced contraction expansion of modified PAN fibres can also be induced electrically in a chemical cell by electrolysis and production of H^+ and OH^- ions.

7.2 Force–Strain Behaviour of Modified PAN

The force–strain behaviour and yield strength of PAN fibre gel are presented with experimental data. Since the molecular structure of PAN fibre is believed to change during the shrinking/swelling process, the mechanical properties of both elongation and contraction states are presented individually. Measured force and strain curves of fibre gel for each state are displayed in Figure 7.3. Initial lengths of fibre gels were an average of 175 mm and 110 mm for the elongated state and contracted state, respectively. The force–strain behaviour of PAN in both states was approximated to be linear. However, as the strain was increased the force curve of the gel showed a slight nonlinear behaviour regardless of the gel state. The resistance force of the fibre gel increased as the strain approached the breaking point. PAN fibre gels in the contracted state had stronger mechanical properties and more flexible elasticity than PAN fibre gels in the elongated state.

7.3 Actuation Properties of Modified PAN

7.3.1 Length-change Characteristics of Modified PAN: Effect of pH Variation

The length change relating to pH variation was investigated and is shown in Figure 7.4. In this experiment, ten samples of raw PAN fibre with lengths of 150 mm were saponified in 1 M LiOH solution for thirty minutes. The fibre-gel samples were then immersed in a pH solution from a level of pH 1. The solution pH was gradually increased by one pH level (shadowed area in Figure 7.4). When the solution reached pH 13 the solution concentration was decreased by one pH level until pH 1 (white area in Figure 7.4). The PAN gel length started to change at pH 10 and a drastic change was observed between pH 12–13 while pH was increasing. The same change was monitored between pH 4–2 while the pH was decreasing.

7.3.2 Generative Force Characteristics: pH-driven and/or Electrically Driven PAN Actuator

Single-strand PAN fibres were used for generative force measurement with both chemically induced and electrochemically induced pH changes adopted. A load cell was used to measure force generation from a single PAN strand fibre (high pH → low pH). A fibre specimen with 100 mm in length was

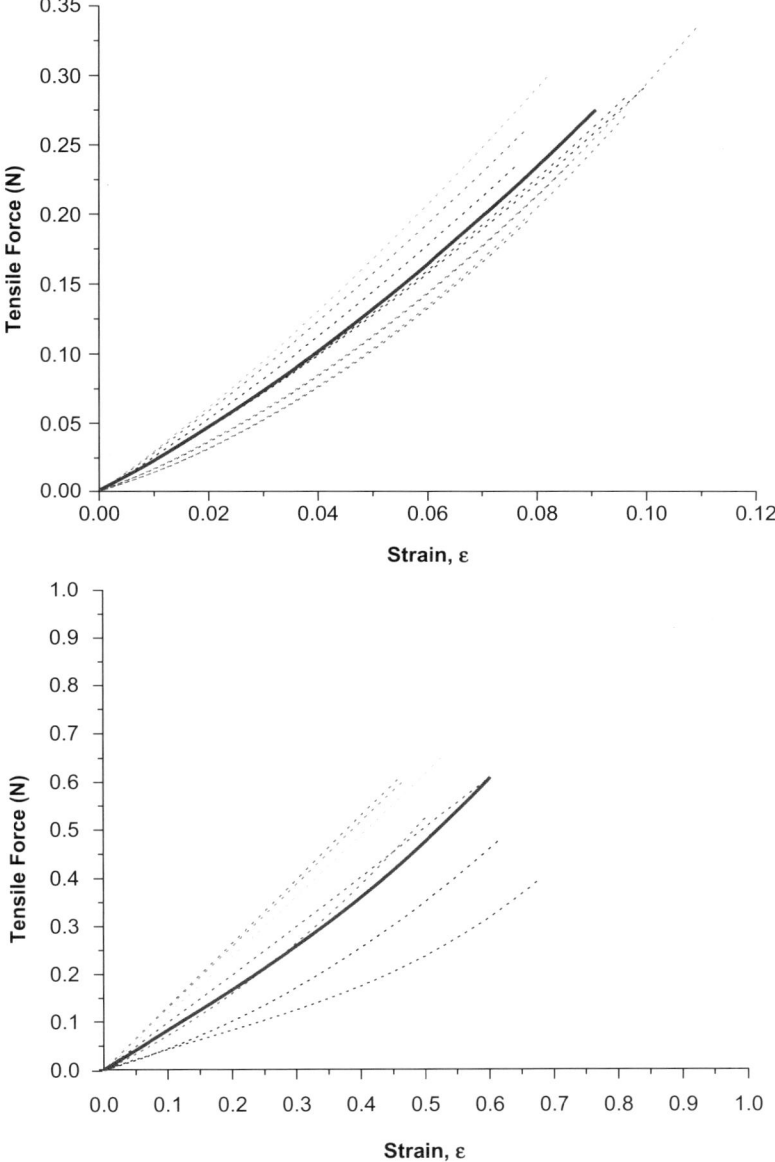

Figure 7.3 Force–strain curve of: (top) elongated and (bottom) contracted state of a single PAN strand. Solid line is the average value of a 10-sample set. Dotted lines show force–strain curves of each sample. Large scattering between different samples is observed. This is probably due to nonuniformity of the fibres. ε=engineering linear strain.[16]

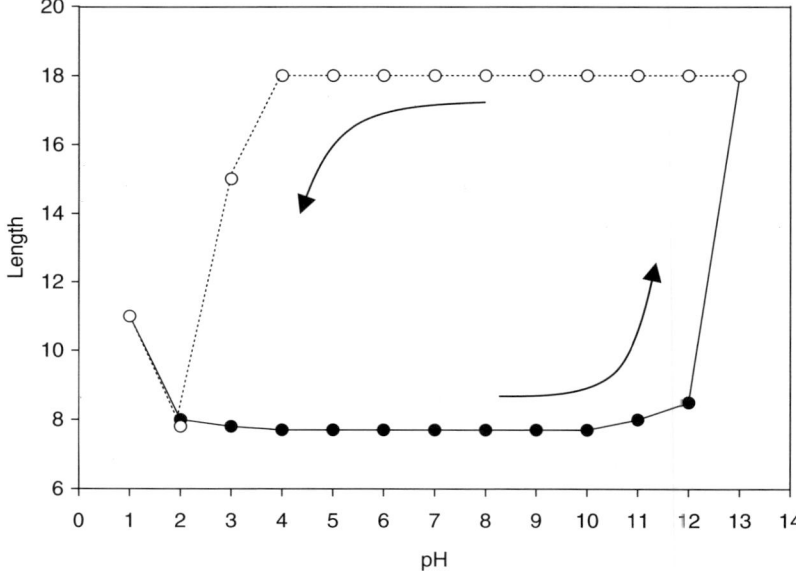

Figure 7.4 Length change of PAN fibre versus pH variation. A large hysteresis is observed when pH is changed form low values to high values and from high values to low values.[16]

prepared. Both ends of a fibre and the ring-type gripper were bonded by epoxy glue. For an electrically driven PAN actuator, stainless steel plate was used as the electrode and the counter electrode for the electrical actuation system while also being used to embrace the 100-mm single-strand fibre with a dc voltage applied between the electrodes. The distance between the electrodes was approximately 15 mm, and 5 V was charged through the electrodes. For the pH activation system a 1 M HCl solution was used, which was poured through a small nozzle until the fibre gel reached the steady state.

Figure 7.5 shows profiles of force development for both actuation systems. Single-strand fibres produced an approximate force of 0.1 N for both activation methods and had similar standard deviation ranges. Force generation reached the steady state within a few seconds in the chemical-activation system. On the other hand, it took approximately 10 min for the force generation to reach a steady state in the electrochemically driven system.

7.3.3 Generative Force Characteristics: Effect of Different Anions

Ions residing in the molecular structure of the polymer and in the solvent are part of the reason why the osmotic pressure swells and shrinks the PAN gel. It was generally noticed that the difference of force generation of the three acidic solutions was very small or negligible.

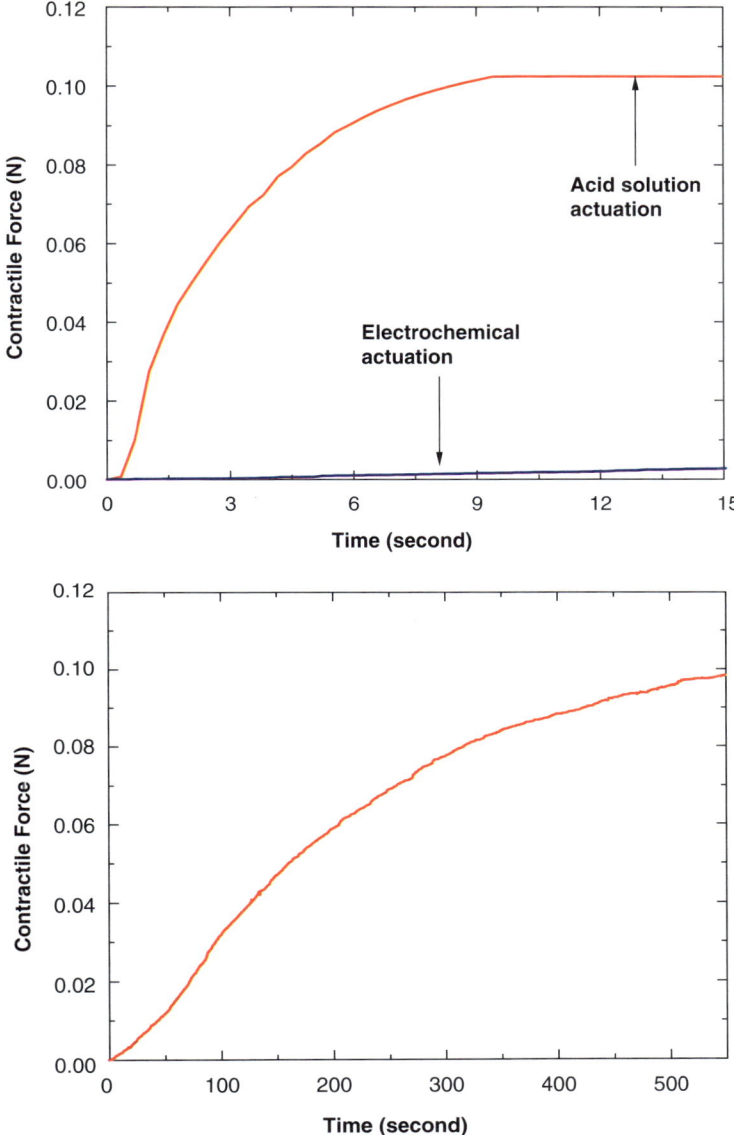

Figure 7.5 Profile of a single-strand force generation in: (top) 1 M HCl -- (bottom) 5 V electric field. Note that the slow electrochemically induced swelling correlates with time dependence of the electrochemically induced pH change.[16]

7.3.4 Generative Force Characteristics: Effect of Acidity

The PAN fibre was saponified in 1 M LiOH solution for 30 min. An HCl solution with different concentrations (0.01 M, 0.1 M, 0.5 M, 1 M and 2 M) was

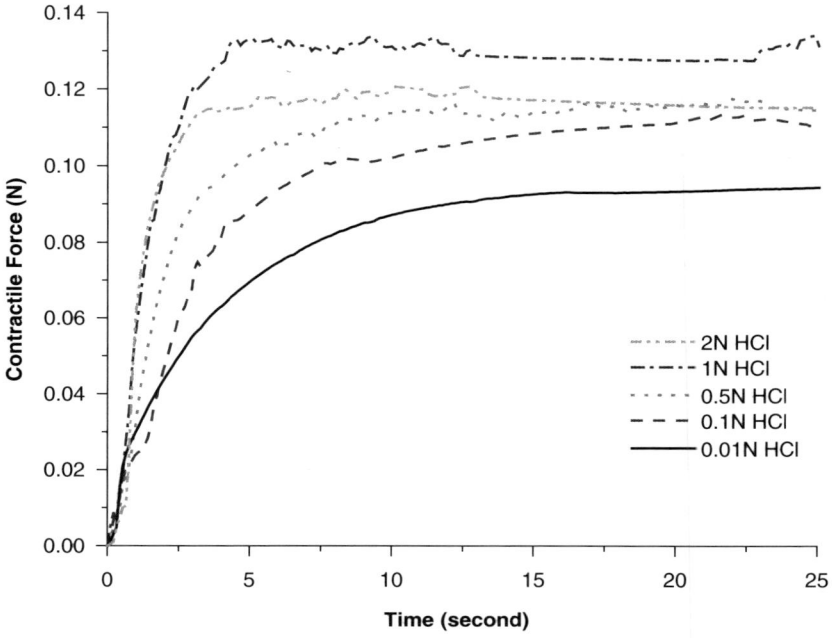

Figure 7.6 Force-generation curves for different concentrations of HCl acid solution (100 mm in length).

prepared. The same fibre gel specimen was repeatedly used throughout the entire test procedure to investigate the force generation as a function of pH difference under the same conditions.

In Figure 7.6, the measured generative force (contractile force) for different acid concentrations is presented. Force generation generally increased as the concentration of acidic solution increased. A decline in force generation was observed when the solution concentration was greater than 1 M. There was a time delay to reach the peak point at low molar concentrations. In high concentrations, such as 1 M and 2 M, the force reached its maximum point almost immediately. Even in 0.01 M concentrations of HCl, the maximum force reached 70% of force generated in the 1 M solution, whose concentration was 100 times stronger. The amount of hydrogen ions present determined the response time of shrinking, but these results suggest that the abundance of hydrogen ions above the critical amount had no significant contribution for force generation.

7.4 Performance of PAN Bundle Artificial Muscle

Chemically induced artificial-muscle systems need an irrigation system to guide acidic and basic solutions to the actuator and to dump the waste liquid out of the actuation system. The dumped hazardous chemicals should also be "collected" for the environment. An electrochemically driven actuation system has been considered as a convenient system due to the complexity of the chemically induced actuation discussed above. This produces less environmental hazards and can be more compact, as well as convenient to control although a chemically induced system has better response time and power generation. This section describes the measured performance of the chemically and electrochemically induced PAN bundle artificial muscles. The use of bundle other than a strand (~ 2000 fibres) introduces the practicality of employing PAN artificial muscles.

7.4.1 Electric-current Effect on Force Generation

The effect of electric current for force generation was investigated. The current density was estimated to be 10 mA/cm^2 and 20 mA/cm^2, respectively. The PAN bundle actuators were under a 10 gm$_f$ pretension stress. The experimental set up

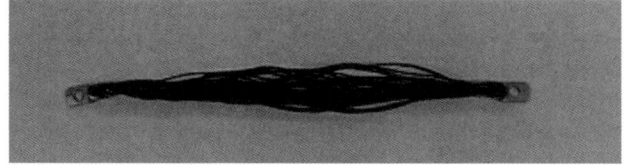

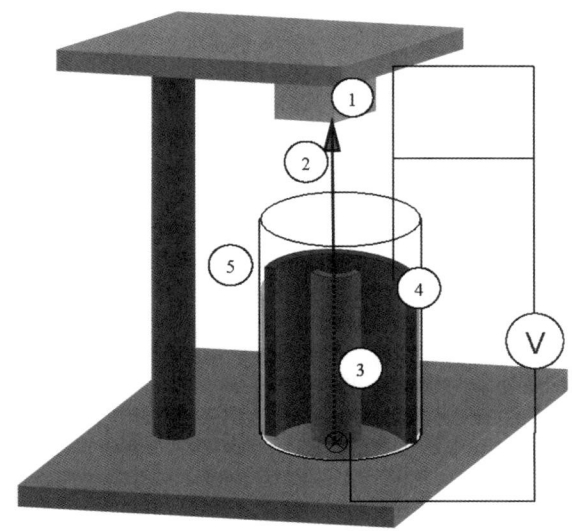

Figure 7.7 Electric activation of PAN bundle actuator test apparatus: (1) the load cell, (2) PAN specimen, (3) the electrode (anode), (4) the electrode (cathode), (5) solution bath ($\Phi 32 \times 15$ mm).[15]

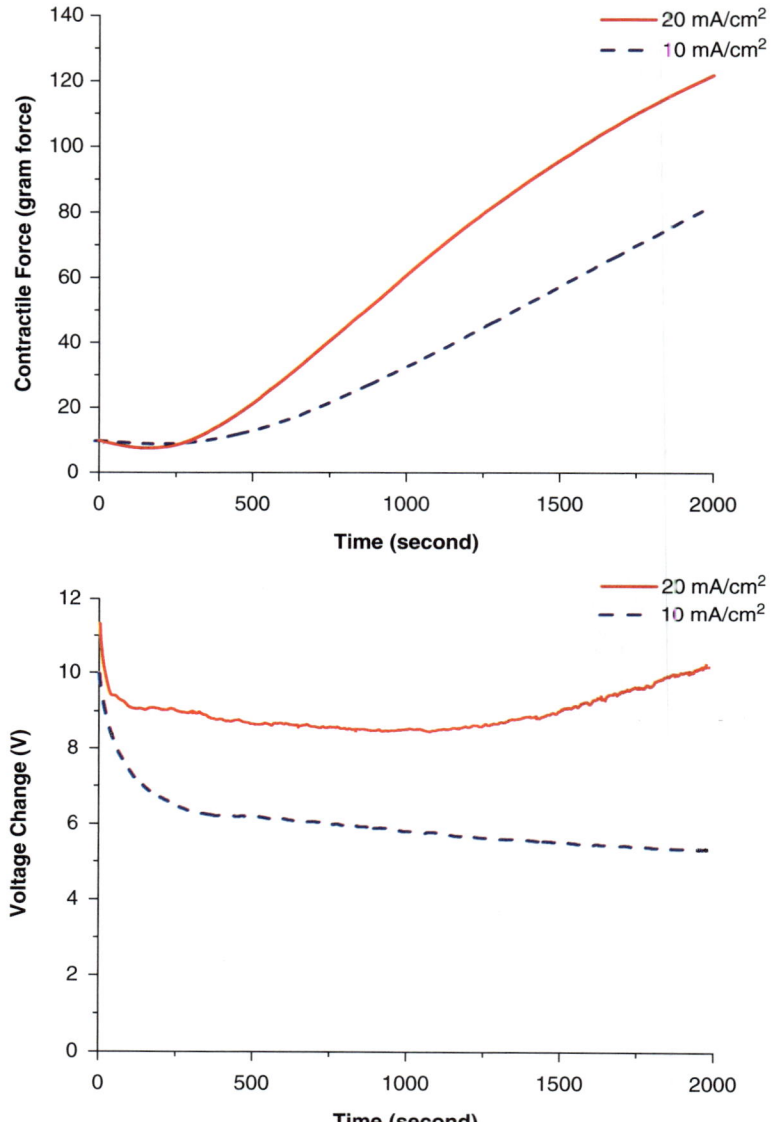

Figure 7.8 Force generation under different current (top), with voltage change (bottom) at 10 g force of pre-tension.[15]

is illustrated in Figure 7.7. Figure 7.8 shows the force generation under varying current densities and voltage changes. There was a 250–300 s time delay before a contractile force was developed, and as the current density increased the time interval decreased. However, it took 30 min for the PAN bundle actuator in

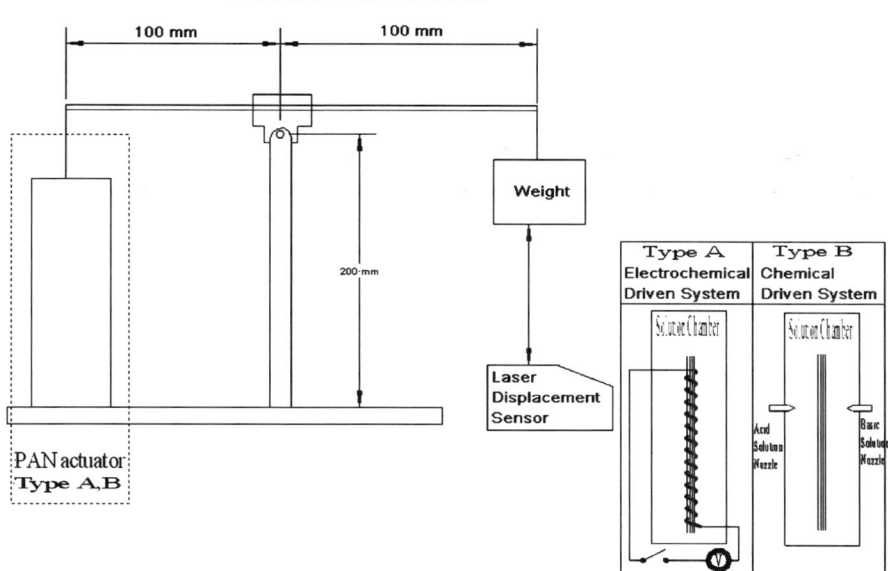

Figure 7.9 PAN bundle muscle crane for work–output measurement.[15]

the electrochemically driven system to generate 80% of the maximum force of the chemical actuation system.

7.4.2 Work Performance

A PAN bundle actuator crane system can be used to measure the "isotonic" work output of a PAN muscle system. Detailed dimensions of the crane are provided in Figure 7.9. The crane system was designed to measure both electrochemical and pH-driven systems. The load was applied using a mass of 25 g, 50 g, 75 g, and 100 g. The electrochemically driven system consisted of titanium electrodes and an ion diaphragm with 20 mA/cm^2 current density. The system was operated in a 500 ppm (parts per million) NaCl solution. For chemical driven system, 1 M HCl solution was used as a stimulus. Work generation under chemical actuating and electrochemical actuating systems is shown in Figure 7.10.

7.5 Summary of Performance Capability of PAN Artificial Muscle

Conventional actuators have been used for decades for various applications. Depending upon the characteristics of the actuators, they have been specified

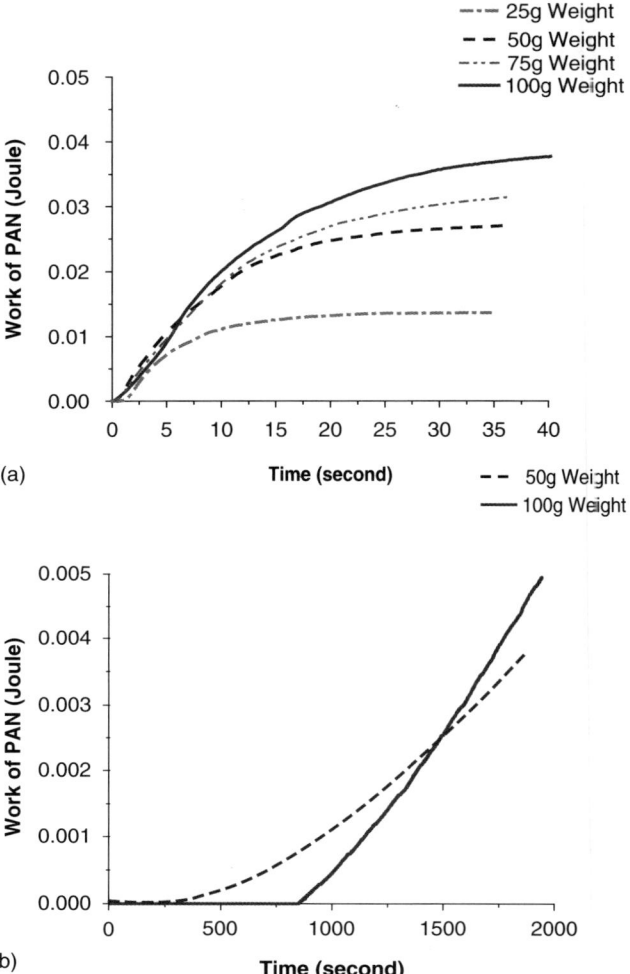

Figure 7.10 Work of chemically driven (top) and electrochemically driven (bottom) PAN artificial-muscle system.[15]

and distinguished for appropriate application fields. Comparing the performance capabilities of PAN actuators with traditional actuators it is necessary to understand and to identify the PAN actuator applications. Performance characteristics of PAN actuators have been interpreted in quantitative form. Performance characteristics of PAN actuators are provided in Figure 7.11, where other conventional actuator technologies are also described.[11] The actuation stress of PAN is in the range of a moving-coil transducer, but the actuation strain is better than a solenoid and similar to natural muscle.

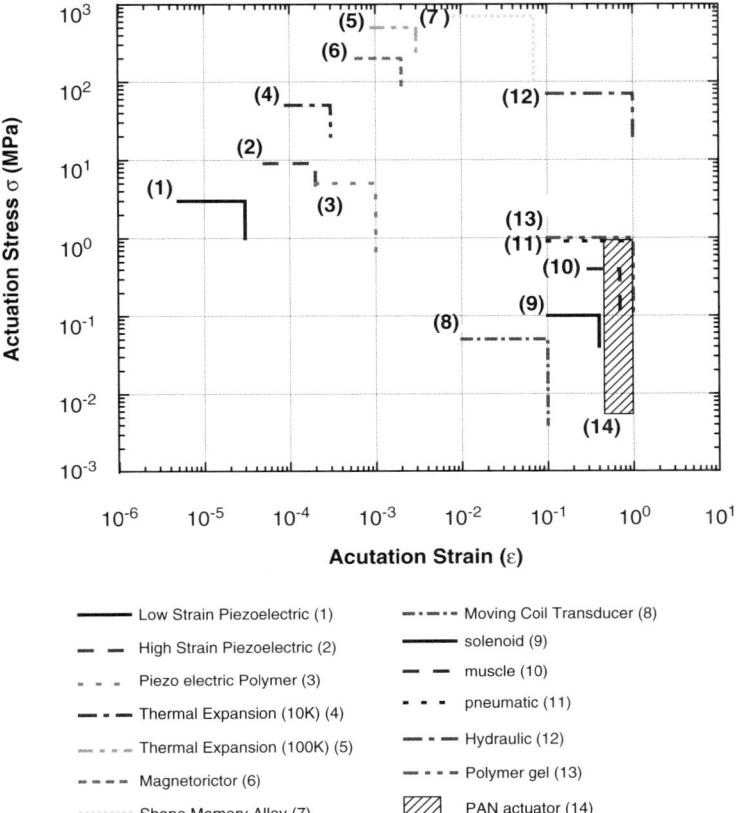

Figure 7.11 Actuation stress (σ), versus actuation strain (ε), for various actuators. Heavy lines indicate upper limits of performance.[15]

References

1. S. Umemoto, N. Okui and T. Sakai, *Polymer Gels,* Plenum Press, New York, 1991, 257–270.
2. T. Umemoto, T. Matsumura, T. Sakai and N. Okui, *Polym. Gels Networks*, 1993, **1**, 115–126.
3. K. Salehpoor, M. Shahinpoor and M. Mojarrad, *Proc. SPIE*, 1996, **2716**, 116–124.
4. H.B. Schreyer, G. Nouvelle, K.J. Kim and M. Shahinpoor, *Biomacromolecules*, 2000, **1**, 642–647.
5. S. Popovic, H. Tamagawa and M. Taya, *Proc. SPIE*, 2000, **3987**, 177–186.
6. Y. Osada, H. Okuzaki and H. Hori, *Nature*, 1991, **355**, 242–244.
7. S.W. Kim, K.S. Lee, I.H. Cho, J.H. Lee, J.W. Lee, Y.K. Lee, K.J. Kim and J.D. Nam, *Polym. Korea*, 2002, **26, No. 4**, 468–476.

8. Y. Li and T. Tananka, *J. Chem. Phys.*, 1989, **90, No. 9**, 5161–5166.
9. P.J. Flory, *Principles of Polymer Chemistry*, Cornell University Press, Ithaca, New York, 1953.
10. M. Doi, M. Matsumoto and Y. Hirose, *Macromolecules*, 1992, **25**, 5504–5511.
11. T. Tanaka, et al., *Nature*, 1987, **325, No. 6107**, 796–798.
12. J.E. Huber, N.A. Fleck and M.F. Ashby, *Proc. Roy. Soc. London*, 1997, **453**, 2185–2205.
13. A. Katchalsky and M. Zwick, *J. Polym. Sci.*, 1955, **16**, 221–233.
14. A. Hickey and N. Peppas, *Polymer*, 1997, **38**, 5931–5936.
15. K. Choè, K.J. Kim, D. Kim, C. Manford, S. Heo and M. Shahinpoor, *J. Int. Mat. Syst. Struct.*, 2006, **17**, 563–576.
16. K. Choè and K.J. Kim, *Sensors and Actuators A*, 2006, **126**, 165–172.

CHAPTER 8
Unimolecular Electronic Devices

ROBERT M. METZGER

Laboratory for Molecular Electronics, Department of Chemistry, Box 870336, University of Alabama, Tuscaloosa, AL 35487-0336, USA

8.1 Introduction

This chapter reviews "molecular electronics", or "molecular-scale electronics", or "unimolecular electronics" (UE):[1] it summarises the last decade of research into molecules that either singly, or in parallel monolayer arrays (one molecule thick), act as either passive or active electronic components. This should lead to electronic devices with dimensions of 1 to 3 nm.

The electronic properties of single molecules are easily studied by spectroscopy: photons can easily interrogate molecules in the gas phase, in a solid, or in solution, while the exact location of the molecule within the sample is only of secondary concern. A disadvantage compared to "photonics" is that an electrical circuit that uses photons as control elements cannot easily be reduced to nm dimensions.

In contrast, touching single molecules electrically and without damage is much more difficult, but potentially rewarding. The goal in UE is to "reach out and touch" individual molecules with electrodes, without damaging them, and exploiting their chemical structure to control the flow of electrical signals. Their chemical structure makes them into "electroactive molecules".

Since molecules can be very small (0.5 to 3 nm), UE may be the *reductio ad absurdum* of inorganic electronics, and may have been presaged by Feynman's talk in 1959 ("there is plenty of room at the bottom"): but Feynman discussed assemblies of atoms, not molecules.[2]

The urge to make ever smaller electronic devices is, in part, driven by Moore's "law",[3] which noted a halving of the distance between components every two years, and a concomitant doubling of the speed of computation in digital circuits: at present, 65-nm design rules are used for 3-GHz computers, and 50-nm design rules exist in research. Going down to 3-nm design rules is very difficult for inorganic electronics and for inorganic metal electrodes, but should be easy for molecules. If UE becomes practical in time, ultrafast molecular-based computing may be reached.

UE started in earnest, when Aviram and Ratner (AR) proposed in 1974 electrical rectification, or diode behaviour, by a single molecule with suitable electronic asymmetry.[4] In the early 1980s, three topical conferences organised by the late Forrest L. Carter sparked some interest, but there were few significant results. In the late 1990s, the United States Defense Advanced Projects Agency sponsored attempts towards molecular integrated circuits and UE. Since about 1995, the push to nanotechnology inspired new and serious efforts in UE. Physicists have rediscovered chemistry, not for the first time! It is theoretically quite clear that molecules can function as electronic components, by using their one-electron-donor or one-electron-acceptor properties; the large technical challenge is how to address them reliably in a nm-scale circuit.

8.2 Donors and Acceptors; HOMOs and LUMOs

One simple way to understand how "electroactive" organic molecules can be used is to tabulate their first adiabatic ionisation potentials I^D (for electron donors D) or their first adiabatic electron affinities A^A (for electron acceptors A), and compare them to the work functions ϕ of inorganic metals that may be used to contact them (Figure 8.1): the work function ϕ represents the energy required to remove an electron from the metal and take it to an infinite distance away in vacuum: ϕ is related to the Fermi level E_F of the metal, and is moderately dependent (0.1 to 0.3 eV) on the Miller indices of the exposed crystallographic face of the metal. There tends to be a mismatch between I_D, A_A, and ϕ, so positive or negative applied potentials are needed to bring the molecular energy levels into resonance with the ϕ, or E_F, of inorganic metal electrodes. Vertical approximations to I_D are easily measured; electron affinities A_A are difficult to measure. Usually, "good" electron donors (relatively low I_D) are poor electron acceptors (have very small A_D), and conversely, "good" acceptors (with large A_A) are difficult to oxidise (large I_A). The semimetal graphite, as the infinite two-dimensional extension of polycyclic aromatic hydrocarbons, is as good a donor as it is an acceptor. Theory yields estimates of I_D and A_A by Koopmans' theorem[5]: the HOMO (highest occupied molecular orbital) level is a vertical approximation to I_D, while the LUMO (lowest unoccupied molecular orbital) level is a vertical approximation to A_A. These approximations ignore electron correlation and Franck–Condon reorganisation.

The practical range of I_D and A_A is limited, because the molecules, and their cations, or anions must be stable in ambient air or solvent: thus, very potent electron donors D or acceptors A can be designed, but they are not stable enough for synthesis, analysis, or assembly.

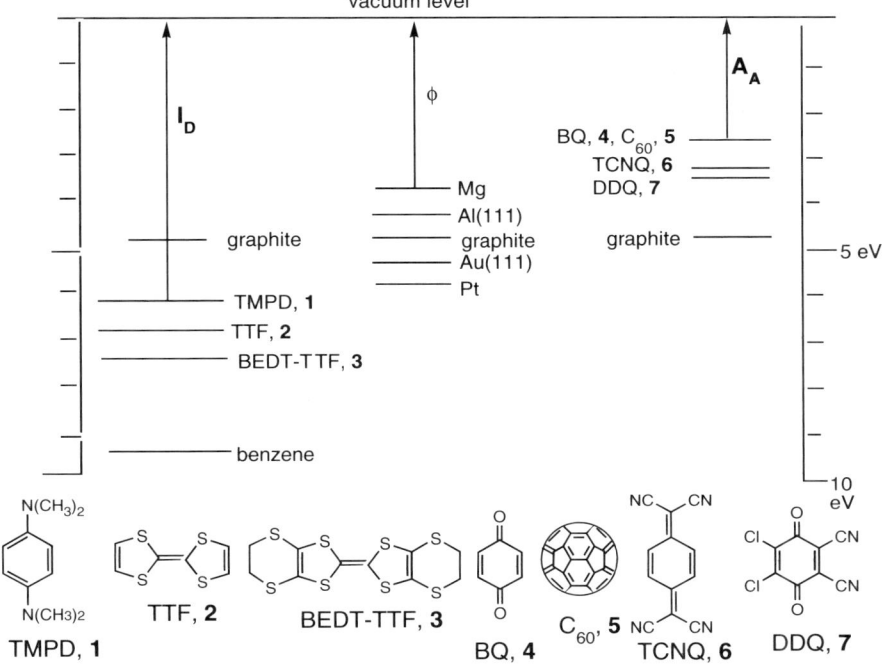

Figure 8.1 Representative one-electron donors D and their ionisation potentials I_D, and one-electron acceptors A and their electron affinities A_A, and metals, and their work functions ϕ.

8.3 Contacts

How one can use an inorganic metal (or semiconducting) electrode to interrogate a molecule? If a molecule gently "touches" a metal surface, then the chemical potentials must become equal across the interface: the resultant band bending is accompanied by a surface dipole of some magnitude, as the chemical potential or Fermi level of the metal and the HOMO of the molecule move to become equal by partial electron transfer at the interface: this is the Schottky barrier.[6] The second, or third, or fourth electrode that would interrogate the molecule (or monolayer of molecules) must also be brought down onto the molecule without heating it, or compressing it: in electrical engineering terminology, one seeks a contact that is "ohmic", *i.e.* obeys Ohm's law,[7] *i.e.* it does not have an energy barrier to electron transfer across it. Such a "gentle" contact is not easily achieved. Luckily, the advent of the scanning tunneling microscope (STM),[8] the atomic force microscope (AFM),[9] and the conducting-tip AFM (CT-AFM)[10] allows close control of the tip-to-substrate distance and of the force with which the tip approaches the molecule.

Molecules deposited on surfaces by physisorption can move after deposition, either to reach a thermodynamic steady state on the surface, or in response to an externally applied field. If one puts a 1-V bias across a monolayer 1 nm thick, the electric field is large: $1\,\text{GV}\,\text{m}^{-1}$, probably large enough to move or reorient molecules around in order to minimise the total energy.

Amphiphilic molecules can be transferred quantitatively from the Pockels–Langmuir (PL) monolayer at the air/water interface, onto a metal or other solid substrate, by the Langmuir–Blodgett (LB) or vertical transfer method,[11,12] or by the Langmuir–Schaefer (LS) or quasihorizontal transfer:[13] the coverage of the surface is well quantified by the transfer ratio = (area covered on the substrate)/(area lost from the PL film). To make the organic molecule amphiphilic, pendant alkyl groups (which yield a hydrophobic end) or pendant carboxylic acid groups (to make a hydrophilic end) or ions (for a hydrophilic end, provided that the substrate provides a suitable counterion) are often necessary. After transfer, molecules may reorganise over time: the kinetic packing of the PL monolayer may relax to a different thermodynamic order.

Molecules can also be covalently bound to certain metal surfaces:[14a] carboxylates onto oxide-covered aluminium, and thiols or thioesters to gold, *etc.*: these are called "self-assembled monolayers" (SAMs), even though the term "self-assembly" is also used for a different purpose in biochemistry (the fitting of a molecule into the active site of an enzyme). The advantage of SAMs is that they are sturdily anchored at the right distance from the metal substrate; the disadvantage is that true perfect monolayer coverage, so easily achieved for LB films, is very difficult to obtain in SAMs.

Molecules with thiol terminations can be bound simultaneously to two or three or more electrodes; so far, this has been done for two electrodes ("break junctions"), but not yet for three.

Making 1 to 3 nm gaps between electrodes is difficult: electron beam lithography can make 50 nm gaps routinely, 20 nm gaps with considerable effort, and smaller gaps with even more effort. Physicists have worked around these restrictions. Two-electrode break junctions were pioneered by Muller[14b] and applied to 1,4-benzenedithiol by Reed and coworkers:[15] a thin Au wire is vapour-deposited onto a flexible substrate, with a narrower region in the centre; the wider parts of the Au wire are anchored under two static supports on the top, then the thinner Au region is piezoelectrically pressed from below, until the Au wire breaks, creating two Au shards with a narrow gap, whose width can be controlled to within less than 0.1 nm by the piezodevice. A 1,4-benzenedithiol solution in benzene, suspended above the break junction, will form many single one-end-only thiolate bonds to Au randomly along the wire, while in the 8-Å gap one (or two, or more) benzenedithiolates will bond simultaneously to both shards. The minimum conductance will be due to a single molecule in the gap.[15]

Inspired by earlier work,[16] Reed and coworkers developed a nanopore technique to study a small assembly of a few hundred to a few thousand molecules.[17]

Nanogaps between electrodes can also be made by controlled electromigration; Au wires can be broken into very sharp tips, if a current is passed through them;[18] it helps a lot if the sample is held at 4.2 K. But this is not easy to control.

Chemists can bridge 50 nm gaps between electrodes by providing, *e.g.*, two 25-nm diameter nanoparticles of Au or Ag, coated with the usual "spinach" of bithiols, and bond them chemically to a 3-nm molecule squeezed between them.

Lindsay and coworkers established by CT AFM that the current–voltage (I–V) curves for octanedithiol, bonded to an Au(111) susbtrate and also bonded to an Au nanoparticle (to make contact easier), fell into several broad families, depending on the force used by the AFM cantilever, and estimated 900 ± 50 MΩ as the resistance per molecule.[19]

Careful attention has focused on the metal/molecule interface. Allara and coworkers established spectroscopically that Ti, when evaporated and deposited atop a SAM on Au, far from being a benign cover layer (potentially oxidised on the surface) in fact interpenetrates within the monolayer.[20] Thus, careful ongoing studies seek to understand what does, or does not happen, at the metal/organic interface, both at zero bias and under applied voltage.

8.4 Two-probe, Three-probe and Four-probe Electrical Measurements

Central to electronics is the I–V measurement, *i.e.* the measurement of the electrical current I through a device, as a function of the electrical potential, or bias, or voltage V placed across it. Electrical devices are most often two-terminal devices (resistors, capacitors, inductors, rectifiers and diodes, negative differential resistance (NDR) devices). Amplification is also possible with Esaki tunnel diodes and NDR devices ("diode logic"), because an input signal, applied across a load of R Ω, placed in series with an NDR device with negative resistance $-R\,\Omega$, provides a zero net resistance at the output, and therefore large signal amplification across the sum of those two resistances. However, difficulties in using organic NDR thiolates at room temperature, have prevented the commercialisation of organic NDR devices as amplifiers. Commercial devices used for amplification are three-terminal devices (bipolar junction transistors (BJT), field effect transistors (FET), vacuum triodes) or four- or five-terminal devices (vacuum tetrodes, vacuum pentodes).

The best way to measure the resistance of a macroscopic device is to use four probes: the outermost two are used to provide a current I from a constant-current source, and the potential drop V between the inner two is measured: the resultant resistance $R = V/I$, after some corrections for geometry, is the true resistance of the device; the contact resistances at the probe/device interfaces cancel out.

For a nanoscopic device (*e.g.* of dimensions 3 nm $\times$ 3 nm $\times$ 3 nm), present technology cannot yet reliably generate four electrodes 3 nm apart. Electron-beam lithography can easily reach 20 nm $\times$ 20 nm $\times$ 20 nm; going below that is difficult.

For two-probe measurements of a two-terminal device, all resistances (measuring instrument-to-first-electrode, first-electrode-to-molecule, molecule-to-second-electrode, and second-electrode-to measuring instrument) are additive.

To minimise extraneous large resistances, droplets of wetting solders, Ag paint, Au paint, or Ga/In eutectic are used. To minimise Schottky-barrier problems, the same metal is used on both sides of the molecule or monolayer. Most metals are covered by an oxide (impervious, or defect ridden, as, *e.g.*, Al). In contrast, gold has no oxide, but has another problem: Au atoms migrate somewhat after deposition, to minimise total energy, and migrate even more under electric field ("electromigration").[21]

Three-electrode measurements have been made, where two electrodes are prepared beforehand, the molecule is placed between them by physisorption or chemisorption, and the third "gate" electrode is an STM or CT-AFM tip. This technique has been used to measure FET behaviour in a monolayer. The electric field for the FET can also be supplied from the gate conductor through the barrier oxide below the molecules being tested.

8.5 Resistors

Molecules can function as resistors. Of course, organic chemists will tell us that saturated straight-chain alkanes will conduct less well than unsaturated poly-alkenes or poly-conjugated aromatic hydrocarbons. In the 1960s Henry Taube proved by kinetic studies of electron-transfer rates between metal ions across alkane ligands occurs more slowly than across unsaturated ligands.[22,23] Confirming this, in 1996 Weiss and coworkers studied the STM currents across a thioalkyl SAM on Au, and found a pronounced conjugation and molecular length dependence of the conductivity.[24]

Ohm's law[7] indicates that the resistance R (Ohms, or Ω) and the conductance G (Siemens = ohm^{-1}, or S) of a device is given by:

$$R = 1/G = V/I \tag{8.1}$$

where V is the applied potential (V), and I is the current (A). This law is valid for macroscopic metals, or for semiconductors at any given temperature, where the resistance is mainly due to scattering off impurities and lattice defects in the material. In semiconductors, the current follows an Arrhenius-like temperature dependence,

$$I = I_0 \exp(-\Delta E/k_B T) \tag{8.2}$$

where ΔE is the activation energy for the dominant carriers (electrons or holes), T is the temperature and k_B is Boltzmann's constant.

For nanoscopic objects the current I is determined by Landauer's formula:[25]

$$I = (2e/h) \int_{-\infty}^{\infty} f_L(\epsilon) - f_R(\epsilon) \, \text{Tr}\{G^a(\epsilon)\Gamma^R(\epsilon)G^r(\epsilon)\Gamma^L(\epsilon)\} d\epsilon \tag{8.3}$$

where e = charge on one electron, h = Planck's constant, ε = energy, $f_L(\varepsilon)$ and $f_R(\varepsilon)$ = Fermi–Dirac distributions in the left and right electrodes, respectively, $G^a(\varepsilon)$ and $G^r(\varepsilon)$ = advanced (and retarded) Green's functions for the molecule,

$\Gamma^R(\varepsilon)$ and $\Gamma^L(\varepsilon)$ = matrices that describe the coupling between molecule and the metal electrodes, and Tr{ } = trace operator. In this formula, the quantum of resistance R_0 and its reciprocal, the quantum of conductance, G_0, are given by Landauer's constant (or the von Klitzing constant,[26] now known to 1 part in 10^9):

$$R_0 = 1/G_0 = h/2e^2 = 12.813 \, k\Omega = 1/(7.75 \times 10^{-5} \, S) \quad (8.4)$$

This does not say that the intrinsic resistance of any molecule is 12.813 kΩ; it says that the resistance of that molecule *plus* the two metallic electrodes is 12.813 kΩ.[26] The *minimum overall* resistance of a molecular wire and its junctions to arbitrary metal electrodes is 25.626 kΩ (assuming two carriers). The conductance *within* the nanowire can be much higher, particularly if there is no scattering ("ballistic" conductance), but the overall conductance is no larger than R_0^{-1}.

The resistance of eqn (8.4) must be divided by a factor N, if N elementary one-dimensional wires, or N molecules, bridge the gap in parallel between the two metal contacts:

$$R_N = h/2e^2 N = (12.91/N) \, k\Omega \quad (8.5)$$

Bulk electrical conductivities range over 25 orders of magnitude (from 1.33×10^{-18} S m^{-1} for fused silica, to 1.56×10^{-3} S m^{-1} for silicon, to 5.5×10^{-5} S m^{-1} for ultrapure "conductivity" water, to 20 S m^{-1} for the quasi-one-dimensional organic metal TTF TCNQ, to about 0.01 S m^{-1} for highly conducting organic polymers, and finally to 6.3×10^7 S m^{-1} for Au, all at room temperature). The conductivity is essentially infinite for superconductors below their critical temperature.

For metal-insulator-metal (MIM) structures, where there is assumed to be a rectangular barrier of energy Φ_B and width d on both sides of the molecule, in the direct tunneling regime $V < \Phi_B e^{-1}$, the Simmons formula[27] can be used:[28]

$$I = e2\pi hd^2)^{-1}\{(\Phi_B - eV/2)\exp -4\pi(2m)^{1/2}h^{-1}\alpha(\Phi_B - eV/2) \\ + (\Phi_B + eV/2)\exp -4\pi(2m)^{1/2}h^{-1}\alpha(\Phi_B + eV/2)\} \quad (8.6)$$

where the dimensionless constant α corrects for a possible nonrectangular barrier, or for the effective mass in place of the true carrier (electron) rest mass m. A fit to the experimental I vs. V curves for a SAM of alkanethiols between Au electrodes in a very small pad of diameter 45 ± 7 nm at 300 K yielded $\Phi_B = 1.37 \pm 0.06$ Volts and $\alpha = 0.66 \pm 0.02$ for n-dodecanethiol, $C_{12}H_{25}SH$, and $\Phi_B = 1.40 \pm 0.04$ V and $\alpha = 0.68 \pm 0.02$ for n-hexadecanethiol, $C_{16}H_{33}SH$.[28] Assuming a molecular cross-sectional area of 23 Å^2 (typical for LB monolayers of alkanes), the circular pad contains at most 300 molecules in parallel; $I = 20$ nA at $V = 0.8$ V for $C_{12}H_{25}SH$ yields an Ohm's law conductance of 2.5×10^{-8} S = $(36 \, M\Omega)^{-1}$, and a specific conductance per molecule of 8.3×10^{-11} S molecule^{-1}. Using the SAM thickness of 14.4 Å for $C_{12}H_{25}S-$ yields 2.5×10^{-8} S/1.44×10^{-9} m = 19.4 S m^{-1} or 0.065 S m^{-1} molecule^{-1}: this may not be a fair use of the data, but it will give us some rough idea.

The range of conductivities between straight-chain hydrocarbons and aromatic hydrocarbons is much smaller than the 25 orders of magnitude mentioned for all bulk materials. Indeed, the "best" (Landauer formula) specific resistance of $2.5616 \times 10^4\,\Omega$ molecule^{-1} is only six orders of magnitude smaller than the estimated $1.2 \times 10^{10}\,\Omega$ molecule^{-1} measured for n-dodecanethiol.[28] A good design for minimising resistances suggests aromatic molecules whose LUMO is low enough to be reached with small biases (<1 V).

Single-wall carbon nanotubes (SWCNT)[29] are very robust "molecules" of pure carbon, which behave either as electrical semiconductors or as quasimetals, depending on the topology of folding.[30,31] Alas, the nanotubes are not yet fully chemically processable. For that we may need defect-free, differently end-derivatised SWCNT, e.g. A_n-SWCNT-B_m, with n polar or formally charged groups A and m polar or oppositely charged groups B, such that the nanotubes can be chemically separated by chromatography by charge, dipole moment, and conductivity: if this can be achieved, then the A_n-SWCNT-B_m would become ideal connectors in UE.

There is also a quantum limit:[32] if an electron is confined to a small dot, i.e. a two-dimensional confined region, or quantum dot, of capacitance C (typically 1 fF); then adding another electron will cost a "charging energy" e^2/C. If ($e^2/2C$) < $k_B T$ (where k_B is Boltzmann's constant, and T is the absolute temperature), then a Coulomb blockade occurs:[32] no more charges can be added, for a threshold voltage $V_{CB} < (k_B T/e)$. This causes a flat region of no current rise in the I–V curve, until $V \geq (k_B T/e)$ (at 300 K, $V_{CB} = 0.026$ V).

8.6 Rectifiers or Diodes

AR[4] proposed a D-σ-A molecular rectifier, with an electron-donor moiety (D), bonded to an electron acceptor moiety (A) through an insulating saturated "σ" bridge; the current, small at negative bias, becomes large at and beyond a threshold positive bias, because at that bias the HOMOs and LUMOs and Fermi levels of the two electrodes start to allow electron transfer to the electrodes, the first highly polar electronic excited state D^+-σ-A^- gets populated, and will decay to the less-polar ground state D^0-σ-A^0 by inelastic tunneling through the molecule.[4] This decay may be enhanced by some intramolecular charge transfer (ICT) or intervalence transfer (IVT) mixing of the donor and acceptor states, i.e. the existence of an extra ICT or IVT absorption band. If the two moieties are too far apart (the σ bridge is too long), then they will not communicate, and no rectification will occur. If they are too close, then a new, single, mixed ground state will form, and the molecule will not rectify. What is the right length for σ? Probably σ should have between 2 and 6 C atoms or their equivalent.

There are three distinct processes for asymmetrical conduction, i.e. rectification, in "metal–organic–metal" (MOM) assemblies. The first is due to Schottky barriers[5] at the "metal/organic" interface(s): these are "S" (for Schottky) rectifiers.[33,34]

The second process arises, if the "chromophore" (*i.e.* the part of the molecule whose molecular orbital must be accessed during conduction) is placed asymmetrically within a "metal–molecule–metal" sandwich, *e.g.* because of the presence of a long alkyl "tail".[35,36] We shall call molecules that rectify by this process "A" (for "asymmetric") rectifiers.[37] The inclusion of LB "tails" causes an "A" contribution.

The third process occurs when the current passing through a molecule, or monolayer of molecules, involves electron transfers between molecular orbitals, whose significant probability amplitudes are asymmetrically placed within the chromophore: this third process may be true "unimolecular rectification", or "U" (for unimolecular) rectification;[35] these "U" rectifiers are what we endeavour to achieve.

The requirements for assembling organic molecules between two inorganic metal electrodes may result in a combination of "A", or "S", and "U" effects. Pure "U" rectifiers are rare.[36]

The electron transport from metal to organic material to metal has received theoretical attention.[35,38] First, asymmetries in current–voltage plots (often ascribed to rectification) also occur if a chromophore is placed asymmetrically within the electrode gap 35 ("A" rectifiers). This has been seen by STM.[39] Second, elastic electron transfer between a metal and a single molecular orbital of a molecule can be expressed by:[38,40]

$$I = I_0\{\tan^{-1}\theta(E_0 + peV) - \tan^{-1}\theta(E_0 - (1-p)eV)\} \quad (8.7)$$

where E_0 is the molecular orbital energy (typically a LUMO or HOMO), V is the applied potential, and p is the fractional distance of the molecule from, say, the left electrode. If the molecule is centred in the gap, then $p = 1/2$. Tunneling across molecules is expected to be approximately exponential, to some power of the potential, so a sigmoidal curve is usually seen, symmetrical about $I = 0$ and $V = 0$.

Rectification has a figure of merit, the rectification ratio (RR), defined as the current at a positive bias V divided by the absolute value of the current at the corresponding negative bias $-V$:

$$\text{RR}(V) = I(V)/|I(-V)| \quad (8.8)$$

Commercial doped Si, Ge, or GaAs pn junction rectifiers have RR between 10 and 100. Between 1982 and 1997 we studied many D-σ-A molecules as potential rectifiers,[41–75] but could not measure reliably their I–V properties. Between 1986 and 1993, Sambles did develop reliable techniques for studying rectification by LB multilayers and even monolayers, by sandwiching them between electrodes of different work functions: Mg on one side, to minimise damage to the film, and noble metals (Ag, Pt) on the other side.[33,34,76–78] To avoid difficulties with potentially asymmetric Schottky barriers, and to avoid the thorny issue of how electron transport occurs between adjacent layers in an LB multilayer, we studied almost exclusively single LB or LS monolayers, and used the same metal (first Al, then Au) on both sides of the monolayer. Since 1997 we have identified eight unimolecular rectifiers (structures **8–15** in Figure 8.2), as LB or

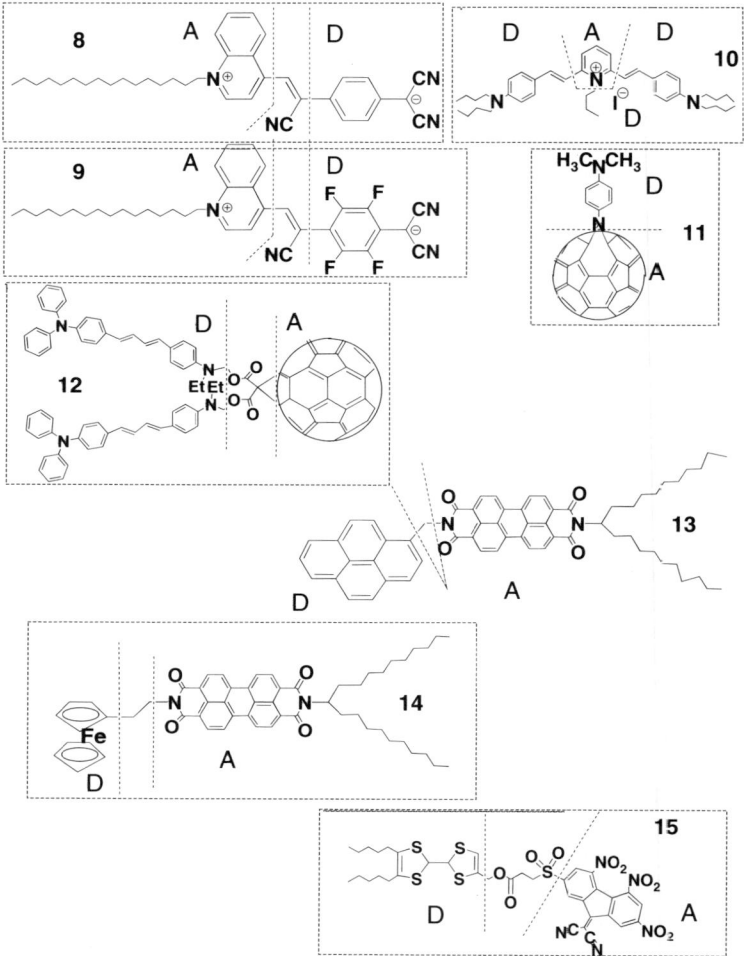

Figure 8.2 Eight unimolecular rectifiers (structures **8–15**).

LS monolayers, either between Al electrodes,[79–81] or between Au electrodes.[82–91] As shown in Figure 8.2, the structures have D and A moieties and all, except for **15**, have pendant alkyl groups for organising the molecules as monolayers. The evaporation of a metal electrode (Al or Au) onto glass, quartz, or very flat Si substrates was routine, as was the transfer of an LB monolayer atop the metal electrode. But depositing the second metal electrode atop the delicate LS or LB monolayer was not routine: a liquid-nitrogen-cooled sample stage in the vacuum evaporator was enough to cool Al vapour upon contact with the cold organic monolayer (at least 50% of the metal–organic–metal "pads" were not electrically shorted).[81] For Au, this was not enough, so the "cold gold" technique was implemented (cooling the Au vapour atoms to room temperature by multiple collisions with Ar vapour)[78,82,83] (see Figure 8.3). Most molecules were studied

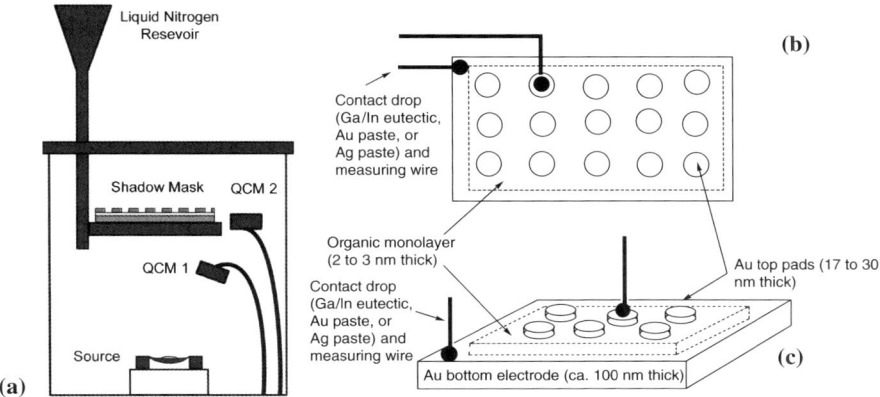

Figure 8.3 (a) Edwards E308 evaporator, with Au source, two quartz crystal thickness monitors (one, QCM1, pointed to Au source, to monitor Au vapour deposition on chamber walls, the other, QCM2, to monitor Au film thickness deposited through a shadow mask atop the organic layer). (b), (c) Geometry of MOM "Au–organic monolayer | Au pads".

at room temperature in a Faraday cage, but **8** was also studied for its rectification for $105\,K < T < 370\,K$.[80] Characteristic I–V curves for them at room temperature are shown in Figure 8.4. Most of these compounds were also studied for their spectroscopic properties (V-UV, IR, grazing-angle IR, spectroscopic ellipsometry, XPS, EPR of their radical ions, surface plasmon resonance, small-angle X-ray scattering).[79,92–94] Efforts were made to identify the molecular mechanisms for the rectification, and to support them by theoretical calculations.[35,38,79,95] The direction of larger electron flow ("forward direction"), is shown by arrows in Figure 8.4, and is in the direction from the electron donor D to the electron acceptor A, as expected. Not all compounds tested rectified,[96–98] because of their chemical structure and/or monolayer assembly. Table 8.1 summarises the characteristics of the measured rectifiers. Our work has been reviewed almost too extensively.[99–136]

Before discussing our own results in detail, we mention some recent contributions of other research groups: (i) Bryce, Petty, and coworkers[137] studied a new D-σ-A compound, inspired by our older effort,[42] containing the D = TTF and the A = TCNQ: this TTF-σ-TCNQ ester gave strong PL films at the air/water interface. But the TCNQ group lies flat, rather than end-on, on the water surface; the LB multilayers were Y-type, rather than Z-type, and rectification could not be observed.[137] (ii) Bryce, Heath, and coworkers found a rectifier in an analogue of **15**.[138] (iii) Ashwell and coworkers studied several zwitterionic systems by scanning tunneling spectroscopy (STS), starting with **8**;[139] for **8**, and for several other zwitterionic systems, the addition of acid could stop, or reverse, the rectification.[140–142] (iv) Yu and coworkers reported rectification by STS, which could be reversed by protonation.[143] (v) Weber, Mayor, and coworkers linked bithiols between mechanically controlled Au break junctions:

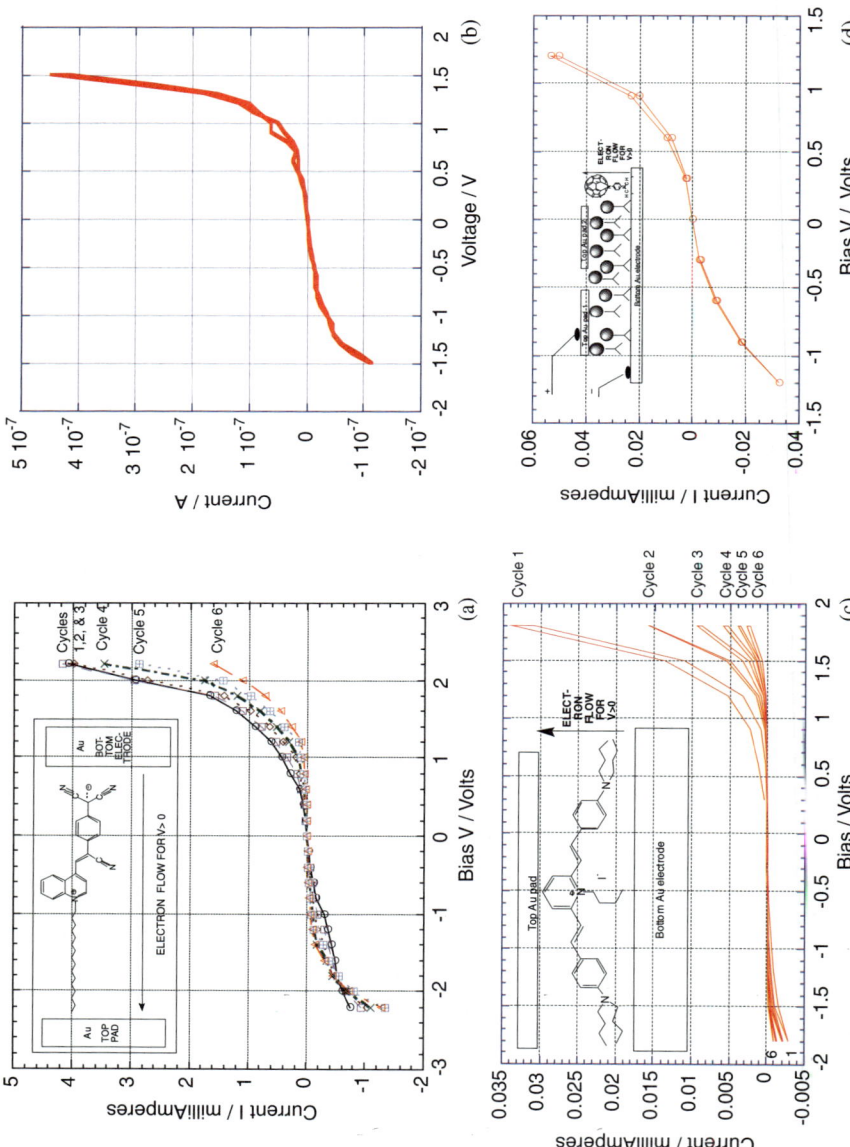

Figure 8.4 The rectification of MOM sandwiches consisting of three elements: (i) a macroscopic bottom Au or Al electrode, (ii) an 0.3 mm² top Al or "cold Au" electrode pad, and, between them, (iii) (a) an LB monolayer of **8**;[83] (b) an LB monolayer of **9**;[91] (c) an LB monolayer of **10**;[84] (d) an LB monolayer of **11**;[85] (e) an LS monolayer of **12**;[87] (f) an LB monolayer of **13**;[88] (g) an LS monolayer of **14**;[88] (h) an LS monolayer of **15**.[90]

Unimolecular Electronic Devices

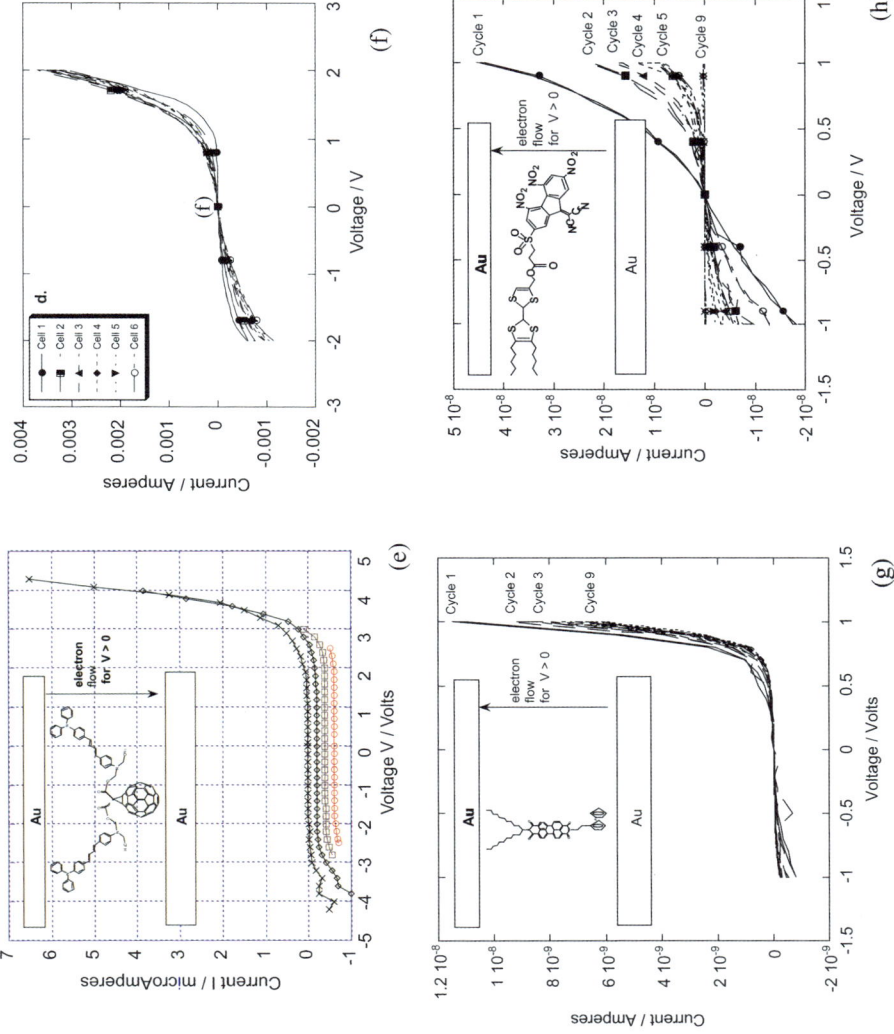

Figure 8.4 (Continued).

Table 8.1 Summary data for eight unimolecular rectifiers **8–15**. All compounds were measured at room temperature in air between Au electrodes inside a Faraday cage (**8** was also measured earlier between Al electrodes at 300 K,[79] also between 105 K and 370 K[80]). The column "# pads" lists how many independent typical MOM pads were discussed in each publication as rectifying (out of hundreds measured).

Str.	Type	Transf. Press. (mN/m)	LB or LS?	RR eqn (8.8)	# pads	Survives cycling?	U, A, or S?	Refs.
8	D^+-π-A^-	20	LB	2–27	16	no	U,A	83
9	D^+-π-A^-	28	LB	3–64	3	no	U,A	91
10	D^+ iodide	22	LB	8–60	24	no	A	84
11	D-σ-A	22	LB	2	1	no	A	85
12	D-σ-A	23	LS	2–16	9	yes	A	87
13	D-σ-A	32	LB	2–5	4	yes	U,A	88
14	D-σ-A	35	LB	28	1	yes	U,A	88
15	D-σ-A	35	LS	3	1	no	U	90

rectification (RRs scattered between 2 and 10) was seen if the molecule consisted of one tetrafluorophenyl group at one end, and a plain phenyl group at the other end, but no rectification was seen if the ends were chemically symmetrical.[144]

The first confirmed rectifier, hexadecylquinolinium tricyanoquinodimethanide **8**,[76–77,79–80,82–83] is a ground-state zwitterion D^+-π-A^-, connected by a twisted "pi" bridge (not a "sigma" bridge): it is sparingly soluble in polar solvents, and has a ground-state static electric dipole moment of **8** of $\mu_{GS} = 43 \pm 8$ Debye at infinite dilution in CH_2Cl_2.[79] The absorption spectrum in solution shows a hypsochromic band, peaked between 600 and 900 nm: this is an intervalence transfer (IVT) or internal charge-transfer band,[79,92] which fluoresces in the near IR.[92] The excited-state dipole moment is $\mu_{ES} = 3$ to 9 Debye.[92] Since the molecule is hypsochromic, the ground state must be D^+-π-A^-, and the first electronically excited state must be D^0-π-A^0; the finite twist angle between the quinolinium ring and the tricyanoquinodimethanide ring allows for an intense IVT band between the D^+ and A^- ends of the molecule: there is an intermolecular aggregate peak at 570 nm[79] polarised in the plane of an LB multiplayer[91] (which was believed to be the IVT peak[79]), and the "real" IVT peak, polarised perpendicular to the monolayer, at 535 nm.[91] A second, transient peak can appear at the air/water interface at 670 nm.[145] Molecule **8** forms amphiphilic Pockels–Langmuir monolayers at the air/water interface, with a collapse pressure of 34 mN m^{-1} and collapse areas of 50 Å^2 at 20 °C,[79] that transfers on the upstroke, with transfer ratios around 100% onto hydrophilic glass, quartz, or aluminium[74,79] or fresh hydrophilic Au,[82,83] but transfers poorly on the downstroke onto graphite, with a transfer ratio of about 50%.[74] The LB monolayer thickness of **8** is 23 Å[79] or 29 Å[83] by X-ray diffraction, 23 Å by spectroscopic ellipsometry,[83] 22 Å by surface plasmon resonance,[79,93] and 25 Å by X-ray photoelectron spectrometry (XPS).[93] With an averaged monolayer thickness of 23 Å and a calculated molecular length of 33 Å, the molecule on Al or Au has a tilt angle of 46° from the surface normal.[79] The XPS spectrum of one monolayer of **8** on Au shows two N(1s) peaks.[93] An angle-resolved XPS spectrum shows that the N atoms of the CN group are closer to the Au substrate than the quinolinium N.[93] The valence-band portion of the XPS spectrum agrees roughly with the density of molecular energy states.[92] The contact angle of a drop of water on fresh "hydrophilic Au" is 40° (it should be 0° if the gold were perfectly hydrophilic); this angle is 92° above a monolayer of **8** deposited on fresh hydrophilic Au (this exposes the nonpolar tail to water).[93] The orientation of **8** is confirmed by a grazing-angle FTIR study of **8** on Al[79] or on Au.[93]

LB monolayers and multilayers of **8** were sandwiched between macroscopic Al electrodes,[79] and later, using the "cold gold" technique,[78] between Au electrodes.[82,83] Between Al electrodes (with their inevitable patchy and defect-ridden covering of oxide), the monolayer has a dramatically asymmetric current. For **8**, RR = 26 at 1.5 V.[79] Assuming a molecular area of 50 Å^2, the current at 1.5 V corresponds to 0.33 electrons molecule^{-1} s^{-1}.[79] The RRs vary from pad to pad, as does the current, because these are all two-probe measurements, with all electrical resistances (Al, Ga/In or Ag paste, wires, *etc.*) in series. As high potentials are scanned repeatedly, the *I–V* curves become less asymmetric; the RRs decrease gradually with repeated cycling of the bias across the monolayer. In the range 105 K < *T* < 390 K, the onset of rectification of **8** between Al electrodes showed no temperature dependence.[80]

With oxide-free Au electrodes, the current through the "Au–monolayer of **8**–Au" pads increased dramatically, as expected, but the asymmetry persisted: the highest current was 90 400 electrons molecule^{-1} s^{-1}.[82,83] The best RR was 27.53 at 2.2 V.[83] Figure 8.4(a) shows how the rectification ratio decreases from cycles 1 to 6. For some cells, the current increases until breakdown occurs; in some cells this happens at 5.0 V, *i.e.* the cells suffer dielectric breakdown only at a field close to 2 GV m^{-1}.[83]

Ashwell and coworkers confirmed that Z-type 30-layer films of **8** rectify between Au electrodes;[145] the currents[145] were three orders of magnitude smaller than those reported for the monolayer.[83]

A tetrafluoro analogue of **8**, *i.e.* molecule **9**, also rectifies, Figure 8.4(b).[91]

The unwelcome gradual decreases in the electrical conductivity and in the RR of an LB monolayer of **8** (from an initial value of 27[79,83] to close to 1 upon repeated cycling) led to combining the LB and SAM techniques, by measuring thioacetyl variants of **8**, which could bind strongly to Au electrodes.[86,89] These variants were synthesized[86,89] with the aim of preparing molecules that (1) form good Langmuir (or Pockels–Langmuir) monolayers at the air/water interface, then (2) bind covalently to an Au substrate after either LB or LS transfer: the good ordering, afforded by the LB technique, should combine with a very sturdy chemical bond to the Au substrate (SAM formation) after LB transfer. The variant of **8**[86] with an undecyl "tail" followed by a thioacetyl termination ("C11 thioacetyl") gave disappointing results: the pressure–area isotherm indicated that the Pockels–Langmuir film collapsed at relatively low surface pressures, compared to **8**, and yielded disordered LB monolayers, with competition between strong physisorption by the dicyanomethide end of the molecule and Au-to-thiolate chemisorption. The monolayer rectified in either direction, depending on where in the LB monolayer, *i.e.* on which molecule ("right side up" or "upside down") the STM tip was probing.[86] Longer variants (C14 and C16 thioacetyl derivatives) did much better.[89]

2,6-Didibutylamino-phenylvinyl-1-butylpyridinium iodide, **10**, forms a Pockels–Langmuir film at the air/water interface, and transfers to hydrophilic substrates as a Z-type multilayer.[84] The monolayer thickness was 0.7 nm by spectroscopic ellipsometry, 1.3 nm by X-ray diffraction, and 1.15 nm or 1.18 nm by surface plasmon resonance at $\lambda = 532$ nm and 632.8 nm, respectively.[84] The films exhibit an absorption maximum at 490 nm (which is slightly hypsochromic in solution), attributable to iodide-to-pyridinium back-charge-transfer, and a second harmonic signal $\chi^{(2)} = 50$ pm V^{-1} at normal incidence ($\lambda = 1064$ nm) and 150 pm V^{-1} at 45°.[84] The rectification shows a decrease of rectification upon successive cycles (Figure 8.4(c)). Some cells have initial RRs as high as 60. The favoured direction of electron flow is from the gegenion to the pyridinium ion, *i.e.* in the direction of "back charge transfer"; the rectification in **2** may be attributed to an interionic electron transfer, or to an intramolecular electron transfer.[84]

Dimethylaminophenylazafullerene, **11**, is a moderate rectifier, but can also exhibit a tremendous but spurious apparent rectification ratio (as high as 20 000),[85] which is probably due to a partial penetration ("electromigration") of Au stalagmites.[85] The azafullerene **11** consists of a weak electron donor (dimethylaniline) bonded to a moderate electron acceptor (N-capped C_{60}), with an IVT peak at 720 nm.[85] The Langmuir film is very rigid, *i.e.* the slope of the isotherm is relatively large. However, the molecular areas are 70 Å^2 at extrapolated zero pressure, and 50 Å^2 at the chosen LB film transfer pressure of 22 mN m^{-1},[85] whereas the true molecular area of C_{60} is close to 100 Å^2. Therefore it is thought that the molecules **3**, transferred onto Au on the upstroke, are somewhat staggered, as is shown in the insert of Figure 8.4(d), with the more hydrophilic dimethylamino group closer to the bottom Au electrode. The film thickness is 2.2 nm by XPS.[85] Angle-resolved N(1s) XPS spectra confirm that the two N atoms are closer to the bottom Au electrode than is the C_{60} cage.[85] One must ignore the *I–V* plots that show large currents due to electromigration. Some cells show a much smaller current, which is slightly rectifying in the forward direction, with RR ≈ 2 (Figure 8.4(d)).[85]

Very sturdy rectification was seen in a Langmuir–Schaefer (LS monolayer of fullerene-bis-(4-diphenylamino-4"-(N-ethyl-N-2"'-hydroxyethylamino-1,4-diphenyl-1,3-butadiene malonate **12** between Au electrodes.[87] Molecule **12** is based on two triphenylamines (two one-electron donors) and a single fullerene (weak one-electron acceptor): a Langmuir–Schaefer monolayer of **12** rectifies (Figure 8.4(e)); RR does not decrease at all upon successive cycling. The monolayer is probably very dense and stiff; so stiff, in fact, that it cannot be transferred onto an Au substrate by the vertical LB process, but adheres to Au, if it is transferred by the horizontal, or Langmuir–Schaefer process.[87]

N-(10-nonadecyl)-N-(1-pyrenylmethyl)perylene-3,4,9,10-bis(dicarboximide), **13**, is a D-σ-A molecule, based on the moderate pyrene donor D, a one-carbon bridge, and the moderate perylenebisimide acceptor A 88: it has a persistent RR (Figure 8.4(f)).[88]

N-(10-nonadecyl)-N-(2-ferrocenylethyl)perylene-3,4,9,10-bis(dicarboximide), **14**, is a D-σ-A molecule, based on the moderate ferrocene donor D, a two-carbon bridge, and the moderate perylenebisimide acceptor A;[88] its has an IVT band that peaks at 595 nm;[88] its Pockels–Langmuir isotherm shows that **14** can be transferred as a monolayer at the fairly high surface pressure of 35 mN m^{-1}, and forms a rectifier, with RR between 25 and 35, which does not change much upon cycling (Figure 8.4(g)).[88]

4,5-Dipentyl-5′-methyltetrathiafulvalen-4′-methyloxy 2,4,5-trinitro-9-dicyano-methylenefluo-rene-7-(3-sulfonylpropionate), **15**, is also a D-σ-A molecule, based on D = tetrathiafulvalene, and A = dicyanomethylenetrinitrofluorene: it has an IVT band maximum at 1220 nm, and was transferred to a fresh hydrophilic Au substrate at a surface pressure of 21 mN m^{-1};[57] its I–V curves (Figure 8.4(h)) show that the RRs decrease upon cycling, and become unity after about 9 cycles of measurement.[90] Similar results were reported for a very closely related molecule.[138]

8.7 Switches

Switches require bistability, *i.e.* the availability of two states, which are both at least kinetically stable. A crystal with bulk bistability is CuTCNQ, which is metastable between its neutral form Cu^{0}TCNQ^0 and its ionic form, Cu$^+$TCNQ$^-$: this allowed for high- and low-voltage conductivities,[146] but despite much work, many publications and patents, it did not become a practical device. An LB monolayer of a bistable 3catenane closed-loop molecule, with a naphthalene group as one "station", and tetrathiafulvalene as the second "station", and a tetracationic catenane hexafluorophospate salt traveling on the catenane, like a "train" on a closed track, was deposited on poly-Si as one electrode, and topped by a 5-nm Ti layer and a 100-nm Al electrode. The current–voltage plot is asymmetric as a function of bias (which may move the train on the track), and a succession of read-write cycles shows that the resistance changes stepwise, as the train(s) move from the lower-conductivity station(s) to the higher-conductivity station(s).[147] Infrared spectroscopy showed that depositing Ti atop the catenane does indeed lead to chemical reaction of Ti with the "top" of the monolayer, but preserves the "working part" of the molecular switch.[148]

8.8 Capacitors

Bistable molecules and unimolecular rectifiers could also be used as capacitors, but this possibility has not received much attention so far.

8.9 Future Flash Memories

A flash memory device has a middle electrode that is not in electrical contact with the outside circuit, so, after a polarisation pulse, charges can be stored for a long time (but not forever) on this middle electrode. Two monolayers of unimolecular rectifiers, separated by a middle "floating" electrode, could be used as flash memory devices, but this has never been tested.

8.10 Field Effect Transistors

A field effect transistor (FET) requires a semiconducting channel connecting source and drain electrodes whose "thickness" can be modified by an applied bias on the gate electrode. For this, any semiconductor will do. Present integrated circuits use FETs preferentially over BJTs because of their ease of fabrication. FET behaviour was observed for LB monolayers and multilayers some time ago.[149] FET behaviour has been observed by STM for a single-walled carbon nanotube curled over parallel Au lines, with the STM acting as a gate electrode,[150] and much work has since been devoted to these FETs. The difficulty of ordering the nanotubes has so far prevented practical use.

8.11 Negative Differential Resistance Devices

Using a "nanopore" technique, molecules of 2'-amino-4-ethynylphenyl-4'-ethynylphenyl-5'-nitro-benzene-1-thiolate, attached to Au on one side and topped by a Ti electrode on the other, exhibit negative differential resistance (NDR);[151] this molecule, when studied by STM, shows time-dependent oscillations in conductance, presumably due to a change in tilt angle of the organothiolate with respect to the Au substrate.[152] However, if the second "top" metal is deposited at room temperature, then evidence of chemical reactions at the open surface of alkoxyorganothiolates (Al, Cu, Ag, and especially the very reactive Ti) or of interpenetration to the bottom of the SAM, close to the Au/S interface (Au), has been presented.[153,154]

8.12 Coulomb-blockade Device and Single-electron Transistor

The organometallic equivalent of a no-gain organometallic single-electron transistor (SET), *i.e.* a Coulomb-blockade device, was realised at 0.1 K with a Co(II) complex, using two electromigrated Au electrodes covalently bonded to the molecule, and a Si gate electrode at 30 nm from the molecule.[155]

8.13 Future Unimolecular Amplifiers

A three-sided molecule, designed to control the current pathway within it by judicious choice of three moieties with different electron affinities and/or ionisation potentials, when covalently bonded to three metal electrodes 3 nm apart, should be the unimolecular equivalent of a bipolar junction transistor.[122,124,130,132] Many suitable molecules can be designed, with endgroups designed for SAM formation with two or three dissimilar metal electrodes, but at present it is highly nontrivial to fabricate, even by electromigration, three electrodes 3 nm apart.

8.14 Future Organic Interconnects

Once a sufficient set of resistors, capacitors, rectifiers, and amplifiers have been demonstrated to work with conventional metal electrodes, one can initiate a new project, of assembling all-organic polymeric electrodes to replace the inorganic metals. This would lead to the all-organic computer! The controlled electrochemical growth of conducting oligomer filaments has already been demonstrated.[156]

Acknowledgements

This work was made possible by the diligence and insight of many colleagues, students, and post-doctoral fellows, to whom I owe a large debt of gratitude. I hope that they had fun working on these problems!

References

1. R.M. Metzger, in R.M. Metzger, P. Day and G. Papavassiliou, ed., *Lower-Dimensional Systems and Molecular Electronics*, NATO ASI Series, 1991, **B248**, 659.
2. R.P. Feynman, in H.D. Gilbert, ed., *Miniaturization*, Reinhold, New York, 1961, p. 282.
3. G.E. Moore, *Electronics*, 1965, **38**, 114.
4. A. Aviram and M.A. Ratner, *Chem. Phys. Lett.*, 1974, **29**, 277.
5. T.C. Koopmans, *Physica*, 1933, **1**, 104.
6. W. Schottky, *Z. Phys.*, 1942, **118**, 539.
7. G.S. Ohm, *Die Galvanische Kette, Mathematisch Bearbeitet*, Riemann, Berlin, 1827.
8. G. Binnig, H. Rohrer, Ch. Gerber and E. Weibel, *Phys. Rev. Lett.*, 1982, **49**, 57.
9. G. Binnig, C.F. Quate and Ch. Gerber, *Phys. Rev. Lett.*, 1986, **56**, 930.
10. U. Dürig, O. Züger and D.W. Pohl, *Phys. Rev. Lett.*, 1990, **65**, 349.
11. K.B. Blodgett, *J. Am. Chem. Soc.*, 1935, **57**, 1007.

12. K.B. Blodgett and I. Langmuir, *Phys. Rev.*, 1937, **51**, 964.
13. I. Langmuir and V.J. Schaefer, *J. Am. Chem. Soc.*, 1938, **60**, 1351.
14. (a) W.C. Bigelow, D.L. Pickett and W.A. Zisman, *J. Colloid Sci.*, 1946, **1**, 513; (b) C.J. Muller, Ph.D. Thesis, University of Leiden, 1991.
15. M.A. Reed, C. Zhou, C.J. Muller, T.P. Burgin and J.M. Tour, *Science*, 1997, **278**, 252.
16. K.S. Ralls, R.A. Buhrman and T.C. Tiberio, *Appl. Phys. Lett.*, 1989, **55**, 2459.
17. C. Zhou, M.R. Deshpande, M.A. Reed, L. Jones II and J.M. Tour, *Appl. Phys. Lett.*, 1997, **71**, 611.
18. H. Park, A.K.L. Lim, A.P. Alivisatos, J. Park and P.L. McEuen, *Appl. Phys. Lett.*, 1999, **75**, 301.
19. X.D. Cui, A. Primak, X. Zarate, J. Tomfohr, O.F. Sankey, A.L. Moore, T.A. Moore, D. Gust, G. Harris and S.M. Lindsay, *Science*, 2001, **294**, 571.
20. T. Tighe, T. Daniel, Z. Zhu, S. Upilli, N. Winograd and D.L Allara, *J. Phys. Chem. B.*, 2005, **109**, 21006.
21. J.R. Black, *IEEE Trans. Electron. Devices*, 1969, **ED–16**, 338.
22. H. Taube and E.S. Gould, *Acc. Chem. Res.*, 1969, **2**, 321.
23. H. Taube, *Angew. Chem. Int. Ed. Engl.*, 1984, **23**, 329.
24. L.A. Bumm, J.J. Arnold, M.T. Cygan, T.D. Dunbar, T.P. Burgin, L. Jones II, D.L. Allara, J.M. Tour and P.S. Weiss, *Science*, 1996, **271**, 1705.
25. R. Landauer, *IBM J. Res. Dev.*, 1957, **1**, 223.
26. K. von Klitzing, G. Dorda and M. Pepper, *Phys. Rev. Lett.*, 1980, **45**, 494.
27. J.G. Simmons, *J. Phys. D*, 1971, **4**, 613.
28. J. Chen, T. Lee, J. Su, W. Wang, M.A. Reed, A.M. Rawlett, M. Kozaki, Y. Yao, R.C. Jagessar, S.M. Dirk, D.W. Price, J.M. Tour, D.S. Grubisha and D.W. Bennett, in M.A. Reed and T. Lee ed., *Molecular Nanoelectronics*, American Scientific Publishers, Stevenson Ranch, CA, 2003, p. 39.
29. S. Iijima and T. Ichihashi, *Nature*, 1993, **363**, 603.
30. R. Saito, M. Fujita, G. Dresselhaus and M.S. Dresselhaus, in L.Y. Chiang, A.F. Garito and D.J. Sandman, ed., *Electrical, Optical and Magnetic Properties of Organic Solid-State Materials*, *Mater. Res. Soc. Symp. Proc.*, 1992, 247, 333.
31. C.T. White, D.H. Robertson and J.W. Mintmire, *Phys. Rev. B*, 1993, **47**, 5485.
32. D.V. Averin and K.K. Likharev, *J. Low-Temp. Phys.*, 1986, **62**, 345–373.
33. N.J. Geddes, J.R. Sambles, D.J. Jarvis, W.G. Parker and D.J. Sandman, *Appl. Phys. Lett.*, 1990, **56**, 1916.
34. N.J. Geddes, J.R. Sambles, D.J. Jarvis, W.G. Parker and D.J. Sandman, *J. Appl. Phys.*, 1992, **71**, 756.
35. C. Krzeminski, C. Delerue, G. Allan, D. Vuillaume and R.M. Metzger, *Phys. Rev. B*, 2001, **64**, # 085405.
36. V. Mujica, M.A. Ratner and A. Nitzan, *Chem. Phys.*, 2002, **281**, 147.
37. M.L. Chabinyc, X. Chen, R.E. Holmlin, H. Jacobs, H. Skulason, C.D. Frisbie, V. Mujica, M.A. Ratner, M.A. Rampi and G.M. Whitesides, *J. Am. Chem. Soc.*, 2002, **124**, 11730.

38. I.R. Peterson, D. Vuillaume and R.M. Metzger, *J. Phys. Chem. A*, 2001, **105**, 4702.
39. A. Stabel, P. Herwig, K. Müllen and J.P. Rabe, *Angew. Chem. Int. Ed.*, 1995, **34**, 1609.
40. L.E. Hall, J.R. Reimers, N.S. Hush and K. Silverbrook, *J. Chem. Phys.*, 2000, **112**, 1510.
41. R.M. Metzger and C.A. Panetta, *J. Phys. Paris. 44 Colloque.C.*, 1983, **3**, 1605.
42. R.M. Metzger and C.A. Panetta, In F.L. Carter, ed., *Molecular Electronic Devices*, Vol. II, Dekker, New York, 1987, p. 5.
43. C.A. Panetta, J. Baghdadchi and R.M. Metzger, *Mol. Cryst. Liq. Cryst.*, 1984, **107**, 103.
44. R.M. Metzger, C.A. Panetta, N.E. Heimer, A.M. Bhatti, E. Torres, G.F. Blackburn, S.K. Tripathy and L.A. Samuelson, *J. Mol. Electron.*, 1986, **2**, 119.
45. R.M. Metzger, C.A. Panetta, Y. Miura and E. Torres, *Synth. Met.*, 1987, **18**, 797.
46. R.M. Metzger and C.A. Panetta, in J.L. Heiras and T. Akachi ed., *Proc. Eighth Winter Meeting on Low-Temperature Physics*, UNAM, Mexico City, 1987, p. 81.
47. E. Torres, C.A. Panetta and R.M. Metzger, *J. Org. Chem.*, 1987, **52**, 2944.
48. R.M. Metzger and C.A. Panetta, in P. Delhaès, M. Drillon, ed., *Organic and Inorganic Lower-Dimensional Materials, NATO ASI Ser*, **B168**, 271 Plenum, New York, 1988.
49. R.M. Metzger, R.R. Schumaker, M.P. Cava, R.K. Laidlaw, C.A. Panetta and E. Torres, *Langmuir*, 1988, **4**, 298.
50. Y. Miura, E. Torres, C.A. Panetta and R.M. Metzger, *J. Org. Chem.*, 1988, **53**, 439.
51. Y. Miura, R.K. Laidlaw, C.A. Panetta and R.M. Metzger, *Acta Crystallogr. C*, 1988, **44**, 2007.
52. R.K. Laidlaw, Y. Miura, C.A. Panetta and R.M. Metzger, *Acta Crystallogr. C*, 1988, **44**, 2009.
53. R.K. Laidlaw, J. Baghdadchi, C.A. Panetta, Y. Miura, E. Torres and R.M. Metzger, *Acta Crystallogr. B*, 1988, **44**, 645.
54. Y. Miura, C.A. Panetta and R.M. Metzger, *J. Liquid Chromat.*, 1988, **11**, 245.
55. R.M. Metzger and C.A. Panetta, *J. Mol. Electron.*, 1989, **5**, 1.
56. R.M. Metzger and C.A. Panetta, *J. Chim. Phys.*, 1988, **85**, 1125.
57. R.M. Metzger and C.A. Panetta, *Synth. Met.*, 1989, **28**, C807.
58. R.M. Metzger, R.K. Laidlaw, E. Torres and C.A. Panetta, *J. Cryst. Spectros, Rces.*, 1989, **19**, 475.
59. R.M. Metzger and C.A. Panetta, in A. Aviram, ed., *Molecular Electronics-Science and Technology* New York Engineering Foundation, New York, NY, 1989, p.293.
60. R.M. Metzger, D.C. Wiser, R.K. Laidlaw, M.A. Takassi, D.L. Mattern and C.A. Panetta, *Langmuir*, 1990, **6**, 350.

61. R.M. Metzger and C.A. Panetta, in R.M. Metzger, P. Day and G.C. Papavassiliou, ed., *Lower-Dimensional Systems and Molecular Electronics, NATO ASI Ser. Ser. B*, **248**, 611 Plenum Press, New York, 1991.
62. R.M. Metzger and C.A. Panetta, in L.Y. Chiang, D.O. Cowan and P. Chaikin, ed., *Advanced Organic Solid State Materials, Mater. Res. Soc. Symp. Proc. Ser.*, 1990, **173**, 531..
63. R.M. Metzger and C.A. Panetta, *New J. Chem.*, 1991, **15**, 209.
64. R.M. Metzger and C.A. Panetta in J.L. Beeby, ed., *Condensed Systems of Low Dimensionality NATO ASI Ser. B*, 1991, **253**, 779..
65. R.M. Metzger and C.A. Panetta, *Synth. Met.*, 1991, **42**, 1407.
66. C.A. Panetta, N.E. Heimer, C.L. Hussey and R.M. Metzger, *Synlett*, 1991, 301.
67. R.M. Metzger, in A. Aviram, ed., *Molecular Electronics-Science and Technology, Am. Inst. Phys. Conf. Proc.*, 1992, **262**, 85.
68. X.-L. Wu, M. Shamsuzzoha, R.M. Metzger and G.J. Ashwell, *Synth. Met.*, 1993, **57**, 3836.
69. P. Wang, J.L. Singleton, X.-L.Wu, M. Shamsuzzoha, R.M Metzger, C.A. Panetta and N.E. Heimer, *Synth. Met.*, 1993, **57**, 3824.
70. R.M. Metzger, in M. Blank, ed., *Electricity and Magnetism in Biology and Medicine*, San Francisco Press, San Francisco, CA, 1993, p. 175.
71. R.M. Metzger, in R.R. Birge, ed., *Molecular and Biomolecular Electronics, Am. Chem. Soc. Adv. in Chem. Ser.*, **240**, 81 American Chemical Society, Washington, DC, 1994.
72. H. Nadizadeh, D.L. Mattern, J. Singleton, X.-L. Wu and R.M. Metzger, *Chem. Mater.*, 1994, **6**, 268–277.
73. R.M. Metzger, *Mater. Sci. Eng. C*, 1995, **3**, 277.
74. R.M. Metzger, H. Tachibana, X. Wu, U. Höpfner, B. Chen, M.V. Lakshmikantham and M.P. Cava, *Synth. Met.*, 1997, **85**, 1359.
75. R.M. Metzger, in H. Sasabe, ed., *Hyper-Structured Molecules I, Chemistry, Physics and Applications*,Gordon & Breach Science Publishers, Amsterdam, 1999, p. 19.
76. G.J. Ashwell, J.R. Sambles, A.S. Martin, W.G. Parker and M. Szablewski, *J. Chem. Soc. Chem. Commun.*, 1990, 1374.
77. A.S. Martin, J.R. Sambles and G.J. Ashwell, *Phys. Rev. Lett.*, 1993, **70**, 218.
78. N. Okazaki and J.R. Sambles, In *Extended Abstracts of the International Symposium on Organic Molecular Electronics, Nagoya, Japan*, 2000, p. 66.
79. R.M. Metzger, B. Chen, U. Höpfner, M.V. Lakshmikantham, D. Vuillaume, T. Kawai, X. Wu, H. Tachibana, T.V. Hughes, H. Sakurai, J.W. Baldwin, C. Hosch, M.P. Cava, L. Brehmer and G.J. Ashwell, *J. Am. Chem. Soc.*, 1997, **119**, 10455.
80. B. Chen and R.M. Metzger, *J. Phys. Chem. B*, 1999, **103**, 4447.
81. D. Vuillaume, B. Chen and R.M. Metzger, *Langmuir*, 1999, **15**, 4011.
82. T. Xu, I.R. Peterson, M.V. Lakshmikantham and R.M. Metzger, *Angew. Chem. Int. Ed.*, 2001, **40**, 1749.
83. R.M. Metzger, T. Xu and I.R. Peterson, *J. Phys. Chem. B.*, 2001, **105**, 7280.

84. J.W. Baldwin, R.R. Amaresh, I.R. Peterson, W.J. Shumate, M.P. Cava, M.A. Amiri, R. Hamilton, G.J. Ashwell and R.M. Metzger, *J. Phys. Chem. B*, 2002, **106**, 12158.
85. R.M. Metzger, J.W. Baldwin, W.J. Shumate, I.R. Peterson, P. Mani, G.J. Mankey, T. Morris, G. Szulczewski, S. Bosi, M. Prato, A. Comito and Y. Rubin, *J. Phys. Chem. B*, 2003, **107**, 1021.
86. A. Jaiswal, R.R. Amaresh, M.V. Lakshmikantham, A. Honciuc, M.P. Cava and R.M. Metzger, *Langmuir*, 2003, **19**, 9043.
87. A. Honciuc, A. Jaiswal, A. Gong, K. Ashworth, C.W. Spangler, I.R. Peterson, L.R. Dalton and R.M. Metzger, *J. Phys. Chem. B*, 2005, **109**, 857.
88. W.J. Shumate, D.L. Mattern, A. Jaiswal, J. Burgess, D.A. Dixon, T.R. White, A. Honciuc and R.M. Metzger, *J. Phys. Chem. B*, 2006, **110**(23), 11146.
89. A. Jaiswal, D. Rajagopal, M.V. Lakshmikantham, M.P. Cava and R.M. Metzger, *Phys. Chem. Chem. Phys.*, in press.
90. W.J. Sumate, Ph. D. dissertation, University of Alabama, 2005.
91. A. Honciuc, A. Otsuka, Y.-H. Wang, S.K. McElwee, S.A. Woski, G. Saito and R.M. Metzger, *J. Phys. Chem B*, 2006, **110**(31), 15085.
92. J.W. Baldwin, B. Chen, S.C. Street, V.V. Konovalov, H. Sakurai, T.V. Hughes, C.S. Simpson, M.V. Lakshmikantham, M.P. Cava, L.D. Kispert and R.M. Metzger, *J. Phys. Chem. B.*, 1999, **103**, 4269.
93. T. Xu, T.A. Morris, G.J. Szulczewski, R.R. Amaresh, Y. Gao, S.C. Street, L.D. Kispert, R.M. Metzger and F. Terenziani, *J. Phys. Chem. B*, 2002, **106**, 10374.
94. F. Terenziani, A. Painelli, A. Girlando and R.M. Metzger, *J. Phys. Chem. B*, 2004, **108**, 10743.
95. O. Kwon, M.L. McKee and R.M. Metzger, *Chem. Phys. Lett.*, 1999, **313**, 321.
96. S. Scheib, M.P. Cava, J.W. Baldwin and R.M. Metzger, *J. Org. Chem.*, 1998, **63**, 1198.
97. T.V. Hughes, B. Mokijewski, B. Chen, M.V. Lakshmikantham, M.P. Cava and R.M. Metzger, *Langmuir*, 1999, **15**, 6925.
98. T. Xu, T.A. Morris, G.J. Szulczewski, R.M. Metzger and M. Szablewski, *J. Mater. Chem.*, 2002, **12**, 3167.
99. R.M. Metzger, B. Chen, D. Vuillaume, U. Höpfner, J.W. Baldwin, T. Kawai, H. Tachibana, H. Sakurai, M.V. Lakshmikantham and M.P. Cava, in L.Y. Chiang, L.R. Dalton, A.Y. Jen, J. Reynolds and M. Rubner ed., *Electrical, Optical and Magnetic Properties of Organic Solid-State Materials IV*, MRS Proc., **488**, Materials Research Society, Pittsburgh, PA, 1998, p. 335.
100. R.M. Metzger, B. Chen, D. Vuillaume, M.V. Lakshmikantham, U. Höpfner, T. Kawai, J.W. Baldwin, X. Wu, H. Tachibana, H. Sakurai and M.P. Cava, *Thin Solid Films*, 1998, **327–329**, 326.
101. S. Scheib, M.P. Cava, J.W. Baldwin and R.M. Metzger, *Thin Solid Films*, 1998, **327–329**, 100.

102. R.M. Metzger and M.P. Cava, in *Molecular Electronics, Science and Technology*, Ann. N. Y. Acad. Sci., 1998, **852**, 95.
103. R.M. Metzger, *Adv. Mater. Opt. Electron.*, 1998, **8**, 229.
104. R.M. Metzger, *Mol. Cryst. Liq. Cryst. Sci. Technol. A*, 1999, **337**, 37.
105. R.M. Metzger, *J. Mater. Chem.*, 1999, **9**, 2027.
106. R.M. Metzger, *J. Mater. Chem.*, 2000, **10**, 55.
107. R.M. Metzger, *Synth. Met.*, 2000, **109**, 23.
108. R.M. Metzger, *Acc. Chem. Res.*, 1999, **32**, 950.
109. R.M. Metzger, B. Chen and J.W. Baldwin, in R. Glaser and P. Kaszinski, ed., *Anisotropic Organic Materials-Approaches to Polar Order*, Am. Chem. Soc. Symp. Proc., 2001, **798**, 50.
110. R.M. Metzger, in S.K. Pantelides, M.A. Reed, J.S. Murday and A. Aviram, ed., *Molecular Electronics*, Mater. Res. Soc. Symp. Proc., **582**, paper H12.2, Materials Research Society, Warrendale, PA, 2001.
111. R.M. Metzger, *Adv. Mater. Opt. Electron.*, 1999, **9**, 253.
112. R.M. Metzger, *Synth. Met.*, 2001, **124**, 107; and in I. Ledoux-Rak, Z. Bao, J. Zyss ed., *Molecular Photonics, from Macroscopic to Nanoscopic Applications*, European Mater. Res. Soc. Symp. Proc.**96** Elsevier, Amsterdam, 2001.
113. R.M. Metzger, in L. Merhari, J.A. Rogers, A. Karim, D.J. Norris and Y. Xia, ed., *Nonlithographic and Lithographic Methods of Nanofabrication – From Ultrahigh – Scale Integration to Photonics to Molecular Electronics"*, Mater. Res. Soc. Symp. Proc., **636**, Materials Res. Soc., Warrendale, PA, 2001, p. D7.8/JJ9.8.1 2001.
114. R.M. Metzger, *J. Macromol. Sci. A.*, 2001, **38**, 1499.
115. R.M. Metzger, *J. Solid State Chem.*, 2002, **168**, 696.
116. R.M. Metzger, in M.A. Reed and T. Lee, ed., *Molecular Nanoelectronics*, American Scientific Publishers, Stevenson Ranch, CA, 2003, p.19.
117. R.M. Metzger, in: *Structural and Electronic Properties of Molecular Nanostructures*, AIP Conf. Proc. **663** AIP Conf. Proc., Melville, NY, 2002, p. 531.
118. R.M. Metzger, *Nanotechnology*, 2003, **13**, 585.
119. R.M. Metzger, *Synth. Met.*, 2003, **137**, 1499.
120. D. Adams, L. Brus, C.E.D. Chidsey, S. Creager, C. Creutz, C.R. Kagan, P. Kamat, M. Lieberman, S. Lindsay, R.A. Marcus, R.M. Metzger, M.E. Michel-Beyerle, J.R. Miller, M.D. Newton, D.R. Rolison, O. Sankey, K.S. Schanze, J. Yardley and X. Zhu, *J. Phys. Chem. B.*, 2003, **107**, 6668.
121. R.M. Metzger, *Chem. Rev.*, 2003, **103**, 3803–3834.
122. R.M. Metzger, in J.R. Reimers, C.A. Picconatto, J.C. Ellenbogen and R. Shashidhar, ed., *Molecular Electronics III*, Ann. N. Y. Acad. Sci., 2003, **1006**, 252.
123. R.M. Metzger, in *Encyclopedia of Supramolecular Chemistry*, Vol. II, p. 1525 Marcel Dekker, New York, NY, 2004.
124. R.M. Metzger, in L. Ouahab and E. Yagubskii ed., *Organic Conductors, Superconductors and Magnets, from Synthesis to Molecular Electronics"*, NATO ASI Ser. II Kluwer, Dordrecht, The Netherlands, 2004, p. 269.

125. D.L. Allara, C.L. McGuinness and R.M. Metzger, in L. Ouahab and E. Yagubskii ed., *Organic Conductors, Superconductors and Magnets, from Synthesis to Molecular Electronics"*, *NATO ASI Ser. II* Kluwer, Dordrecht, The Netherlands, 2004, p. 295.
126. R.M. Metzger, *Electrochem. Soc. Interf.*, 2004, **13**, 40.
127. R.M. Metzger, in M. Razeghi and G.J. Brown, ed., *Quantum Sensing and Nanophotonic Devices" SPIE* **5359** 153 SPIE – The International Society for Optical Engineering, Bellingham, WA, 2004.
128. R.M. Metzger, *Chem. Record*, 2004, **4**, 291.
129. R.M. Metzger, in J. Kahovec, ed., *Electronic Phenomena in Organic Solids, Macromol. Symposia*, **212**, 63 Wiley-VCH, Weinheim, Germany, 2004.
130. R.M. Metzger, in *Proc. of the First Conference on Foundations of Nanoscience, Self-Assembled Architectures and Devices FNANO04* Sciencetechnica, 2004, p. 151.
131. I.R. Peterson and R.M. Metzger, *IEE Proc. Circ. Dev. Syst.*, 2004, **151**, 452.
132. R.M. Metzger, in G. Cuniberti, G. Fagas and K. Richter, ed., *Introducing Molecular Electronics, Springer Lecture Notes on Physics*, **680** Springer, Berlin Heidelberg New York, 2005, p. 313.
133. R.M. Metzger, *Colloids Surf. A*, 2006, **285**, 2.
134. R.M. Metzger, *Anal. Chim. Acta*, 2006, **568**, 146.
135. R.M. Metzger, *Chem. Phys.*, 2006, **326**, 176.
136. R.M. Metzger, in *Multifunctional Conducting Molecular Materials*, G. Saito, F. Widl, R.C. Haddan, K. Tanigoki, T. Endki, H.E. Katz and M. Maesato (eds.), Royal Society of Chemistry, Cambridge, p. 238.
137. D.F. Perepichka, M.R. Bryce, C. Pearson, M.C. Petty, E.J.L. McInnes and J.P. Zhao, *Angew. Chem. Int. Ed.*, 2003, **42**, 4636.
138. G. Ho, J.R. Heath, M. Kontratenko, D.F. Perepichka, K. Arseneault, M. Pézolet and M.R. Bryce, *Chem. Eur. J.*, 2005, **11**, 2914.
139. G.J. Ashwell, W.D. Tyrrell and A.J. Whittam, *J. Mater. Chem.*, 2003, **13**, 2855.
140. G.J. Ashwell, A. Chwialkowska and L.R. Herrmann High, *J. Mater. Chem.*, 2004, **14**, 2848.
141. G.J. Ashwell and M. Berry, *J. Mater. Chem.*, 2005, **15**, 108.
142. G.J. Ashwell, W.D. Tyrrell and A.J. Whittam, *J. Am. Chem. Soc.*, 2005, **126**, 7102.
143. G.M. Morales, P. Jiang, S. Yuan, Y. Lee, A. Sanchez, W. You and L. Yu, *J. Am. Chem. Soc.*, 2005, **127**, 10456.
144. M. Elbing, R. Ochs, M. Keotopp, M. Fischer, C. von Hänisch, F. Weigend, F. Evers, H.B. Weber and M. Mayor, *Proc. Natl. Acad. Sci. US*, 2005, **102**, 8815.
145. G.J. Ashwell and G.A.N. Paxton, *Aust. J. Chem.*, 2002, **55**, 199.
146. R.S. Potember, T.O. Poeher and D.O. Cowan, *Appl. Phys. Lett.*, 1979, **34**, 405.
147. C.P. Collier, G. Mattersteig, E.W. Wong, K. Beverly, J. Sampaio, F.M. Raymo, J.F. Stoddart and J.R. Heath, *Science*, 2000, **289**, 1172.

148. E. DeIonno, H.-R. Tseng, D.D. Harvey, J.F. Stoddart and J.R. Heath, *J. Phys. Chem. B*, 2006, **110**, 7609.
149. J. Paloheimo, P. Kuivalainen, H. Stubb, E. Vuorimaa and P. Yli-Lahti, *Phys. Lett.*, 1990, **56**, 1157.
150. S.J. Tans, M.H. Devoret, H. Dai, A. Thess, R.E. Smalley, L.J. Geerligs and C. Dekker, *Nature*, 1997, **386**, 474.
151. J. Chen, M.A. Reed, A.M. Rawlett and J.M. Tour, *Science*, 1999, **286**, 1550.
152. Z.J. Donhauser, B.A. Mantooth, K.F. Kelly, L.A. Bumm, J.D. Monnell, J.J. Stapleton, D.W. Price,Jr., A.M. Rawlett, D.L. Allara, J.M. Tour and P.S. Weiss, *Science*, 2001, **292**, 2303.
153. B.C. Haynie, A.V. Walker, T.B. Tighe, D.L. Allara and N. Winograd, *Appl. Surf. Sci.*, 2003, **203–204**, 433.
154. A.V. Walker, T.B. Tighe, O.M. Cabarcos, M.D. Reinard, B.C. Haynie, S. Uppili, N. Winograd and D.L. Allara, *J. Am. Chem. Soc.*, 2004, **126**, 3954.
155. J. Park, A.N. Pasupathy, J.I. Goldsmith, C. Chang, Y. Yaish, J.R. Petta, M. Rinkoski, J.P. Sethna, H.D. Abruña, P.L. McEuen and D.C. Ralph, *Nature*, 2002, **417**, 722.
156. H. He, J. Zhu, N.J. Tao, L.A. Nagahara, I. Amlani and R. Tsui, *J. Am. Chem. Soc.*, 2001, **123**, 7730.

CHAPTER 9
Piezoelectric Ceramics as Intelligent Multifunctional Materials

A. YOUSEFI-KOMA

School of Mechanical Engineering, College of Engineering, University of Tehran, Tehran, Iran

9.1 Introduction

Piezoelectricity is an electromechanical phenomenon derived from the Greek word "piezein" for "press" meaning "pressure electricity".[1] The piezoelectric effect of natural crystals was discovered by the Curie brothers in 1880. Rochelle salts and tourmaline were the best-known piezoelectric materials at that time. An extensive research and material development led to the present highly effective piezoelectric ceramic materials with adjustable parameters optimally adapted to customer requirements.

Although various applications of piezoelectric ceramics such as sonar and phonograph pickups have been realised since 1930, it is less than 30 years that piezoelectric ceramics have been employed as structural actuators/sensors in the development of intelligent structures. Forward[15] investigated the application of piezoelectric ceramics as passive dampers in mechanical systems. One of the first successful applications of piezoelectric materials in structural dynamics and control was introduced by Forward *et al*.[16] In their study piezoelectric sensors and actuators were used for vibration control of a flexible beam. Since then the application of piezoelectric ceramics as multifunctional intelligent material systems has grabbed the attention of researchers (Bailey and Hubbard[1], de Luis and Crawley[11]). Most of these efforts were funded by the Department of Defense (DoD), the Air Force, the Ballistic Missile Defense Organization (BMDO), the Strategic Defense Initiative Organization (SDIO). The Defense Advanced Research Project Agency (DARPA) got involved in smart structures and material systems, particularly piezoelectric ceramics, in 1993 by recognising this field as a breakthrough technology for the defence and aerospace industries. The Navy has also supported smart structures research by funding several projects on smart actuator materials specially ceramics, and single-crystal piezoelectric ceramics (Sater et al.[42]). This chapter briefly reviews the concept of piezoelectricity with emphasis on piezoelectric ceramics. Analytical and numerical modeling of piezoelectric ceramics is then discussed. Finally, some of the major applications of piezoelectric ceramics and their commercial products are introduced.

9.2 Piezoelectricity

The microscopic origin of the piezoelectric effect is the displacement of ionic charges within a crystal structure (Janos and Hagood[30]). Piezoelectric materials produce an electric field when subject to a mechanical strain, and if an electric field is applied to them a deformation results. Thus, piezoelectric behaviour can be manifested in two distinct ways. The first one is called the "direct" piezoelectric effect that occurs when a piezoelectric material becomes electrically charged when subjected to a mechanical stress (Cady[6]). This property can be used in sensors that measure strain, strain rate, forces, pressure, and vibration. The second property is called the "converse" piezoelectric effect that occurs when the piezoelectric material becomes strained when placed in an electric field. This effect can be employed to develop mechanical actuators to generate strain, forces, and movement. The certain features of piezoelectric materials such as fast response, high bandwidth, and strain-rate measurement have made them the most popular intelligent materials in the area of smart structures.

After the discovery of the piezoelectric effect by Jacques and Pierre Curie, Valsek observed the hysteresis loop of the polarisation with respect to an applied electrical field along and axis of a Rochelle salt crystal in 1921. In 1935 Busch and Scherrer in Switzerland discovered ferroelectricity in potassium dihydrogen phosphate, which was the second family of piezoelectric crystals. One of the most significant discoveries was the poling process in polycrystalline piezoelectric ceramics in 1946. Until then the piezoelectric effect was only observed on single-crystal materials. However, single-crystal materials are generally more expensive to produce and limited in size. More accessible polycrystalline materials fueled many new applications (Janos and Hagood[30]).

9.3 Piezoelectric Ceramics

Since 1917 piezoelectric ceramics have found some commercial applications as transducers such as ultrasonic submarine detectors. Scientists then started developing new piezoelectric ceramics to replace natural piezoelectric materials such as tourmaline and quartz. Piezoelectric ceramics were developed from high-quality barium-titanate ceramics used in capacitors. These ceramics are

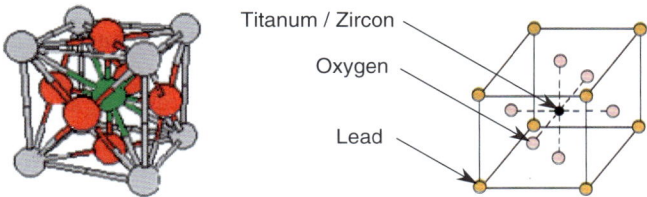

Figure 9.1 Atomic structure of lead zirconate titanate (PZT) crystal.

mostly oxide materials on the basis of lead oxide, zirconate oxide and titanate oxide. Metal oxides of the elements lithium, magnesium, zinc, nickel, manganese, niobium, antimony or strontium are added for the appropriate adjustment or stabilisation of the material parameters. Piezoelectric ceramics include several ceramics such as lead zirconate titanate (PZT), lead metaniobate (LMN), lead titanate (LT), and lead magnesium niobate (PMN). Among these, PZT has been the most extensively produced and employed piezoelectric ceramics. A schematic of the atomic structure of a PZT crystalline is shown in Figure 9.1. The physical properties of the resulting PZT compound can be controlled by different composition ratios of the basic materials, grinding duration, calcinations, shaping and sintering.

The crystalline structure of a PZT is derived from the mineral perovskite ($CaTiO_3$). This structure is formed above the Curie temperature from regular oxygen octahedrons with titanate and zirconium ions in the centre. Below the Curie temperature a distortion of this structure occurs and then this leads to the formation of metal oxygen dipoles. Initially, the directions of the single dipoles are distributed randomly in the crystalline, thus, no polarisation can be measured. A strong electric field applied during the process causes an alignment of single dipoles in each crystalline along the direction of the applied field. After the electric field has been removed the polarisation alignment is maintained. A typical procedure of PZT manufacturing process is illustrated in Figure 9.2. Piezoelectric ceramics are produced in different configurations such as plates, discs, fibres, tubes, and ribbons (Figure 9.3).

9.4 Piezoelectric Ceramic Actuators

Piezoelectric ceramics have been extensively used as actuators in intelligent material systems. Conventional servovalve/hydraulic actuators suffer from various limitations such as multiple energy conversions (mechanical, hydraulic, electrical), large number of parts, *i.e.* potential failure sites and large weight penalty, high vulnerability of the hydraulic pipes network, frequency bandwidth limitation (Giurgiutiu[20]). Piezoelectric ceramic actuators overcome these limitations by integration with the host structure as embedded or bonded actuators, direct conversion of electrical energy to high-frequency linear motions and elimination the need for hydraulic power systems. Electrical

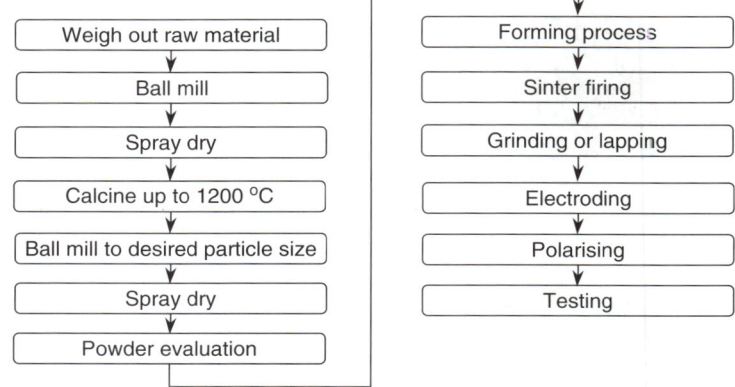

Figure 9.2 PZT manufacturing process.

Figure 9.3 Piezoelectric ceramics (From Sensor Technology Limited and CeraNova Corp.).

energy is easier to transmit and electrical lines are much less vulnerable than hydraulic pipes. However, these *piezoceramic* actuators have their own drawbacks such as small strokes (Giurgiutiu et al.[21]).

Piezoelectric ceramic actuators are used for various functions, including control and damping of mechanical vibrations, tool adjustment and control, micropumps, mirror positioning, wave generation, structural deformation, engineering inspection systems and scanning electron microscopes. When a voltage is applied to piezoelectric actuators, they produce small displacements with a high force capability. Their key features include compactness and low weight, fast response, displacement proportional to the applied voltage, large force, and broad operating temperature range.

The most common piezoelectric ceramic actuators include ceramics patches, stacks, and piezoelectric fibre actuators. For example, QuickPack piezoelectric ceramic actuator was first introduced by Active Control Experts (ACX, a division of CYMER) as a commercially available PZT actuator patch that includes all the wiring to the electrodes (Figure 9.4). A piezoelectric ceramic

Figure 9.4 Piezoelectric ceramic Quickpack (from ACX, a division of CYMER).

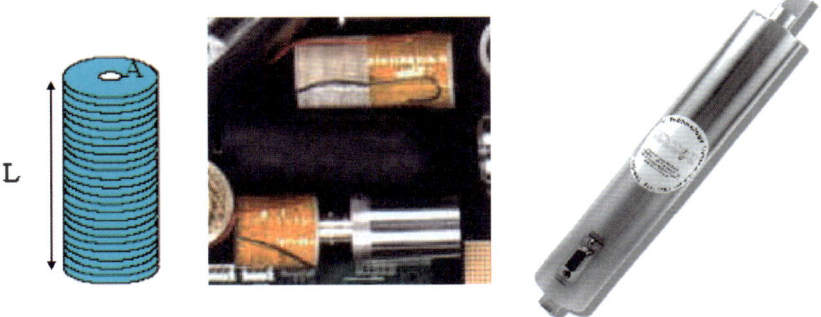

Figure 9.5 Piezoelectric ceramic stack actuator (from Sensor Technology Limited).

stack actuator is made of several ceramic discs that are usually assembled as a cylindrical actuator. The common feature of stack actuators is that many thin layers of piezoelectric ceramics, typically PZT are glued or cofired together with an electrode between each layer. This arrangement allows the mechanical displacement to sum in series, while the electrical properties remain in parallel (Figure 9.5). Stack actuators provide large stroke forces in a broadband frequency. These actuators are studied extensively in the literature (Flint et al.[14]). In order to have flexible piezoelectric ceramic actuators with high actuation capability an active fibre composite (AFC) actuator made of piezoelectric ceramic fibres was developed at MIT (Rodgers and Hagood[41]). The PZT–fibre piles have continuous aligned PZT fibres in an epoxy layer, and polyimide/copper electrode films. The electrode films are etched into an interdigitated pattern that affects electric field along the fibre direction, thus activating the primary piezoelectric effect (Figure 9.6).

9.5 Modeling

Piezoelectric ceramics can be modeled either numerically or analytically using the basic constitutive equations. A nonlinear modeling of piezoelectric ceramics has been studied (Chan and Hagood[7]). The mathematical modeling of a nonlinear laminated anisotropic piezoelectric structure has also been discussed (Tzou et al.[52]). In that study a generic theory is proposed and its nonlinear thermo-electromechanical equations have been derived based on the variational

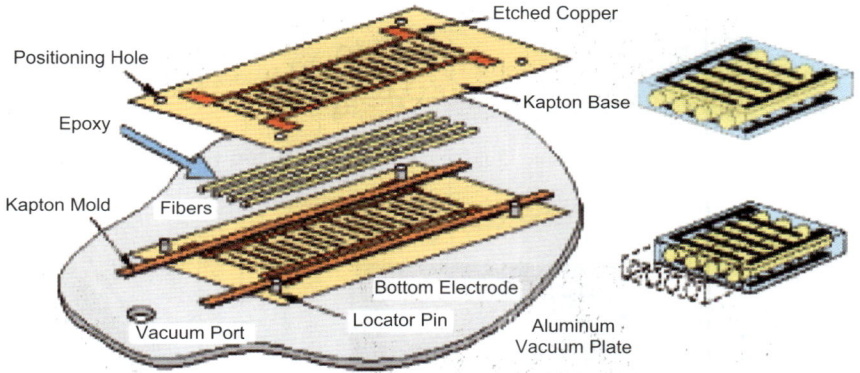

Figure 9.6 Piezoelectric ceramic fibre actuator, Active Fiber Composite (AFC).

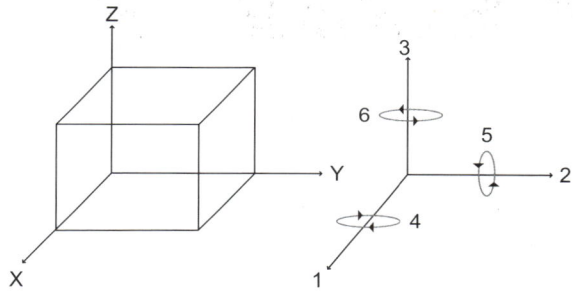

Figure 9.7 A block of piezoelectric element and its corresponding axes and directions.

principal. The heat-dissipation problem associated with piezoelectric ceramic materials is addressed by Zhou et al.[61]. With a reasonable approximation piezoelectric ceramics are assumed to respond linearly with very little hysteresis. Consider a piezoelectric ceramic block as shown in Figure 9.7. The constitutive equation of the piezoelectric ceramic block can be expressed as follows (Berlincourt, et al.[5], Jaffe, et al.[29], and Zelenka[60]).

$$\{S\} = [S_E]\{T\} + [d]^T\{E\} \tag{9.1}$$

$$\{D\} = [d]\{T\} + [\varepsilon_T]\{E\} \tag{9.2}$$

where $[S_E]$ is the adiabatic compliance matrix, $[d]$ is the adiabatic PZT coupling matrix, $[\varepsilon_T]$ is the adiabatic dielectric constant matrix including permittivity components, $\{S\}$ is the strain vector, $\{T\}$ is the stress vector, $\{D\}$ is the charge density vector, and $\{E\}$ is the electric field vector. Equation (9.1) relates the applied electric field to the strain in the piezoelectric ceramic (actuator

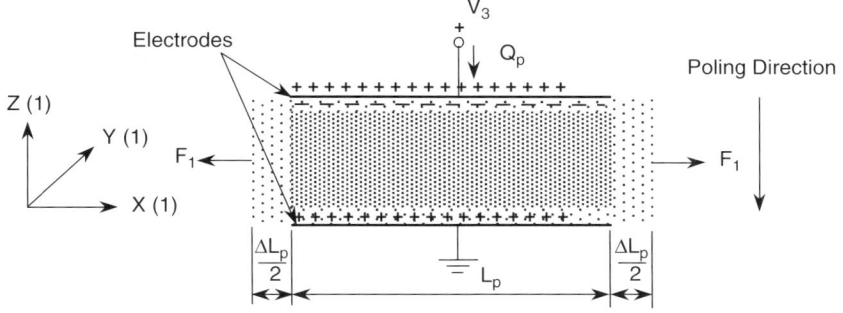

Figure 9.8 PZT element and its poling direction.

equation). On the other hand, eqn (9.2) relates the electrical charge to the stresses and strains in the piezoelectric ceramic (sensor equation). Assume that axis 1 is aligned in the direction in which the piezoelectric element is uniaxially stretched during its manufacturing process, and axis 3 has two properties. First, it is parallel to the direction of the piezoelectric element thickness and second, the positive direction points in the opposite direction of the electric field used in the manufacturing process to pole piezoelectric element (Figure 9.8). In this case Sesslar[44] showed that eqn (9.1) can be written as

$$\underbrace{\begin{Bmatrix} \varepsilon_1 \\ \varepsilon_2 \\ \varepsilon_3 \\ \gamma_{23} \\ \gamma_{31} \\ \gamma_{12} \end{Bmatrix}}_{\{S\}} = \underbrace{\begin{bmatrix} S_{11} & S_{12} & S_{13} & 0 & 0 & 0 \\ S_{21} & S_{22} & S_{23} & 0 & 0 & 0 \\ S_{31} & S_{32} & S_{33} & 0 & 0 & 0 \\ 0 & 0 & 0 & S_{44} & 0 & 0 \\ 0 & 0 & 0 & 0 & S_{55} & 0 \\ 0 & 0 & 0 & 0 & 0 & S_{66} \end{bmatrix}}_{[S_E]} \underbrace{\begin{Bmatrix} \sigma_1 \\ \sigma_2 \\ \sigma_3 \\ \tau_{23} \\ \tau_{31} \\ \tau_{12} \end{Bmatrix}}_{\{T\}} + \underbrace{\begin{bmatrix} 0 & 0 & d_{31} \\ 0 & 0 & d_{31} \\ 0 & 0 & d_{33} \\ 0 & d_{15} & 0 \\ d_{15} & 0 & 0 \\ 0 & 0 & 0 \end{bmatrix}}_{[d]^T} \underbrace{\begin{Bmatrix} E_1 \\ E_2 \\ E_3 \end{Bmatrix}}_{\{E\}}$$

(9.3)

Besides analytical models introduced in the literature, several finite-element (FE) models have also been developed (Sheta *et al.*[46]). The electric field components, E_k, are related to the electrostatic potential φ as

$$E_k = -\nabla_k \varphi, \quad k = 1, 2, 3 \qquad (9.4)$$

Using the variational method the linear dynamic equation of the electro-mechanical model of the piezoelectric ceramic that is integrated within an elastic structure is written as follows.

$$\begin{bmatrix} M_{uu} & 0 \\ 0 & 0 \end{bmatrix} \begin{Bmatrix} \ddot{u} \\ \ddot{\varphi} \end{Bmatrix} + \begin{bmatrix} K_{uu} & K_{u\varphi} \\ K_{\varphi u} & -K_{\varphi\varphi} \end{bmatrix} \begin{Bmatrix} u \\ \varphi \end{Bmatrix} = \begin{Bmatrix} F \\ G \end{Bmatrix} \qquad (9.5)$$

where u, φ, F, and G represent the nodal displacement, electric potential, force, and applied charge vectors, respectively. K_{uu} and M_{uu} are the stiffness and mass matrices of the original structure respectively. $K_{\varphi\varphi}$ represents the stiffness matrix of the piezoelectric ceramic, and $K_{u\varphi}$ is the coupled stiffness between electrical field and elastic strain. These matrices are given as

$$[K_{u\varphi}] = \int_v B_u^T e B_\varphi dv \tag{9.6}$$

$$[K_{\varphi\varphi}] = \int_v B_\varphi^T \varepsilon B_\varphi dv \tag{9.7}$$

where ε is the dielectric matrix and v is the volume. The other matrices are given as

$$B_u = \begin{bmatrix} \partial/\partial x & 0 & 0 \\ 0 & \partial/\partial y & 0 \\ 0 & 0 & \partial/\partial z \\ \partial/\partial y & \partial/\partial x & 0 \\ 0 & \partial/\partial z & \partial/\partial y \\ \partial/\partial z & 0 & \partial/\partial x \end{bmatrix} [N_u]^T \tag{9.8}$$

$$B_\varphi = \begin{Bmatrix} \partial/\partial x \\ \partial/\partial y \\ \partial/\partial z \end{Bmatrix} [N_\varphi]^T \tag{9.9}$$

$$[e] = [d][S_E]^{-1} \tag{9.10}$$

where N_φ and N_u are the shape functions for the electric and displacement fields, respectively.

Defining the following matrices

$$\tilde{u} = \begin{Bmatrix} u \\ \varphi \end{Bmatrix}, \quad \tilde{M} = \begin{bmatrix} M_{uu} & 0 \\ 0 & 0 \end{bmatrix}, \quad \tilde{K} = \begin{bmatrix} K_{uu} & K_{u\varphi} \\ K_{\varphi u} & -K_{\varphi\varphi} \end{bmatrix}, \quad \tilde{F} = \begin{Bmatrix} F \\ G \end{Bmatrix} \tag{9.11}$$

From eqn (9.5), the final electromechanical equation of the integrated intelligent structure within a piezoelectric ceramic is written as

$$\tilde{M}\ddot{\tilde{u}} + \tilde{K}\tilde{u} = \tilde{F} \tag{9.12}$$

Thus, integration of piezoelectric ceramic element into an FE model introduces an additional nodal-dependent variable representing the nodal electrostatic potential, φ, and also modifies the mass and stiffness matrices as described in eqn (9.11). Presently, several commercial FE software packages, such as ANSYS, also have particular electromechanical elements to model piezoelectric ceramics.

As a simple case consider a thin and long 1D piezoelectric ceramic block with the thickness of t in which the electric field is applied along axis 3 and the strain developed along axis 1 (Figure 9.8), the maximum actuation strain, free strain, due to an electric voltage V applied to a piezoelectric element is

$$\varepsilon_{max} = \Lambda = d_{31}\left(\frac{V}{t}\right) \tag{9.13}$$

Table 9.1 Typical PZT properties.

Description	Constant	Value	Unit
Volume density	ρ	7350.0	kg/m^3
Elastic module	E	71.4×10^9	Pa
Piezoelectric strain constant	d_{31}	200.0×10^{-12}	m/V
Electric permittivity	ε	150.4×10^{-10}	F/m

Where

$$d_{31} = \text{Piezoelectric coupling (strain) coefficient} = \frac{\text{Strain developed along axis 1}}{\text{Electric field applied along axis 3}} \quad (9.14)$$

A list of typical PZT piezoelectric ceramic properties is presented in Table 9.1.

Piezoelectric ceramics modeling is often performed as an integrated part of flexible structures dynamic modeling. As an alternative to the commonly used modal method, a wave model of flexural structures can be obtained directly from the partial differential equations describing the structural dynamics in the frequency domain. This type of model does not depend on modes and is therefore free of considerations of modal truncation and related difficulties. This is a promising approach for many types of flexible structures such as beams, which can be viewed as flexural waveguides. This method has been addressed in a number of textbooks (*e.g.* Graff[22] and Humar[28]), with interest motivated by the wide range of applications. For instance, Wang and Huang[54] performed an analytical and numerical study of wave propagation induced by piezoelectric ceramic actuators for monitoring the surface damage of structures. Dynamic wave modeling of flexible structures can also be used to design controllers based on wave absorption. Fuller *et al.*,[18] performed an experimental investigation on simultaneous active control flexural and extensional waves in elastic beams. Matsuda *et al.*[34] used the transfer matrix method to show that elastic motion is a superposition of waves traveling in a flexible structure. The approach diagonalised the unity transfer matrix into real-Jordan form. In the following two sections dynamic modeling of piezoelectric ceramics as sensors and actuators are presented.

9.5.1 Sensors

One of the first models of piezoelectric sensors was introduced by Lee *et al.*[32] Dynamic modeling of very thin piezoelectric sensors with uniform strain distribution for strain and strain-rate measurements of flexible beam vibrations was studied by Vaz.[53] A generalised formulation for piezoelectric strain and strain rate sensors with a strain-assumed linear is derived by Yousefi-Koma and Vukovich.[58] Piezoelectric elements can be used either as strain or strain-rate sensors. The output voltage of a piezoelectric sensor bonded onto a beam is proportional to the spatial integral of strain, and its output current is

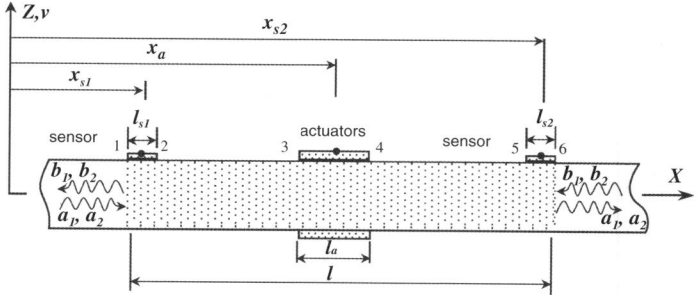

Figure 9.9 Input and output waves of the sensing area.

proportional to the spatial integral of strain rate. For a piezoelectric sensor located between x_i and x_j (Figure 9.9), the output voltage and current are

$$V_s = \frac{E_s d_{31s} W_s}{C_s} \int_{x_i}^{x_j} \varepsilon \, dx \qquad (9.15)$$

$$I_s = (E_s d_{31s} W_s) \int_{x_i}^{x_j} \frac{d\varepsilon}{dt} dx \qquad (9.16)$$

Where E_s, W_s, d_{31s}, and C_s are elastic modulus, width, strain constant, and electric capacitance of the piezoelectric sensor, respectively, and ε represents the strain.

The transverse vibration of a flexible beam can be described in terms of flexural waves. Using an Euler–Bernoulli model of the beam, vibration in the frequency domain can be written as a *"wavefunction"*

$$v(x, \omega) = a_1(x, \omega) + a_2(x, \omega) + b_1(x, \omega) + b_2(x, \omega) \qquad (9.17)$$

Subscript 1 represents traveling waves and subscript 2 denotes evanescent waves for which amplitudes decay exponentially with x. a_1 and a_2 represent waves traveling in the positive x direction, and b_1 and b_2 denote waves in the negative direction. These waves are given by:

$$\begin{array}{ll} a_1(x,\omega) = A_1 e^{-i\mu x}, & a_2(x,\omega) = A_2 e^{-\mu x} \\ b_1(x,\omega) = A_1 e^{i\mu x}, & b_2(x,\omega) = A_2 e^{\mu x} \end{array} \qquad (9.18)$$

where

$$\mu = \sqrt{(\omega/\alpha)}, \qquad \alpha = \sqrt{(E_b I_b)/\rho_b} \qquad (9.19)$$

A_1, A_2, B_1, and B_2 are constants that can be obtained from the boundary conditions. E_b, I_b, ρ_b are elastic modulus, moment of inertia of the cross section, and mass per unit length of the flexible beam, respectively, and ω is frequency.

A "*wave state vector*", **w**, and a "*physical state vector*", **s**, at position x_i are defined as (Pines and von Flotow[39])

$$\mathbf{w}_i \equiv [a_1(x_i, \omega) \quad a_2(x_i, \omega) \quad b_1(x_i, \omega) \quad b_2(x_i, \omega)]' \tag{9.20}$$

and

$$\mathbf{s}_i \equiv [v(x_i, \omega) \quad \theta(x_i, \omega) \quad M(x_i, \omega) \quad V(x_i, \omega)]' \tag{9.21}$$

Superscript ' represents the transpose of a vector or matrix. θ, M, and V represent slope, bending moment, and shear force of the flexible beam respectively. Defining the coefficient matrix **H** as

$$\mathbf{H} = \begin{bmatrix} 1 & 1 & 1 & 1 \\ -i\mu & -\mu & i\mu & \mu \\ -E_b I_b \mu^2 & E_b I_b \mu^2 & -E_b I_b \mu^2 & E_b I_b \mu^2 \\ -i E_b I_b \mu^3 & E_b I_b \mu^3 & i E_b I_b \mu^3 & -E_b I_b \mu^3 \end{bmatrix} \tag{9.22}$$

the physical state vector can be written with respect to the wave state vector as

$$\mathbf{s}_i = \mathbf{H}\,\mathbf{w}_i \tag{9.23}$$

Defining the dimensionless parameter, k_{ij}, as

$$k_{ij} = (x_j - x_i)\mu \tag{9.24}$$

the spatial transition matrix, T_{ij}, given by

$$T_{ij} = \begin{bmatrix} e^{-ik_{ij}} & 0 & 0 & 0 \\ 0 & e^{-k_{ij}} & 0 & 0 \\ 0 & 0 & e^{ik_{ij}} & 0 \\ 0 & 0 & 0 & e^{k_{ij}} \end{bmatrix} \tag{9.25}$$

relates the wave state vectors at positions x_i and x_j as

$$\mathbf{w}_j = T_{ij}\mathbf{w}_i \tag{9.26}$$

It can be shown[55] that these two equations can be written with respect to the wave state vectors as follows:

$$V_s = -\mathrm{sgn}(z)\frac{t_b E_s d_{31s} W_s}{2 E_b I_b C_s}\left(M_o\sqrt{\frac{\alpha}{\omega}}\right)[-i \ -1 \ i \ 1]\{\mathbf{w}_j - \mathbf{w}_i\} \tag{9.27}$$

and

$$I_s = -\mathrm{sgn}(z)\frac{t_b E_s d_{31s} W_s}{2 E_b I_b}(M_o\sqrt{\alpha\omega})[1 \ -i \ -1 \ i]\{\mathbf{w}_j - \mathbf{w}_i\} \tag{9.28}$$

where

$$M_o \equiv E_b I_b \mu^2 \tag{9.29}$$

9.5.2 Actuators

Two different approaches, *i.e.* equilibrium equations and energy method, have mainly been employed in the literature to derive piezoelectric actuators (Crawley and de Luis[10], Vaz,[53] Yousefi-Koma[59]). A pair of piezoelectric ceramics bonded on the top and bottom surfaces of the flexible beam is used as the actuator. The forces induced by bonded piezoelectric actuators can be modeled as a pair of moments at the end points of the piezoelectric actuators. Neglecting the geometric stiffness effects of the piezoelectric actuator on the beam (these effects are later included in the dynamic model for control implementation) and assuming a linear strain distribution over piezoelectric actuator thickness, the endpoint moment induced by an applied voltage V to the piezoelectric actuator can be written as

$$M_a = (E_a W_a t_a t_b) \frac{\kappa(1+\tilde{t})}{\kappa + 6(1 + 2\tilde{t} + \frac{4}{3}\tilde{t}^2)} \Lambda \qquad (9.30)$$

where E, W, and t denote elastic modulus, width, and thickness, and subscripts b and a represent the flexible beam and the piezoelectric actuator, respectively. The stiffness ratio, κ, thickness ratio, $\tilde{t}$, and the strain induced by the piezoelectric effect, Λ, are

$$\kappa \equiv \frac{(EWt)_b}{(EWt)_a}, \quad \tilde{t} \equiv \frac{t_a}{t_b}, \quad \text{and } \Lambda = \frac{d_{31a}}{t_a} V \qquad (9.31)$$

The induced bending moment, M_a, can be incorporated as the boundary condition in the dynamic equation. Using eqns (9.20)–(9.26), the induced moment can be related to the waves at the end points of the actuator as follows.

$$[\mathbf{I} \quad -\mathbf{T}_{ij}] \begin{bmatrix} w_j^+ \\ w_i^- \end{bmatrix} = \Omega_{ij} M_a \qquad (9.32)$$

where

$$\Omega_{ij} = (-\mathbf{T}_{ij} + \mathbf{I}) \mathbf{H}^{-1} [0\ 0\ 1\ 0]' \qquad (9.33)$$

Superscripts – and + denote left and right neighbourhoods at the specific positions, respectively.

9.6 Applications

To date, multifunctional piezoelectric ceramics have been employed in various areas of applications for performance enhancement such as vibration/acoustic control, shape control, structural health monitoring, micropositioning, fast valves and nozzles, transducers, gas ignitions, shock absorbers in luxury cars, active engine mounts, spacecraft jitter reduction, and sports. Several research projects have been conducted to demonstrate the application of these materials.

The following is a description of some of the selected major applications in various areas of research and industry.

9.6.1 Vibration/Acoustic Control

Vibration/acoustic control is the dominant area of piezoelectric ceramic applications. Several vibration/acoustic applications have been investigated including tail buffet control, wing flutter control, aircraft interior noise reduction, vibration cancellation in sport equipments such as ski, bat, and snow board. For instance, intelligent constrained layers using piezoelectric ceramic have been used for bending vibration control of Euler–Bernoulli beams (Shen et al.[45]).

For high-performance twin-tail aircraft such as the F/A-18 and F-15, buffet-induced tail vibrations occur when unsteady pressures associated with separated flow, or vortices, excite the vibration modes of the vertical fin structural assemblies. This is a significant problem particularly for the F/A-18 aircraft that requires frequent inspection to prevent catastrophic failure. Lazarus et al.[31]) studied the application of active piezoelectric actuators on an F/A-18 vertical tail. Their results showed that a decrease of more than 60% on the first bending mode amplitude could be introduced by a smart control system using piezoelectric ceramic actuators. In a recent project on tail buffet by NASA Langley, Air Force Research Laboratory (AFRL), and Daimler Chrysler, called Actively Controlled Response of Buffet Affected Tails (ACROBAT), surface-bonded piezoelectric ceramic wafer actuators were attached to the port vertical tail of a 1/6 scale wind-tunnel model of an F/A-18, while an active rudder control was employed on the starboard vertical tail. A single-input single-output (SISO) control system was implemented resulting in 60% reduction of power spectrum density (PSD) peak at the first bending mode and 19% reduction of RMS value at the root for an angle of attack of $37°$ (Moses[36]). As a follow on to ACROBAT, a blended system approach was adopted for the Scaling Influences Derived from Experimentally Known Impact of Controls (SIDEKIC) project (Moses[37]), where the active rudder was employed to reduce the first bending resonance ($\sim 17\,Hz$) and piezoelectric ceramic wafers were used for suppression of the first torsional resonance ($\sim 58\,Hz$).

In order to examine the feasibility of the application of piezoelectric ceramics for buffet suppression on a full-scale F/A-18, a program was initiated under The Technical Cooperation Program (TTCP) with the collaboration of Canada, US and Australia. A dynamically scaled model of a vertical tail was first fabricated in the National Research Council Canada (NRC) to investigate the feasibility of active vibration control system with piezoelectric ceramic actuators. A total of 12 pairs of piezoelectric ceramics were bonded on both sides of a flexible fin shown in Figure 9.10. An FE model of the intelligent fin with piezoelectric ceramics was developed for control development (Figure 9.11). An active control system based on system identification technique suppressed vibrations significantly (Yousefi-Koma et al.[55]). The full-scale aircraft was

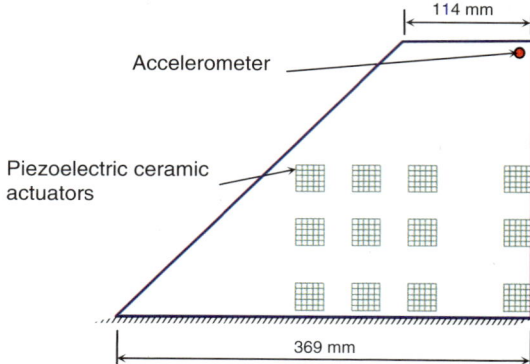

Figure 9.10 Smart fin with piezoelectric ceramics.

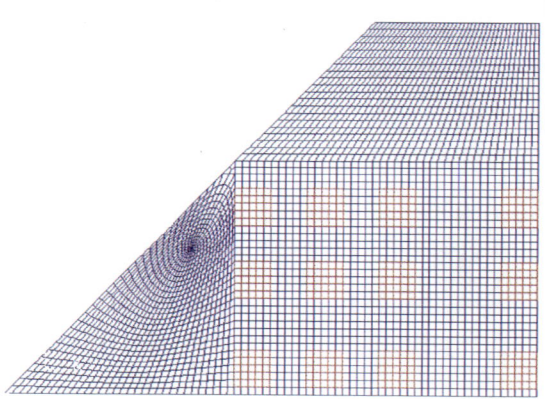

Figure 9.11 FE model of the smart fin.

tested in the International Follow-On Structural Test (IFOST) Program rig in Australia (Chen et al.[8], Spangler and Jacques[44], Hopkins et al.[27], and Nitzsche et al.[38]). The starboard fin of the aircraft was instrumented with piezoelectric actuators over a wide area on both sides of the fin. Results demonstrated that the active control system was effectively able to suppress the buffet response of the vertical fin at high angle of attack. Amplitude reductions of up to 60% at the nominal flight configuration and close to 10% at the worst case were demonstrated.

At particular flight conditions of an aircraft, a large vibration (flutter or limit cycle oscillation) caused by aeroelastic forces may result in serious safety and/or flight control problems. In fighter jets, the underwing weapons may intensify this phenomenon. The feasibility of employing the piezoelectric ceramics to control panel flutter was investigated by Suleman and Venkayya[51]. An analytical and experimental investigation of flutter suppression of a fixed wing by

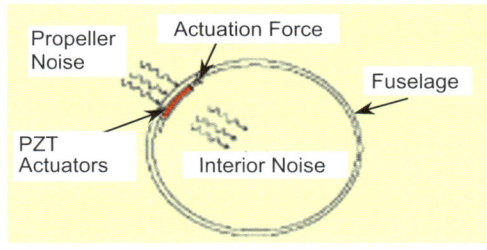 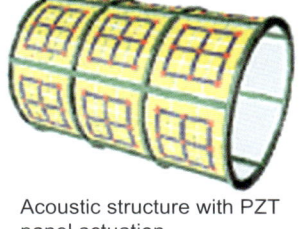

Figure 9.12 Structural acoustic control of fuselage using piezoelectric ceramics.

piezoelectric actuators was performed by Heeg[25] with the corresponding experimental studies performed at NASA Langley Research Centre (LaRC) under the Flutter Research and Experimental Device (FRED) project. The Piezoelectric Aeroelastic Response Tailoring Investigation (PARTI) (Heeg and McGowan[26], and McGowan et al.[35]) was conducted at MIT with NASA support to investigate aircraft wing-flutter suppression using piezoelectric ceramic actuators. A 12% increase in flutter damping and 75% reduction in root bending moment caused by gust were achieved.

Undesired vibration in turbine engine components such as stator and rotor blades, lubrication units, and casing reduces engine performance and life. Lucent Technologies and Pratt & Whitney have designed and applied an active vibration control system for engines (Finn[13]) to demonstrate that vibration reduction may increase engine durability and reduce the maintenance and service cost. Grewal et al.[23] employed an adaptive feedforward control system to reduce the aircraft cabin noise in turboprop aircrafts using bonded piezoelectric ceramic actuators. Fripp, et al.[17] have also studied application of piezoelectric ceramics in noise control of aircraft fuselages during operation. Applications of PZT actuators have also been studied at MIT for helicopter fuselage noise cancellation (Figure 9.12). A comprehensive literature survey in this area has been done by Yousefi-Koma and Zimcik.[57]

9.6.2 Rotor-blade Flap

Chen and Chopra[9] described the construction of a composite blade with diagonally oriented PZT wafers embedded in the fibreglass skin. Significant twist response was measured when excitation was close to resonance frequencies. Servoflap concepts have been investigated as an alternate approach to achieving induced-strain rotor-blade actuation using piezoelectric ceramics. Bimorph piezoelectric ceramic actuators were used in early servo-flap experiments targeting the Boeing CH-47D helicopter (Spangler and Hall.[48] Prechtl and Hall[40] built a mechanically amplified ISA flap actuator (X-frame actuator) using a piezoelectric ceramic stack for CH-47D rotor-blade model. The X-frame actuator occupies the leading edge part of airfoil. Active trailing-edge flaps and active tip-twist blade models were tested in the hover stand (Spencer, et al.[49]). Suppression of 1/rev vibrations and induction of 2/rev oscillatory loads were successfully demonstrated. A full-scale implementation of Smart Materials Actuation Rotor Technology (SMART) started in 1993 at Boeing Mesa. A prototype flap actuator with a two-stage amplification and biaxial operation was constructed and tested (Straub[50]). The SMART program consisted of Boeing, MIT, UCLA, and the University of Maryland. Support to the program has also been offered by NASA Ames, Rockwell Science Centre, and TRS Ceramics, Inc. A new displacement amplification principal, the double X-frame (2X) concept has been adopted, fabricated, and tested (Hall et al.[24]). Up to 3° flap authority was predicted with a high-voltage stack (TRS and RSC piezoelectric ceramic stack), and 4° with a single-crystal stack (Straub[50]).

9.6.3 Adaptive Structural Shape Control

Barrett[3] built an electrically active torque plate consisting of a metallic substrate and diagonally attached PZT wafers. Twisting of the torque plate is created by activation of the PZT elements with polarities in opposing phase on the top and bottom surfaces. Flight testing of the model was successfully performed. Active fibre composites (AFC) were incorporated into the construction of a 1/6th scale CH-47D blade model for wind-tunnel testing at Boeing Helicopters (Rodgers and Hagood[41]). Using AFC at the tip of the blade alternates the aerodynamic shape of the blade and consequently changes the aerodynamic forces to reduce the vibration. A wind-tunnel test has been performed for a model-scaled blade control. The progressive design and miniaturisation of microaerial vehicles (MAV) has been severely limited due to the size and mass of traditional actuators and electronic components. A MAV with all moving intelligent surfaces was fabricated and tested by Barrett et al.[2] This model used a piezoelectric stability augmentation system. This high-bandwidth, high-authority control stabilator was the key enabling technology for the aircraft. A 30% scaled model of an unmanned combat air vehicle (UCAV) was fabricated at NASA Langley Research Centre. This model was equipped with conventional electric motors, SMA, and piezoelectric ceramic ultrasonic actuators as well as strain, pressure, force, and acceleration sensors. Two wind-tunnel tests were conducted. High-rate, large-deflection, conformal-trailing-edge control was successfully demonstrated under realistic flight conditions (Bartley-Cho et al.[4] and Scherer et al.[43]).

9.6.4 Structural Health Monitoring

Recent advances in intelligent piezoelectric sensors/actuators, data acquisition and electronics capabilities suggest the feasibility of piezoelectric ceramics applications to health monitoring. Using piezoelectric ceramic excitation and sensing capabilities the load history can be recorded online. This information can then be used to detect the amplitude and location of damage on the

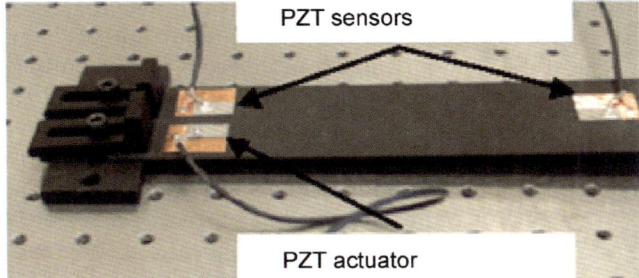

Figure 9.13 Structural health monitoring using piezoelectric ceramics.

structure that simplifies inspection and reduces the lifecycle monitoring cost significantly (Figure 9.13). Fedele et al.[12] used two arrays of piezoelectric ceramic sensors and actuators. Piezoelectric actuators introduced minor excitations to the system and piezoelectric sensors monitored the response. They were able to find the locations and magnitudes of the flaws by frequency-domain comparison of the original and damaged structure responses. A similar study was performed by Lichtenwalner et al.[33] on a composite flexible beam on the MD-900 rotor system. Piezoelectric ceramic transducers acted as both sensors and actuators. The statistical analysis of changes in transfer functions could determine the locations and magnitude of damages.

9.6.5 Compact Hybrid Actuators

Piezoelectric ceramic materials are employed to develop new types of electro-mechanical actuators and devices that take advantage of the high energy density of these materials. It is aimed to cross multiple domains to create new classes of hybrid actuators with significantly higher performance than electromagnetic motors, centralised pneumatic or hydraulic systems. Kinetic Ceramics Inc. has developed a prototype PZT pump operating at 10 kHz with an output pressure of 11 MPa. The valve functions by generating/collapsing a bubble that blocks an orifice. A new design of inchworm motor using multi-layer piezoelectric ceramic stack for guidance of next-generation missiles has been introduced by Burleigh Instruments, Inc. 136 N of continuous force with 60 min of operation is achieved. The actual speed is 26.5 mm/s with less than 1 nm resolution.

9.7 Commercial Products

Since the discovery of the unique properties of piezoelectric ceramics, they have been employed in various commercial products. Piezoelectric ceramics have been used in sonar system for ships and submarines as transmitters. This application is still considered as the main market for these materials. One of the first novel success stories was fabrication of smart ski by Active Control Experts (ACX, a division of CYMER) in which PZT patches were used as a passive shunt to suppress mechanical vibrations (Figure 9.14). The same idea was then applied to fabricate smart snowboards and smart bats (Figures 9.15

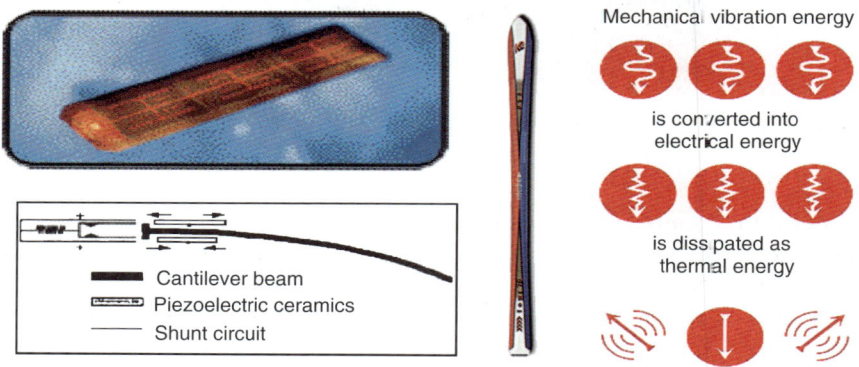

Figure 9.14   Smart ski (from ACX, a division of CYMER).

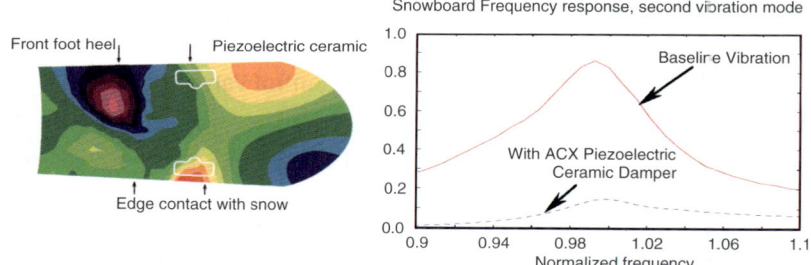

Figure 9.15   Smart snowboard (from ACX, a division of CYMER).

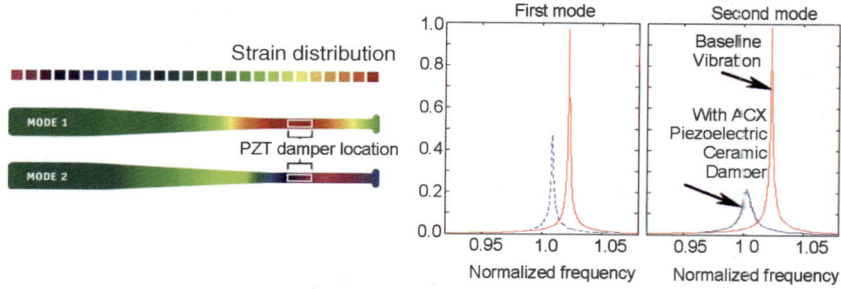

Figure 9.16   Smart bat (from ACX, a division of CYMER).

and 9.16). In passive shunt technique mechanical vibration energy transforms to electrical energy and is then dissipated as thermal energy to the environment. As seen in Figures 9.15 and 9.16, using embedded PZT patches in snowboards and bats the vibration peaks are reduced dramatically.

A piezoelectric ceramic stack actuator is also used in fuel-injection systems (Figure 9.17). These materials have found some medical applications such as audio pickup for heart monitoring (Figure 9.18). High-frequency valves are

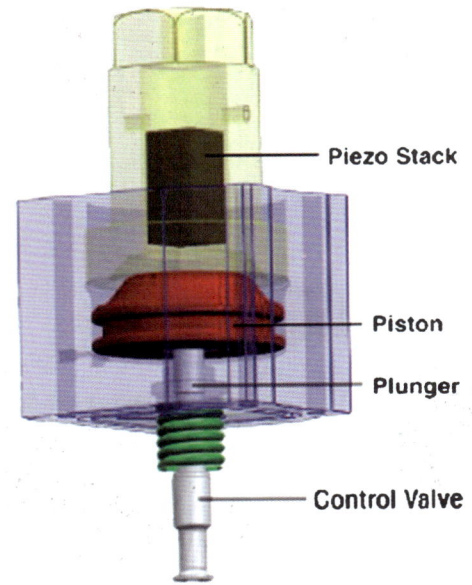

Figure 9.17 Fuel-injection system.

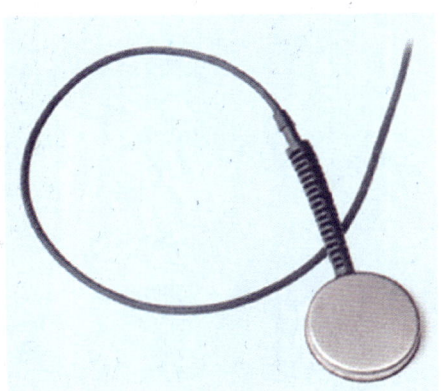

Figure 9.18 Piezoelectric heart audio pickup.

also fabricated using bimorph piezoelectric ceramics (Figure 9.19). A commercial miniature-size ultrasonic motor has been developed using PZT patches assembled on a disc (Figure 9.20). The smallest of these ultrasonic, piezoelectric motors is about the size of a grain of rice. They are 1.8 mm in diameter and

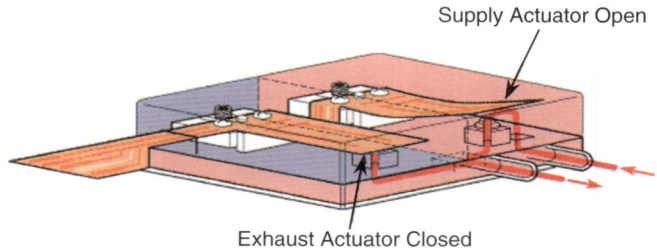

Figure 9.19   Piezoelectric valve.

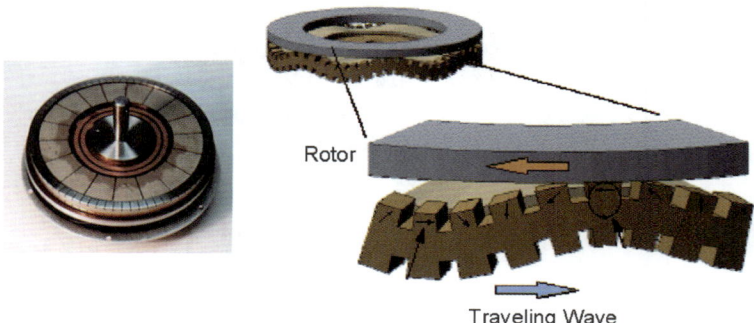

Figure 9.20   Piezoelectric ceramic ultrasonic motors.

Figure 9.21   Advanced 6D micropositioning systems (from Physik Instrumente (PI)).

Piezoelectric Ceramics as Intelligent Multifunctional Materials 251

4 mm long. The motor is capable of micro motion for very precise positioning. The technology provides a high power-to-weight ratio without the hoses, compressors, and other supporting hardware associated with hydraulic and pneumatic devices.

There are many other commercial products based on piezoelectric ceramics. For instance, Physik Instrument (PI) has developed several intelligent systems

Figure 9.22 Active optics: piezo-steering mirrors (from Physik Instrumente (PI)).

Figure 9.23 Microscopy (from Physik Instrumente (PI)).

Figure 9.24 Micromanufacturing, microassembly (from Physik Instrumente (PI)).

for various applications such as micropositioning systems, active optics, microscopy, and micromanufacturing using PZT ceramics (Figures 9.21–9.24).

References

1. T. Bailey and J.E. Hubbard, *J. Guid. Cont. Dynam.*, 1985, **8**, pp. 605–611.
2. R.M. Barrett, R.S. Gross and F.T. Brozoski, *Proceedings of 36th Structures, Structural Dynamics and Materials Conference*, AIAA-95-1081, AIAA, Reston, VA, 1995, pp. 2289–2296.
3. R. Barrett, *Smart Mater. Struct.*, 1993, **2**, 55–65.
4. J.D. Bartley-Cho, D.P. Wang and M.N. West, *Proceedings of SPIE Conference on Smart Structures and Material Systems: Industrial and Commercial Applications of Smart Structures Technologies*, 4698, Bellingham, WA, 2002, pp. 53–63.
5. D.A. Berlincourt, D.R. Curran and H. Jaffe, *Physical Acoustics Principles and Methods*, ed. W.P. Mason, Academic Press, New York, 1964, pp. 169–270.
6. W.G. Cady, Dover Publications, Inc., Mineola, New York, USA, 1964.
7. K.H. Chan and N.W. Hagood, *Proceedings of the SPIE Conference on Smart Structures and Materials: Smart Structures and Intelligent Systems*, Orlando, Florida, Feb. 14–16, 1994, pp. 195–205.
8. Y. Chen, V. Wickramasinghe and D.G. Zimcik, *Proceedings of SPIE Conference on Smart Structures and Material Systems*, 2006.
9. P.C. Chen and I. Chopra, *SPIE publication*, 1997, **3041**, 217–229.
10. E.F. Crawley and J. de Luis, *AIAA J.*, 1987, **25**, No. 10, 1373–1385.

11. J. De Luis and E.F. Crawley, *Proceedings of the 31st AIAA/ASME/ASCE/AHS/ASC Structures, Structural Dynamics and Materials Conference*, 1989, pp. 2340–2350.
12. P. Fedele, G. Cafasso and L. Leece, *7th International Conference on Adaptive Structures*, Technomic Publishing Co. Inc.Lancaster, Pa, 1997, pp. 232–242.
13. E.D. Finn, Lowering the Volume on Helicopters, *Aerospace America*, 1997, 24–25.
14. E.M. Flint, C.A. Rogers and C. Liang, *Adaptive Structures and Composite materials: Analysis and Application AMSE International Mechanical Engineering Congress and Exposition*, Chicago, Illinois, Nov. 6–11, 1994, pp. 201–210.
15. R.L. Forward, US Patent 4, 158,787, June 19, 1979.
16. R.L. Forward, C.J. Swigert and M. Obal, *The Shock and Vibration Bulletin, Part 4 Damping and Machinery Dynamics*, May 1983, **Bulletin 53**, 51–61.
17. M. Fripp, D. O'Sullivan, S. Hall, N.W. Hagood and K. Lilienkamp, *Proceedings of SPIE Smart Structures and Materials*, 1997, 3041.
18. C.R. Fuller, G.P. Gibbs and R.J. Silcox, *J. Intell. Mater. Syst. Struct.*, 1990, **1**, 235–247.
19. M.V. Gandhi and B.S. Thompson, Chapman and Hall, London, UK, 1992.
20. V. Giurgiutiu, *J. Intell. Mater. Syst. Struct.*, 2000, **11**, 525–544.
21. V. Giurgiutiu, A. Craig and C.A. Rogers, *J. Intell. Mater. Syst. Struct.*, 1997, 8.
22. K.F. Graff, *Wave Motion in Elastic Solids*, Ely House, London, Clarendon Press, Oxford University Press, 1975.
23. A. Grewal, D.G. Zimcik and B. Leigh, *J. Aircraft*, 2001, **38**(1), 164–173.
24. S.R. Hall, T. Tzianetopolou, F. Straub and H. Ngo, *Proceedings of SPIE Smart Structures and Materials: Smart Structures and Integrated Systems*, 2000, **3985**, 26–37.
25. J. Heeg, NASA Technical paper 3241, 1993.
26. J. Heeg and A.R. McGowan, *Proceedings of SPIE Conference on Smart Structures and Material Systems: Industrial and Commercial Applications of Smart Structures Technologies*, **2447**, Bellingham, WA, 1995, pp. 2–13.
27. M. Hopkins, D. Henderson, R. Moses, T. Ryall, D. Zimcik and R. Spangler, *Proceedings of SPIE Conference on Smart Structures and Material Systems: Industrial and Commercial Applications of Smart Structures Technologies*, **3326**, Bellingham, WA, 1998, pp. 27–33.
28. J.L. Humar, *Dynamics of Structures, Prentice Hall International Series in Civil Engineering and Engineering Mechanics*, Englewood Cliffs, NJ, 2002.
29. B. Jaffe, R. Cook and H. Jaffe, *Piezoelectric Ceramics*, Academic Press, New York, NY, 1971.
30. B. Z. Janos and N. W. Hagood, Internal report # AMSL 00-4, Active Materials and Structures Laboratory, MIT, 2000.
31. K.B. Lazarus, E. Saarmaa and G.S. Agnes, *Proceedings of SPIE Conference on Smart Structures and Material Systems: Industrial and Commercial*

Applications of Smart Structures Technologies, **2447**, Bellingham, WA, 1995, pp. 179–192.
32. C.K. Lee, T.C. O'Sullivan and W.W. Chiang, *Proceedings of the 32nd AIAA/ASME/ ASCE/AHS/ASC Structures, Structural Dynamics and Materials Conference*, 1991, pp. 2197–2207.
33. P. Lichtenwalner, J.P. Dunne, R.S. Becker and E.R. Baumann, *SPIE Conference on Smart Structures and Material Systems, Industrial and Commercial Applications of Smart Structures Technologies*, **3044**, Bellingham, WA, 1997, pp. 186–194.
34. K. Matsuda, Y. Kanemitsu and S. Kijimoto, *J. Sound Vib.*, 1998, **216**(2), 269–279.
35. A.R. McGowan, J. Heeg and R.C. Lake, *Proceedings of the 37th Structures Structural Dynamics and Materials Conference*, AIAA-96-1511, AIAA, Reston VA, 1996, pp. 1722–1732.
36. R.W. Moses, *Proceedings of SPIE Conference on Smart Structures and Material Systems: Industrial and Commercial Applications of Smart Structures Technologies*, 3044, Bellingham, WA, 1997, pp. 87–98.
37. R.W. Moses, *Proceedings of the 40th Structures, Structural Dynamics and Materials Conference*, AIAA-99-1318, AIAA, Reston, VA, 1999, pp. 1034–1042.
38. F. Nitzsche, D.G. Zimcik and K. Langille, *Proceedings of the 38th Structures, Structural Dynamics and Materials Conference*, Adaptive Structures Forum, AIAA-97-1386, AIAA, Reston, VA, 1997, pp. 1467–1477.
39. D.J. Pines and A.H. von Flotow, *J. Sound Vib.*, 1990, **142, No. 3**, 391–412.
40. E.F. Prechtl and S.R. Hall, *Proc. SPIE*, 1997, **3041**, 158–182.
41. J.P. Rodgers, N.W. Hagood and D.B. Weems, *Proceedings of SPIE Smart Structures and Materials*, 1996, 2717.
42. J.M. Sater, C.R. Crowe, R. Antcliff and A. Das, "An Assessment of Smart Air and Space Structures: Demonstration and Technology (IDA)" paper P-3552, September 2000.
43. L.B. Scherer, C.A. Martin, B. Sanders, M. West, J. Florance, C. Wieseman, A. Burner and G. Fleming, *Proceedings of SPIE Conference on Smart Structures and Material Systems: Industrial and Commercial Applications of Smart Structures Technologies*, **4698**, Bellingham, WA, 2002, pp. 64–75.
44. G.M. Sesslar, *J. Acoust. Soc. Am.*, 1981, **70, No. 6**, 1596–1608.
45. I.Y. Shen, W. Guo and Y.C. Pao, *Adaptive Structures and Composite Materials: Analysis and Application, AMSE International Mechanical Engineering Congress and Exposition*, Chicago, Illinois, Nov. 6–11, 1994, pp. 133–143.
46. E.F. Sheta, R.W. Moses and L.J. Huttsell, *J. Sound Vib.*, 2006, **292**(3–5), 854.
47. R.L. Spangler and R.N. Jacques, *Proceedings of the 40th Structures, Structural Dynamics and Materials Conference*, AIAA-99-1316, Reston, VA, 1999, pp. 1023–1033.
48. R.L. Jr. Spangler and S.R. Hall, Report # SSL 1-89, SERC 14-90, MIT Space Engineering Research Center, MIT, Cambridge, Massachusetts 02139, Jan, 1989.

49. M.G. Spencer, R.M. Sanner and R.I. Chopra, *Proceedings of SPIE Smart Structures and Materials: Smart Structures and Integrated Systems*, 2000, **3985**, pp. 38–49.
50. F.K. Straub, *8th ARO Workshop on Aeroelasticity of Rotorcraft Systems*, Penn State University, Oct 18–20, 1999.
51. A. Suleman and V.B. Venkayya, *Proceedings of the 6th AIAA/USAF/NASA/ISSMO Symposium on Multidisciplinary Analysis and Optimization*, Bellevue, Washington, 1996, pp. 141–151.
52. H.S. Tzou, Y. Bao and R. Ye, *Proceedings of the SPIE Conference on Smart Structures and Materials: Smart Structures and Intelligent Systems*, Orlando, Florida, Feb. 14–16, 1994, pp. 206–214.
53. A.F. Vaz, *9th CASI Symposium on Aerospace Structures and Materials*, Ottawa, Canada, 1996.
54. X.D. Wang and G.L. Huang, *J. Intell. Mater. Syst. Struct.*, 2001, **12**(2), 105–115.
55. A. Yousefi-Koma and G. Vukovich, *Jnl. Vib. Ctrl.*, 2005, **11**(8), 1043.
56. A. Yousefi-Koma, Y. Chen and D.G. Zimcik, *Proceedings of the 7th International Workshop on Smart Mater. Struct.*, CANSMART, Montreal, Canada, Oct. 2004, pp. 7–16.
57. A. Yousefi-Koma and D.G. Zimcik, *Can. Aero. Space J.*, 2003, **9, No. 4**, 163–172.
58. A. Yousefi-Koma and G. Vukovich, *Can. Aero. Space J.*, 1999, **45, No. 4**, 379–389.
59. A. Yousefi-Koma, "Active Vibration Control of Smart Structures Using Piezoelements", Ph.D. Thesis, Carleton University, Ottawa, Canada, 1997.
60. J. Zelenka, *Piezoelectric Resonators and Their Applications*, Elsevier Science Publishing Co. Inc., New York, 1986.
61. S. Zhou, C. Liang and C.A. Rogers, *Adaptive Structures and Composite materials: Analysis and Application, AMSE International Mechanical Engineering Congress and Exposition*, Chicago, Illinois, Nov. 6–11, 1994, pp. 183–191.

CHAPTER 10
Ferroelectric Relaxor Polymers as Intelligent Soft Actuators and Artificial Muscles

Q. M. ZHANG,[1] BAOJIN CHU[1] AND Z.-Y. CHENG[2]

[1] Electrical Engineering Department and Materials Research Institute, The Pennsylvania State University, University Park, PA 16802, USA
[2] Materials Research and Education Center, Auburn University, Auburn, AL 36849-5341, USA

10.1 Introduction

As the best-known ferroelectric polymers, poly(vinylidene fluoride) (PVDF) and its copolymer with trifluoroethylene (TrFE) have been widely used in piezoelectric sensors and transducers.[1,2] On the other hand, the relatively low piezoelectric coefficients, low piezoelectric strain, and low elastic-energy-density limit their applications as actuator materials. Recently, we demonstrated that by proper defects modification, PVDF-based polymers can be converted from a piezoelectric polymer to an electrostrictive polymer with a very large electrostrictive strain. Furthermore, these are the first ferroelectric polymer known to exhibit the relaxor ferroelectric phenomenon. In this chapter, we will review the experimental results related to the ferroelectric behaviour and electromechanical responses in this class of relaxor ferroelectric polymers.

Presented in Figure 10.1 is a phase diagram for the normal ferroelectric P(VDF-TrFE) copolymer.[3] In the ferroelectric phase, the copolymer is in the all-*trans* conformation, resulting in a unit cell with a large lattice constant along the polymer-chain direction and a smaller unit-cell dimension perpendicular to the chain.[4] In the paraelectric phase, however, where the crystallites adopt conformations containing a mixture of *trans* and *gauche* bonds, the unit-cell dimension along the chain direction is significantly shortened, while the cell dimension perpendicular to the chain expands (see Figures 10.2 and 10.3).[5] As a result,

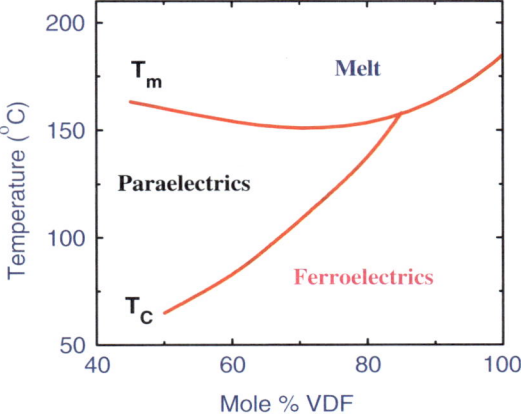

Figure 10.1 Phase diagram of PVDF and P(VDF-TrFE) polymers showing a ferroelectric–paraelectric transition that signals a change from a ferroelectric (polar) phase to a paraelectric (nonpolar phase).[3]

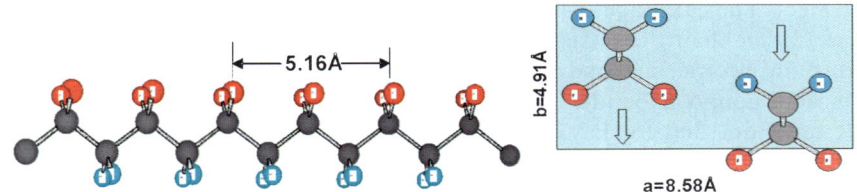

Figure 10.2 A view of ferroelectric phase having lattice dimensions of $a = 8.58$ Å, $b = 4.91$ Å and a chain direction (or fibre axis) of 2.58 Å.

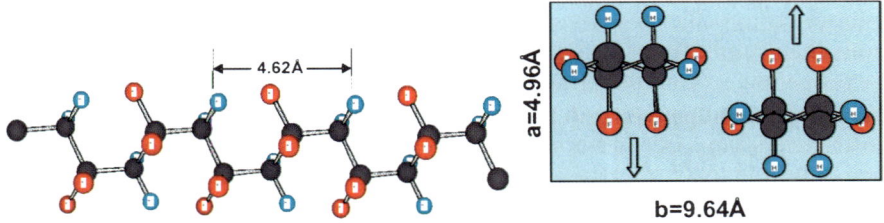

Figure 10.3 A view of paraelectric phase having lattice dimensions of $a = 4.96$ Å, $b = 9.64$ Å and a chain direction (or fibre axis) of 4.62 Å.

when the copolymer goes through the ferroelectric to paraelectric (F–P) phase transition, there is a large lattice constant change, as illustrated in Figure 10.4, which is for the 65/35 mol% copolymer.[6] The results here indicate that by utilising the molecular conformation change between the polar and nonpolar conformations, a large strain response can be achieved in PVDF-based

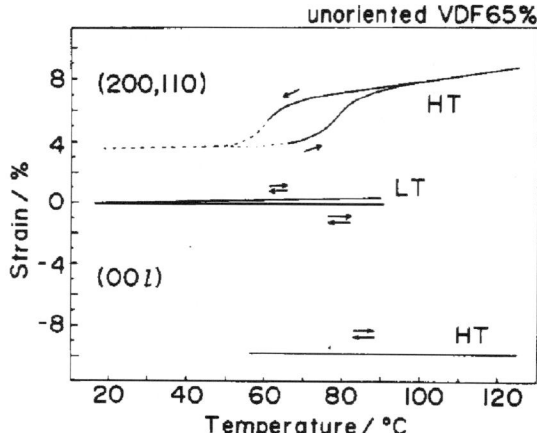

Figure 10.4 As a result of the copolymer going through the F–P transition, there is a large lattice constant change for the P(VDF-TrFE) 65/35 mol% copolymer.[6]

polymers. The challenge is how to alter the energy levels so that the nonpolar phase rather than the polar phase is stable at room temperature and how to realise an electric field induced reversible conformation change between the polar and nonpolar bonds.

It turns out that for P(VDF-TrFE) copolymer at compositions with VDF/TrFE ratio at below 70/30 mol%, one can employ strategies such as high-energy electron irradiation treatment or adding bulky third monomer to form random terpolymer to convert the room-temperature phase into a nonpolar phase to exhibit high electrostriction.[7–10] This chapter reviews the results related to the electromechanical response, microstructures, relaxor ferroelectric dynamics, as well as the microdevice performance of the electrostrictive PVDF-based polymers.

10.2 High-energy Electron-irradiated Copolymer (HEEIP)

10.2.1 Microstructures of HEEIP

Presented in Figure 10.5 is the change of the polarisation hysteresis loop measured at room temperature for the P(VDF-TrFE) 68/32 mol% copolymer with the irradiation dose.[8] With increased dosage, the near-square polarisation hysteresis loop, characteristic of a normal ferroelectric material, is transformed to a slim polarisation loop (at 75 Mrad). At very high dose (175 Mrad), the polymer becomes a linear dielectric in which the crystallinity is near zero. The differential scanning calorimetry (DSC) data also reveal the disappearance of the ferroelectric transition for the copolymers irradiated with 75 Mrad dose or

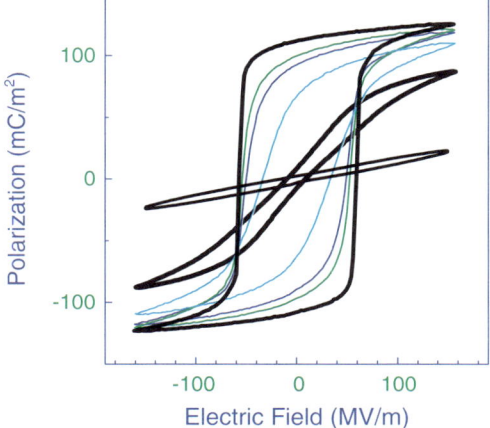

Figure 10.5 Polarisation hysteresis loops at room temperature for a P(VDF-TrFE) 68/32 mol% copolymer with different doses of irradiation, from high to low: 0, 20, 35, 50, 75, and 175 Mrad.[8]

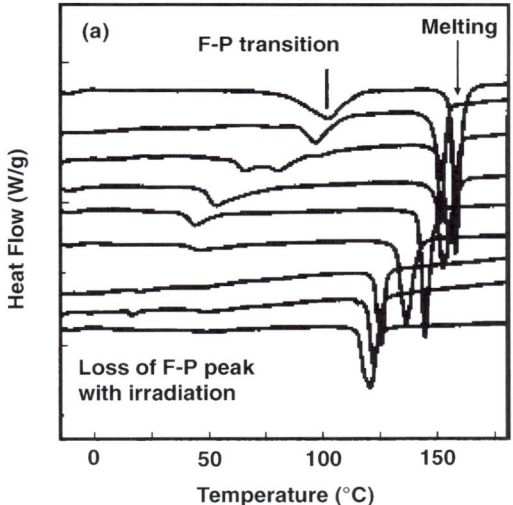

Figure 10.6 DSC data for an irradiated P(VDF-TrFE) 68/32 mol% copolymer with different doses, from top to bottom: 0, 10, 20, 35, 50, 65, 75, 85, and 100 Mrad.[8]

higher, as shown in Figure 10.6.[8] The DSC results indicate that the broad dielectric constant peak observed in Figure 10.7 does not correspond to the ferroelectric phase transition. When cooled under zero electric field, the copolymer irradiated with 75 Mrad or higher dose remains in the nonpolar phase. All these results indicate that the high-energy electron-irradiated copolymers are converted to a ferroelectric relaxor.

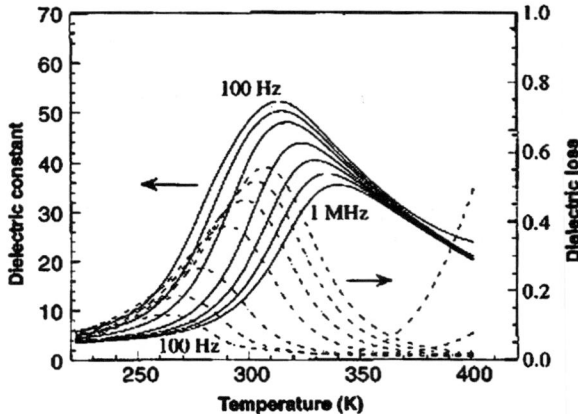

Figure 10.7 The dielectric constant and loss of the high-energy electron-irradiated copolymer.

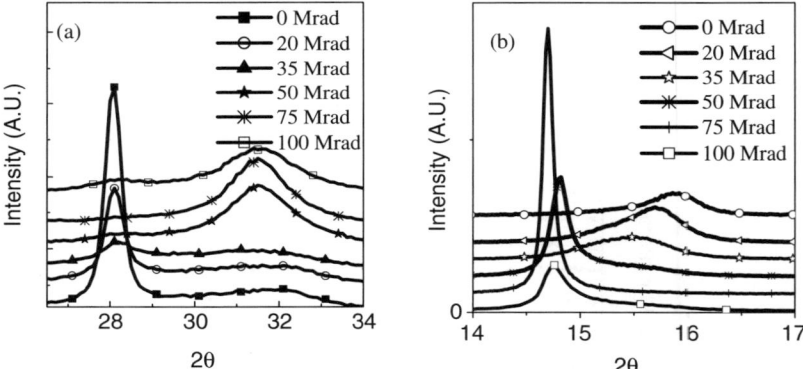

Figure 10.8 X-ray diffraction data for a P(VDF-TrFE) 68/32 mol% copolymer with different irradiation doses (measured at room temperature): (a) Reflection associated with (001) peak and (b) reflection associated with (110)/(200) peak.[8]

The microstructure change of the irradiated copolymer with irradiation dose was also investigated by Cheng et al.[8,11] The wide-angle X-ray diffraction data taken at the angular range near the (200/110) reflection and (001) reflection are shown in Figure 10.8. As can be seen from the evolution of the (001) peak, with increased dose, the polar-phase peak intensity decreases while the nonpolar peak intensity increases. In spite of the change of the peak intensity, the peak position of the two phases does not show a change with dose, suggesting that the transformation from the normal ferroelectric to the relaxor ferroelectric with the irradiation dose is a first-order process. In contrast, both the peak intensity and peak position of (200/110) peak change continuously with dose, indicating a strong intrachain coupling and consequently a distortion of the

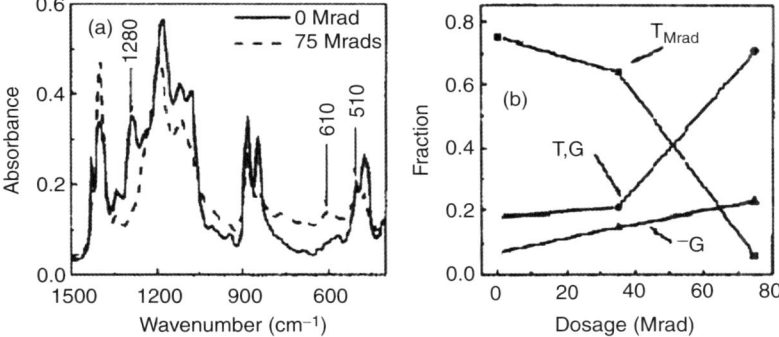

Figure 10.9 FT-IR data of a P(VDF-TrFE) 68/32 mol% copolymer at room temperature: (a) comparison of spectra for copolymer before and after irradiation and (b) fraction of different conformations versus irradiation dose.[8]

lattice constant in the two phases due to this coupling. The FT-IR data also show a change of the molecular conformation from the polar (all-*trans* conformation) to the random nonpolar conformations (*trans-gauche*, TGTG', and T$_3$GT$_3$G') as shown in Figure 10.9.[8] There are three FT-IR absorbance peaks used here to follow the change of the three main conformations with dose. The one at 1288 cm^{-1} is from the all-*trans* conformation ($T_{m>4}$), 614 cm^{-1} is for the TGTG', and 510 cm^{-1} is for the T$_3$GT$_3$G'.[12] The fraction F_i of each conformation is calculated following the relation

$$F_i = \frac{A_i}{A_1 + A_2 + A_3} \quad (10.1)$$

where A_i is the absorbance of these conformations.

A common feature in the inorganic relaxor ferroelectric is the presence of the so-called nanopolar regions, the population of which increases as the temperature is reduced.[13,14] For the irradiated copolymers, one indication of the existence of these polar regions is the change of the molecular conformations with temperature. An increase in the local "nanopolar" region with reduced temperature should be reflected in an increase in the absorbance of the all-*trans* conformation in the FT-IR data. As expected, for the nonirradiated copolymer as well as for the copolymer irradiated with lower doses to become a pure ferroelectric relaxor, the polar-order regions increase with reduced temperature due to the presence of the normal ferroelectric phase in these copolymers.[8] In contrast, for the 68/32 mol% copolymer irradiated with 75 Mrad (a pure ferroelectric relaxor), as shown in Figure 10.10,[8] there is essentially no change of the molecular conformation with reduced temperature, revealing that there are no local "nanopolar" regions developed in the HEEIP relaxor ferroelectric polymer. It should be noted that the relaxor ferroelectric terpolymers, which will be presented later, show quite different behaviour.

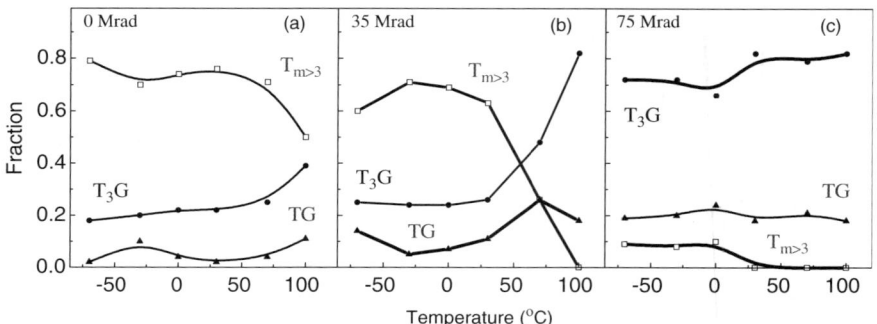

Figure 10.10 Fraction of different conformations versus temperature for a P(VDF-TrFE) 68/32 mol% copolymer with different doses: (a) dose = 0 Mrad, and (b) dose = 35 Mrad and (c) dose = 75 Mrad.[8]

^{19}F nuclear magnetic resonance (NMR) spectroscopy was used to probe the structure of P(VDF-TrFE) before and after the high-energy electron irradiation to elucidate the chemistry associated with the development of outstanding electroactive properties.[15] In addition to the crosslinking, it was shown that about 1 mol% of CF_3 pendant groups was formed for copolymers irradiated with a 75-Mrad dose. Several CF_3 end groups were also observed in the irradiated copolymer. These defects are likely to cause lattice distortion, reduction of the crystallite size, as well as the reduction of the crystallinity, all of which will affect and reduce the macroscopic ferroelectric ordering in the copolymer.

10.2.2 Electromechanical Responses of HEEIP

One approach to significantly increase the electrostrictive response with a high force level and high elastic-energy density in polymers is to work with polymers with a higher dielectric constant. As has been shown in Figure 10.7, HEEIP relaxor ferroelectric polymer shows a dielectric constant of near 60 at room temperature and, consequently, an electrostrictive strain of ∼5% under a 150 MV/m field has been demonstrated, as shown in Figure 10.11(a).[7,16,17] The plot of strain vs. P^2 yields a straight line, indicating the response is electrostrictive in nature [($S_3 = Q_{33}P_3^2$, Figure 10.11(b)]. For the irradiated copolymer, Q_{33} is found in the range between −4 to −15 m^2/C (the electrostrictive M coefficient is greater than 2×10^{-18}),[18] depending on the sample-processing conditions. The strain response does not change appreciably with temperature, as suggested by the plot in Figure 10.11(c).

The basis for such a large electrostriction in the irradiated P(VDF-TrFE) copolymers is the large change in the lattice strain as the polymer goes through the transformation between the ferroelectric and paraelectric phase. Figure 10.12 presents the wide-angle X-ray scattering data obtained on the irradiated copolymer when subject to different external fields.[11,19] This shows that the X-ray peaks change with external field, reflecting that there is a local field-induced

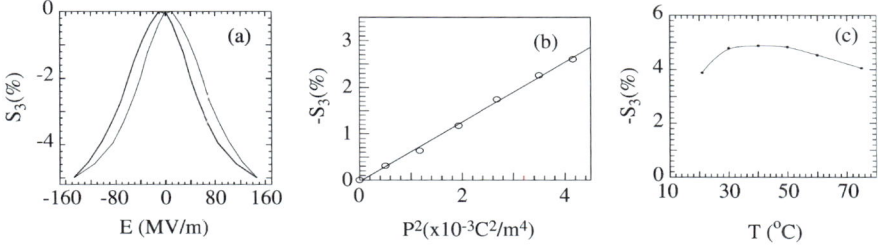

Figure 10.11 A high electrostrictive strain (S_3 or longitudinal strain) about -5% under 150 MV/m field from a high-energy electron-irradiated P(VDF-TrFE) copolymer.

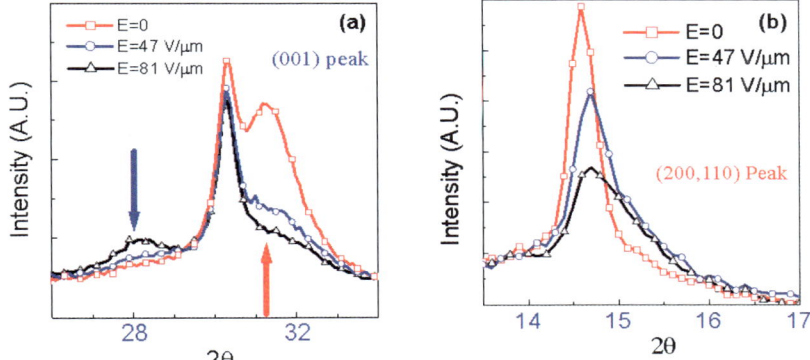

Figure 10.12 X-ray diffraction pattern measured at room temperature as a function of applied electric field for irradiated film. (a) the (001) diffraction observed using transmission scan (peak at $2\theta \sim 30.2°$ is from gold electrodes) and arrows indicate the polar and nonpolar phase X-ray peaks and (b) the (110)/(200) diffraction peak obtained using a reflection scan.

conformation change in the polymer. The data confirm that it is the reversible field-induced transformation between the polar and nonpolar conformations that is responsible for the observed large electromechanical responses in the HEEIP relaxor ferroelectric polymers.

Of special interest is the finding that in the electron-irradiated P(VDF-TrFE) copolymers, large anisotropy in the strain responses exists along and perpendicular to the chain direction, as can be deduced from the change in the lattice parameters between the polar and nonpolar phases. Therefore, the transverse strain can be tuned over a large range by varying the film-processing conditions. For unstretched films, the transverse strain is relatively small ($\sim +1\%$ at ~ 100 MV/m) while the amplitude ratio between the transverse strain and longitudinal strain is less than 0.33.[20] This feature is attractive for devices utilising the longitudinal strain such as ultrasonic transducers in the thickness mode, and actuators and sensors making use of the longitudinal electromechanical responses

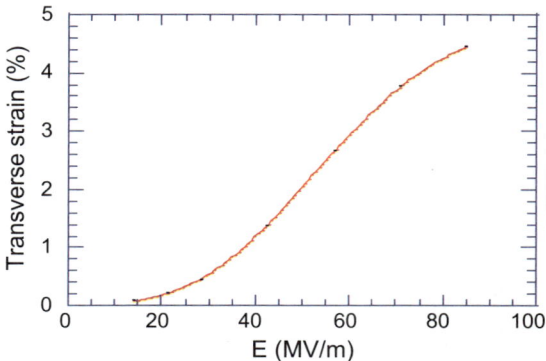

Figure 10.13 For stretched films, a large transverse strain (S_1) along the stretching direction can be achieved.[19]

of the material. For example, with a very weak transverse electromechanical response in comparison with the longitudinal, one can significantly reduce the influence of the lateral modes on the thickness resonance and improve the performance of the thickness transducer. On the other hand, for stretched films, a large transverse strain (S_1) along the stretching direction can be achieved as shown in Figure 10.13, where a transverse strain of about $+4.5\%$ was observed in the irradiated copolymer under an electric field of 85 MV/m.[19]

It is also interesting to note that the strain along the thickness direction (parallel to the electric field) is always negative for PVDF and its copolymers regardless of the sample processing conditions.[20] In fact, this is a general feature for a system in which the polarisation response originates from the dipolar interaction and is true for all polymeric piezoelectric and electrostrictive responses.[21] The sign of the strains perpendicular to the applied-field direction will depend on the sample-processing conditions. For the anisotropically stretched films discussed here, the electric-induced strain along the stretching direction, which is perpendicular to the applied field, is positive, whereas in the direction perpendicular to both stretching and applied field directions, the strain is negative. For unstretched samples that are isotropic in the plane perpendicular to the applied field, the strain component in the plane is an average of the strains along the chain (positive) and perpendicular to the chain (negative) and is, in general, positive.[20]

For electrostrictive materials, the electromechanical coupling factor (k_{ij}) has been derived by Hom et al.[22] based on the consideration of electrical and mechanical energies generated in the material under the external field:

$$k_{3i}^2 = \frac{kS_i^2}{s_{ii}^D \left[P_E \ln\left(\frac{P_S+P_E}{P_S-P_E}\right) + P_S \ln\left(1 - \left(\frac{P_E}{P_S}\right)^2\right) \right]} \qquad (10.2)$$

where $i=1$ or 3 correspond to the transverse or longitudinal direction (for example, k_{31} is the transverse coupling factor), s_{ii}^D is the elastic compliance under constant polarisation, and S_i and P_E are the strain and polarisation responses, respectively, for the material under an electric field E. The coupling factor depends on E, the electric field level. In eqn (10.2), it is assumed that the polarisation-field (P–E) relationship follows approximately

$$|P_E| = P_S \tanh(k|E|) \qquad (10.3)$$

where P_S is the saturation polarisation and k is a constant.

The electromechanical coupling factors for the irradiated copolymers calculated based on eqn (10.2) are shown in Figure 10.14.[19,20] Near room temperature and under an electric field of 80 MV/m, k_{33} can reach more than 0.3, which is comparable to that obtained in a single-crystal P(VDF-TrFE) copolymer. More interestingly, a k_{31} of 0.65 can be obtained in a stretched copolymer, which is much higher than the values measured in unirradiated P(VDF-TrFE) copolymers.[23] These results are also verified by recent resonance studies in these polymers.

In a polymer, there is always a concern about the electromechanical response under high mechanical load; that is, whether the material can maintain high strain levels when subject to high external stresses. Figure 10.15(a) depicts the transverse strain of a stretched and irradiated 65/35 copolymer under a tensile stress along the stretching direction and the longitudinal strain of unstretched and irradiated 65/35 copolymer under hydrostatic pressure.[24-26] As can be seen from the figure, under a constant electric field the transverse strain increases initially with the load and reaches a maximum at a tensile stress of about 20 MPa. Upon a further increase of the load, the field-induced strain is reduced. One important feature revealed by the data is that even under a tensile stress of 45 MPa, the strain generated is still nearly the same as that without load, indicating that the material has a very high load capability. Shown in Figure 10.15(b) is the longitudinal strain under hydrostatic pressure. At low electric fields, the strain does not change much with pressure, while at high fields it shows increase with pressure.

Recently, it was also reported that high electrostrictive strain can also be obtained in the high energy proton irradiation on P(VDF-TrFE) 56/44 mol%

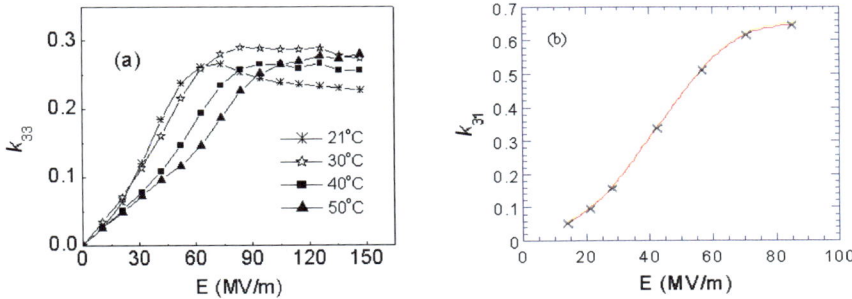

Figure 10.14 The electromechanical coupling factors for irradiated copolymer.

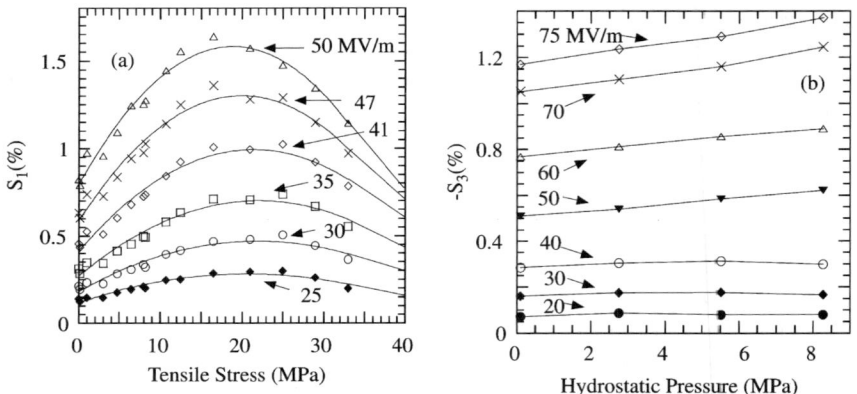

Figure 10.15 (a) The transverse strain (amplitude) of a stretched and irradiated P(VDF-TrFE) 65/35 mol% copolymer under a tensile stress along the stretching direction[24,26] and (b) the longitudinal strain of an unstretched and irradiated P(VDF-TrFE) 65/35 mol% copolymer under hydrostatic pressure.[25]

copolymer.[27] The electrostrictive coefficient M_{33} measured from the proton irradiated copolymer films is about $-1.83 \times 10^{-18}\,\mathrm{m^2/V^2}$, comparable to those in the high-energy electron-irradiated copolymers. The irradiated copolymer also exhibits ferroelectric relaxor behaviour. One principal effect of the high energy irradiation is network formation (crosslinking).[15,28] Casalini and Roland[29] used a free-radical crosslinker and coagent to form a three-dimensional network of P(VDF-TrFE) copolymer with high crosslink density and showed that in such a P(VDF-TrFE) network, a thickness strain of more than -10% can be induced under a field of 9 MV/m when measured under dc condition. By fitting the data with an $S = M\,E^2$ relation, where S is the thickness strain and E is the applied field, the coefficient $M = 1.6 \times 10^{-15}\,\mathrm{m^2/V^2}$, much higher than these obtained in the high-energy irradiated copolymers. The authors suggested that the large strain is from the rotation of the small ferroelectric crystals, created due to the high density of crosslinking that greatly reduces the crystallite sizes, under an external field.

10.3 Electrostrictive Responses and Relaxor Ferroelectric Behaviour in P(VDF-TrFE)-based Terpolymers

10.3.1 The Electromechanical Response in P(VDF-TrFE)-based Terpolymers

Although high-energy irradiations can be used to convert the normal ferroelectric P(VDF-TrFE) into a relaxor ferroelectric with high electrostriction, the irradiation also introduces many undesirable defects to the copolymer, such as the formation of crosslinkings, radicals, and chain scission.[15,28] From the basic ferroelectric response point of view, the defects modification of the ferroelectric properties can also be realised by introducing randomly in the polymer chain a third monomer, which is bulkier than VDF and TrFE. Furthermore, by a proper molecular design that enhances the degree of molecular-level conformational changes in the polymer, the terpolymer can exhibit a higher electromechanical response than the high-energy electron-irradiated copolymer, as indeed been observed in the terpolymer of P(VDF-TrFE-CFE) (CFE: chlorofluoroethylene).[10,19] The high electrostrictive response was also observed in several other ferroelectric fluoroterpolymers such as P(VDF-TrFE-CTFE) (CTFE: chlorotrifluoroethylene).[9,12,30–32] To facilitate the discussion and comparison with the irradiated P(VDF-TrFE) copolymer, the composition of the terpolymers (use P(VDF-TrFE-CFE) as an example) are labeled as VDF_x-$TrFE_{1-x}$-CFE_y, where the mole ratio of VDF/TrFE is $x/1-x$ and y is the mol% of CFE in the terpolymer.

Presented in Figure 10.16(a) is the thickness strain (*i.e.* longitudinal strain or S_3) of the P(VDF-TrFE-CFE) terpolymer at composition of 68/32/9 mol% and a thickness strain of more than 7% can be reached, which is by far the highest among all the high-energy electron-irradiated copolymers and terpolymers investigated (measured at frequencies equal to or higher than 10 Hz).[10,19] The transverse strain response of this terpolymer was also characterised. Presented in Figure 10.16(b) is the transverse strain S_1 from an uniaxially stretched film measured along the stretching direction and the Figure 10.16(c) is the transverse strain measured from a nonstretched film, which is smaller than that from the uniaxially stretched films. For the uniaxially stretched films, the transverse strain perpendicular to the stretching direction is quite small and for stretched films, this strain is negative. Using the relationship

$$S_v = S_3 + S_1 + S_2 \qquad (10.4)$$

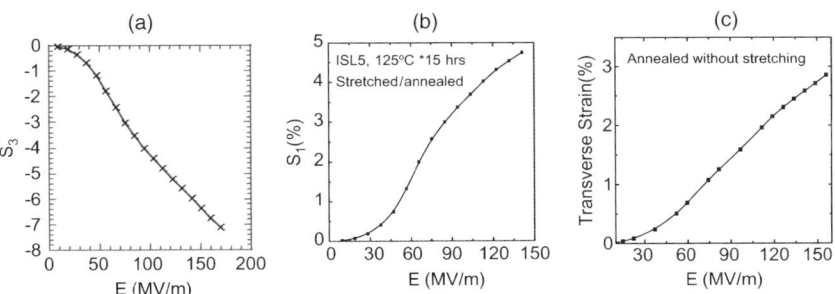

Figure 10.16 (a) The field-induced longitudinal strain (S_3), (b) transverse strain for uniaxially stretched terpolymer measured along the stretching direction, and (c) transverse strain (S_1) for the unstretched P(VDF-TrFE-CFE) 68/32/9 mol% as a function of applied field. The measurement frequency is in the 1 to 10 Hz range.[19]

where S_1 and S_2 are the transverse strains, the volume strain can be determined and it was found that for the terpolymer of P(VDF-TrFE-CFE), the volume strain is very small (on the order of 10% of the thickness strain S_3). This is different from the irradiated copolymers, which exhibit quite high volume strain and suggests different electrostrictive strain-generation mechanisms in these two classes of polymers. Recently, in a systematic study of recrystallisation in the irradiated P(VDF-TrFE) copolymers, it is found that the interfacial layer between the crystals and amorphous in these electrostrictive polymers play an important role in the electromechanical performance.[33] Therefore, it is suggested that the interfacial layer has a significant contribution to the strain response observed in the terpolymers.

In addition to the terpolymer of P(VDF-TrFE-CFE), ferroelectric fluoroterpolymers with other termonomers such as CTFE (CTFE: chlorotrifluoroethylene), HFP (hexafluoropropylene), and CDFE (chlorodifluoroethylene) have also been investigated.[9,30,31,32,34] Among them, P(VDF-TrFE-CTFE) was investigated in some details regarding their electromechanical response and the results show that although very large thickness strain was reported when measured at near dc (static) condition (strain of 1.2% under 20 MV/m, for example, corresponding a electrostrictive coefficient $M = 3 \times 10^{-17} \, m^2/V^2$, $S = ME^2$), the strain response measured at non-dc condition is lower than that from the terpolymer of P(VDF-TrFE-CFE).[9,32]

10.3.2 The Microstructure and Ferroelectric Relaxor Behaviour of P(VDF-TrFE-CFE) Terpolymers

The effect of the termonomer CFE on the ferroelectric response of the terpolymers was recently examined. In analogy to the irradiated copolymers, for terpolymers with DVF/TrFE mole ratio below 70/30 mol%, increasing the CFE mol% can eventually lead to the complete conversion of the terpolymer into the ferroelectric relaxor. This evolution process can be monitored from the X-ray, DSC, as well as FT-IR measurements. For example, from the peak area of the polar (normal ferroelectric) and nonpolar (relaxor phase) peaks of the X-ray data, the polar phase fraction in the terpolymer can be deduced and is shown in Figure 10.17. This fraction is a function of CFE content for a series of P(VDF-TrFE-CFE) terpolymers with DVF/TrFE ratio in 64/36 mol% to 75/25 mol% range.[35] The result indicates that with about 8 mol% of CFE, the terpolymers in this VDF/TrFE ratio range can be completely converetd into a ferroelectric relaxor with nondetectable normal ferroelectric fraction. In contrast to the relaxor ferroelectrics based on the irradiated copolymer, in the relaxor terpolymer there exists a significant fraction of all-*trans* conformation at room temperature even though the X-ray data and DSC data all show that there is no macroscopic ferroelectric phase. Presented in Figure 10.18 is the evolution of different conformation fractions for the relaxor terpolymer P(VDF-TrFE-CFE) 68/32/9 mol% as a function of temperature.[36] With reduced temperature from 150°C (above the melting), there is a gradual increase of the all-*trans*

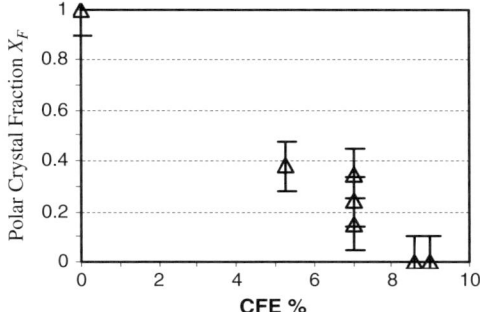

Figure 10.17 The polar-phase fraction in the terpolymer versus CFE molar percentage.

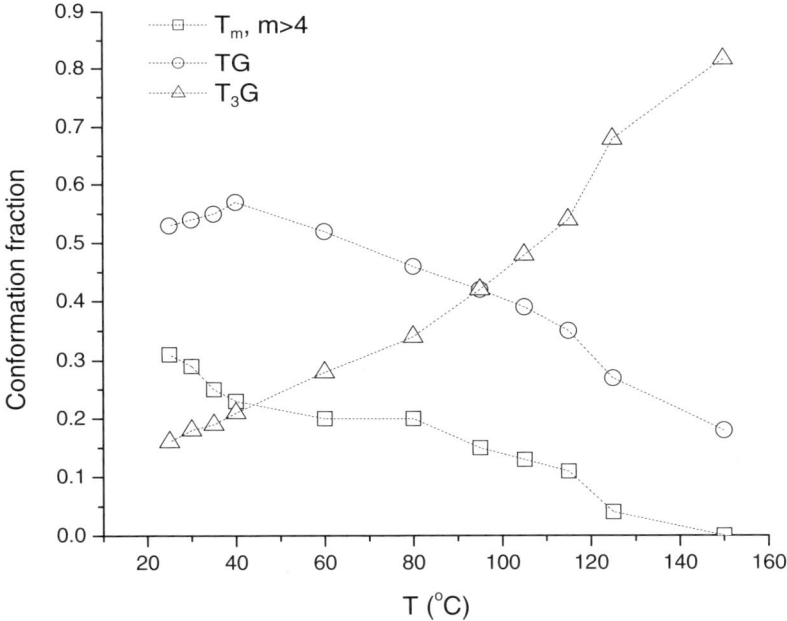

Figure 10.18 $T_{m>4}$, TGTG′, and T_3GT_3G' conformation fractions for a terpolymer of 68/32/9 VDF/TrFE/CFE. Lines are drawn to guide the eyes.

conformation ($T_{m>4}$) which suggests that there exists local nanopolar-regions at room temperature for these terpolymers. It also interesting to note that the fraction of TGTG' also increases with reduced temperature, while the fraction of T_3GT_3G' reduces with reduced temperature. The change of the molecular conformations with temperature for the terpolymer here is certainly quite different from that observed in the high-energy electron-irradiated copolymers, in spite of the fact that both relaxor polymers exhibit very similar macroscopic responses.

The dynamic processes in the relaxor P(VDF-TrFE-CFE) terpolymers as well as the high-energy electron-irradiated P(VDF-TrFE) copolymers were investigated by Bobnar et al.[37,38] and it was found that the dynamic processes in these relaxor ferroelectric polymers are very similar to those in the classical inorganic relaxors such as lead magnesium niobate. The results show that the broad dispersive dielectric maximum is a result where, as in dipole glasses, the dielectric constant ε_1 (the linear dielectric constant) at a certain temperature which depends on the experimental time scale, *i.e.* frequency, starts to deviate from its static value (Figure 10.19). The temperaure dependence of the static dielectric constant is also shown in Figure 10.19. Due to the conduction in the terpolymer, the static dielectric constant can only be measured at temperatures below 240 K. The rise of the dielectric constant at high temperature (for the static dielectric constant as well as the one at 20 Hz and 100 Hz) is caused by the conduction in the sample. The inset to Figure 10.19 shows that the characteristic relaxor

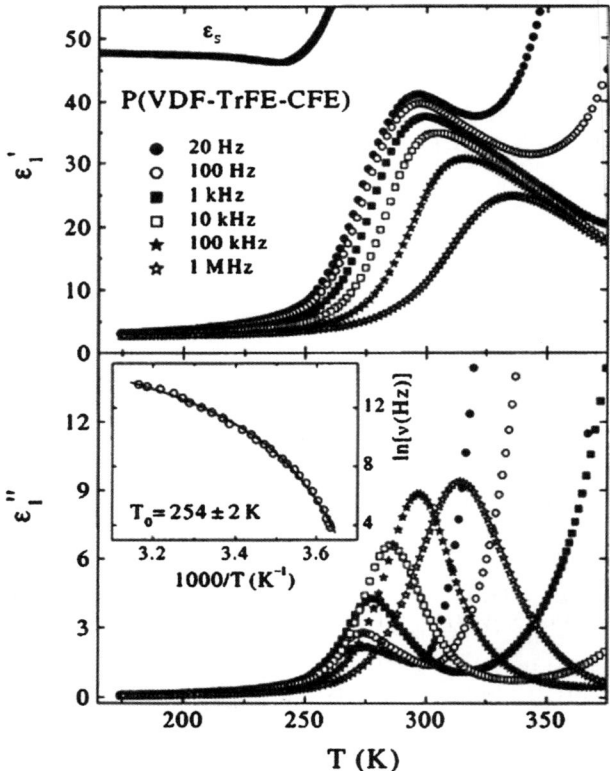

Figure 10.19 Temperature dependence of a real ε_1' and an imaginary ε_1'' linear dielectric constant measured at several frequencies for the relaxor P(VDF-TrFE-CFE) terpolymer; the static dielectric constant ε_s is also shown. The insert shows that the charcteristic relaxation time follows the Vogel–Fulcher law.[37]

frequency, determined from peaks in the imaginary dielectric constant ε_1'', follows the Vogel–Fulcher (V–F) law $f=f_0 \exp[-U/k(T-T_0)]$ with the Vogel–Fulcher temperature $T_0 = 254$ K for the P(VDF-TrFE-CFE) 68/32/9 mol% terpolymer. It is also found that the dielectric behaviour of the irrdaiated P(VDF-TrFE) copolymers follows the V–F law.[39] This indicates the similarity between these relaxor ferroelectric polymers and the inorganic relaxor ferroelectrics. However, it is very interesting to note that the value of f_0 obtained in these relaxor ferroelectric polymers is much smaller than that from the inorganic relaxor ferroelectrics.

The temperature dependence of the third-order nonlinear dielectric constant ε_3 and dielectric nonlinearity a_3 ($=\varepsilon_3/\varepsilon_0^3 \varepsilon_1^4$, where ε_0 is the vacuum permittivity) indicates a crossover from decreasing paraelectric-like to increasing glass-like temperature behaviour (Figure 10.20). Such a behaviour is in accordance with the prediction of the spherical random-bond–random-field model of relaxor ferroelectrics.[40] Although one standard way to analyse the dielectric data is to fit to the Cole–Cole plot, this procedure can not provide direct and

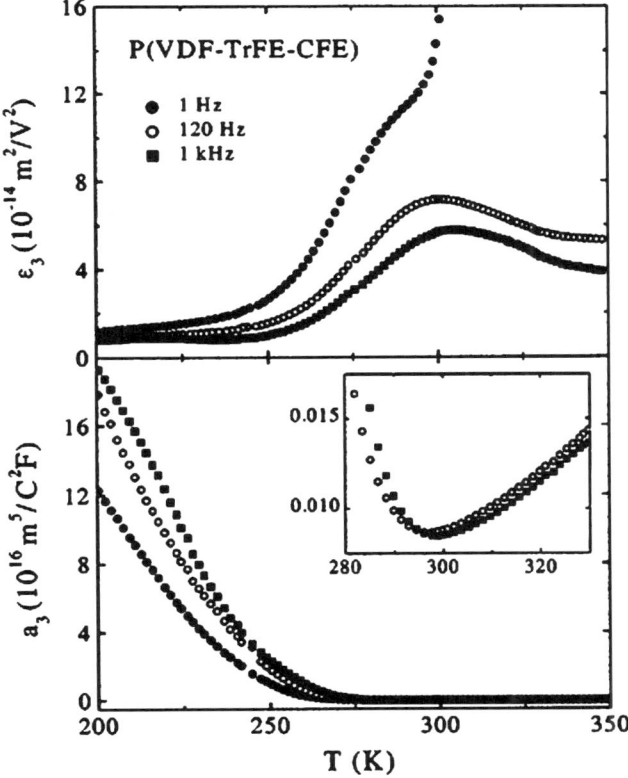

Figure 10.20 Temperature dependence of the 3rd-order nonlinear dielectric constant ε_3 and dielectric nonlinearity a_3 at three frequencies for the relaxor P(VDF-TrFE-CFE). The insert shows the paraelectric-to-glass crossover.[37]

independent information about the actual relaxation spectrum. Furthermore, this method is not suitable when the relaxation spectrum becomes extremely polydispersive in the relaxor with decreasing temperature and when the freezing temperatrure is approached. The information on the relaxation spectrum and thus on the dynamic processes can be directly extracted using the so-called temperature–frequency plot.[41] By varying the reduced dielectric constant

$$\delta = \frac{\varepsilon'_1 - \varepsilon_\infty}{\varepsilon_s - \varepsilon_\infty} = \int_{z1}^{z2} \frac{g(z)\mathrm{d}z}{1 + (\omega/\omega_a)^2 \exp(2z)} \quad (10.5)$$

between the values 1 and 0 different segments of the relaxation spectrum $g(z)$ are being probed. The temperature–frequency plot for the terpolymer P(VDF-TrFE-CFE) 68/32/9 mol% is shown in Figure 10.21(a). The solid lines through the $d = 0.95, \ldots, 0.7$ data are fits to the V–F law and the freezing

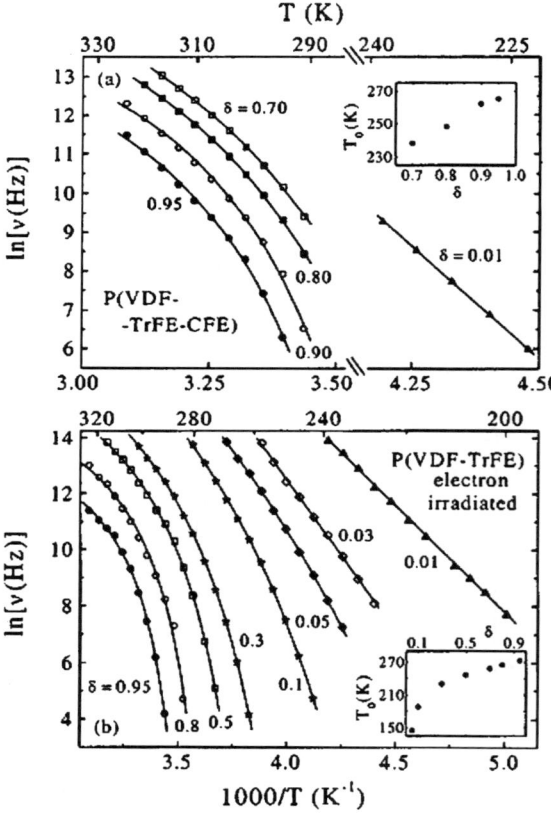

Figure 10.21 Temperature–frequency plot for several fixed values of the reduced dielectric constant d in (a) a P(VDF-TrFE-CFE) terpolymer and (b) the high-energy electron-irradiated P(VDF-TrFE) copolymer.[37]

temperature of the system is determined by the divergence of the longest relaxation time, i.e. $T_f = T_0(\delta \rightarrow 1) = 269$ K, which is higher than the V–F temperature determined from the characteristic relaxation time in Figure 10.21 because the bulk of the relaxation times remains finite below T_f. The temperature–frequency plot for the irradiated P(VDF-TrFE) 68/32 mol% is also shown in Figure 10.21(b). The high-frequency components (curves with small δ) remain active down to temperatures much below T_f and follow the Arrhenius law $f = f_0 \exp(-E/kT)$.

10.4 Performance of Microelectromechanical Devices

Several microelectromechanical devices, which make use of the large strain and elastic energy density of the electrostrictive PVDF-based polymers, have been investigated.[42–44] In this section, the results of a polymeric microactuator (PMAT) will be reviewed. To fully use the properties of the stretched electrostrictive P(VDF-TrFE) copolymer film, the PMAT configuration adopted for this investigation is shown schematically in Figure 10.22.[43,44] The active polymer, which is a film ($\sim 10\,\mu$m thickness) of uniaxially stretched HEEIP, is bonded to an inactive polymer film using Spurr epoxy to form a unimorph actuator. In this design, the inactive polymer is of the same material as the active polymer except there are no electrodes on the film. The elastic modulus along the uniaxial stretching direction is 1 GPa and is about 3 times larger than that in the perpendicular direction. In the PMAT, the actuation is generated by the strain along the stretching direction (the x-direction in Fig. 22) and the unimorph actuator is bonded to the silicon wafer with Spurr epoxy at the two ends ($x = \pm L_0/2$). Under an electric field, the expansion of the active polymer generates an upward motion of the PMAT (the z-direction in Figure 10.22). The two ends of the PMAT along the y-direction are mechanically free to maximise the actuation. Sputtered Au electrodes are used and a 0.5-mm unelectroded margin is provided to prevent electric breakdown (Figure 10.22(b)). It is expected that this unelectroded margin along the y-direction will cause some reduction of the actuator strain S_1 of the active polymer.

Two groups of actuators were fabricated with different L_0 (the actuation length along the x-direction): $L_0 = 1$ mm (PMAT1) and $L_0 = 0.5$ mm (PMAT0.5). The width w (along the y-direction, Figure 10.22(a)) of PMAT1 is 4.5 mm and for PMAT0.5, $w = 3$ mm. The inactive polymer film thickness t_n for PMAT1 group is about the same as that (t_a) of the active film and for PMAT0.5, t_n is about half of t_a. The actuator pattern on the silicon wafer was fabricated using the standard bulk silicon micromachining technique (KOH anisotropic wet-etching) as described in references.[42–46] The displacement response was measured by a laser vibrometer (Polytec PI, Inc, model OFV-511), which has a beam size of 5 μm and an upper measuring frequency of 100 kHz. An HP 4192 impedance analyser was used for electric impedance characterisation. To provide high dc bias voltages in the impedance measurement, a special blocking circuit was designed.

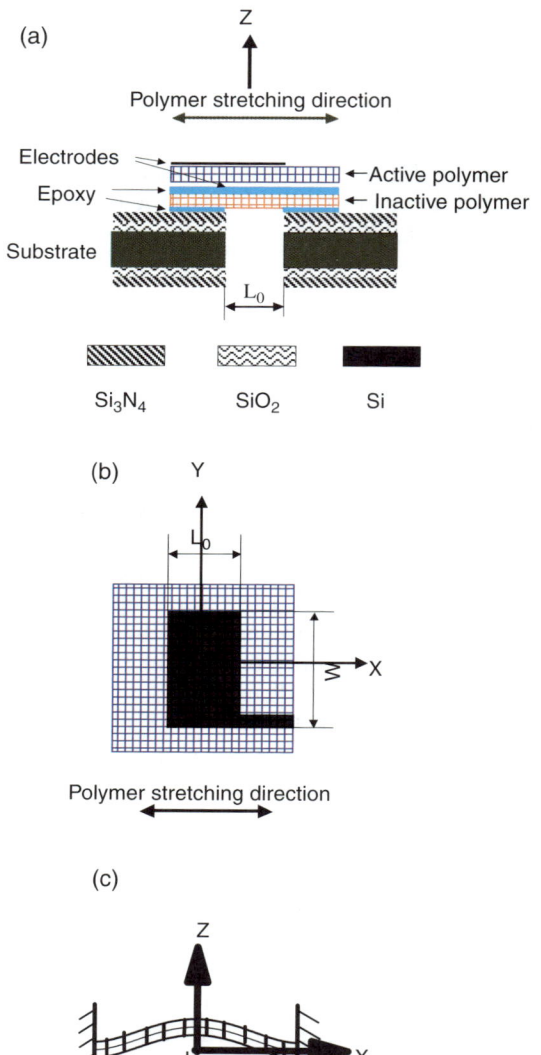

Figure 10.22 (a) Schematic of the PMAT investigated. (b) The electrode pattern of the PMAT active polymer (black area) where the electrode width along the x-direction is the same as the device actuation length L_0 in the same direction and along the y-direction, the unelectroded margin width is 0.5 mm to prevent the breakdown at the edges. (c) Schematic of the actuation response of the current PMAT under external fields: because of $S_1 > 0$, the actuator moves upward from the neutral position $(z=0)$.[43,44]

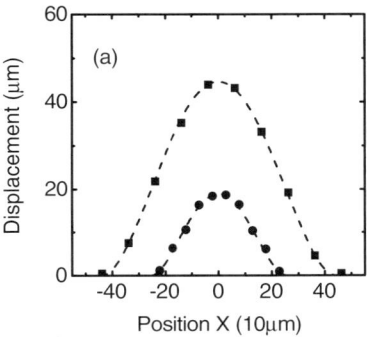
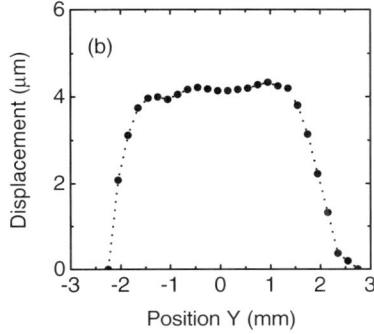

Figure 10.23 (a) Displacement profile for a PMAT1 (measured at 100 Hz under 100 V/μm field) and PMAT0.5 (measured at 50 Hz under 65 V/μm field) along the x-direction where the dots are measured data and solid curves are fitting from eqn (10.1). (b) Displacement profile for a PMAT1 along the y-direction (measured under 52 V/μm dc bias field and 100 V (rms value) ac field at 100 Hz).

For the PMAT investigated here, the actuation displacement along the z-direction as a function of the coordinate x can be expressed as:[47]

$$z = a\left(\left(\frac{L_0}{2}\right)^2 - x^2\right)^2 \qquad (10.6)$$

where the two ends of the actuator are fixed at $x = \pm L_0/2$ as shown in Figure 10.22(c). The constant a in the equation is determined by the effective strain S_e of the polymer films in the x-direction, i.e.

$$\int_{-L_0/2}^{L_0/2} \left(\left(\frac{dz}{dx}\right)^2 + 1\right)^{1/2} dx = (1 + S_e)L_0 \qquad (10.7)$$

which is based on the geometric constraint that the total length of the actuator after the actuation should be equal to the polymer film length after the strain S_e. Because of the inactive polymer layer, S_e is different from the strain S_1 of the active polymer measured in the free condition. From the force balance condition in the x-direction, S_e is related to S_1 as,

$$E_a t_a (S_1 - S_e) = E_n t_n S_n \qquad (10.8)$$

which leads to $S_e = S_1/(1 + k)$, where $k = (E_n t_n)/(E_a t_a)$ is a measure of the clamping effect due to inactive layer (clamping ratio). To include the contribution from the metal electrodes and Spurr epoxy bonding layer, k should be modified to $k = (\Sigma E_{ni} t_{ni})/(E_a t_a)$, where E_{ni} and t_{ni} are the elastic modulus and thickness of ith inactive layer, respectively. It should be noted that in the derivation, we omitted the mechanical clamping effect at the two fixed ends

($x = \pm L_0/2$) which will reduce S_1 even if $k = 0$. All the analysis is based on the assumption that $L_0 \gg t$, where t is the total thickness of the actuator.

To evaluate the performance of the PMAT, we first present in Figure 10.23(a) the displacement profiles measured on a PMAT1 and a PMAT0.5 as a function of x and along the line of $y = 0$ (see Figure 10.22(b)). The solid lines are the fitting from eqn (10.1), which match the experimental results (dots) quite well, indicating that the unimorphs function properly. In addition, the displacement profile as a function of y along the $x = 0$ line for a PMAT1 was also characterised and the results are shown in Figure 10.23(b). The electrode width along the y-direction is $w = 4.5$ mm for PAMT1 and the width at 90% of actuation level is about 3.5 mm, indicating the clamping effect due to the unelectroded margins.

The displacement versus applied electric field was measured for a PMA1 and the result is presented in Figure 10.24(a). The results derived from the eqns (10.6) to (10.8) based on S_1 acquired from an active film in the free condition are also presented in the figure for comparison. The experimental data points follow the curve with clamping ratio $k = 3$ very well, which seems to be reasonable considering the fact that besides the inactive layer ($k = 1$), there are Spurr epoxy layers (the elastic modulus $G = 5$ GPa and thickness ~ 1 μm), Au electrodes (200 nm total thickness and elastic modulus ~ 100 GPa), unelectroded margins, as well as the clamping effect from the two fixed ends. The maximum displacement of the unimorph actuator reaches 60 μm, corresponding to $S_e = 0.85\%$ in the films. Such a high displacement output of the PMAT is due to the high strain level S_1 in the active polymer. By reducing k to equal 1, the displacement output can be increased. It should be pointed out that in the current PMAT design, only the top polymer film is active, resulting in an

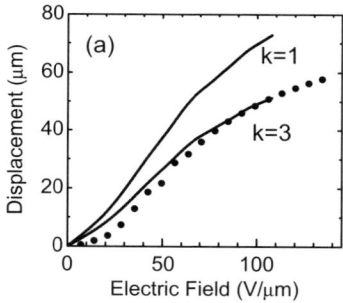
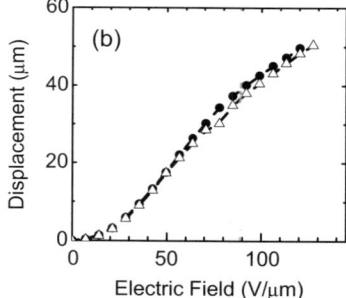

Figure 10.24 (a) Displacement at the center of a PMAT1 as a function of the applied electric ac field (10 Hz) where dots are the experiment data and solid curves are the calculation based on the eqns (10.6) to (10.8) and the data on S_1 measured at the free condition of the active polymer at 10 Hz. In the figure, the calculated curves with the clamping ratio of $k = 1$ and $k = 3$ are shown. The model curve is in good accord with the experimental data. (b) Comparison of the displacement at the center of a PMAT measured in air (solid circles) and in silicon oil (open triangles) as a function of applied electric field amplitude (100 Hz). There is very little reduction of the displacement due to the fluid loading.

upward motion of the actuator. If needed, one can alternatively drive the two polymer layers, which will result in the motion of the actuator in both upward and downward directions to double the displacement output.

The frequency dependence of a PMAT1 was also characterised. Presented in Figure 10.25(a) is the normalised displacement output measured in the frequency range from 10 Hz to 100 kHz (the upper limit of the optic probe). In this measurement, a dc bias field of 55 V/μm was applied so that an effective piezoelectric state was induced. As can be seen, over four frequency decades, the displacement voltage ratio (DVR) decreases from 31 nm/V to 18 nm/V. The results demonstrate two main features of the PMAT investigated: (a) it possesses a high DVR, and (b) it is capable of operating over a broad frequency range. At 86 kHz, a resonance was observed. Interestingly, when the same PMAT was operated in a liquid (silicon oil), the DVR remains nearly the same in the same frequency range. This result demonstrates the high load capability of the PMAT, which is due to the high elastic energy density of the active material utilised. In the fluid medium, the resonance is shifted to lower frequency (13.2 kHz) due to the fluid loading, which should follow (if the fluid loading does not change the mode shape of the actuator):[48]

$$\frac{f_i|_{\text{in fluid}}}{f_i|_{\text{vacuum}}} = 1/(1 + 2A_p/M_p)^{1/2} \tag{10.9}$$

where M_p is the mass of the resonator plate and A_p accounts for the fluid-loading effect. For the fundamental resonance of the PMAT, it can be derived that $(f_i|_{\text{in fluid}})/(f_i|_{\text{in vacuum}}) = 1/6.12$,[48,49] which is very close to the measured ratio of $13.2/86 = 1/6.5$. For comparison, the electric impedance data measured under the same dc bias field for the PMAT1 in air is presented in Figure 10.25(b) where

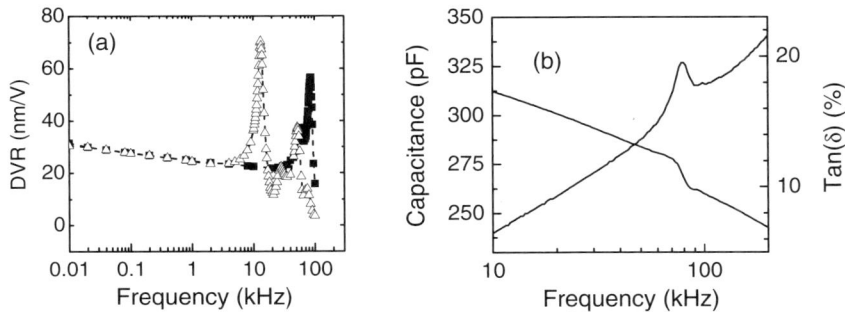

Figure 10.25 (a) The displacement voltage ratio (DVR) of a DC field biased (55 MV/m) PMAT1 as a function of ac field frequency (the ac field is 0.64 V/μm). The measurement was carried out both in air (solid dots) and in silicone oil (open triangles). Due to the fluid loading, the resonance frequency is shifted to lower frequency. In the data taken from silicone oil, higher harmonics were also observed. (b) Electric impedance data of the same PMAT1 measured in air under the same applied DC and ac electric field. A strong electromechanical resonance was observed.

a strong electric resonance was also observed in the same frequency range, indicating the strong electromechanical coupling of the device.

The displacement outputs measured under high electric field in air and in silicon oil are compared in Figure 10.25(a) and again, there is very little reduction of the displacement in the silicon oil compared with that in air.

10.5 Summary

In summary, it is demonstrated that the stable phase of PVDF-based ferroelectric polymers at room temperature can be converted from the normal ferroelectric (corresponding to piezoelectric polymer) to nonferroelectric (corresponding to electrostrictive polymer) by different approaches, such as high-energy irradiation to P(VDF-TrFE) copolymer and adding bulky termonomers, such as CFE and CTFE, to form a random terpolymer. The modified polymers exhibit a high dielectric constant and the characteristics of the relaxor ferroelectrics, such as a strong dielectric dispersion in the radio-frequency region, a broad dielectric peak as a function temperature, the shift of the dielectric constant (real and imaginary parts) maximum temperature with frequency, a slim polarisation hysteresis loop at the dielectric constant peak. These polymers represent the first known relaxor ferroelectric polymers. More importantly, these relaxor ferroelectric polymers exhibit high electromechanical responses, such as a high electric-field induced electrostrictive strain ($>7\%$) with an elastic modulus and a high electromechanical coupling factor ($k_{33} \sim 0.3$ and $k_{31} \sim 0.65$) over a broad temperature range. In addition, these electrostrictive polymers exhibit a highly tunable anisotropy in the strain response. For example, the transverse strain can vary from one third up to the same as the longitudinal strain. Additionally, these electrostrictive polymers possess a high load capability.

The microstructure of the relaxor polymers was investigated. Although the dielectric behaviour and electromechanical performance are the same for all the relaxor polymers, the microstructure, especially the mechanisms responding to high electromechanical performance, and morphology are different. For the irradiated copolymers, it is found that the electrostrictive strain response originates from the electric-field induced conformation change between the nonpolar bonds and polar bonds. As a result, the irradiated copolymer also exhibits a large volume strain. For the terpolymers, it is found that at room temperature the molecular conformations are high in the T_3GT_3G' as well as all-*trans* bonds and that the volume strain is very small. Therefore, it is suggested that the there exists local nanopolar regions in the terpolymer and that the interfacial layer in the terpolymers plays an important role in the electromechanical response.

Taking advantage of the high electromechanical response of the electrostrictive fluoropolymers, several microdevices have been developed. The performance of these devices confirms the high electromechanical performance and high load capability of the polymers. Furthermore, the results show that the relaxor polymers maintain high electromechanical responses at frequencies of more than 10^5 Hz.

Acknowledgement

The financial support from NIH under Grant No. R21EY016799-01 is greatly appreciated. Thanks also due to 3M for a nontenured faculty award to ZY.

References

1. H.S. Nalwa, *Ferroelectric Polymers: Chemistry, Physics and Applications*, Marcel Dekker, New York, 1995.
2. T.T. Wang, J.M. Herbert and A.M. Glass, *The Applications of Ferroelectric Polymers*, Blackie and Son, New York, 1988.
3. A.J. Lovinger and T. Furukawa, *Ferroelectrics*, 1983, **50**, 227.
4. R. Hasegawa, Y. Takahashi, Y. Chatani and H. Tadokoro, *Polymer J.*, 1972, **3**, 600.
5. Y. Takahashi, Y. Matsubara and H. Tadokoro, *Macromolecules*, 1983, **16**, 1588.
6. K. Tashiro, S. Nishimura and M. Kobayashi, *Macromolecules*, 1990, **23**, 2802.
7. Q.M. Zhang, V. Bharti and X. Zhao, *Science*, 1998, **280**, 2101.
8. Z.-Y. Cheng, D. Olson, H.S. Xu, F. Xia, J.S. Hundal, Q.M. Zhang, F.B. Bateman, G.J. Kavarnos and T. Ramotowski, *Macromolecules*, 2002, **35**, 664.
9. H. Xu, Z.-Y. Cheng, D. Olson, T. Mai, Q.M. Zhang and G. Kavarnos, *Appl. Phys. Lett.*, 2001, **78**, 2360.
10. F. Xia, Z.-Y. Cheng, H. Xu, H. Li, Q.M. Zhang, G. Kavarnos, R. Ting, G. Abdul-Sedat and K.D. Belfield, *Adv. Mater.*, 2002, **14**, 1574.
11. Z.-Y. Cheng, Z. Li, Y. Ma, Q.M. Zhang and F.B. Bateman, in *Nano- and Microelectromechanical Systems NEMS and MEMS and Molecular Machines*, ed. A.A. Ayon, T. Buchheit, D.A. LaVan and M. Madou, Materials Research Society Series, Warrendale, PA, 2002, Vol. 734, p.A.2.5.1.
12. N.M. Raynolds, K.J. Kim, C. Chang and S.L. Hsu, *Macromolecules*, 1989, **223**, 1092–1100.
13. L.E. Cross, *Ferroelectrics*, 1994, **151**, 305.
14. Z.-Y. Cheng, R.S. Katiyar, X. Yao and A. Guo, *Phys. Rev. B*, 1997, **55**, 8165.
15. P. Mabboux and K. Gleason, *J. Fluorine Chem.*, 2002, **113**, 27.
16. X. Zhao, V. Bharti, Q.M. Zhang, T. Ramotowski, F. Tito and R. Ting, *Appl. Phys. Lett.*, 1998, **73**, 2054.
17. Z.-Y. Cheng, V. Bharti, T.-B. Xu, H. Xu, T. Mai and Q.M. Zhang, *Sens. Actuators A. Phys.*, 2001, **90**, 138.

18. Z.-Y. Cheng, V. Bharti, T. Mai, T.B. Xu, Q.M. Zhang, T. Ramotowski, K.A. Wright and R. Ting, *IEEE Trans. Ultrason. Ferro. Freq. Cont.*, 2000, **47**, 1296.
19. C. Huang, R. Klein, F. Xia, H.F. Li, Q.M. Zhang, F. Bauer and Z.-Y. Cheng, *IEEE Trans. Dielec. Elec. Insulat.*, 2004, **112**, 299–311.
20. Z.-Y. Cheng, T.-B. Xu, V. Bharti, S. Wang and Q.M. Zhang, *Appl. Phys. Lett.*, 1999, **74**, 1901–1903.
21. Y.M. Shkel and D.J. Klingenberg, *J. Appl. Phys.*, 1998, **83**, 415.
22. C. Hom, S. Pilgrim, N. Shankar, K. Bridger, M. Masuda and S. Winzer, *IEEE Trans. Ultrason. Ferro. Freq. Cont.*, 1994, **41**, 542–551.
23. K. Omote, H. Ohigashi and K. Koga, *J. Appl. Phys.*, 1997, **81**(6), 2760.
24. V. Bharti, Z.-Y. Cheng, S. Gross, T.-B. Xu and Q.M. Zhang, *Appl. Phys. Lett.*, 1999, **75**, 2653.
25. S.J. Gross, Z.-Y. Cheng, V. Bharti and Q.M. Zhang, *Proc. IEEE 1999 Int. Symp. Ultrasonics*, 1999, 1019–1024.
26. Z.-Y. Cheng, V. Bharti, T. -B. Xu, S. Wang, Q.M. Zhang, T. Ramotowski, F. Tito and R. Ting, *J. Appl. Phys.*, 1999, **86**, 2208.
27. S. Guo, X.-Z. Zhao, Q. Zhuo, H.L.W. Chan and C.L. Choy, *Appl. Phys. Lett.*, 2004, **84**, 3349.
28. V. Bharti, H. Xu, G. Shanthi, Q.M. Zhang and K. Liang, *J. Appl. Phys.*, 2000, **87**, 452–461.
29. R. Casalini and M. Roland, *Appl. Phys. Lett.*, 2001, **79**, 2627.
30. T. Chung and A. Petchsuk, *Macromolecules*, 2002, **35**, 7678.
31. J.T. Carrett, C.M. Roland, A. Petchsuk and T.C. Chung, *Appl. Phys. Lett.*, 2003, **83**, 1190.
32. G.S. Buckley, C.M. Roland, R. Casalini, A. Petchsuk and T.C. Chung, *Chem. Mater.*, 2002, **14**, 2590–2593.
33. Z.M. Li, S.Q. Li and Z.-Y. Cheng, *J. Appl. Phys.*, 2005, **971**, 014102.
34. Z.M. Li, Y.H. Wang and Z.-Y. Cheng, *Appl. Phys. Lett.*, 2006, **88**, 062904.
35. R. Klein, F. Xia, Q.M. Zhang and F. Bauer, *J. Appl. Phys.*, 2005, **97**, 094105.
36. R. Klein, MS Thesis, The Pennsylvania State University, 2004.
37. V. Bobnar, B. Vodopivec, M. Kosec, A. Levstik, B. Hilczer and Q.M. Zhang, *Macromolecules*, 2003, **36**, 4436.
38. V. Bobnar, B. Vodopivec, A. Levstik, Z.-Y. Cheng and Q.M. Zhang, *Phys. Rev. B*, 2002, **67**, 94205.
39. Z.-Y. Cheng, Q.M. Zhang and F.B. Bateman, *J. Appl. Phys.*, 2002, **92**, 6749.
40. V. Bobnar, Z. Kutnjak, R. Pirc, R. Blinc and A. Lovstik, *Phys. Rev. Lett.*, 2000, **84**, 5892.
41. Z. Kutnjak, C. Filipic, A. Levstik and R. Pirc, *Phys. Rev. Lett.*, 1993, **70**, 4015.
42. Feng Xia, Srinivas Tadigadapa and Q.M. Zhang, *Sens. Actuators A*, 2006, **125**, 346.
43. T.B. Xu, PhD Thesis, Penn State University, 2001.
44. (a) T.B. Xu, Z.-Y. Cheng and Q.M. Zhang, *Appl. Phys. Lett.*, 2002, **80**, 1082; (b) T.-B. Xu and Ji Su, *J. MEMS*, 2005, **14**, 539.

45. B. Rashidian and M.G. Allen, *Proc. ASME DSC*, 1991, **32**, 171.
46. J.-H. Mo, A.L. Robinson, D.E. Fitting, F.L. Terry and P.L. Carson, *IEEE Trans. Electron Devices*, 1990, **37**, 134.
47. M. Madou, *Fundamentals of Microfabrication*, CRC Press, New York, 1997, p. 145.
48. R.D. Blevins, *Formulas for Natural Frequency and Mode Shape*, Krieger Publishing Company, Florida, 1995, p. 413.
49. R.A. Walsh, *Electromechanical Design Handbook*, McGraw-Hill, New York, 2000, p. 5.34.

CHAPTER 11
Magnetic Polymeric Gels as Intelligent Artificial Muscles

MIKLÓS ZRÍNYI

Department of Physical Chemistry, Budapest University of Technology and Economics, HAS-BME Laboratory of Soft Matters, H-1521, Budapest, HUNGARY

11.1 Introduction

Motility is one of the most important living phenomena. The most conspicuous natural device that confers motility is the muscle. From an engineering point of view, muscles are soft and wet mechanical transducers, capable of performing their functions by quick and reversibly shortening in a process called unidirectional contraction. At present there are several adaptive (smart, intelligent) materials that can actuate or alter their properties in response to a changing environment. Among them, mechanical actuators have been the subjects of much investigation in recent years. They undergo a controllable change of shape due to some physical effects and can convert energy (electrical, thermal, chemical) directly to mechanical energy. This can be used to do work against load.

Today's robots can perform quick and precise movements, and as long as they are limited to one specific function. Nevertheless, they cannot perform diverse and flexible movements like living organisms. Flexible movement means two things: soft touching and quick responses to various situations. These two requirements can be fulfilled by soft and wet elastic materials, which possess intelligence on material level.

Certain polymer gels represent one class of actuators that have the unique ability to change elastic and swelling properties in a reversible manner.[1-3] These wet and soft materials offer lifelike capabilities for the future direction of technological development.

Since the muscle tissue consists of 80% multicomponent aqueous solution (water) and 20% proteins that comprise the elastic material, it is not surprising that efforts to develop artificial muscle have been devoted to polymer gels. Polymer gels are unique smart materials in the sense that no other class of materials can be made to respond to so many different stimuli like polymer gels. Both the stimuli and the response can be quite diverse. Volume phase transition in response to infinitesimal change of external stimuli like pH, temperature, solvent composition, electric field, and light has been observed in various gels.[2–4] Their application in devices such as actuators, controlled delivery systems, sensors, separators and artificial muscles has been suggested and are in progress.

For developing an artificial muscle it is important to induce quick and reversible shortening or lengthening (shape change) in response to a changing environment. Attempts at developing stimuli-responsive gels for technological purposes are complicated by the fact that structural changes, like shape and swelling degree changes that occur in smart gels, are kinetically restricted by the collective diffusion of chains and the friction between the polymer network and the swelling agent. This disadvantage often hinders the effort of designing optimal gels for different applications. This does not apply to more developed electrically activated polymers.

In order to accelerate the response of an adaptive gel to stimuli, the use of magnetic-field-sensitive gels (ferrogels) has been developed in our laboratory.[5–12]

11.2 Ferrogel as a New Type of Responsive Gel

A ferrogel is a chemically crosslinked polymer network swollen by a ferrofluid. A ferrofluid, or a magnetic fluid, is a colloidal dispersion of monodomain magnetic particles.[13] Their typical size is about 10 nm and they are superparamagnetic. In the ferrogel, the fine, distributed magnetic particles are located in the swelling liquid and attached to the flexible network chains by adhesive forces. The solid particles are the elementary carriers of a magnetic moment. In the absence of an applied field the moments are randomly oriented, and thus the gel has no net magnetisation. As soon as an external field is applied, the magnetic moments tend to align with the field to produce a bulk magnetic moment. With ordinary field strengths, the tendency of the dipole moments to align with the applied field is partially overcome by thermal agitation, such as the molecules of paramagnetic gas. As the strength of the field increases, all the particles eventually align their moments along the direction of the field, and as a result, the magnetisation saturates. If the field is turned off, the magnetic dipole moments quickly randomise and thus the bulk magnetisation is again reduced to zero. In a zero magnetic field a ferrogel presents a mechanical behaviour very close to that of a swollen network filled with nonmagnetic colloidal particles.

In this chapter, chemically crosslinked polyvinyl alcohol (PVA) hydrogel filled with magnetite particles is reported. PVA is a neutral water-swollen polymer that reacts under certain conditions with glutaraldehyde (GDA) resulting in chemical crosslinkages between PVA chains. The crosslinking density can be conveniently varied by the amount of GDA relative to the vinyl alcohol [VA] units of PVA chains. A ferrogel is characterised by the following symbol: sample name/polymer concentration (given by wt%) at which the crosslinks were introduced/the ratio of vinyl alcohol units to the crosslinking molecules/concentration of magnetite particles (given by wt%). For example, a typical ferrogel, FG/6.3/300/4.25, was prepared at a solution of 6.3 wt% of PVA; the ratio of vinyl alcohol monomer units to the crosslinker molecules (GDA) is 300; the magnetite content at preparation is 4.25 wt%.

In a uniform magnetic field a ferrogel experiences no net force. When it is placed into a spatially nonuniform magnetic field, forces act on the magnetic particles, and the magnetic interactions are enhanced. The stronger field attracts the particles, and due to their small size and strong interactions with molecules of dispersing liquid and polymer chains they all move together. Changes in molecular conformation can accumulate and lead to shape changes. The magnetic field drives and controls the displacement and the final shape is set by the balance of magnetic and elastic interactions. The force density, f_m on a piece of magnetic gel can be written as

$$\mathbf{f}_m = \mu_0 (\mathbf{M} \nabla) \mathbf{H} \qquad (11.1)$$

where μ_0 is the magnetic permeability of vacuum, M represents the magnetisation and $\nabla \mathbf{H}$ takes into account the gradient of magnetic field, H. It should be kept in mind, that the magnetic force density vector varies from point to point in accordance with the positional dependence of product $(M \nabla) \mathbf{H}$.

The orientation of $\mathbf{f}_m$ is parallel to the direction of magnetic field. In a nonaccelerating system the force density manifests itself as a stress distribution, which must be balanced by the network elasticity. A completely balanced set of forces is in this respect equivalent to no external force at all. However, they affect the gel internally, tending to change its shape or size or both. In general, the deformation induced by magnetic field cannot be considered homogeneous, since the driving force $(M \nabla) \mathbf{H}$ varies from point to point in space. However, one can find a special distribution of magnetic field, where the deviation from the homogeneous case is not significant as described in Ref. 15. (Figure 11.1)

In this case, the condition for uniaxial deformation of a ferrogel cylinder can be written as follows:[6,8]

$$\lambda_H^3 - \beta(H_h^2 - H_m^2)\lambda_H - 1 = 0 \qquad (11.2)$$

where λ_H denotes the deformation ratio due to field induced strain. The parameter β is defined as

$$\beta = \frac{\mu_0 \chi}{2G} \qquad (11.3)$$

where χ stands for the initial susceptibility of the magnetoelast, H_h and H_m represent the magnetic field strength at the bottom and the top of a suspended ferrogel cylinder, respectively. Equation (11.2) can be considered as a basic equation for describing the unidirectional magnetoelastic properties. It says that if we suspend a magnetic composite in a nonhomogeneous magnetic field

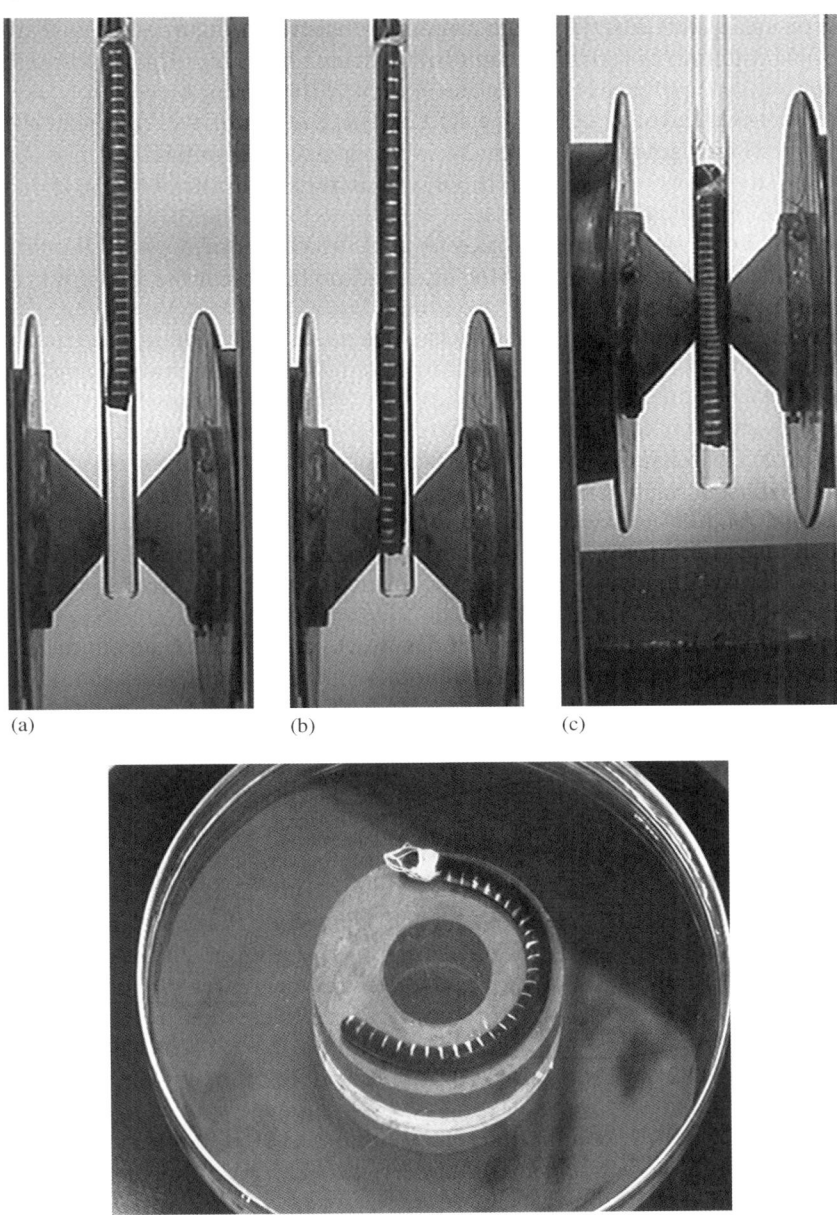

Figure 11.1 Shape distortion of ferrogels due to nonuniform magnetic field (a) no external magnetic field; (b) the maximal field strength is located under the lower end of the gel; (c) the maximal field strength is focused in the middle of the gel along its axis; (d) bending induced by a permanent magnet.

in such a way that $H_h > H_m$ then elongation occurs. In the opposite case when the field intensity is higher on top of the gel, i.e. $H_h < H_m$ eqn (11.2) predicts unidirectional compression. The magnetic-field-sensitive polymer gels can be made to bend and straighten, as well as elongate and contract repeatedly many times without damaging the gel. The response time to obtain the new equilibrium shape was found to be less than a second and seems to be independent of the size of the gel.

It must be mentioned that all the shape changes reported here are completely reversible. Since the magnetic field can be created by electromagnets, it is easy to achieve dynamical conditions by a modulated current intensity. We applied stepwise and sine-wave modulation by a function generator in the frequency range of 0.01–100 Hz. A cylindrical-shaped magnetoelast characterised by a height of 8 mm and radius of 4.5 mm was put onto the upper surface of a standing electromagnet as shown in Figure 11.2. It can be seen that the magnetoelastic response time is rather quick, one cycle requires half a second. We have to mention that up to 40 Hz the magnetic stimulus and the elastic response are strongly coupled. Neither phase shift, no significant mechanical (or magnetic) relaxation takes place.[12]

We have studied the dependence of elongation on the steady current intensity required by the electromagnets to produce an external magnetic field.

A cylindrical gel tube was suspended in water to prevent evaporation of swelling liquid and to balance the weight of gels by the buoyancy. The experimental arrangement is shown in Figure 11.3. The position of the top surface of the ferrogel was fixed by a rigid nonmagnetic copper thread. This was connected to a force-measuring unit in order to discover the force developed in the ferrogel due to a nonuniform magnetic field. The steady current intensity flowing through the electromagnets was varied in order to produce the external magnetic field. The field gradient around the ferrogel was parallel to the axis of the gel tube. The intensity of the current was varied between 0 and 8 A by an electronic device and the voltage was kept constant (20 V). The highest field intensity was 800 mT between the poles of electromagnets, and it disappeared within 200 mm. The distance of the lower end of the undeformed gel measured from the $z = 0$ axis is denoted by z_o.

The elongation of the ferrogel was monitored by a CCD camera, which allowed us to determine the displacement of the lower end of the gel with an accuracy of 10^{-2} mm. Both the elongation and also the force due to magnetic interactions have been measured.

Figure 11.2 Snapshot of shape change of a magnetoelast due to modulated magnetic field. The frequency of the field is 40 Hz.

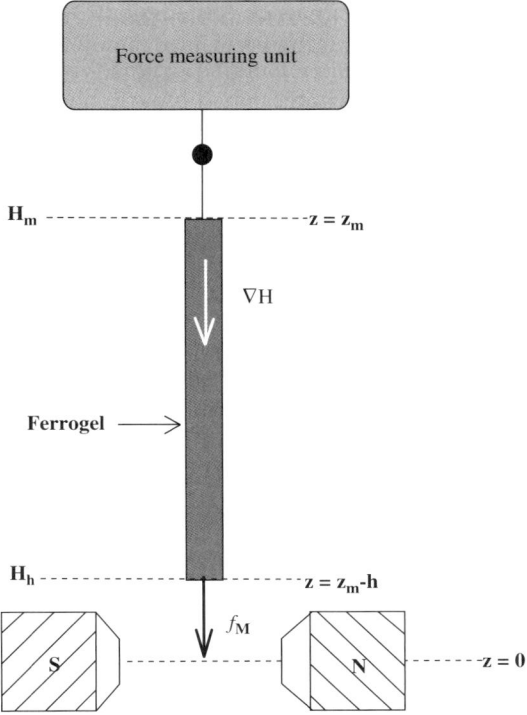

Figure 11.3 Schematic diagram of experimental setup to study the magnetoelastic properties of ferrogels.

Figure 11.4A shows the effect of a magnetic field on the deformation of ferrogel. The relative displacement is plotted against the steady current intensity. It can be seen that the displacement of the lower end of the gel – due to the magnetic force – is rather significant. A large magnetostriction takes place. In some cases we were able to produce an elongation of 40% of the initial length by applying a nonuniform magnetic field. It may be seen that at small current intensities the displacement slightly increases. However, at a certain current intensity a comparatively large, abrupt elongation occurs. This noncontinuous change in the size of the ferrogel appears within an infinitesimal change in the steady current intensity. Further increase in the current intensity results in another small extension. We have found that by decreasing the current a contraction takes place. Similarly to the extension, the measurement of the contraction was found to have a noncontinuous dependence on current intensity. It is worth mentioning that a significant hysteresis characterises the extension–contraction processes as seen in Figure 11.4. Not only the relative displacement, but also the measured force shows similar dependence. By variation of the experimental conditions, we have found a crossover between continuous and discontinuous shape transitions. The crossover between continuous and noncontinuous transitions seems to be determined by the position

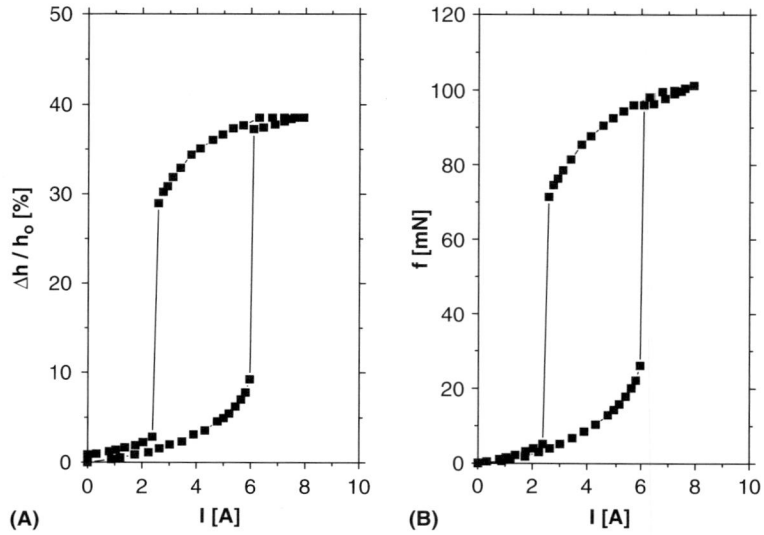

Figure 11.4 Noncontinuous elongation of a ferrogel FG/5.72/300/4.95 The initial length of the gel, h_o is 163 mm, $z_0 = 54.2$ mm, (A) relative displacement as a function of current; (B) force as a function of current.

of gel in the nonuniform magnetic field. It is worth mentioning that the discrete shape transition occurred within a time interval of one second, independently from the gel size.

11.3 Interpretation of the Abrupt Shape Transition

It is possible to interpret the abrupt shape transformation and hysteresis phenomena on the basis of eqn (11.2). In order to find the dependence of elongation on the steady current intensity, first we have to relate the magnetic field strengths, H_h and H_m to the steady current intensity. Let us assume that the magnetic field strength varies along the gel axis as

$$H(z) = H_{max}h(z) \qquad (11.4)$$

where H_{max} represents the maximal field strength at the position $z = 0$ and $h(z)$ is a unique function characterising the experimental arrangement (geometry of poles and gap distance).

It was found that the z-direction distribution of magnetic field strength can be satisfactorily approximated by the following forms:

$$h(z) = \begin{cases} 1 - kz^2 & \text{if } |z| < \delta \\ (1 - k\delta^2)\,e^{-\gamma(|z|-\delta)} & \text{if } |z| \geq \delta \end{cases} \qquad (11.5)$$

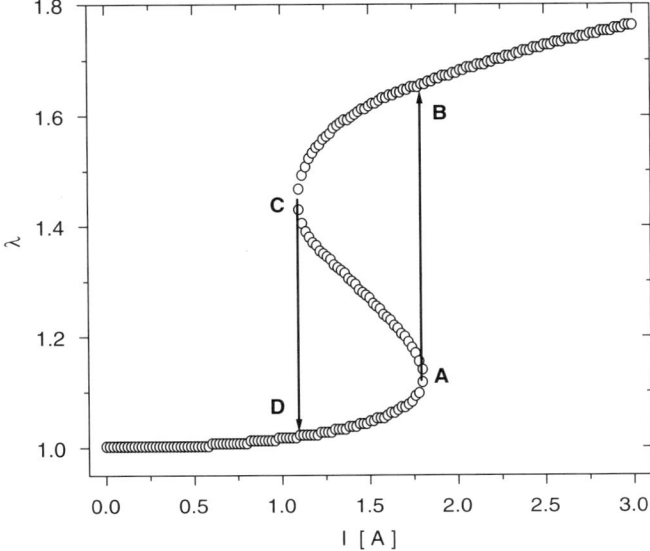

Figure 11.5 Dependence of deformation ratio, λ on the steady current intensity as calculated on the basis of eqn (11.2). For the calculation $h_0 = 10$ cm $z_0 = 5$ cm and $\gamma = 0.4$ were used.

where γ is a characteristic constant describing the exponential decay of field strength at larger distances, δ means the radius of poles and the constant $k = \frac{\gamma}{2\delta + \gamma \delta^2}$ was determined by taking into account the same slope of $h(z)$ curves at distance $r = \delta$, where the functions (in eqn (11.5)) approach each other.

According to the Biot–Savart law H_{max} can be written as $H_{max} = k_I I$, where k_I means a proportionality factor and I is the steady current intensity. It must be mentioned that the value of parameter k_I strongly depends on the quality of the electromagnet. The numerical solution of eqn (11.2) provides the λ dependence, which is shown in Figure 11.5.

In contradistinction to the state of equilibrium the internal parameter (deformation ratio) is not a single-valued function of the external parameter (current intensity) that is metastability develops with an abrupt nonequilibrium transition.

The transition from A → B takes place at a different value of current intensity from that for C → D. An area ABCD remains as a kind of hysteresis loop. This cycle depicted in Figure 11.5 is time independent and may be repeated several times. Here, a macroscopic energy barrier is provided by magnetic interactions and this energy barrier compels the ferrogel to go around it a higher or lower current than the equilibrium value.

We have also studied the influence of parameters z_0 on the shape of the $\lambda_H(I)$ dependence. These results are summarised in Figure 11.6. One can see that with increasing z_0 both the measure of abrupt transition and hysteresis increases. The parameter γ, which controls the gradient of the field strength, also plays an

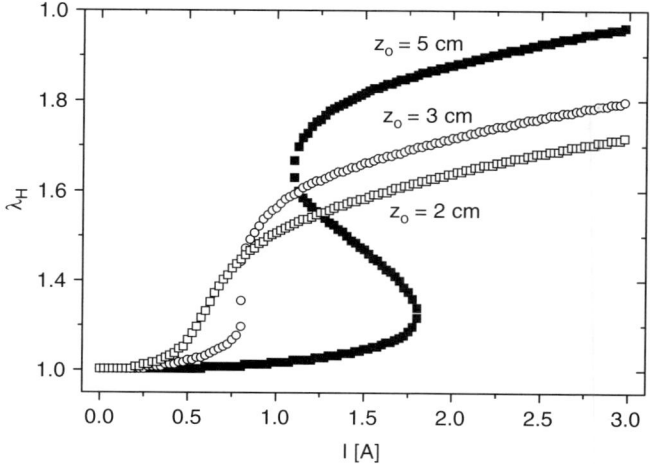

Figure 11.6 Dependence of deformation ratio, λ_H on the steady current at different initial position, z_0 of the gel.

essential role. The necessary condition for realising an abrupt transition requires a value of $\gamma > 0.15$. Below this value no discontinuous transition occurs if $h_0 = 10$ cm and $z_0 = 5$ cm.

11.4 Nonhomogeneous Deformation of Ferrogels

If the magnetic force density, $(M \nabla) H$ varies significantly from point to point in space then the deformation of ferrogels cannot be considered as homogeneous.

Let us consider a very simple physical situation similar to our unidirectional experiments discussed in the previous section. Namely, a long and thin ferrogel cylinder is suspended in water vertically. The magnetic field is induced by a solenoid-based electromagnet placed under the gel. The axis of the gel cylinder (z) is parallel with the magnetic field and its gradient. In this case, the deformation of the gel is uniaxial and can be considered as one-dimensional.

The governing equation for this situation describing the displacement of each point of the gel along the z-axis is the following second-order, nonlinear ordinary differential equation:[15]

$$G\left(\frac{d^2 u_z(Z)}{dZ^2} + \frac{2}{(du_z(Z)/dZ)^3}\frac{d^2 u_z(Z)}{dZ^2}\right) + M(u_z(Z))\frac{dH(u_z(Z))}{dZ} = 0 \quad (11.6)$$

where $u_z(Z)$ represents the displacement given in the reference – undeformed – configuration, G is the shear modulus of the gel, M denotes the magnetisation, and H stands for the magnetic field strength.

The magnetisation, M of a ferrogel can be described by the Langevin function. Assuming the magnetisation of individual particles in the gel to be equal to the saturation magnetisation of the pure ferromagnetic material, the magnetisation M, of ferrogel, in the presence of an applied field can be expressed as:[13]

$$M = \Phi_m M_s L(\xi) = \Phi_m M_s \left(\coth \xi - \frac{1}{\xi} \right) \quad (11.7)$$

where Φ_m stands for the volume fraction of the magnetic particles in the whole gel, and ξ of the Langevin function $L(\xi)$ is defined as

$$\xi = \frac{\mu_0 m H}{k_B T} \quad (11.8)$$

where m is the giant magnetic moment of nano-sized magnetic particles, k_B denotes the Boltzmann constant and T stands for the temperature. According to eqn (11.7) the magnetisation of a ferrogel is in direct proportion to the concentration of magnetic particles and their saturation magnetisation.

Based on eqn (11.6) with boundary conditions $u_z(0) = 0$, $t(Z_m) = 0$, we have calculated the unidirectional deformation of a ferrogel cylinder. This is shown in Figure 11.7. On the left-hand side the magnetic field strength along

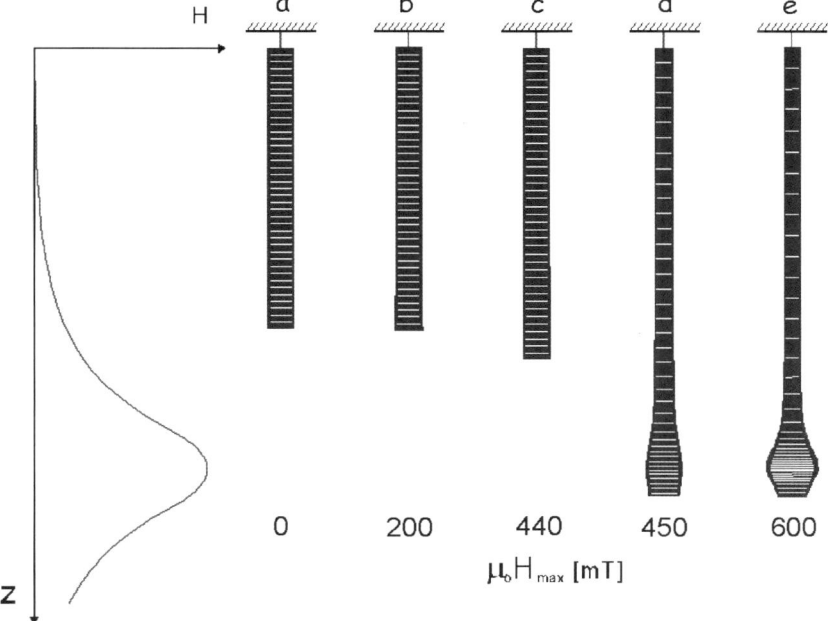

Figure 11.7 Schematic representation of the uniaxial deformation of a ferrogel cylinder calculated numerically from eqns (11.6)–(11.8). The external magnetic-field distribution described by eqns (11.4) and (11.5) is also indicated in the figure. The gel on the left-hand side is undeformed (B = 0). The rest two represent the abrupt transition within a slight increase of the field intensity.

the z-axis is plotted. The distribution of the field we employed in the calculations was similar to that in real experiments. As one can see, the gel elongates eventually as the magnetic-field intensity increases. At a certain field intensity, the gel falls abruptly into a new equilibrium position similar to what we observed experimentally. The white lines on the gel body demonstrate the nonhomogeneity of the deformation. Different distance between adjacent lines indicates different degree of deformation. The high degree of nonhomogeneity is easily seen.

At lower field intensities (b and c) the upper part of the gel elongates to a greater extent than the lower part, whereas at higher field intensities (d and e) the lower part of the gel contracts while the upper part elongate.

In order to test the validity of our model we compared theoretical calculations with the results of unidirectional experiments. In Figure 11.8 the strain of the bottom end of the gel is plotted against the maximum field intensity. Both experimental and calculated points are shown. The calculated points fit quite well to the measured ones, indicating that our model is able to reproduce not only the noncontinuous characteristics of the deformation process, but also provides accurate, realistic numerical values.

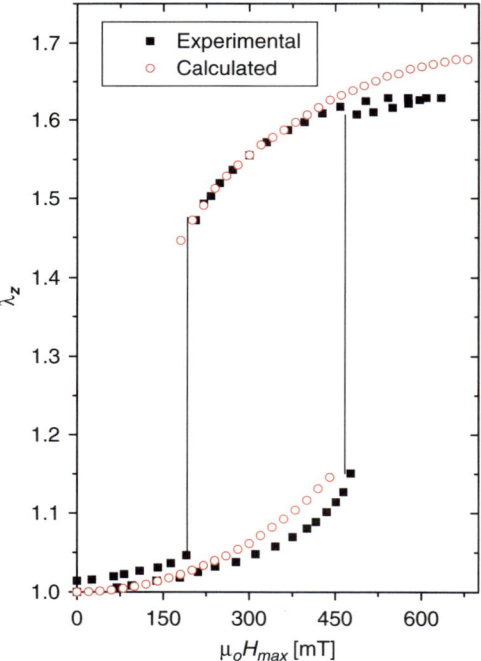

Figure 11.8 Uniaxial elongation of a ferrogel cylinder. The points represent the displacement of the bottom end of the gel. The blank points were calculated on the basis of eqn (11.6) with magnetic-field distribution given by eqns (11.4)–(11.7), respectively.

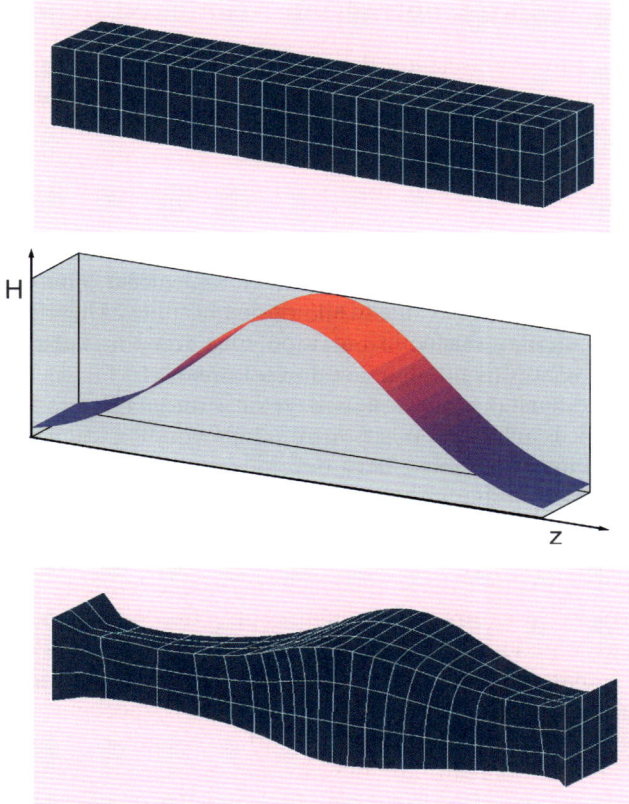

Figure 11.9 FEM calculation of the deformation of a ferrogel block in a uniaxial magnetic field at different field intensities. The magnetic field strength has a Gaussian distribution along the block as shown in the figure.

Nowadays, the finite element method (FEM) is prevalent in solid mechanics and provides a useful tool to study realistic, three-dimensional nonlinear deformations. The FEM system MARC[15] has been chosen for the calculations, because it is developed for modeling nonlinear mechanical deformations and it can be extended with the afore-mentioned user subroutines. As an illustrative example of three-dimensional deformation of ferrogels, in Figure 11.9 we present the deformation of a ferrogel block whose ends are fixed to a wall. The magnetic field is uniaxial and has a Gaussian distribution along the gel axis as shown in the middle of the figure.

Looking at the pictures one may associate with the motion of a living worm. This indicates that the deformation and motion of ferrogels has a close relation with the one of simple living organisms. This is due to the complexity and nonhomogeneity of the magnetic-field-induced deformations.

11.5 Muscle-like Contraction Mimicked by Ferrogels

Forces internal to the muscle are derived from a special mechanism, which is designed to transform chemically bound energy into mechanical work or locomotion. During muscular contraction one end of the muscle remains fixed, while the other end moves towards to origin. Since the volume of muscle remains essentially unaltered, it also experiences an increase in diameter.

In the analyses of the mechanical behaviour of an isolated muscle it is customary to introduce two types of strains. Under load condition the muscle is first stretched passively before being stimulated to contract. This process is then followed by active contraction due to shortening of fibres. Therefore under load condition the muscular thickening may be considered as net effects of both passive and active strains taking place simultaneously.

When contraction takes place against a resistance, force is generated and mechanical work is released. When contraction occurs against the maximum resistance, maximum muscular force develops, but no measurable change in muscle length, as well as mechanical work, can be observed. This isometric contraction represents the largest force that a muscle can actively generate. Under no-load conditions the stimulation of muscle will cause it to contract to its smallest length without development of any active tension.

The passive–active nature of muscular contraction cannot be realised by a traditional elastic material of which deformation depends only on the surface traction. In a material designed for artificial muscle application there must be a physical or chemical process that can generate mechanical stress inside the material independently on acting surface traction. This controllable "body force" is able to determine the rate and measure of contraction even if the material is preloaded and it is needed to make the material capable of mimicking the passive and active mechanism of muscle.

Development of an artificial muscle must face the task of reproducing at least two main characteristics of real muscle fibres, namely the high and fast contractility. It is also important to have a reliable control system. Electric or magnetic fields are the most practical stimuli from the point of view of signal control.

In order to mimic muscular contraction we have studied the unidirectional shortening of ferrogel samples excited by nonhomogeneous magnetic field. A cylindrical gel sample was suspended in water vertically between planar parallel poles of an electromagnet. The position of the top surface of the gel was fixed and the highest field strength was located at this point.

The steady current intensity in the solenoid-based electromagnet was varied in order to produce a different magnetic field distribution. The maximum magnetic field strength (magnetic flux density) of 300 mT developed at the upper part of the gel and disappeared within 120 mm along the axis of cylindrical gel sample. It is worth mentioning that 300 mT is a field strength that is less than the field strength measured at the surface of common permanent magnets. Due to the field gradient directed from bottom to top along the gel axis, contraction occurs. Figure 11.10 shows that under no-load conditions

Figure 11.10 Magnetic-field-induced contraction.

stimulation of magnetic gel will cause it to contract to its smaller length without development of any mechanical tension.

The measure of this atomic (no-load) contraction strongly depends on the structure of the magnetic gel as well as on the distribution of magnetic field along the axis of the gel.

Since the highest magnetic-field strength can be controlled by the intensity of the steady current flowing through the electromagnet, it is possible to realise different degree of contraction as shown in Figure 11.9.

The contractile activity of magnetic gels can be used to lift a load that is to produce work. A nonmagnetic load (lead) of variable mass was connected to the lower end of the gel. A CCD camera that provides us the displacement of the lower part of the gel with an accuracy of 0.01 mm monitored the contraction of the gel.

In the presence of a load the gel elongates (passive strain). When a magnetic field is created, as a consequence, a contraction (active strain) takes place. If the mass of the load is not too high, the net effect is a significant contraction.

We have studied the active deformation of a magnetic-field-sensitive gel fibre under a load, which has a mass 20-fold larger then the mass of the gel itself.

This gel contains 0.39 g polymer and 0.31 g magnetite. In the absence of the field the gel elongates. The magnetic stimulus results in a large decrease in the length. When the field is turned off, the gel stretches again.

We have determined the work done by the ferrogel as a function of load. These results can be seen in Figure 11.11.

11.6 Control of Pseudomuscular Contraction

We consider here a vertically suspended cylindrical gel fibre by means of which a nonmagnetic load can be lifted up. First the gel is preloaded with a

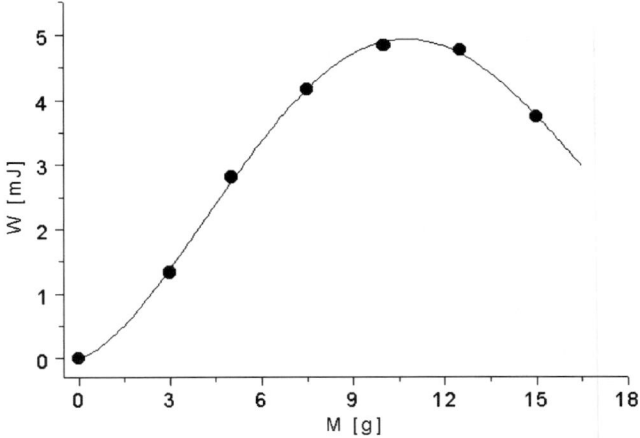

Figure 11.11 Mechanical work released by a ferrogel as a function of load. Solid line is a guide for the eye.

mass, M. As a consequence a passive strain λ_M develops. The magnitude of this extension can be calculated on the basis of rubber elasticity theory.[6] The Neo-Hookean law can give the mechanical stress:

$$\sigma_n = G(\lambda_z - \lambda_z^{-2}) \qquad (11.9)$$

where the nominal stress, σ_n is defined as the ratio of the equilibrium elastic force to the undeformed cross-sectional area of the sample. If applying a load having a mass of M deforms the gel, then the nominal stress can be written as

$$\sigma_n = \frac{Mg}{a_0} \qquad (11.10)$$

where g denotes the gravitational constant.

It is advisable forward to rearrange eqn (11.9) in order to find a solution for the deformation ratio, λ_M due to the load.

$$\lambda_M^3 - \alpha \lambda_M^2 - 1 = 0 \qquad (11.11)$$

The dimensionless quantity, α includes the ratio of nominal stress, σ_n and the elastic modulus, G of ferrogels:

$$\alpha = \frac{\sigma_n}{G} = \frac{Mg}{a_0 G} \qquad (11.12)$$

Here, a_0 denotes the undeformed cross-sectional area of the gel at rest. Eqn (11.11) tells us how the deformation ratio depends on the applied mechanical stress and the modulus. It is obvious that in the presence of a load, the pendant gel elongates, that is $\lambda_M > 1$.

When an external magnetic field is applied, the strain will be changed to $\lambda_{M,H}$, which is the overall strain. Let us consider homogeneous deformation and a linear relationship between the magnetisation and the magnetic-field strength. This latter assumption is valid at small magnetic field intensities. In this case a Taylor series expansion of the Langevin distribution function in eqn (11.7) yields: $L(\xi) \cong \xi/3$. This results in a linear relationship between magnetisation and field intensity:

$$M \simeq \chi H, \; H \to 0 \qquad (11.13)$$

where χ denotes the initial susceptibility, which is defined as

$$\chi = \Phi_m M_S \frac{m}{3k_B T} \qquad (11.14)$$

It was found for ferrogels: $M_S m/3k_B T = 0.338$.

The magnetic force can be determined by using eqn (11.1). On the basis of the additivity of mechanical and magnetic stress it is possible to derive the following equation:[16]

$$\lambda_{M,H}^3 - \alpha \lambda_{M,H}^2 - \beta(H_h^2 - H_m^2)\lambda_{M,H} - 1 = 0 \qquad (11.15)$$

where H_h and H_m denote the magnetic field strength at the bottom and the top of a ferrogel fibre, respectively. The parameter β can be considered as the stimulation coefficient defined as:

$$\beta = \frac{\mu_0 \chi}{2G} \qquad (11.16)$$

Equation (11.15) says that if we suspend a ferrogels fibre in a nonhomogeneous magnetic field in such a way, that $H_h > H_m$, then elongation occurs $\lambda_{M,H} > \lambda_M > 1$. When the field is turned off the load is lifted up. In the opposite case $H_h < H_m$ work is released when the magnetic field is applied $\lambda_M > 1$, but $\lambda_{M,H} < \lambda_M$.

The displacement of the load can be expressed as

$$\Delta h = h_0 (\lambda_{M,H} - \lambda_M) \qquad (11.17)$$

We have found that the displacement of the end of the gel is much more significant at elongation than at contraction. An expression for the mechanical work can be derived as follows:

$$W = -Mg\Delta h = -Mgh_0(\lambda_{M,H} - \lambda_M) \qquad (11.18)$$

where h_0 denotes the undeformed length of the magnetic gel fibre. As a representative example, we have calculated the mechanical work as a function of load. These results are shown in Figure 11.12.

It is obvious that the mechanical work strongly depends on the applied load. One can see that at small loads the work increases with the load. $M=18\,g$ represents the highest mechanical work. It is also seen, that above a certain value, if the load is comparatively heavy, the work decreases with increasing mass. Similar results have been found for mechanical work produced by

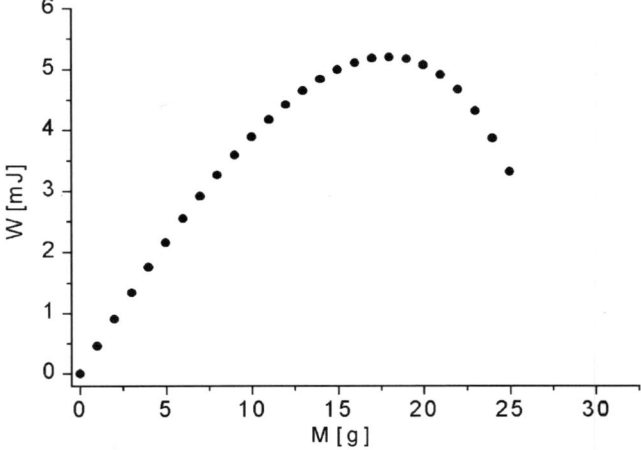

Figure 11.12 The mechanical work as a function of load calculated on the basis of eqns (11.12)–(11.14).

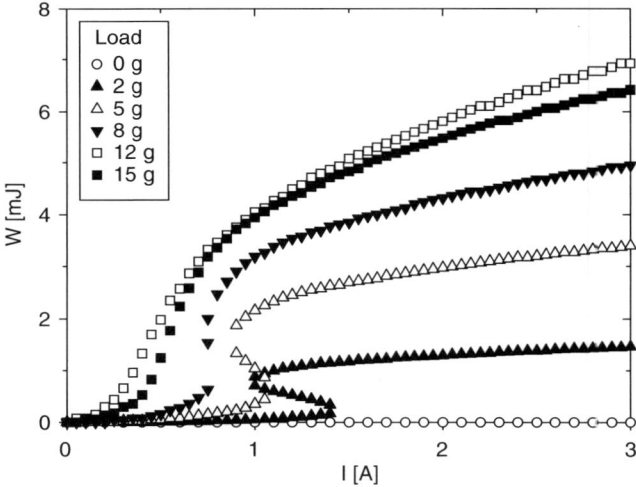

Figure 11.13 The mechanical work at different applied loads as calculated on the basis of eqn (11.15). The mass of the load is indicated on the figure.

swelling.[14] Any other experimental situation can be studied with aid of eqns (11.15)–(11.18). Figure 11.13 shows the dependence of mechanical work as a function of steady current intensities.

Two kinds of behaviour can be seen in the figure. When the applied load is small ($M < 8g$) then at a certain current intensity abrupt transition occurs. Below this transition point the work is negligible. If the mass of load exceeds $8\,g$ the mechanical work increases with the increase of the steady current. This result concludes that higher field gradient results in higher mechanical work. One can also see that at small loads the work increases with the mass of loads. It is also seen, that above a certain value, if the load is comparatively heavy, the work decreases with increasing mass.

11.7 Future Aspects

The ability of magnetic-field-sensitive gels to undergo a quick controllable change of shape can be used to create an artificially designed system possessing sensor and actuator functions internally in the gel itself. The peculiar magneto-elastic properties may be used to create a wide range of motion and to control the shape change and movement that are smooth and gentle similar to that observed in muscle. Thus, application of magnetic-field-sensitive gels as a soft actuator for robots and other devices has special interest. Unlike in metallic machine systems devices made of gels work without noise, heat evolution and exhaustion. An understanding of magnetoelastic coupling in gels will hasten the gel engineering to switches, sensors, micromachines, biomimetic energy-transducing devices and controlled delivery systems. If the magnetic field is created inside the gel by incorporated small powerful electromagnets and the field is coordinated and controlled by a computer, then the magnetic-field-sensitive gel may be used as an artificial muscle.

It must also be finally mentioned that ferroelectric polymers are different from magnetic gels or ferrogels as described in Ref. 17.

Acknowledgements

This research was supported by the Széchenyi NRDP No. 3/043/2001 and the Hungarian National Research Fund (OTKA, Grant No. T038228). This research is sponsored by NATO's Scientific Division in the framework of the Science for Peace Programme.

References

1. W. Kuhn, B. Hargitay, A. Katchalsky and H. Eisenberg, *Nature*, 1950, **165**, 514.
2. Y. Osada, *Adv. Polym. Sci.*, 1987, **82**, 1.
3. D. De Rossi, K. Kawana, Y. Osada and A. Yamauchi, *Polymer Gels Fundamentals and Biomedical Applications*, Plenum Press, New York, 1991.
4. T. Tanaka, *Science*, 1982, **218**, 467.
5. M. Zrínyi, L. Barsi and A. Büki, *Polym. Gels Networks*, 1997, **5**, 415–427.
6. M. Zrínyi, L. Barsi and A. Büki, *J. Chem. Phys.*, 1996, **104**(20), 8750–8756.
7. D. Szabó, L. Barsi, A. Büki and M. Zrínyi, *Model. Chem.*, 1997, **134(2–3)**, 155–167.

8. L. Barsi, A. Büki, D. Szabó and M. Zrínyi, *Prog. Colloid Polym. Sci.*, 1996, **102**, 57–63.
9. M. Zrinyi, *Trends Polym. Sci.*, 1997, **5(9)**, 280–285.
10. M. Zrínyi, L. Barsi, D. Szabó and H.-G. Kilian, *J. Chem. Phys*, 1997, **108(13)**, 5685–5692.
11. D. Szabó, G. Szeghy and M. Zrinyi, *Macromolecules*, 1998, **31**, 6541–6548.
12. M. Zrinyi, D. Szabo and H.-G. Kilian, *Polym. Gels Networks*, 1999, **6**, 441.
13. R.E. Rosenweig, *Ferrohydrodynamics*, Cambridge University Press, Cambridge, London, New York, New Rochelle, Melbourne, Sydney, 1985.
14. M. Zrínyi and F. Horkay, *J. Intell. Mater. Syst. Struct.*, 1993, **4(2)**, 190–201.
15. M. Zrínyi, D. Szabó, G. Filipcsei and J. Fehér, in: *Polymer Gels and Networks*, ed. Y. Osada and A.R. Khokhlov, Marcel Dekker Inc., New York, 2002.
16. M. Zrínyi, D. Szabó and L. Barsi, in: *Polymer Sensors and Actuators*, ed. Y. Osada and De Rossi, Springer Verlag, Berlin, 1999.
17. H.S. Nalwa, *Ferroelectric Polymers: Chemistry, Physics and Applications*, Marcel Dekker, New York, 1995, also look at Q.M. Zhang and Baojin Chu's chapter in this book on ferroelectric relaxors.

CHAPTER 12
Intelligent Materials: Shape-Memory Polymers

MARC BEHL,[1] ROBERT LANGER[2] AND ANDREAS LENDLEIN[1]

[1] Institute of Polymer Research, GKSS Research Center Geesthacht, Kantstr. 55, D-14513, Teltow, Germany
[2] Institute Professor, Massachusetts Institute of Technology, 45 Carleton Street, Cambridge, MA 02139, USA

12.1 Introduction

Shape-memory polymers can be deformed and fixed in a temporary shape. They only recover their original, permanent shape when exposed to an appropriate stimulus.[1] Therefore, shape-memory polymers are stimuli-sensitive materials and belong to the group of smart polymers. The shape change from the temporary to the permanent shape is predefined and determined by the mechanical deformation, which lead to the temporary shape. Shape-memory polymers reported so far use either heat or light as stimulus. Furthermore, the thermally induced shape-memory effect can be actuated indirectly by irradiation with IR light, application of electric current, exposure to alternating magnetic fields or immersion in water.

The shape-memory effect is not an intrinsic material property, but results from the combination of the polymer's molecular architecture and the resulting polymer morphology in combination with a tailored processing and programming technology. The shape-memory polymer is initially formed into its permanent shape by conventional processing methods, *e.g.* extruding or injection moulding. In a second step, which is called the programming process, the polymer sample is deformed and fixed in an individual temporary shape. The permanent shape recovers from the temporary shape upon application of an external stimulus. The cycle of programming and recovery can be repeated several times. Temporary shapes in subsequent cycles can be different. When compared to shape-memory metallic alloys, polymers allow significantly higher deformations between temporary and permanent shapes and can be programmed in a shorter time interval. As the shape-memory effect relies on the molecular architecture of the polymer and does not require specific repeating units intrinsic material properties can be adjusted in a wide range by variation of different molecular parameters such as type of monomers and composition. In this way material properties can be adjusted to the needs of specific applications which span various areas, *e.g.* intelligent fabrics,[2] heat-shrinkable tubes for electronics or films for packaging,[3] sun sails in space crafts,[4] self-disassembling mobile phones[5] and intelligent medical devices[6] and implants for minimally invasive surgery.[7,8]

Shape-memory polymers are polymer networks containing molecular switches that can be actuated by external stimuli.[1,9] In these networks, the netpoints determine the permanent shape. If they are chemical in nature, the crosslinks are covalent bonds. If they are physical in nature, the netpoints consist of intermolecular interactions like van-der-Waals interactions or hydrogen bonds. The netpoints are crosslinking chain segments. These are flexible as long as the working temperature is higher than the thermal transition temperature of the chain segments. In this case the polymer network behaves entropy elastic.

Physical crosslinking is reached in polymers morphology consisting of at least two segregated domains as realised, *e.g.*, in block copolymers. The phase providing the highest thermal transition T_{perm} acts as the netpoint. The polymer network will display shape-memory functionality, if the material can be temporarily fixed in a deformed state under environmental conditions relevant for the particular application. For that purpose the recoiling of the chain segments, which were oriented under external stress, has to be prevented reversibly. This is achieved by introducing additional reversible netpoints as molecular switches. By vitrification or crystallisation of side chains or the chain segments themselves, physical crosslinking is obtained. Reversibly covalent crosslinking for fixing the temporary shape is obtained by the attachment of functional groups to the chain segments. These functional groups must be able to react reversibly with each other forming a covalent bond by control of an external stimulus.

The introduction of such functional groups being able to undergo photoreversible reactions enabled the extension of the shape-memory technology to light as stimulus.[10] Other stimuli, like electrical current or alternating magnetic fields, are based on the thermally induced shape-memory effect, as these stimuli are used for indirectly heating of the material.

In this review actual developments in the field of thermally induced shape-memory polymers as well as the extension of the shape-memory technology to other ways of actuation are presented. Finally, multifunctionalised polymers are introduced, in which different functionalisations are combined. Special emphasis is placed on biodegradable shape-memory polymers having a high potential in biomedical applications.

12.2 Thermally Induced Shape-memory Polymers

12.2.1 General Concept and Characterisation of Shape-memory Effect

The thermally induced shape-memory effect results from the temporary fixation of the material in a deformed state in a temperature range relevant for the intended application. In phase-segregated linear block copolymers one phase acts as the netpoint determining the permanent shape and an other phase as switch (Figure 12.1).[9] The network chains are working as molecular switches, as their flexibility is a function of temperature. Above the thermal transition T_{trans} related to the switching phase the chain segments are flexible, while below T_{trans} the flexibility of the segments is at least partly limited. The recovery of the permanent shape is driven by entropy elasticity as the oriented switching segments are gaining entropy by recoiling.

Shape-memory polymers can be divided into two classes according to T_{trans}, which can be either a melting temperature T_m or glass-transition temperature T_g. While the latter are extending over a broad temperature interval, melting temperatures show in most cases a relatively sharp transition.

In covalently crosslinked shape-memory polymers a maximum weight content of switching segments can be realised, as the permanent shape is fixed by covalent netpoints. In thermoplasts a certain minimum of hard segments is required to form a sufficient number of physical crosslinks at temperatures between T_{trans} and T_{perm}, stabilising the permanent shape.

The shape-memory effect can be quantified in cyclic thermomechanical tests using a tensile tester equipped with a thermochamber.[11] A single cycle includes the programming of a test piece and recovery of its permanent shape. The permanent shape of the test piece is heated up to a temperature T_{high} above T_{trans}. Then it is deformed to ε_m leading to an orientation of the polymer chains. Afterwards, the working temperature is lowered $T_{low} < T_{trans}$ resulting in a fixation of the temporary shape. Finally, the sample is reheated to T_{high}

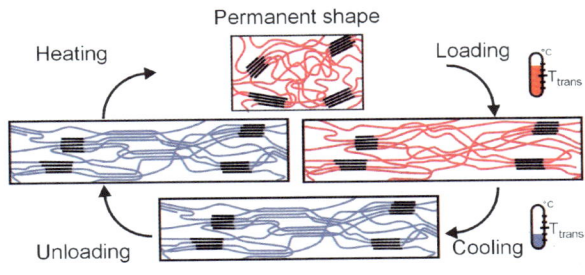

Figure 12.1 Molecular mechanism of the thermally induced shape-memory effect for a linear block copolymer. T_{trans} = thermal transition temperature related to the switching phase.

resulting in a contraction of the polymer chains and recovery of the permanent shape. Various test programs have been defined differing in the programming procedure (cold drawing at $T < T_{trans}$ or temporary heating of the polymer to a temperature $T > T_{trans}$) or in the control mode of the tensile tester (stress or strain controlled). Other test parameters of the programming procedure, which can be varied in the tests, are the lower (T_{low}) and upper (T_{high}) working temperature of the thermochamber or the number of cycles (N).

The data obtained from these tests enable the determination of shape-memory properties including the strain fixity rate R_f, the strain recovery rate R_r and the switching temperature T_{switch}. R_f describes the ability of the switching segment to fix a mechanical deformation in the form of an elongation to ε_m applied in the programming process resulting in the temporary shape

$$R_f(N) = \frac{\varepsilon_u(N)}{\varepsilon_m} \tag{12.1}$$

R_r quantifies the ability of the material to memorise its permanent shape (eqn (12.2)). It results from the ratio of the change in strain recorded during shape-memory effect $\varepsilon_m - \varepsilon_p(N)$ and the change in strain in the course of programming given by $\varepsilon_m - \varepsilon_p(N-1)$

$$R_r(N) = \frac{\varepsilon_m - \varepsilon_p(N)}{\varepsilon_m - \varepsilon_p(N-1)} \tag{12.2}$$

12.2.2 Thermoplastic Shape-memory Polymers

Examples of linear block copolymers with $T_{trans} = T_m$, which are reviewed in Ref. 1, include polyurethanes, polyetherester and ABA triblock copolymers. In ABA triblock copolymers of polytetrahydrofuran (B block) and poly(2-methyl-2-oxazoline) (A block) polytetrahydrofuran acts as switching segment ($T_m = 20$–40 °C) and poly(2-methyl-2-oxazoline) forms the phase related to T_{perm}. The polyetherester consist of poly(ethyleneterephthalate) as the hard segment and crystallisable poly(ethylene glycol) as the switching segment ($T_m = 40$–60 °C).[12–14] In the copolyesterurethane example poly(ε-caprolactone) is forming the switching phase ($T_m = 44$–55 °C) and the polyurethane is forming the hard segment.[11,15,16] In polyurethanes containing mesogenic groups[17,18] poly(ε-caprolactone) is used as switching segment ($T_m = 38$–59 °C) while the segments from mesogenic diols and 4,4-methylene bis(phenyl isocyanate) are forming the phase related to T_{perm}.

Microphase separation could be investigated by scanning electron microscopy for block copolymer with *trans*-poly(isoprene) switching segments and polyurethane hard segment.[19] In the block copolymer the urethane segments are capable of assembling into spherical domains. T_{trans} is a melting transition around 60 °C related to the *trans*-poly(isoprene) segments. The polyurethane segments act as physical crosslinks of the block copolymer. For switching segment contents of 70 wt% R_rs of 85% and R_fs close to 100% have been determined at $\varepsilon_m = 100\%$. A switching segment content of 35 wt% at $\varepsilon_m = 50\%$ resulted in R_rs of only 35% but with similar R_fs.

Physical crosslinking with a thermal transition relying on a glass-transition temperature $T_{trans} = T_g$ is realised in linear block copolymers. Examples are polyurethanes whose switching domains are in most cases mixed phases, containing segments of polytetrahydrofuran or poly(ethylene adipate).[1] The quality of phase separation between polytetrahydrofuran and polyurethane-segments depends strongly on the molecular weight of the poly(tetrahydrofuran)diols used as educt or reactant. Polyurethanes of 4,4-methylene bis(phenyl isocyanate) containing two poly(tetrahydrofuran)diols differing in their molecular weight ($M_n = 1000$ or 1800 g·mol^{-1}) were investigated.[20] Additionally, the arrangement of the blocks in the linear polymer was varied between a more random and a blocky sequence structure. A random arrangement results in a single T_g associated to a mixed phase, while in a blocky arrangement two T_gs are observed. The mechanical properties of these polyurethanes with two types of polytetrahydrofuran showed higher stresses and strains at break (ε_b) than polyurethanes with only one kind of polytetrahydrofuran or blends of polytetrahydrofuran and polyurethane. R_fs and R_rs of more than 90% have been reported for polymers with a blocky sequence structure. The influence of the molar ratio of the two poly(tetrahydrofuran)diols on the ε_b was investigated. The highest R_fs and R_rs were reported for a polymer of 70/30 wt% (low molecular weight/high molecular weight) polytetrahydrofuran. This has been attributed to a restructuring process during strain.

Polyurethanes of 4,4-methylene bis(phenyl isocyanate), 1,4-butanediol and polytetrahydrofuran were processed via electrospinning into nanofibres with fibre diameters between 800 nm and 2 µm having an average diameter of 1.2 µm.[21] The shape-memory functionality on this scale could be demonstrated for nonwoven made of these fibres.

Poly(ketone-co-alcohol) with shape memory can be obtained by polymer analogous reaction from polyketones by partial reduction with $NaBH_4/THF$.[22] Polyketones are synthesised by polymerisation of propene, hex-1-ene or propene and ethene with carbon monoxide catalysed by late transition metal complexes. By adjusting the amount of $NaBH_4/THF$ the degree of reduction could be adjusted, which was shown to be directly related to the glass-transition temperature, polarity and mechanical properties of the polymer. The most promising material was a partly reduced poly(ethylene-co-propene-co-carbonoxide). This terpolymer presented a phase-separated morphology with hard microcrystalline ethylene/CO rich segments within a softer amorphous polyketone ethylene-propene/CO matrix. The crystalline domains of this thermoplastic material work as physical crosslinkers resulting in an elastic behaviour above T_g as the switching temperature is given by the glass transition ($T_{trans} = T_g$). Changing the chemical composition of the polymeric material by reduction allows control of the T_g from below room temperature to 75 °C. A melting of polymer has not been detected but in thermal analysis degradation began at 350 °C. R_r between 90 and 95% were quantified.

12.2.3 Covalently Crosslinked Shape-memory Polymers

Covalently crosslinked networks can be prepared following two different synthesis strategies:[1,9] Crosslinking of linear or branched polymers or co-polymerisation/polycocondensation of monomers with tri- or higher functional crosslinkers.

Crosslinking is realised in radiation-crosslinked polyethylene,[3] and its co-polymers.[23,24] Another crosslinking method is thermal vacuum dehydrochlorination of poly(vinyl chloride)[25] and subsequent crosslinking in an HCl atmosphere. Poly[ethylene-*co*-(vinylacetate)] can be chemically crosslinked by the thermally induced radical initiator dicumylperoxide.[26] The same radical initiator is applied for crosslinking of semicrystalline polycyclo-octene obtained by ringopening methathesis polymerisation of cyclo-octene.[27] The degree of crosslinking is controlled by radical initiator content. Crystallinity decreases with increasing crosslinking density. The shape-memory effect relies on the melting of the crystallites, which depends on the *trans*-vinylene content. For the pure polycyclo-octene with 81% *trans*-vinylene content a melting temperature of 60 °C was measured. Shape recovery of these materials occurred within 0.7 s at 70 °C.

The second synthesis strategy of shape-memory polymer networks is realised by copolymerisation of monofunctional monomers with low molecular weight or oligomeric crosslinkers. An example are copolymers from stearyl acrylate, methacrylate and N,N'-methylenebisacrylamide as crosslinker.[28] The switching phase is formed by crystalline domains of the stearyl side chains. Multi-phase copolymer networks are also obtained by radical copolymerisation of poly(octadecyl vinyl ether)diacrylates or –dimethacrylates with butyl acrylate as comonomer.[29,30] The phase separation of this multiphase copolymer results in crystallisation of the octadecyl side chains, which act as switching segments.

Covalent crosslinks have been introduced in a polyurethane by adding trimethylol propane to the reaction mixture.[31] The polyurethane, which is synthesised following the prepolymer method, consists of two different segments. The polyurethane segment is formed from isophorone di-isocyanate and polytetrahydrofuran and a polyurethane segment from toluene di-isocyanate and 1,3-butanediol. Depending on the segment that is prepared in the pre-polymer reaction, where the trimethylol propane is added, crosslinks are preferentially formed in this specific segment. T_{trans} of the switching segment forming domains is a T_g between 10 and 60 °C. For certain polymer compositions a second, weak transition could be observed around 100 °C in dynamic mechanical analysis. Crosslinks, which are located in the polyurethane segment, resulted in lower fixities of the temporary shape and better recovery of the permanent shape as compared to polymers with crosslinks placed in the polyetherurethane segments having the same composition. On the other hand R_r increased with decreasing polyurethane content when crosslinks were in the polyetherurethane segment.

In liquid-crystalline elastomer networks, the thermal transition of the liquid crystalline moieties enables thermally induced switching of the shape-memory

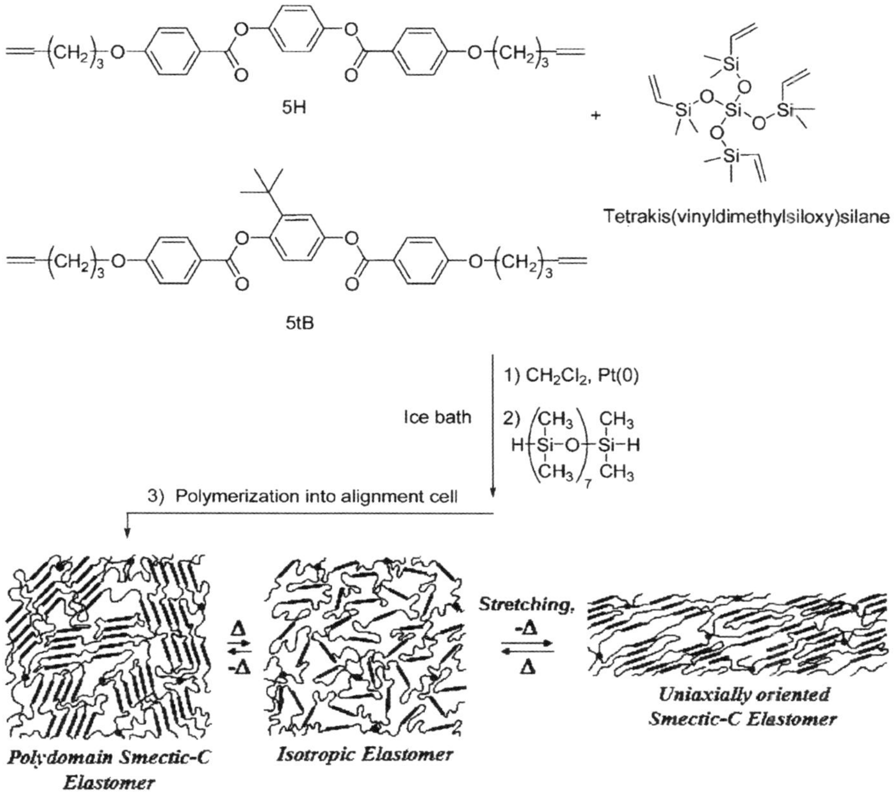

Figure 12.2 Synthesis and molecular mechanism of shape-memory effect in liquid-crystalline smectic-C elastomers. Polydomain smectic-C elastomer undergoes isotropisation upon heating through the clearing transition, allowing for subsequent stretching and fixing of a secondary uniaxially oriented smectic-C elastomer.

effect.[32] Tetrafunctional silanes, working as netpoints, have been coupled to oligomeric silanes working as spacers, to whom two distinct benzoate-based mesogenic groups had been attached, forming a main chain smectic-C elastomer [Figure 12.2].

The networks are heated to the isotropic state of the elastomer. Afterwards, a temporary shape can be programmed by stretching or twisting and subsequent cooling, below the clearing transition (I-SmC) of the smectic-C mesogens. Upon reheating over this clearing transition the permanent shape, acquired during the crosslinking process, can be recovered. By introduction of two mesogens differing in the clearing transition temperature (a difference of 140 °C between the clearing transition of the two mesogens) and variation of the spacer length, the smectic-isotropic clearing temperature could be adjusted between 0 and 90 °C, which is superimposed by the T_g that varies between −17 and 50 °C.[33]

12.2.4 Composites from Shape-memory Polymers and Particles

Composites from shape-memory polymers and particles were developed to enhance mechanical properties. Reinforcement of thermoset epoxy resin using microscale particles (carbon fibres) increased stiffness and reduced recoverable strain levels.[34,35] Nanoscale reinforcements lead to sustained increase in modulus and strength, which is explained by better interface properties.[36,37] Stiffness and elastic modulus of shape-memory polymer composites of thermoset epoxy resins reinforced with SiC nanopowder increased dramatically. This increase was a function of the SiC content.[38] The same study showed that the T_gs of the reinforced and the pure matrix polymer are similar, while the former is slightly higher than the latter. In nanocarbon-filled polyurethanes a contrary tendency was found. Here, a lowering of the glass-transition temperature was observed.[39] High particle contents in shape-memory nanocomposites, such as weight fractions of 30 wt% SiC and above, lead to agglomeration of particles. The increase of stiffness and modulus could be confirmed in subsequent investigations either with SiC particles in epoxy-based systems,[40] or in polyurethanes reinforced with glass fibres.[41] An R_r of 20–40% in the first and of 90% in the fifth cycle at $\varepsilon_m = 100\%$ for filler contents between 0 and 30 wt% were reported.

12.2.5 Indirect Actuation of Thermally Induced Shape-memory Effect in Polymers

Indirect actuation of the thermally induced shape-memory effect can be realised by two different strategies: One possibility is the triggering of thermally induced shape-memory effect by indirect heating, *e.g.* warming by irradiation with light. The other strategy is lowering of T_{trans} by diffusion of a low molecular weight molecule into the polymer bulk acting as plasticiser. In this way, the thermally induced shape-memory effect can be triggered while the temperature of the sample remains constant.

Thermally induced shape-memory polymers can be heated by illumination with infrared light instead of increasing the environmental temperature. This indirect light activation has been used in laser-activated medical devices for the mechanical removal of thrombi[42,43] Thermal conductivity and heat capacity limit the ability of a shape-memory polymer to quickly actuate upon exposure on a heat-transferring fluid or infrared light source.[44] The heat transfer can be enhanced by addition of conductive fillers such as conducting ceramics, carbon black or carbon nanotubes.[44–46] Incorporation of carbon nanotubes in shape-memory polymers enhances not only stress and fixity, but also the photoresponsiveness because of an enhanced photothermal effect.[47] The increase in stress and fixity in these polyurethanes could also be achieved by the incorporation of carbon black. But carbon-black-reinforced materials showed only 25–30% recovery, while in carbon nanotube reinforced materials ratios of almost 100% were possible. These enhanced recovery results from a synergism

Figure 12.3 Electroinduced shape-recovery behaviour of polyurethanes–multiwall carbon nanotube composites. The sample undergoes the transition from temporary shape (linear, left) to permanent (helix, right) within 10 s when a constant voltage of 40 V is applied.

between the anisotropic carbon nanotubes and the crystallising switching segments of the polyurethane.

A certain amount of conductivity in polyurethane shape-memory polymers was realised by incorporation of carbon nanotubes.[48] Application of an electric current heats the sample because of the high ohmic resistance of the composite. In Figure 12.3 the shape recovery of a helix on application of a constant voltage (40 V) within 10 s is shown. In order to enhance the interactions between the carbon nanotubes and the shape-memory polymer matrix, the surface of the nanotubes was treated with acid.[48] However, at the same time the conductivity of the materials decreased. This has been attributed to increased defects in the lattice structure of the carbon–carbon bonds in the nanotube. Potential applications for this kind of shape-memory composites are electroactive actuators in microaerial vehicles.[48]

Composites of shape-memory thermoplasts and magnetic nanoparticles enable remote actuation of the thermally induced shape-memory effect in magnetic fields.[49] The heating of the sample is realised by inductive heating of the nanoparticles in an alternating magnetic field ($f = 258$ kHz, H = 30 kA m^{-1}). Magnetic nanoparticles consisting of iron(III)oxide cores in a silica matrix were incorporated in a biodegradable multiblock copolymer of PDC, consisting of poly(p-dioxanone) as hard segment and poly(ε-caprolactone) as switching segment, and an aliphatic polyetherurethane TFX, from methylene bis(p-cyclohexyl isocyanate), butanediol and poly(tetramethylene glycol). While TFX has an amorphous switching phase, PDC has a crystallisable switching segment. The elastic properties of TFX were improved by incorporation of the particles. The contrary effect has been observed for PDC. In both systems the shape-memory effect could be triggered by an alternating magnetic field. Exemplarily the recoiling of a corkscrew-like spiral of the composite TFX100 is shown (Figure 12.4). The R_rs obtained by indirect heating were comparable to those reached by increasing the environmental temperature. Potential applications for magnetic-field-driven shape-memory polymers are smart implants or instruments, which perform mechanical adjustments in a noncontact mode.

The T_g of commercially available shape-memory polyurethanes (MM3520 from Mitsubishi Heavy Industries) can be significantly lowered by immersion in water.[50,51] The polymer in its temporary shape was immersed in water, and recovers its shape, as T_g is lowered from 39 °C to a temperature below ambient temperature. The shape recovery of such a polyurethane from a circular

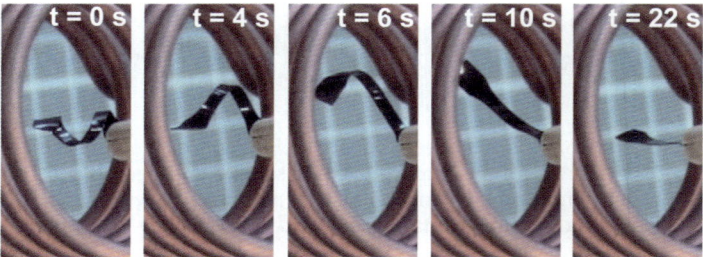

Figure 12.4 Magnetic-field-driven actuation of a thermoplastic shape-memory composite consisting of iron (III) oxide particles in a silica matrix and the TFX polymer. Electroactive shape-recovery behaviour of polyurethanes–multiwall carbon nanotube composites. The sample undergoes the transition from temporary shape (linear, left) to permanent (helix, right) within 10 s when a constant voltage of 40 V is applied.

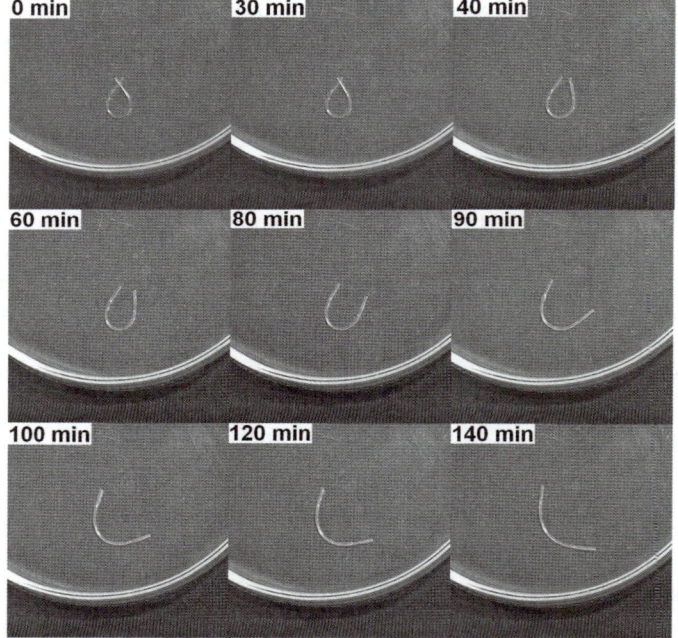

Figure 12.5 Water-driven actuation of a shape-memory effect: The shape-memory polyurethane in circular temporary shape is immersed into water. After 30 min the recovering of the linear permanent shape starts.

temporary shape to a linear shape is presented in Figure 12.5. In the polymer two types of adsorbed water have to be considered, bound and free water.[52] While the effect of free water in the polymer is negligible, bound water acts as a plasticiser and lowers the T_g significantly. The effect of lowering the glass-transition temperature by immersion in water could also be confirmed in composites of polyurethane and carbon nanotubes.[52] Increasing weight content of carbon nanopowder results in less sensitivity to water.[52]

12.3 Light-induced Shape-memory Polymers

Light-induced shape-memory polymers with reversibly reacting molecular switches were introduced in Ref. [10]. Cinnamic acid (CA) or cinnamyliden acetic acid (CAA) moieties work as light-triggered molecular switches (Figure 12.6).

Upon irradiation with UV light of $\lambda > 260$ nm a 2 + 2 cycloaddition reaction occurs between two of these light-sensitive moieties, forming covalent crosslinks by formation of cyclobutane rings. Upon irradiation with UV light of $\lambda < 260$ nm these crosslinks are reversibly cleaved. In these photoresponsive shape-memory polymers the permanent shape is determined by an amorphous permanent polymer network.

Two kinds of light-induced shape-memory polymers have been prepared: a graft polymer and an interpenetrating polymer. In the grafted polymer, the CA molecules are grafted onto the permanent polymer network resulting in a polymer architecture illustrated in Figure 12.7. The grafted polymer is obtained by copolymerisation of n-butylacrylate, hydroxyethyl methacrylate and ethyleneglycol-1-acrylate-2-CA with poly(propylene glycol)-dimethacrylate ($M_n = 560$ g mol^{-1}) as crosslinker.

The second polymer is an interpenetrating polymer network, formed by loading a permanent polymer network of butylacrylate with 3 wt% poly(propylene glycol)-dimethacrylate ($M_n = 1000$ g mol^{-1}) as crosslinker with 20 wt% star-poly(ethylene glycol) endcapped with CAA terminal groups (SCAA) by swelling in a 10 wt% SCAA chloroform solution. After removal of the solvent an opaque yellow film is obtained. In both light-sensitive polymers, the grafted

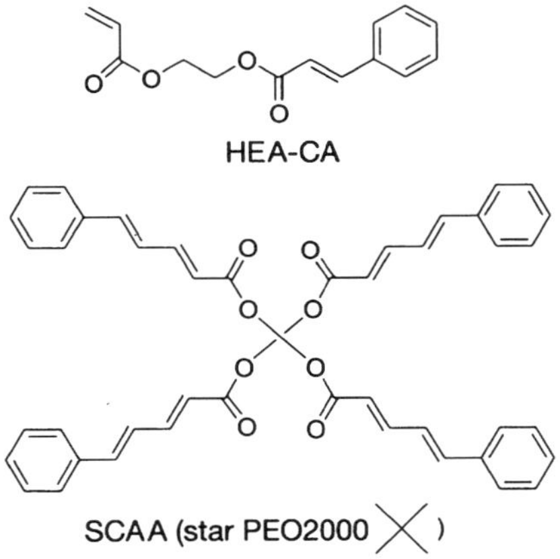

Figure 12.6 Chemical structure of the light-sensitive cinnamic acid and cinnamyliden acetic acid molecular switches.

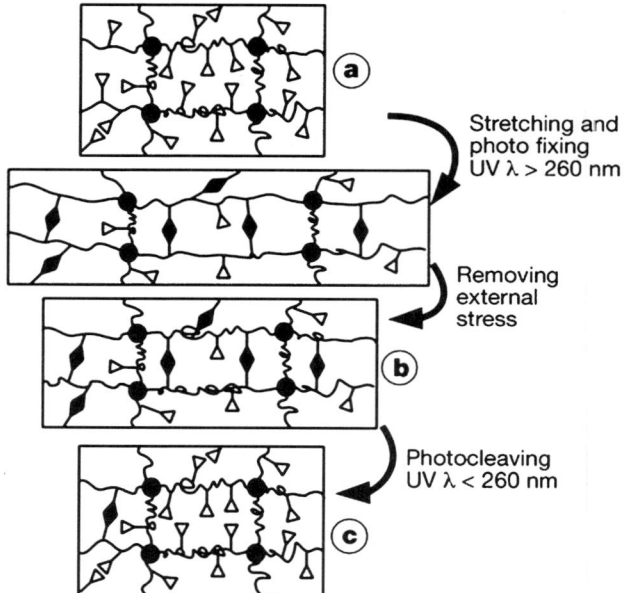

Figure 12.7 Molecular mechanism of shape-memory effect of the grafted polymer network: the chromophores (open triangles) are covalently grafted onto the permanent polymer network (filled circles, permanent crosslinks), forming photoreversible crosslinks (filled diamonds); fixation and recovery of the temporary shape are realised by UV light irradiation of suitable wavelengths.

and the interpenetrating polymer network, the permanent shape is given by the permanent network.

In the programming cycle, as illustrated schematically in Figure 12.7 for the grafted polymer network, the polymer network is stretched first, resulting in a strain of the coiled polymer segments (a). In a second step, the network is irradiated with UV-light $\lambda > 260$ nm, creating new netpoints, which fix the elongated form in a temporary shape (b). Upon irradiation with UV light of $\lambda < 260$ nm the crosslinks are reversibly cleaved and the permanent shape can recover (c). The grafted polymer revealed in the fifth cycle R_fs of max. 52% and R_rs of max. 95%, while the interpenetrating network showed in the third cycle 33% R_f and 98% R_r. As in these light-induced shape-memory polymers the constrain of external heating is circumvented, possible applications were seen in the medical field.

12.4 Multifunctional Polymers with Shape-memory Effect

Multifunctionality is the unexpected combination of material functionalisation.[53] An example of multifunctionality is the combination of biodegradability and thermally induced shape-memory functionality. Especially in biomedical application shape-memory materials would benefit from biodegradability.[54] Bulky implants could be inserted in compressed temporary shape into the human body by minimally invasive surgery. Upon application of the external stimulus they turn into their application relevant shape. Another application for biodegradable shape-memory polymers are intelligent sutures for wound closure,[7] which are able to apply a defined stress to the wound lips, when the shape-memory effect is actuated. In both cases a follow-on surgery to remove the implant is not necessary, as materials degrade within a defined time interval.

Thermoplastic shape-memory polymers with $T_{trans} = T_m$ have been obtained as linear block copolymers from oligo(ε-caprolactone)diols and oligo(p-dioxanone)diols and di-isocyanate as the junction unit.[7] Here, the oligo(ε-caprolactone) acts as the switching segment, while oligo(p-dioxanone) forms the phase related to T_{perm}. R_fs between 98 and 99.5% and R_rs of 98 to 99% in the third cycle were reported for these polymers. In aqueous buffer solution (pH 7) relative mass loss around 50% were found for a block copolymer with 42 wt% oligo(p-dioxanone) content after 270 days. Another example of $T_{trans} = T_m$ are linear multiblock copolymers with poly(l-lactide) and poly(glycolide-co-caprolactone)-segments.[55] Here, the degradability of the switching segment is enhanced by copolymerisation of ε-caprolactone with diglycolide resulting in formation of easily hydrolysable ester bonds. T_{trans} is superimposed by the T_g of the poly(l-lactide) segments. R_fs of 99% and R_rs of 99.6% at ε_m of 200% in the third cycle were determined. A mass loss of 50% in phosphate buffer solution (pH 7.4) was observed at least after 154 days.

Covalently crosslinked biodegradable shape-memory polymers with $T_{trans} = T_g$ were obtained by polyaddition of multifunctional alcohols with di-isocyanates.[56] Star-shaped co-oligoesters, synthesised by ring-opening copolymerisation of rac-dilactide and diglycolide with 1,1,1-tris(hydroxymethyl)ethane or pentaerythrite as initiators, were crosslinked with an isomeric mixture of 1,6-diisocyanato-2,2,4-trimethylhexane and 1,6-diisocyanato-2,4,4-trimethylhexane. This resulted in formation of an amorphous polymer with a T_g acting as T_{trans}. R_rs up to 97% and R_fs higher than 99% in the fifth cycle were reported. Hydrolytic degradation in aqueous buffer solution (pH 7) resulted in a relative mass loss of around 50% after 80 days and complete degradation after 150 days.

12.5 Conclusion and Outlook

The field of shape-memory materials is rapidly developing. While fundamental research is focusing on new types of shape-memory effects the technology platform of existing materials is moving towards some highly sophisticated applications. Potential applications can be seen in various areas of everyday life, from switches, sensors, fabrics or intelligent packaging to self-repairing autobodies. Other application areas include the aerospace industry and reusable composite tooling.[57] In particular, the new methods of stimulation like light or magnetic fields will open additional fields for applications. Here the area of active medical devices and implants seems to be promising[54] and first applications were demonstrated.[42,43] Requirements for functionalised polymers are determined by specific applications and can be very complex. Therefore, the trend in actual shape-memory research is set in the direction of multi-functionalised materials.[53]

References

1. A. Lendlein and S. Kelch, *Ang Chem.-Int. Ed.*, 2002, **41**(12), 2034–2057.
2. J. Hu, X. Ding, X. Tao and J. Yu, *J. Dong Hua University (Eng. Ed.)*, 2002, **19**, 3.
3. A. Charlesby, *Atomic Radiation and Polymers*, Pergamon Press, Oxford, 1960, 198–257.
4. D. Campbell, M.S. Lake, M.R. Scherbarth, E. Nelson and R.W. Six, *Elastic Memory Composite Material, An Enabling Technology For Future Furlable Space Structures. in 46th AIAA/ASME/ASCE/AHS/ASC Structures, Structural Dynamics, and Materials Conference*, 2005.
5. H. Hussein and D. Harrison. *Investigation into the use of engineering polymers as actuators to produce 'automatic disassembly' of electronic products, in Design and Manufacture for Sustainable Development* 2004, T. Bhamra and B. Hon (eds.), Wiley-VCH, Weinheim, 2004.
6. H.M. Wache, D.J. Tartakowska, A. Hentrich and M.H. Wagner, *J. Mater. Sci.: Mater. Med.*, 2003, **14**(2), 109–112.
7. A. Lendlein and R. Langer, *Science*, 2002, **296**(5573), 1673–1676.
8. A. Metcalfe, A.C. Desfaits, I. Salazkin, L. Yahia, W.M. Sokolowski and J. Raymond, *Biomaterials*, 2003, **24**(3), 491–497.
9. A. Lendlein and S. Kelch, *Shape-Memory Polymers*, in *Encyclopedia of Polymer Science and Technology*, John Wiley & Sons, New York, 2002, 125–136.
10. A. Lendlein, H.Y. Jiang, O. Jünger and R. Langer, *Nature*, 2005, **434**(7035), 879–882.
11. B.K. Kim, S.Y. Lee and M. Xu, *Polymer*, 1996, **37**(26), 5781–5793.
12. X.L. Luo, X.Y. Zhang, M.T. Wang, D.H. Ma, M. Xu and F.K. Li, *J. Appl. Polym. Sci.*, 1997, **64**(12), 2433–2440.
13. M.T. Wang, X.L. Luo, X.Y. Zhang and D.Z. Ma, *Polym. Adv. Technol.*, 1997, **8**(3), 136–139.
14. M.T. Wang and L.D. Zhang, *J. Polym. Sci. Part B-Polym. Phys.*, 1999, **37**(2), 101–112.
15. Z.L. Ma, W.G. Zhao, Y.F. Liu and J.R. Shi, *J. Appl. Polym. Sci.*, 1997, **63**(12), 1511–1515.
16. F.K. Li, J.N. Hou, W. Zhu, X. Zhang, M. Xu, X.L. Luo, D.Z. Ma and B.K. Kim, *J. Appl. Polym. Sci.*, 1996, **62**(4), 631–638.
17. H.M. Jeong, J.B. Lee, S.Y. Lee and B.K. Kim, *J. Mater. Sci.*, 2000, **35**(2), 279–283.

18. H.M. Jeong, B.K. Kim and Y.J. Choi, *Polymer*, 2000, **41**(5), 1849–1855.
19. X. Ni and X. Sun, *J. Appl. Polym. Sci.*, 2006, **100**, 879–885.
20. J.W. Cho, Y.C. Jung, Y.C. Chung and B.C. Chun, *J. Appl. Polym. Sci.*, 2004, **93**(5), 2410–2415.
21. D.I. Cha, H.Y. Kim, K.H. Lee, Y.C. Jung, J.W. Cho and B.C. Chun, *J. Appl. Polym. Sci.*, 2005, **96**(2), 460–465.
22. D. Perez-Foullerat, S. Hild, A. Mucke and B. Rieger, *Macromol. Chem. Phys.*, 2004, **205**(3), 374–382.
23. G. Kleinhans and F. Heidenhain, *Kunststoffe*, 1986, **76**, 1069–1073.
24. G. Kleinhans, W. Starkl and K. Nuffer, *Kunststoffe*, 1984, **74**, 445–449.
25. V. Skakalova, V. Lukes and M. Breza, *Macromol. Chem. Phys.*, 1997, **198**(10), 3161–3172.
26. F.K. Li, W. Zhu, X. Zhang, C.T. Zhao and M. Xu, *J. Appl. Polym. Sci.*, 1999, **71**(7), 1063–1070.
27. C.D. Liu, S.B. Chun, P.T. Mather, L. Zheng, E.H. Haley and E.B. Coughlin, *Macromolecules*, 2002, **35**(27), 9868–9874.
28. Y. Kagami, J. Gong and Y. Osada, *Macromol. Rapid Commun.*, 1996, **17**(8), 539–543.
29. E.J. Goethals, W. Reyntjens and S. Lievens, *Macromol. Symp.*, 1998, **132**, 57–64.
30. W.G. Reyntjens, F.E. Du Prez and E.J. Goethals, *Macromol Rapid Commun.*, 1999, **20**(5), 251–255.
31. S.H. Lee, J.W. Kim and B.K. Kim, *Smart Mater. Struct.*, 2004, **13**(6), 1345–1350.
32. I.A. Rousseau and P.T. Mather, *J. Am. Chem. Soc.*, 2003, **125**(50), 15300–15301.
33. I.A. Rousseau, H.H. Qin and P.T. Mather, *Macromolecules*, 2005, **38**(10), 4103–4113.
34. K. Gall, M. Mikulas, N.A. Munshi, F. Beavers and M. Tupper, *J. Intell. Mater. Sys. Struct.*, 2000, **11**(11), 877–886.
35. C. Liang, C.A. Rogers and E. Malafeew, *J. Intell. Mater. Sys. Struct.*, 1997, **8**(4), 380–386.
36. B.J. Ash, R. Stone, D.F. Rogers, L.S. Schadler, R.W. Siegel and B.C. Benicewicz, *T. Apple. Investigation into the thermal mechanical behaviour of PMMA/alumina nanocomposites in Filled and Nanocomposite Polymer Materials*, Material Research Society, Boston, 2001, KK2.10.1.
37. S.K. Bhattacharya and R.R. Tummala, *J. Electron. Packag.*, 2002, **124**(1), 1–6.
38. K. Gall, M.L. Dunn, Y. Liu, D. Finch, M. Lake and N.A. Munshi, *Acta Mater.*, 2002, **50**(20), 5115–5126.
39. B. Yang, W.M. Huang, C. Li and J.H. Chor, *Eur. Polym. J.*, 2005, **41**(5), 1123–1128.
40. Y. Liu, K. Gall, M.L. Dunn and McCluskey, *Mech. Mater.*, 2004, **36**(10), 929–940.
41. T. Ohki, Q.Q. Ni, N. Ohsako and M. Iwamoto, *Compos. Part A-Appl. Sci. Manuf.*, 2004, **35**(9), 1065–1073.

42. D.J. Maitland, M.F. Metzger, D. Schumann, A. Lee and T.S. Wilson, *Lasers Surg Med.*, 2002, **30**(1), 1–11.
43. W. Small, T.S. Wilson, W.J. Benett, J.M. Loge and D.J. Maitland, *Opt. Exp.*, 2005, **13**(20), 8204–8213.
44. C.D. Liu and P.T. Mather, *ANTEC*, 2003, 1962.
45. F.K. Li, L.Y. Qi, J. Yang, M. Xu, X.L. Luo and D.Z. Ma, *J. Appl. Polym. Sci.*, 2000, **75**(1), 68–77.
46. M.J. Biercuk, M.C. Llaguno, M. Radosavljevic, J.K. Hyun, A.T. Johnson and J.E. Fischer, *Appl. Phys. Lett.*, 2002, **80**(15), 2767–2769.
47. H. Koerner, G. Price, N.A. Pearce, M. Alexander and R.A. Vaia, *Nature Mater.*, 2004, **3**(2), 115–120.
48. J.W. Cho, J.W. Kim, Y.C. Jung and N.S. Goo, *Macromol. Rapid Commun.*, 2005, **26**(5), 412–416.
49. R. Mohr, K. Kratz, T. Weigel, M. Lucka-Gabor, M. Moneke and A. Lendlein, *PNAS*, 2006, **103**(10), 3540–3545.
50. B. Yang, W.M. Huang, C. Li, C.M. Lee and L. Li, *Smart Mater. Struct.*, 2004, **13**(1), 191–195.
51. W.M. Huang, B. Yang, L. An, C. Li and Y.S. Chan, *Appl. Phys. Lett.*, 2005, **86**(11), 114105.
52. B. Yang, W.M. Huang, C. Li, L. Li and J.H. Chor, *Scr. Mater.*, 2005, **53**(1), 105–107.
53. A. Lendlein and S. Kelch, *Functionally Graded Materials Viii*, 2005, **492–493**, 219–223.
54. F. El Feninat, G. Laroche, M. Fiset and D. Mantovani, *Adv. Eng. Mater.*, 2002, **4**(3), 91–104.
55. C.C. Min, W.J. Cui, J.Z. Bei and S.G. Wang, *Polym. Adv. Technol.*, 2005, **16**(8), 608–615.
56. A. Alteheld, Y.K. Feng, S. Kelch and A. Lendlein, *Ang. Chem.-Int. Ed.*, 2005, **44**(8), 1188–1192.
57. M.C. Everhart and J. Stahl, *High strain fibre reinforced reusable shape memory polymer mandrels. in International SAMPE Symposium and Exhibition (Proceedings)*, 2005, Long Beach, CA.

CHAPTER 13
Shape-Memory Alloys as Multifunctional Materials

L. MCDONALD SCHETKY

Memry Corporation, Bethel, Connecticut, USA

13.1 Introduction to Shape-memory Alloys

The shape-memory effect is observed in alloys that exhibit a thermoelastic martensite transformation.[1-7] In this type of transformation the martensite forms and disappears with a falling and rising temperature. In certain alloys, an intermetallic phase exists that will undergo a displacive shear-like transformation to martensite when cooled below a critical temperature, referred to as M_S. The transformation is complete when the specimen reaches a temperature designated as M_f, and the specimen is then considered to be in the martensitic state. If the specimen in this condition is deformed, it can be restored to its original shape by heating. The transformation starts at a temperature designated as A_S, and full recovery is achieved when the temperature reaches a higher temperature designated as A_f. The critical temperatures are dependent on the alloy composition and its thermomechanical processing. The first-order martensite transformation exhibits hysteresis, as shown in Figure 13.1, which also shows schematically the shape-recovery process.

The shape-memory phenomenon was first described by Chang and Read[1] in the AuCd alloy. Since then many other shape-memory alloys have been discovered, however, until the discovery of the effect in the equiatomic NiTi intermetallic by Buehler[2] the shape-memory effect was an academic curiosity.

In addition to forming martensite by cooling, the transformation can take place isothermally by the application of sufficient stress; this is referred to as stress-induced martensite. When the stress is removed the martensite reverts to the parent phase, and due to the very high strains that result from this type of transformation it is known as superelasticity. This is shown in Figure 13.2 where the shape-memory effect is illustrated by the right-hand curve showing that the strain generated by an external stress is fully recovered when the

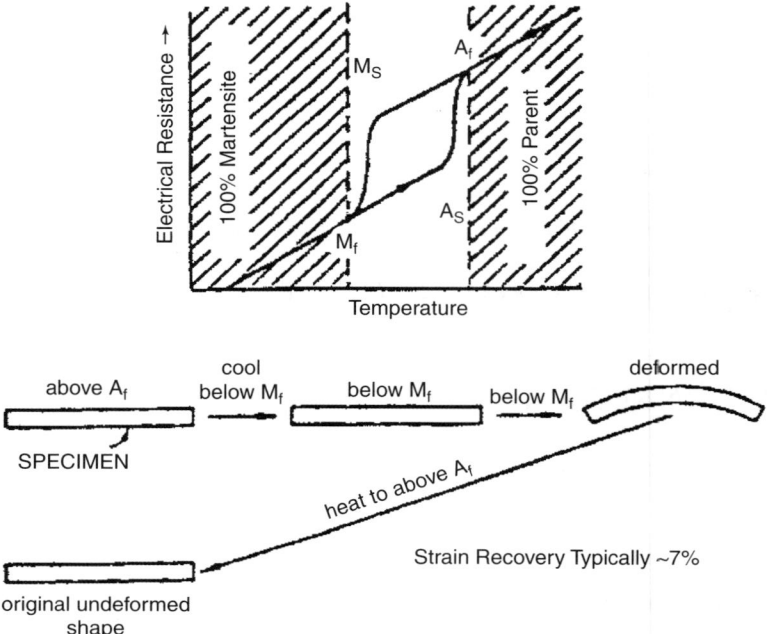

Figure 13.1 The shape-memory effect.

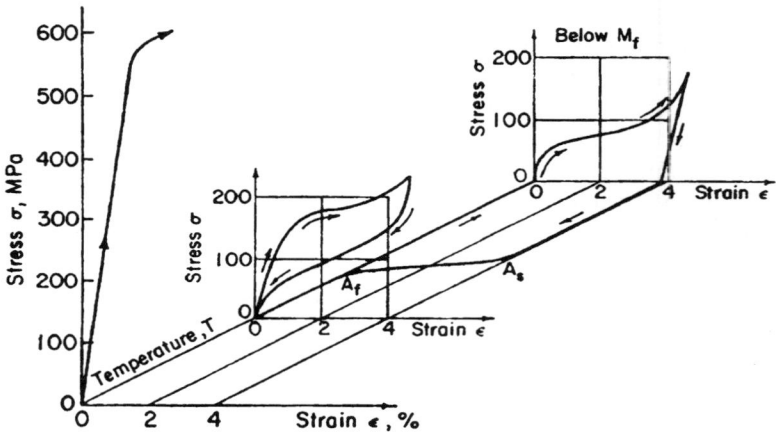

Figure 13.2 The stress–strain–temperature plot for shape memory and superelasticity.

specimen is heated above the A_f. When the strain is induced at temperatures above the A_f the strain is recovered when the stress is removed, as shown in the middle curve. At higher temperatures the alloy behaves like a conventional alloy shown on the left-hand curve.

Since the boundary between individual martensite variants and the twin boundaries move readily under small stress, if subjected to a vibratory stress the frictional loss owing to the oscillation of these boundaries gives rise to high damping. Each of these three properties, shape memory, superelasticity, and damping, have applications that range from simple fasteners to sophisticated actuators in intelligent structures and devices.

The martensite transformation is a first-order displacive transformation in which a body-centred cubic phase on cooling transforms by a shearing mechanism to martensite that is both twinned and ordered. This same transformation is the strengthening mechanism used when steel is heat treated; however, the shape-memory effect does not occur in steel. since when the martensite in steel is heated it tempers and changes its crystal structure rather than reverting back to the parent phase as in a shape-memory alloy. The progress of the martensite transformations can be followed by measuring properties such as length change or resistivity as a function of temperature; a typical shape-memory alloy hysteresis curve is shown in Figure 13.3.

The superelastic behaviour of shape-memory alloys has really three forms: linear superelasticity, work-hardened superelasticity and pseudoelasticity. When martensite is cold worked by about 10% it develops what is referred to as linear superelasticity. This is shown in Figure 13.4(a), and is the form most useful in certain medical device applications. If further worked, the stress–strain curve shows a narrower hysteresis and a smaller residual strain, as shown in Figure 13.4(b).

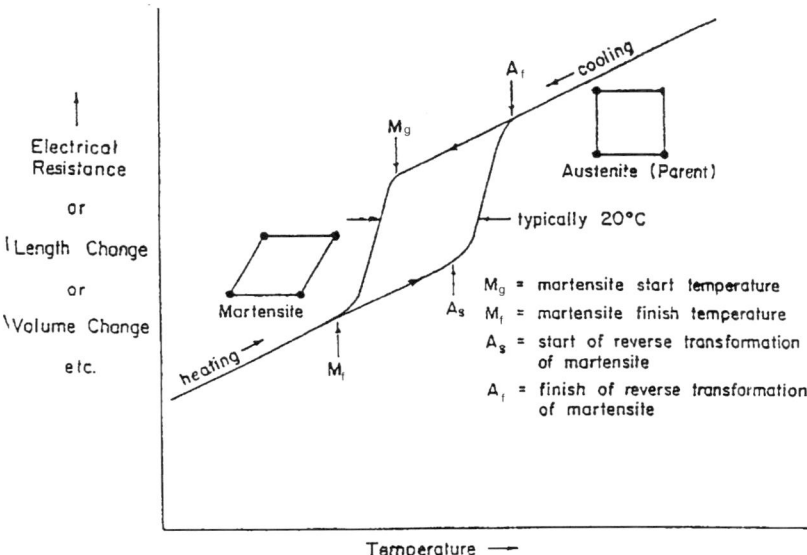

Figure 13.3 Differential scanning calorimeter trace of martensite transformation.

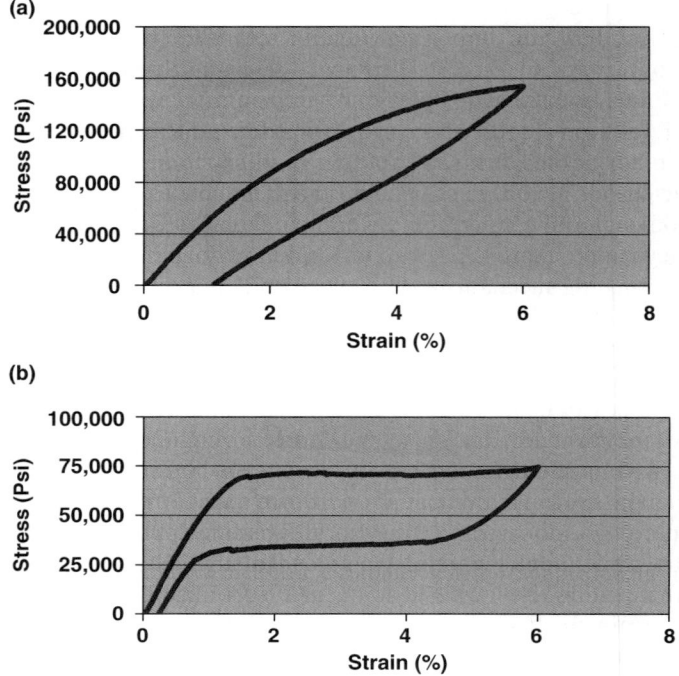

Figure 13.4 a) Superelasticity or pseudoelasticity. b) Linear superelasticity in work-hardened martensite.

13.2 Shape-memory Alloy Applications

13.2.1 Couplings

The first successful commercial shape-memory device was a NiTi tubular coupling used for the joining of high-pressure hydraulic control lines in military aircraft. Prior to the introduction of these couplings leakage of the lines at joints was common; no leaks occurred after this innovation. The principal considerations in designing this type of coupling are the maximum and minimum temperature to which the coupling will be exposed. At excessive temperature, the alloy can exhibit stress relaxation and thermodynamic instability, and the low limit is defined by the point where the coupling may loosen. A typical tubular coupling is machined from bar stock with an inner diameter that is smaller than the outer diameter of the tube to be joined. The coupling is then cooled to the martensitic state and then expanded to a diameter that will allow it to be slipped over the tube ends to be joined. When the coupling is heated to its recovery temperature an interference fit is created. For aircraft couplings, which in flight may see temperatures of $-50\,°C$ the alloy composition must be adjusted to give a transformation temperature below that figure. In practice,

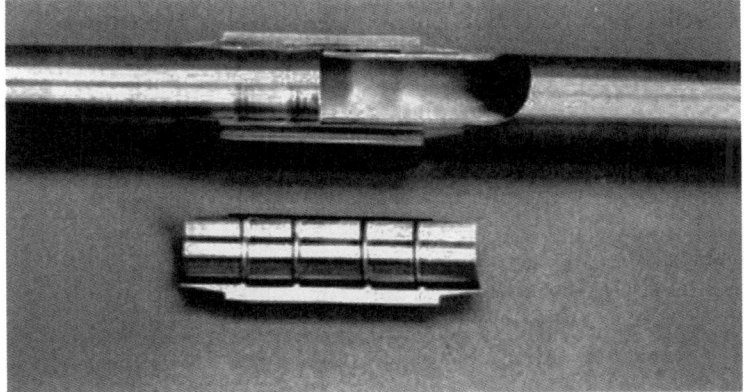

Figure 13.5 Aircraft hydraulic-line coupling.

the coupling is expanded in liquid N_2. To avoid the problem of maintaining the coupling at this cryogenic temperature until installed, a modified NiTi was developed with an addition of Nb that produces a very wide hysteresis in the alloy. This makes it possible to carry out the expansion at low temperature and then store at room temperature; recovery is achieved at 100 °C. Tubular couplings using the NiTiNb alloy have been developed for larger pipes and Russian applications have been reported for pipes as large as 200 cm in diameter. A typical aircraft tube coupling is shown in Figure 13.5.

13.2.2 Seals

Rings made by butt welding NiTi wire offer a very versatile device for a wide range of sealing problems. Using the NiTiNb alloy these rings can be expanded and then stored until installed. One of the large volume applications is the use of the ring to bond the backbraid sheath on cables to the end of an electrical connector. These multipronged connectors may have as many as 100 pins, and the braid seal must be strong enough to allow the plug to be detached without loss of the braid-joint integrity. The military specification for this type of seal requires integrity at service temperatures from −51 °C to 71 °C. Other applications for ring seals are for the hermetic sealing of electronics and precision device packaging. The relatively low temperature of shape recovery mitigates against damage to the device. Another application for ring seals is to position accurately and preload a bearing or a gear to a shaft. The ring seals attaching a backbraid and an electronic container seal are shown in Figure 13.6.

A more recent development in the application of shape-memory sealing devices is for diesel-engine injector fabrication. The injectors have two parallel cylindrical internal cavities that must be connected by a hole that requires drilling through the outer injector body and through the web between the two cylindrical cavities. One cavity supplies the fuel to the second cavity that applies

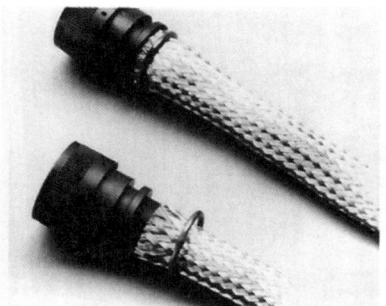

Figure 13.6 A backbraid attachment and an electronic package seal.

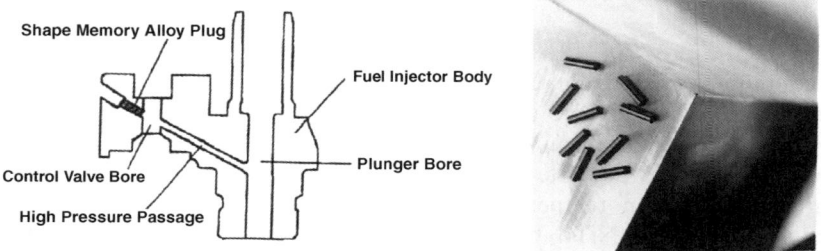

Figure 13.7 Diesel-injector seal and the hole-sealing pins.

the injection pressure. In the past, the outer hole was sealed by silver brazing a plug. Unfortunately, there were unacceptable field failures of this joint. An additional objection to the brazing operation was that this was carried out when the injector was close to its very high precision final dimensions, and the high temperature of brazing could result in some distortion. The solution was to create a NiTiNb prestrained pin which was driven into the hole in the injector body and then heated to recover the pin prestrain. These seals, shown in Figure 13.7, have never experienced a field failure and can support pressures as high as 32 Ksi.

13.2.3   Electrical Connectors

Digital electronics has placed an ever-increasing demand for specialised connectors, both on circuit boards and on wires and fibre-optic carriers. The clock speeds of computers now exceed 100 MHz and at these frequencies there are problems of coupling of signals from one channel to the next, referred to as crosstalk. Other problems that need to be solved relate to signal velocity, noise and switching delays. The correction of these problems requires connectors with high contact force to minimise contact resistance and reaction. A

Figure 13.8 ZIF connector.

shape-memory connector has been developed that provides these features and is called a zero-insertion-force connector (ZIF). The ZIF connector is opened by the shape-memory effect to allow for easy insertion of the circuit board or flexible cable and then closes to provide a high contact force. This type of application makes use of the fact that the modulus of elasticity of martensite is about 300% lower than the modulus of the parent phase. The contacts of the connector are fabricated from a gold-plated phosphor bronze "C"-shaped piece that is held closed by a shape-memory rectangular clip. When the connector is cooled by a blast of CO_2 the spring contact becomes stronger than the confining force of the shape-memory clip and as a result opens for easy insertion of the required conductor. Once insertion is complete the ZIF warms and the shape memory clip reverts to the parent phase and develops sufficient force to firmly close the connector contacts. A typical ZIF connector is shown in Figure 13.8.

This principal is also exploited for single-wire or fibre-optic connectors to provide a vibration-resistant joint. For the case of fibre optics the major requirement is the precise alignment of the axis of the mating optical fibres since any misalignment results in an unacceptable attenuation of the light signal. An additional benefit is that the coupling provides a hermetic seal to prevent the intrusion of moisture.

13.2.4 Virtual Two-way Actuation Using One-way NiTi Shape-memory Alloys

The ZIF connector described above is an example of a two-way device. In that case the objective is to open and close the contacts. In other devices the requirement is for a controlled motion to occur as a result of a change in the ambient temperature or by heating by some external means such as I^2R, hot air or hot fluid. These devices involve the operation of a shape-memory spring whose motion is constrained by a spring load, a dead load or some other mechanical means. The most common configuration of a two-way actuator is a shape-memory spring that is opposed by a conventional spring. When the shape-memory spring is heated its modulus increases and it overcomes the biasing spring and delivers motion in one direction. On cooling, the bias spring overcomes the now lower-modulus shape-memory spring and motion occurs in the opposite direction The force output in both directions is reduced by the residual force in either the bias spring or the shape-memory spring. If the bias spring is made of a shape-memory alloy whose heating and cooling is out of phase with the heating and cooling of the main shape-memory spring, a higher force can be achieved in both directions due to the out-of-phase modulus change in the two spring elements. A commercial example of this type of actuator is used in the automatic transmission fluid temperature controller in a Mercedes-Benz automobile. This device allows full fluid flow only when the transmission temperature reaches its design temperature. Although many other automotive applications have been proposed and many patents issued, a barrier to their application lies in the fact that NiTi has an operating limit of about 90 °C and under the hood temperatures in warm climates can easily reach 125 °C. There have been a number of alloy compositions proposed for higher temperature actuation, however, to date none have proven acceptable in either performance or cost.

13.2.5 Nonbiased Safety Devices

Shape-memory alloy actuators have been developed for a variety of safety systems where the actuator performs the opening or closing of a valve in response to a temperature rise. Domestic gas shut-off valves using a shape-memory spring that when heated by a fire will push forward a ball to create a seal have been developed. European telephone systems require a lightning-protection device, and in one example shape-memory fingers make contact for normal telephone reception, but in the event of a nearby lightning strike the fingers curl to interrupt the current flow.

A successful device for protecting individuals from hot-water scalding in a shower is provided by the Memrytec® valve illustrated in Figure 13.9. If the water from a shower, or in other versions from a sink or tub, suddenly rises to a dangerous level due to a system pressure or temperature change, the shape-memory leaf spring activates a diaphragm valve, shutting off the water flow. A small trickle of water is allowed to flow to drain the hot water, and once normal temperature is restored the valve opens and allows normal water flow.

Other similar shape-memory-actuated safety valves have been developed to close off industrial gas flow in the event of fire. In this case the valve is placed in the supply line of an air-operated control valve and when the safety valve

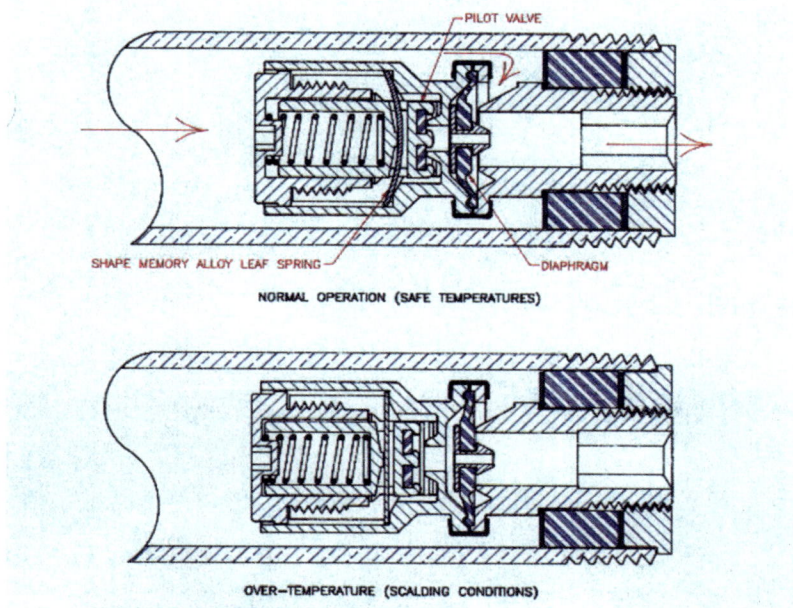

Figure 13.9 The Memrytec antiscald shower valve.

actuates it opens and stops the air flow causing the valve to immediately shut. This is similar to the brake system on a railroad where the loss of air causes the brakes to activate and stop the train. The air valve was first used in semiconductor high-temperature diffusion processors that employ gases that are either poisonous or pyrophoric, thus, in the event of a fire it is imperative to shut off all gas flow. Electric circuit breakers can also be considered as a class of safety devices. The usual tripping mechanism in circuit breakers is a bimetal strip that operates by the bending of a beam due to differential expansion of the two-layer device. The forces and motions obtainable from a bimetal strip are much smaller than can be produced by an equivalent-sized shape-memory actuator.

In larger circuit breakers a series of latches operating in cascade may be required in order to have a larger force available for the opening of the main contacts. Since the opening temperature for the circuit breaker is critical they will often require calibration. Using a shape-memory alloy with its larger available force can eliminate this calibration.

13.2.6 Thermal Interrupter

The lithium-ion battery is the preferred energiser for many consumer products such as hearing aids, computers, cameras, and cell phones. In use, there are conditions that can lead to a rapid increase in the cell temperature, thermal

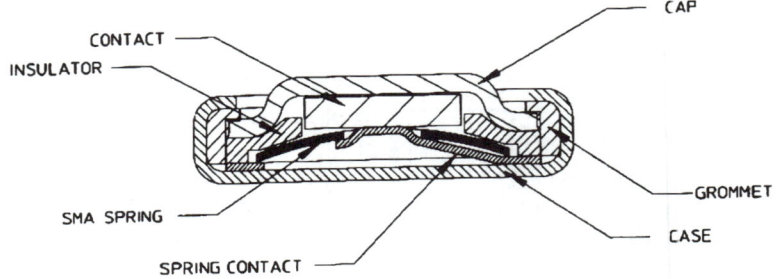

Figure 13.10 Lithium-ion battery thermal protector.

Figure 13.11 Superelastic eyeglass frames showing bending tolerance.

runaway. The major causes of thermal runaway are too rapid a rate of recharging, overcharging or short-circuit during use. A shape-memory spring operating against a spring contact interrupts the current flow in the event of a runaway; this is shown in Figure 13.10.

13.2.7 Eyeglass Frames

Eyeglass frames are worldwide a multibillion dollar business that is heavily influenced by fashion. The large tortoise frames of the past are now metal framed with narrow aspect; unfortunately, these frames are easily damaged. The use of superelastic NiTi for frame components has grown rapidly, the selling feature being their resistance to damage from accidental bending. The shape-memory nose piece is joined to the main frame by a swaged mechanical joint. This is the preferred method of joining this alloy since it is not readily welded. The ear pieces are treated in a similar manner. The ability to severely bend a frame is shown in Figure 13.11. This application has grown to be an

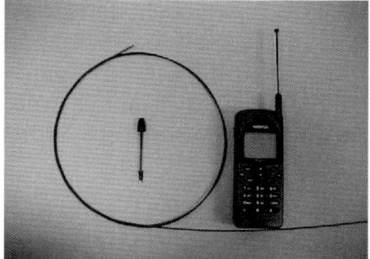 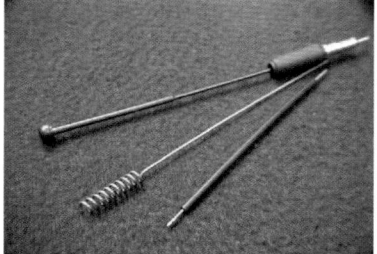

Figure 13.12 Cellular-phone antenna.

important market for NiTi, exceeded in importance only by the medical device field.

13.2.8 Cellular-phone Antennas

Although changes in cell-phone design have reduced this particular market, NiTi is still broadly used because of its kink resistance and durability. A common mode of antenna damage was catching in a car window or door. The NiTi wire is usually coated with a plastic coating such as PVC. A typical antenna is illustrated in Figure 13.12.

13.2.9 Home Appliances

One of the first applications for shape-memory alloys to achieve high volume is a small copper-based shape-memory alloy spring that controls the operation of an electric tea kettle. The spring activates the circuit breaker and shuts off the kettle when it starts to produce steam. It also protects the kettle if it accidentally runs dry. Certain refrigerator models use a shape-memory actuator to open and close a duct connecting the freezer section and the main chamber. One of the major American appliance manufacturers employs a shape-memory actuator to control the door in a self-cleaning oven. The Japanese have demonstrated a variety of appliance applications: a shape-memory control for the louvres in a window-type air conditioner, a control for correctly timing a rice cooker and a thermostatic control valve for a shower.

13.3 Medical Applications

The range of applications described were developed over the period from the mid 1960s when the NiTi shape-memory alloys were discovered to the 1990s. By far the major current market for shape-memory alloys has resulted from the recognition that the NiTi family of alloys have excellent biocompatibility and mechanical properties such as superelasticity that uniquely suite them for a wide range of medical devices. The usage of NiTi has shifted to the point where

30% of NiTi consumption is applied to medical instruments and prostheses; this represents 70% of the total market value for NiTi alloys. First applied to orthodontic archwires, NiTi alloys now have a firm place in laparoscopic surgical devices, cardiovascular devices and in orthopaedics and gastroenterology. Most of these applications utilise either wire or thin-wall tube, although a third form is gaining in importance, thin-film devices.

13.3.1 Orthodontics and Dental Procedures

NiTi alloys have been used in orthodontic procedures since the early 1970s, the first application being the archwires that are attached to the teeth by small clips and provide the force for realignment. The first archwires were fabricated from cold-worked NiTi that has a good combination of low modulus and spring-back. The preferred wire is now pseudoelastic NiTi that provides the desired low and constant force during the realignment process. The low force is an advantage since it moves the teeth without damage to the periodontal ligament or the adjoining tissue. Stainless steel wire that was the common archwire material in the past has two major disadvantages: a high modulus and creep under load. The latter characteristic requires that the dentist initially over-tighten the wire to compensate for the loosening that will occur. Overtightening causes patient discomfort and requires more frequent periodic retightening; in addition the high force causes damage to the proximal tissues. The NiTi archwire operates in the temperature region where stress-induced martensite forms that provides for a low and constant force during the realignment process. The net result is a reduction in the time required to straighten the teeth. Orthodontic wires used for correcting teeth alignment are heat treated to a "U" arch form using a composition that will be pseudoelastic at mouth temperature, 37 °C. In addition to the use of NiTi alloys in archwires they are also used in endodontic files. These small sharply pointed files are used to clean out the tooth cavity in a root-canal procedure. The high flexibility of the NiTi file and its resistance to kinking admirably suite this application.

Modern dental procedures increasingly use tooth implants rather than false teeth plates or bridges. The implant is permanent and esthetically a great improvement over previous tooth-replacement techniques. NiTi components are used in tooth implants and earlier trials in Japan involved thousands of patients with no evidence of biological reaction after 10 years of implantation.

13.3.2 Superelastic Medical Devices

Guide wires are used to introduce catheters into the body and provide a means for getting the catheter to a desired location. Previous guide wires were fabricated from stainless steel that can kink as it is pushed through a convoluted vein making withdrawal very difficult. Superelastic NiTi provides the ideal solution since it is very resistant to kinking and with its low modulus can be moved around the many curves in a blood vessel without danger of damage.

These catheter introducers are now the system of choice for cardiology procedures such as angioplasty and radiology diagnostics. In some cases the tip of the guide wire is plated with a noble metal to provide radiopacity and make localisation by X-ray easier. The mechanical characteristics that recommend superelastic NiTi for guide-wire application are also exploited for athroscopic instruments such as probes and needles used in joint repair. Superelastic wires heat treated to provide a "U" curved tip are used in locating a tumour in the breast after it has been identified by mammography. Such lesions and tumours are difficult for the surgeon to distinguish from the surrounding tissue. The probe, called the Homer Mammalok®, is put into a sheath and while observing the tumour by radiography the needle is inserted to its location and the sheath is then withdrawn resulting in the tip assuming its curved shape that retains its position. The surgeon can then follow the probe to the tumour location and excise it.

13.3.3 Cardiovascular Stents

Stents (endoprostheses) are medical implants for dilating and maintaining proper opening of previously occluded anatomic passages such as a vein or artery. The initial opening of the clogged vessel is by angioplasty, which involves threading a catheter to the position of the occlusion and then pushing through the catheter a small tube with an expandable balloon at its end which, when inflated by air pressure, pushes the occlusion or clot back against the vessel wall. A superelastic stent, which has been collapsed and then inserted into a tube, is pushed in through the catheter to this same location and when pushed out of the confining sheath expands to contact and support the vessel wall. Stents are fabricated by a variety of techniques, wire formed into a tubular shape with radial and longitudinal flexibility, stents that are simply coiled wire as in a spring, and stents produced by laser cutting a complex pattern in a thin wall tube. In all cases the NiTi used is in the superelaastic condition. Typical laser cut stents are illustrated in Figure 13.13.

The same balloon that is used in angioplasticy can be used to expand a stent, which in that case is stainless steel. The superelastic stent is called a self-expanding stent since it will spontaneously expand when released from its confining sheath. NiTi stents are primarily used for peripheral applications such as: abdominal aortic aneurism (AAA) stent grafts, iliac, renal, subclavian, femoral, popliteal, carotid, *etc.* The AAA stents is fabricated by forming and joining wire to produce a scaffold which then has Dacron fabric sewn onto the wire form to produce a pressure-tight tube. The AAA stent graft may be straight, or as in Figure 13.14, branched to repair both the femoral arteries as well as to repair the aortic aneurism.

The critical characteristics required of a cardiovascular stent are: low profile, flexibility, trackability, predictable deployment positioning, mechanical biocompatibility, *i.e.* (surface and edge geometry and nature of moving parts), radiopacity and side-branch accessibility. The earlier stents suffered from

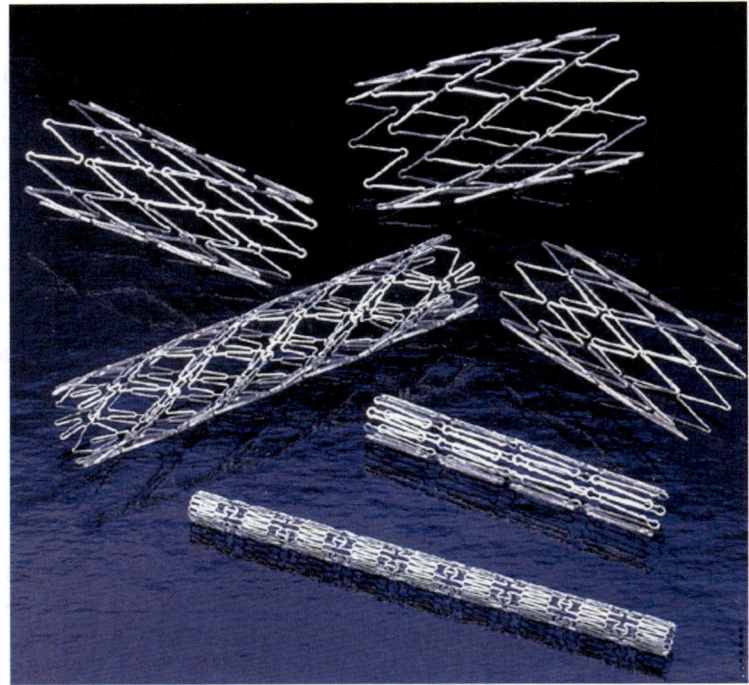

Figure 13.13 Laser-cut tubular stents.

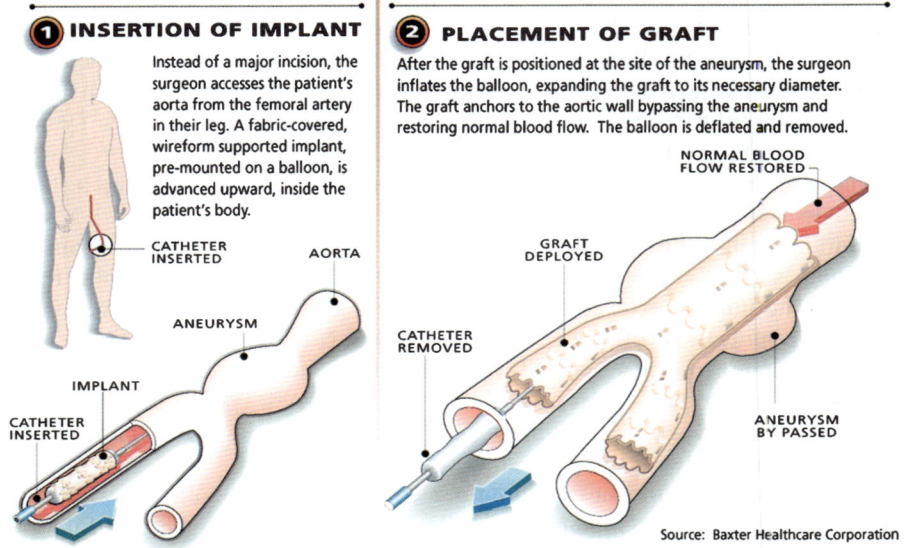

Figure 13.14 AAA stent graft deployment.

restenosis, that is, after a period of time tissue reaction at the interface of the vessel wall and the stent would cause the vessel to again occlude. The solution to this problem was to develop stent coatings that prevented the inflammation and tissue reaction from occurring. This is now the standard stent configuration and is referred to as a drug-eluting stent.

13.4 Engineering Applications

13.4.1 Adaptive Structures

Adaptive structures, also called intelligent or smart materials, refers to the various materials systems that automatically or remotely alter their dynamic characteristics or their geometry to meet their intended performance. By integrating the sensors and actuators into the structural system, typically a composite material, control of shape, vibration and acoustic behaviour can be effected. In addition to active control, passive control of system damping can be achieved in these structures. The sensors employed include piezoelectric ceramics, piezoelectric polymer films, ferroelectrics and fibre optics. For producing the stress-induced changes in the dynamic behaviour of a composite the actuators are either embedded in the composite or surface mounted. In general, piezoelectric actuators are used where small strains and high frequencies are appropriate, while shape-memory actuators are employed where high strains at lower frequency are required. When the shape-memory wires are embedded in the composite three control paradigms are possible: active property tuning, active strain-energy tuning and shape control. Active property tuning relies on unstrained NiTi wires embedded in the composite that, when heated above the transformation temperature, undergo a dramatic increase in modulus of elasticity, on the order of 300%, and the yield stress of the alloy changes by a factor of ten. The change in stiffness of the composite due to the modulus shift of the embedded wires results in a change in the modal response of the structure, and therefore a change in its acoustic emission. The magnitude of the modal shift is proportional to the volume per cent of the embedded wires.

In active strain-energy tuning the embedded wires are prestrained prior to being embedded, and by restraining the wires in a jig, the prestrain is preserved during the elevated temperature curing of the composite. When the wires are heated the recovery stress developed at the composite matrix/wire interface changes the energy balance, which results in a shift in modal response. In a typical case, in response to a vibration detected by a sensor, a signal is sent to a microprocessor that in turn energises the shape-memory actuator wires that shifts the modal response and reduces the amplitude of the vibration or acoustic signature.

Embedded actuators can also be used to change the shape of a structure, for example the curvature of a panel. For shape change the prestrained wires are embedded at a distance from the neutral axis of the panel so that when energised the resulting tensile force produces a local moment that will cause out-of-plane bending and a shape change. Even more control can be obtained by attaching the wire ends to the surface of the component and guides are placed at discrete intervals along the wire. When shape recovery is effected the curvature is defined by the spacing of the chords, each tangent to the curvature of that section of the composite.

In applying shape control there is an inevitable conflict, on the one hand structures are designed to provide load-carrying capabilities that implies some degree of stiffness. On the other hand, the objective of geometrical adaptability requires flexibility of the structure. Inducing shape change in a stiff structure with classical actuator devices requires very large actuator forces, but also produce very high stresses and strains that compromise structural integrity. The conflict is particularly evident for aircraft structures that place high demands on stiffness and low weight as well as structural reliability.

A solution to the conflict is found in using a solid-state actuator such as shape-memory alloys. In contrast with most "classical" actuators, solid-state actuators are strain induced. That means that strains and not forces are primarily supplies by the actuator principle. Producing deformation is a "natural" process for strain-induced actuators and does not require mechanical loading of a passive structure. If work is to be produced during deformation, the actuator material is able to supply forces due to its own stiffness. Integrating strain-induced actuators into a structure would realise adaptability without loading the passive material beyond its limits.

A second advantage of strain-induced actuators, which makes them especially attractive for aircraft structures, is their very high specific performance. In Figure 13.15 a comparison is made of the energy-to-volume ratio of various strain-induced actuators with the value for an electromagnetic actuator. It is evident that shape-memory alloys outperform all other actuators on the basis of specific work output.

These alloys also have another interesting feature, their ability to operate with very large strains, making them suitable for integral hinges allowing high displacement without moving parts.

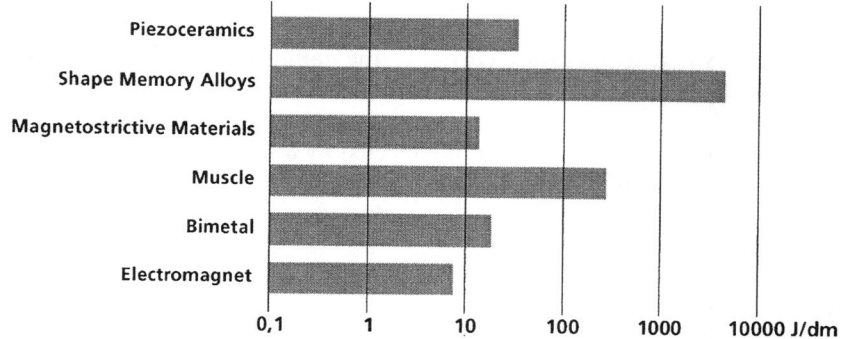

Figure 13.15 Comparison of the energy-to-volume ratio of various actuators.

At the German Aerospace Center in Braunschweig, Germany, research has concentrated on the design of high-lift devices (wings) with variable camber using shape-memory alloy actuators. Several designs have been explored using shape-memory wires attached to model airfoil section in the trailing end of the foil to effect a change in camber. They are also exploring the use of shape-memory actuators to induce a local thickness increase in the wing section, in effect a "bump" that will provide shock control in transonic flow.

A different approach to wing geometry control is being pursued by an American aerospace company, the use of shape-memory torsion tubes to cause wing twist.

The change in airfoil profile required in going from subsonic to supersonic flight has involved compromises in performance. An approach to solving this vexing problem has been to cause the wing to twist, thus presenting a different leading edge shape. The twist is accomplished by a prestrained NiTi cylinder attached to a major spar and, by means of an attached rod, extending to a short spar at the wing tip. The recovery of the prestain by heating the cylinder can produce a twist of the wing extending from the main spar to the wing tip of several degrees. The restoring force is the torsional stiffness of the wing itself. The concept has been demonstrated on a quarter-scale wing with excellent results. Full scale wind-tunnel tests are now planned.

13.4.2 Structural Damping

We mentioned in the introduction that shape-memory alloys exhibit very high damping due to the mobility of the martensite and twin boundaries. Conventional alloys such as steels, copper-based alloys and aluminium alloys exhibit a specific damping capacity (SDCs) of 0.5 to 1.5%, and grey cast iron of the type used for machine tool beds may have SDC values in the range of 10 to 12%. Shape-memory alloys typically have SDC values above 40%. The damping in these alloys is stress dependent, not surprising since the stress to cause motion of the variant boundaries must be above some critical value for each alloy system. Unlike polymeric and rubber materials the damping in shape-memory alloys is frequency independent. Material damping is not easy to incorporate in a structural component since the stress levels that accompany even very high acoustic outputs are quite low. The solution is to use these damping materials for control of vibration at its source thus isolating the vibrating member from adjacent structures.

A number of studies of seismic vibration control have been made using what is called base isolation. In this application large blocks of CuZnAl shape-memory alloy are placed at the interface of a structural column and the concrete foundation. Preliminary tests are quite promising. In a recent European Union supported study a five-storey-building model was subjected to simulated seismic vibrations using the worlds largest shake table located in Greece. Each floor of the model had diagonal stiffener elements that had shape-memory dampers attached. Accelerometers on each floor recorded the floor motion. The results were impressive with reductions in the displacement of each floor ranging from a factor of 30% to 60%. As a result of these tests it was decided to use a similar approach in the repair of the Basilica of the Shrine of Assisi in Italy that had recently been almost destroyed by an earthquake. NiTi tendons were embedded in the rebuilt dome structure with the hope that the dome will survive future seismic events. Studies are also underway to install NiTi tendons in various important ancient statuary to protect them from earthquake shock and vibration.

13.4.3 High-force Devices

Shape-memory devices have been developed as a substitute for explosives in building demolition where danger to nearby structures or to personnel is inherent in the use of explosives. These same actuators have been used in quarries to split blocks of granite and marble. These high-force devices are based on the ability of NiTi in constrained recovery of strain to produce a stress of 415 MPa, thus, with a cross-sectional area of 5 cm^2, a force of approximately 25 t can be delivered. In application, a cylinder or series of cylinders of the required cross section in the martensitic state are compressed in a press. This requires only one tenth the stress that the same deformation would require if the part was in the austenitic or parent phase condition. In use, the prestrained cylinder or set of cylinders are placed in a confining frame and when electrically heated generate a force that can be over 100 t. Another high-force actuator application is for wheel-pullers to remove a gas turbine wheel from its shaft with minimum possibility of damage.

An interesting application that has proven very successful is a shape-memory device replacing an explosive bolt. Explosive bolts are used in a broad variety of space launch procedures, and to separate stages once the launch has been achieved. There is always apprehension that, prior to launch, an electrical storm could induce current in these bolts and cause premature firing. Explosive bolts are also used in underwater systems in petroleum and petrochemical equipment, and in nuclear power plants. Using a short shape-memory tube that has been prestrained in compression, a notched bolt is assembled so that, on shape recovery by activating an electric heater the shape-recovery force causes the bolt to fracture at the notch. The Frangibolt™ is illustrated in Figure 13.16 and is an interesting solution to an industrial problem through the use of shape-memory alloys.

13.4.4 Jet-engine and Other Aeronautical Applications

A number of suggestions for the replacement of hydraulic actuators on jet engines have been made, but, as in the case of automotive applications, the temperature requirements cannot be met by the available NiTi alloys. Recent research using first-principles computer design of alloys has led to a novel strengthening mechanism for NiTi using nanoparticle dispersions. Although

Shape-Memory Alloys as Multifunctional Materials 335

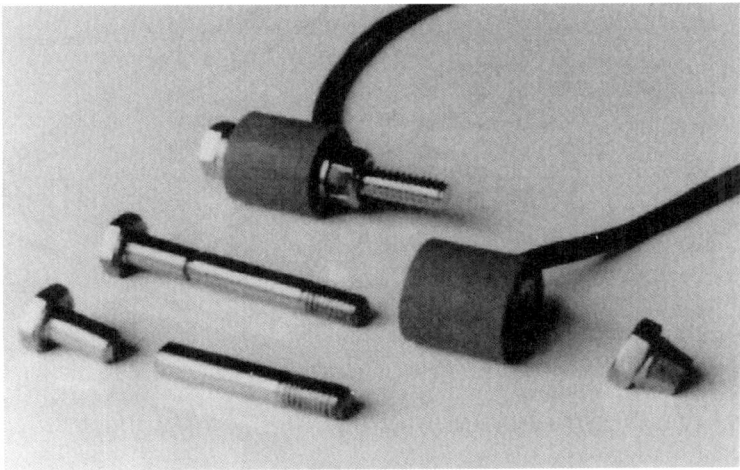

Figure 13.16 The Frangibolt™ components.

the goal was to improve strength and fatigue properties, a surprising increase in the transformation temperatures was also observed. Further larger-scale trials are underway to verify these promising results.

One application that is being studied does not require a high-temperature alloy but utilises small shape-memory blades that rise and fall to control air flow in the annular space between the inner and outer cowl of a high-bypass jet engine. Another application in aircraft that does not require high-temperature performance is helicopter rotor-blade control.

Helicopter rotor blades have intrinsic small differences in dimensions that when installed on the helicopter rotor shaft can cause one blade to track at a higher or lower plane than the other blades. To balance the individual blades is a tedious procedure involving the bending up or down of an aluminium tab attached to the trailing edge of the blade. The pilot must take-off and observe the way in which the blades are tracking, then land and have a technician correct an errant blade by bending the tab. This process is iterated until all blades fly in the same plane, a prerequisite to minimising vibration. A solution to the problem has been the development of a shape-memory actuator that the pilot can command to cause each blade to change track by remotely moving a hinged trailing edge tab. The weight and space requirements for the actuator make for a demanding design. The actuator specifications required a device with a maximum weight of 500 g, an envelope of 2.5 cm diameter × 18 cm length. The required output motion is ±7.5° with an accuracy of ±0.1°. The force output was estimated at 0.60 N m, and the force required to hold a position in flight was estimated to be 1.9 N m. The actuator would be subjected to vibration levels of as high as 828 G and operate in a temperature range of −50 °C to 70 °C. The actuator design was based on two parallel torsion tubes operating out of phase to produce motion in either direction. The NiTi alloy

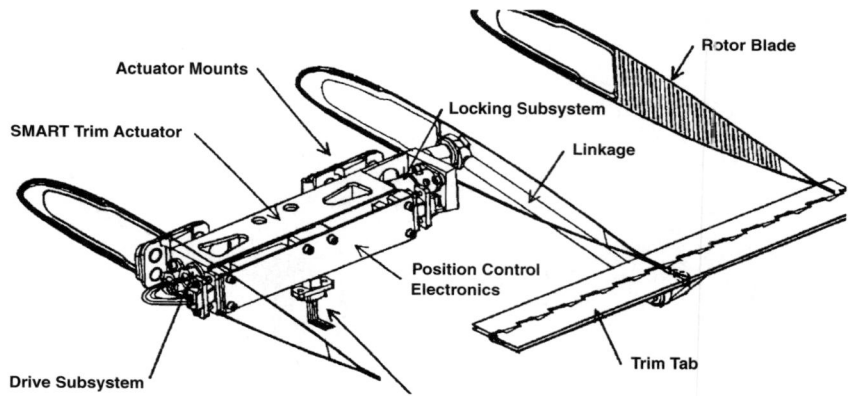

Figure 13.17 SMART trim actuator system.[3]

chosen was the NiTi10%Cu. for its cyclic stability and low hysteresis. The tubes were produced by gun drilling a solid rod and had dimensions of 7.6 cm length with a 6.3 mm diameter and a wall thickness of 0.5 mm. The tubes were heated by a cartridge heater that fitted into the tube bore. When the actuator had moved the tab to its desired location it was locked in place by a shape-memory-actuated sleeve fitted to the output shaft. The position of the output shaft was measured by a Hall-effect sensor that was activated by a magnet fitted to the output crank. In final form the pilot will observe the tracking of the blades and be able to adjust their path by calling for a change in the tab angle. The trim tab actuator mounted to the blade main spar is shown in Figure 13.17.

13.5 Thin-film and Porous Devices

The actuators and devices discussed in the forgoing sections have employed shape-memory alloy wire, strip, rod and tubular forms. A more recent approach to fabrication involves the sputter deposition of thin films of shape-memory alloy onto a substrate, usually silicon. The techniques employed in semiconductor circuit fabrication are used to shape these thin films to create unique, small devices. One of the first thin-film shape-memory devices was a miniature valve that has dimensions in the millimeter range. NiTi is sputtered on to a silicon substrate and then photoetched to form the active diaphragm for a valve. The thin film is released from the silicon by backetching, leaving a thin film supported around its edges by the silicon wafer. The minivalve is assembled with an "O" ring and a biasing element and two etched top and bottom plates as illustrated in Figure 13.18.

The valve has the performance and pressure capabilities of an electrically operated valve an order of magnitude greater in size and weight. Another thin-film application is a diaphragm pump, again using semiconductor techniques of sputter deposition and photoetching. The thin film has electrical contacts

Shape-Memory Alloys as Multifunctional Materials 337

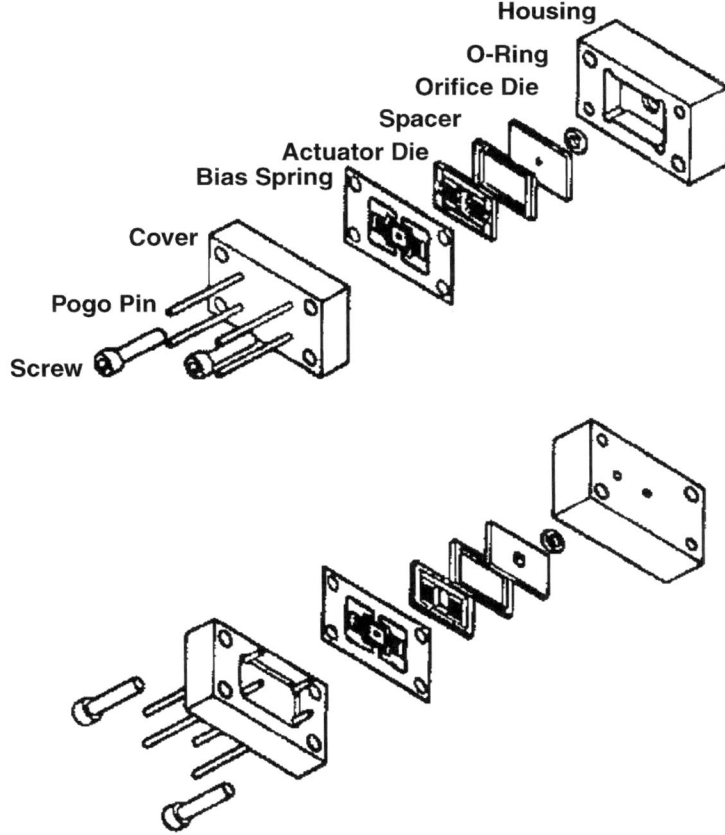

Figure 13.18 Exploded view of thin-film valve.

bonded to two edges and has integral inlet and outlet ports. These miniature pumps can produce pressures as high as 100 psi (0.048 kg/mm) at frequencies as high as 100 Hz. Prototype pumps have been used for hydraulic controls and very small versions have been proposed for implantable pumps for drug delivery in patients requiring a constant controlled drug as well as for the delivery of insulin for diabetics.

A variety of techniques are available to produce porous bodies of NiTi alloy, powder metallurgy techniques and ignition. Ignition involves the mixing of elemental nickel and titanium powders that are compressed in a die to a required shape. One edge of the compact is heated to the temperature at which the mix undergoes a vigorous exothermic reaction, referred to as the ignition. The reaction propagates throughout the specimen leaving a fused porous structure. Since the oxide level in such a structure is quite high, the ductility of the porous body is not high. One of the applications investigated is to use the materials as an energy absorber in military vehicles. A medical group is exploring the use of porous NiTi bodies for spinal disc replacement since there is sufficient pseudoelastic behaviour to provide the required degrees of spinal-bend motion.

References

1. L.C. Chang and T.A. Read, *Trans AIME*, 1951, **191**, 47.
2. W.J. Beuhler, J.V. Gilfrich and R.C. Wiley, *J. Appl. Phys.*, 1963, **34**, 1467.
3. D.K. Kennedy, *J. Intell. Mater. Sys. Struct.*, 2004, **15–4**, 235.
4. L. McDonald Schetky, *Sci. Am.*, 1979, **241, 5**, 74–82.
5. L. Yahia, ed., *Shape Memory Implants*, Springer-Verlag, Berlin, 2000.
6. Y.Y. Chu and L.C. Zhao, ed., *SMST-SMM 2001: Shape Memory Materials and Its Applications: Proceedings of the International Conference on Superelastic Technologies and Shape Memory Materials*, 2001.
7. C.T. Liu, H. Kunsmann, K. Otsuka and M. Wuttig, ed., Shape-Memory Materials and Phenomena — Fundamental Aspects and Applications, *Mater. Res. Soc. Symp. Proc.*, 1992, **648**, 246.

CHAPTER 14
Magnetorheological Materials and their Applications

XIAOJIE WANG AND FARAMARZ GORDANINEJAD

Department of Mechanical Engineering, University of Nevada, Reno, Nevada 89557, USA

14.1 Introduction

Magnetorheological (MR) and electrorheological (ER) fluids are suspensions of magnetisable or polarisable micrometre-size particles in a base liquid. Both fluids are controllable since their rheological properties change dramatically, reversibly and repeatedly in response to an applied magnetic or electric field, respectively. The rheological controllability of these fluids provides an efficient way to design simple and fast electromechanical systems for actuation, valving and motion control.[1–3]

New and emerging applications of MR fluid have received considerable attention in recent years; for example, shock absorbers/dampers for shock and vibration mitigation, and clutches and rotary brakes for torque transfer.[4–7] Research and development has demonstrated that the performance of MR fluids may inherently be better suited to meet the design requirements of most devices than ER fluids.

There have been numerous publications reviewing the progress of the ER fluid and related materials from basic chemistry and physical mechanism to device applications.[1–3,8–16] However, review articles on the MR technology are limited.[17–21]

The focus of this chapter is on a review of MR fluids and their applications including background information, material properties, modeling techniques, and a review of certain MR fluid devices. In addition, a brief overview of recent research on the MR elastomers, which are the solid counterparts of MR fluids, is presented.

14.2 Historical Perspective

The initial discovery and development of MR fluids and devices are dated to over 50 years ago.[22,23] In 1948, Rabinow at the National Bureau of Standards published his research work on a new type of magnetic fluid and new devices utilising this fluid.[23] The fluid he referred to was a mixture of "finely divided iron" with a liquid such as oil. Rabinow also described the fundamental phenomenon underlying MR fluid that is capable of solidifying under an applied magnetic field. He presented a comprehensive experimental study on the design and test of various size clutches for different MR fluids with various magnetic powder mixtures.[23] A typical MR fluid used in the clutch consisted of carbonyl iron and silicone oil, petroleum oil or kerosene.[23,24] Besides the clutch applications, Rabinow also proposed several potential applications for MR fluids such as in hydraulic actuators, dampers, and servovalves.

In roughly the same time period, Winslow[25,26] published his investigations on the properties and applications of an ER fluid, which was analogous to an MR fluid. He discovered that certain types of particles suspended in a base carrier fluid, would produce an "oil-occluding fibrous mass" when acted upon by an electric field and noted an "intrinsically reversible" nature of the fluid.[26] Winslow also described and demonstrated the physical phenomena of fibration in his work.[26] He built and tested several prototype devices using ER fluids. In addition, Winslow mentioned Rabinow's work on MR fluids at the National Bureau of Standards, and discussed some advantages of ER fluids over MR fluids.[26]

Rabinow and Winslow's discoveries soon inspired research activities in MR and ER fluids and devices. However, these devices did not enjoy rapid commercialisation success in the years following their developments, because most of the early MR/ER fluids suffered from an irreversible aggregation of suspended particles and thermal instability.[2] Moreover, in the early stages, in addition to problems due to material properties, computers were at the early stage of development. For automatic control, MR/ER devices need a computer-chip interface. The lack of such technology in the 1950s delayed the interest in these materials for more than thirty years.

A revival of interest in ER fluids began in the middle to late 1980s, spurred by the discovery of water-free ER fluids with an improved thermal and electrical stability.[8–12] Subsequently, research on MR fluids attracted more attention, since MR fluids possess some advantages over ER fluids. These advantages include, high yield stress, a wide range of operating temperature and activation using electromagnets driven by low-voltage power supplies.[4–7,27–30]

In the last 15 years, there have been significant efforts to develop novel MR fluids and devices. There are currently several commercial MR fluids available, and some MR fluid-based devices, such as MR-fluid shock absorbers, have already been commercialised for use in a semiactive suspension system of land vehicles[30–32] and seismic application.[33]

14.3 Magnetorheological Materials

14.3.1 Magnetorheological Fluids

In recent years, efforts have been undertaken to develop MR fluids with higher yield strength, and better stability and durability, which have led to several experimental and commercial MR fluids and MR-fluid devices.

The major components of a typical MR fluid are magnetisable particles, carrier fluids, and additives. The selection of MR particulate materials is limited to a number of magnetisable elements and alloys. To obtain a stronger MR effect, particles with high saturation magnetisation are desirable. Typically, the carbonyl iron is widely used as MR particulate materials for its high saturation magnetisation (about 2.1 T), low remnant magnetisation, and its ready availability.[27,34] A scanning electron micrograph (SEM) of carbonyl iron particles with 97.5% purity used in preparing some MR fluids is shown in Figure 14.1.

Fe-Co alloys with the highest known saturation magnetisation of ~2.4 T have been reported[35] for preparing a MR fluid with much higher strength. Some other magnetic materials such as iron alloys,[36] nickel zinc ferrite,[37] Co-phthalocyanine/Fe nanocomposite[38] and ultrafine, multidomain manganese zinc ferrite[39] have also been used to prepare MR fluids.

The particle size and volume fraction affect the rheological properties of MR fluids at both the "on" (magnetic field is applied) and "off" (magnetic field is not applied) states. The higher volume fraction of ferrous particles can increase the strength of a MR fluid, but, it can also increase its "off-state" viscosity, as well as its sedimentation velocity. The optimum particle–volume fractions in most MR fluids are between 30% and 50%.[40] The size of the particles used for practical MR fluids is in the range of 1–10 μm. It should be noted that for the same volume fraction, the larger particles have a more pronounced MR effect.

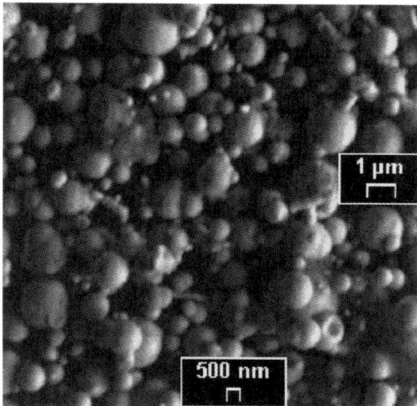

Figure 14.1 SEM image of carbonyl iron particles with 97.5% purity (Zoom × 3500).[34]

However, a large particle size ($>10\,\mu m$) is undesirable since the particles could settle faster in a fluid medium.

Overall, small particles may seem to be more effective since Brownian motion prevents sedimentation. However, a strong Brownian motion also prevents the formation of the MR structures; thus reduces MR effects. Lemaire et al.[41] studied the influence of the particle size on the rheology of MR fluids. They found that for small particles ($<1.0\,\mu m$), where the ratio λ of magnetostatic energy to thermal energy is much larger than unity ($\lambda \approx 10^2$–10^3), the apparent yield stress strongly increases with the size of the particles. Only for the very high values of λ ($\lambda \approx 10^9$) is there no dependence on the size of the particles. de Gans et al.[42] found that for small particles there is a strong increase of MR material properties with particle size. At a certain particle size and field strength, a crossover occurs to a regime where there is only very limited size dependence.

The effect of mixing particles with different sizes on the rheological properties of MR suspensions has also been investigated.[43–47] Kormann et al.[43] reported that the use of nanoparticles suspended in polar liquids could make a stable MR fluid, but the yield stress is very low (~ 6 kPa) and it is highly sensitive to high temperatures. Foister[44] improved the yield stress of a MR fluid by mixing two different sizes of iron particles (average diameters $1.25\,\mu m$ and $7.9\,\mu m$) for the same particle-volume fraction. Such bidisperse suspensions also keep a low off-state viscosity. The smaller particles help the larger particles to form more chain-like structures in bidisperse suspensions than that of in monodisperse suspensions. As a result, the yield stress may be enhanced.[47]

The carrier fluids function as a continuous medium where the magnetic particles can be suspended. Since the MR effect is not sensitive to the dielectric properties of the carrier fluid, the choices of the carrier fluids are flexible. Silicone, synthetic, semisynthetic, mineral, and lubricating oils, other polar organic liquids, and water have all been reported to have been used as carrier fluids.[4,17,19,27,40] Rankin et al.[48] and Park et al.[49] have investigated the effects of a viscoelastic medium as a continuous phase on the rheological properties and stability of MR fluids. They found that the viscoelastic carrier could prevent the particle settling and the MR effects of the fluids subject to a moderate magnetic field could still be significant. Ginder et al.[50] introduced a ferrofluid-based MR fluid to obtain a synergistic rheological effect.

A MR polymer gel (MRPG), which utilises polymeric gels as the carrier fluid, has been developed by Fuchs et al.[34,51,52] These composite polymeric fluids permit control of viscosity, provide high yield stress and exhibit low particle-settling behaviour through different combinations of resins and crosslinkers. In addition to modification of the carrier fluid, the polymer gel, which includes crosslinked copolyimide (CCPI) and solvent N-octylpyrrolidone, also changes the surface properties of the ferrous particles; thus reducing the particle settling and improving particle redispersion.[52] The particle settling rate is shown in Figure 14.2. An increase in the polymer weight ratio of CCPI to solvent (OP) 1:40 to 1:20 reduces the settling. This indicates that the dispersion stability is improved by increasing the concentration of CCPI.

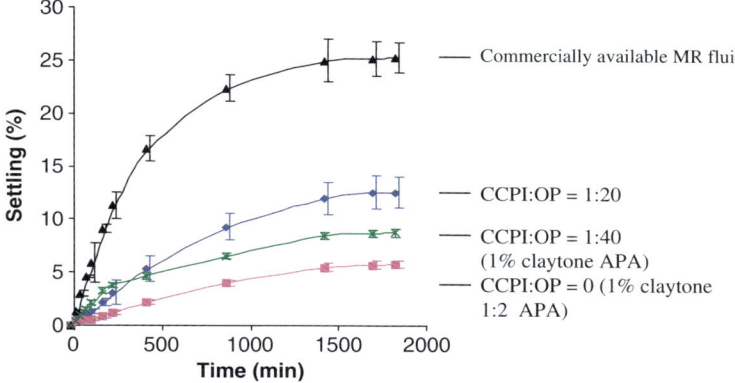

Figure 14.2 Settling curves for the MRPGs.[52]

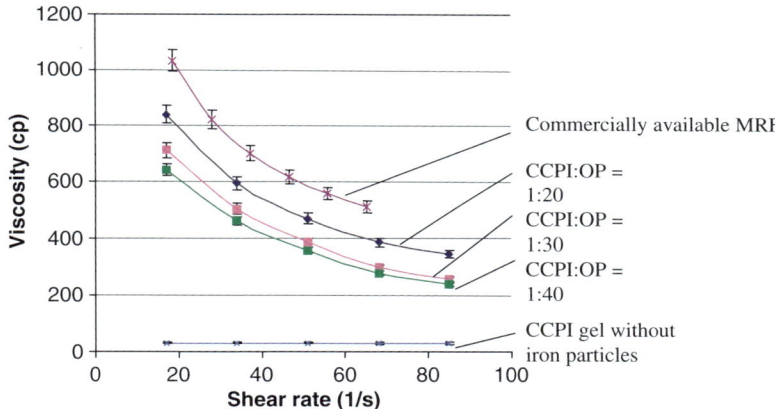

Figure 14.3 Apparent viscosity of MPRGs as a function of shear rate.[52]

A criterion for achieving good particulate dispersion is to ensure that the polymer coats the particle surface homogeneously. If the particle coating is complete, settling should be reduced. Figure 14.3 shows the dynamic yield stresses of two different MRPGs and commercially available MRF as function of applied magnetic-field strength. The off-state viscosity of MRPGs is also presented in Figure 14.3. As can be seen in Figure 14.3, the MRPGs' viscosity can be easily adjusted by controlling the concentration of CCPI in the carrier medium. It is speculated that higher polymer concentrations of MRPGs (CCPI:OP = 1:20) not only improve the particle sedimentation but also increase the dynamic yield stress, while keeping a relative low off-state viscosity.

In general, a fairly robust carrier fluid should be temperature stable, with a high boiling point, nonreactive, and nontoxic. The carrier fluid is an element that contributes to the MR fluid stability and redispersibility. The carrier fluids with low vapour pressure experience fewer sealing problems. The temperature dependence of the carrier fluid's viscosity is the dominating factor in the operating range limitation of the MR fluids. Typical carrier fluids are silicone oil or hydrocarbon oil, since they exhibit many desirable properties for MR-fluid applications.

Fluid additives including colloidal particles and ionic and nonionic surfactants have been used to enhance the performance of MR fluids. Some nanostructured materials such as silica, fibrous carbon,[17,53] magnetic metallic[17,19,40,54] and various polymers[55] added to the MR fluids maintain a coating on the particles. The coating can improve the redispersibility of MR fluid and prevent particle abrasion.[19,43,55] Other additives, for example, the ferromagnetic Co-gamma-Fe_2O_3 and CrO_2 particles were added as stabilising and thickening agents in the carbonyl iron suspension within the viscoelastic medium.[49]

Besides the yield strength and particle settling, durability and lifetime of MR fluids are key parameters that should be considered in practical applications.[24] It has been found that the off-state viscosity of MR fluid would increase dramatically and irreversibly for a long period of operation under high shear and magnetic field. This thickening problem is due to the separation of tiny fragments from the surface of the particles; thus causing particle agglomeration. A proper surface-coating technique would provide a solution to protect the corrosion of iron particles and enhance the fluid durability.[24]

14.3.2 Magnetorheological Elastomers

Magnetorheological elastomers (MREs) are solid-state analogues of MR fluids, which are obtained by dispersing micrometre-size magnetisable particles in elastomers or rubber-like solids. MREs can undergo reversible and rapid (on the order of a few milliseconds) changes when subject to an applied magnetic field.

MR fluids have controllable yield stress, while MR elastomers have initially nonzero shear and tensile moduli, which dramatically increase by an applied magnetic field. Typically, the magnetisable particles are mixed into the liquid-state elastomer or rubber-like matrix. While curing, a magnetic field is applied on the mixture to form chain or columnar structures before solidification. The particles within elastomers are locked into place and are very sensitive to a magnetic field, thus resulting in a field-responsive modulus for the MRE[56] Figure 14.4 presents a schematic of the structure of a typical MRE. The magnetic-field curing strength affects the MRE's maximum performance capability. Figure 14.5 shows scanning electron microscopy (SEM) images of two specimens. In Figure 14.5(a), the particles are not oriented at cure, and in Figure 14.5(b), the particles are oriented at cure using a magnetic field. As can be seen from Figure 14.5(b), column structures of ferrous particles are formed parallel to the magnetic-field direction.

Magnetorheological Materials and their Applications

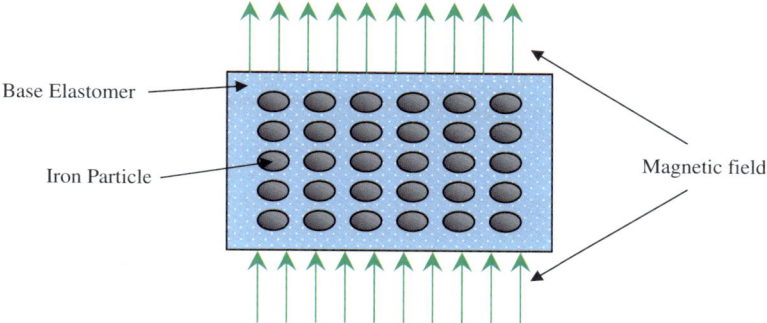

Figure 14.4 Schematic of the chain-like structure of a MR elastomer.

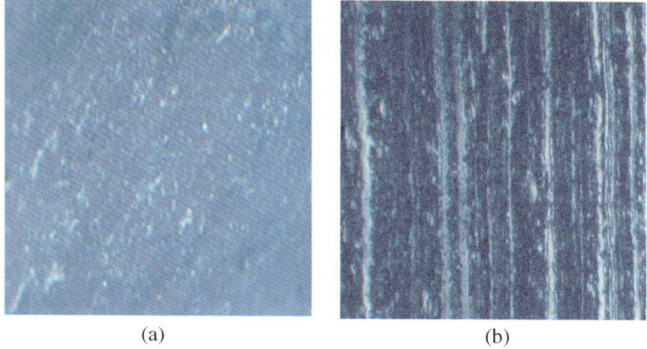

Figure 14.5 Microscopic SEM images of a UNR MRE with 5% weight of iron particles. (a) nonoriented particles during cure with magnification of 110, and (b) oriented particles during cure with magnification of 55.

Similar to MR fluids, the field-dependent properties of MREs result from the magnetic dipole interactions between particles within the host materials. Most models developed to understand the chain behaviour of MR fluids could be used to predict the behaviour of MR elastomers. Jolly et al.[57,58] employed a simple dipole model of particle interactions within a chain to predict the shear modulus of a MR elastomer. Davis[59] borrowed a model from his prior theoretical work on MR fluids to analyse the properties of MREs in zero fields and at saturation.

Several experimental and theoretic studies have demonstrated that the percentage increase of modulus for some MREs with rubber-like matrix is in the range of 30–50%.[57–60] Figure 14.6 shows the effects of magnetic field on storage and shear moduli at various frequencies. As can be seen from Figure 14.6, for the matrix materials considered in this study, the variation of storage modulus is much higher than the loss modulus. Note that the loss modulus shows considerable rate sensitivity at low frequency and less at high frequency.

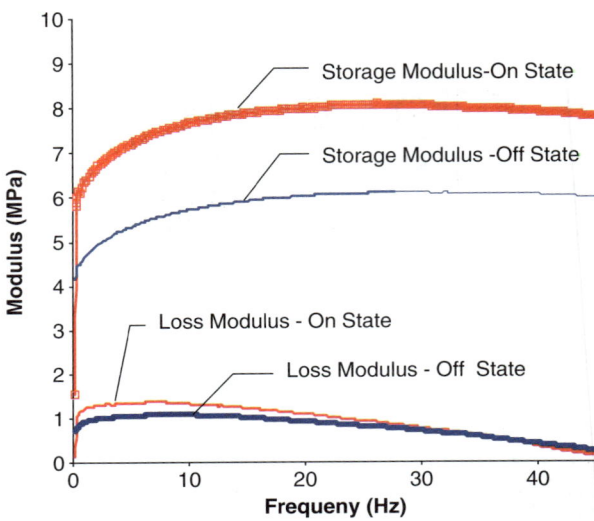

Figure 14.6 Effects of magnetic field and frequency on storage and loss moduli for 60% (by weight) iron particle.

However, the type of elastomer and iron particle volume fraction, determines the variation of loss and storage moduli. Figure 14.7 shows the effects of weight percentage of iron particles and frequency on storage modulus for the off-state and on-state (0.1 T). The results indicate that higher storage moduli can be obtained at an iron particle concentration of about 60 wt%, while maintaining good controllable force range (*i.e.* the range between off and on states).

Jolly *et al.*[58] tested three samples of MR elastomers containing 10%, 20%, and 30% iron particles by volume with a double-lap shear method. They found that the change in modulus under a magnetic field increase with increasing volume percentage of iron. Demchuk and Kuzmin[60] reported that the optimum value of the volume fraction of the magnetisable particles is in the range of 27–30%. They also found that the larger the filled particles, the larger the modulus of MR elastomer under an applied field, however, at zero field, this situation is reverse. Davis's theoretical work shows that the increase in shear modulus for typical elastomer could reach 50% of the zero-field modulus under an applied magnetic field.[59] The optimum particle volume fraction for the maximum fractional change in modulus is about 27%. The loss factor of a MR elastomer is weakly field dependent at low strains (about 1%), while for strains greater than 1%, the increase in loss factor is on the order of 10–15%.[60–62]

There are several types of MREs that differ in their compositions and the way they are processed. The viscoelastic properties of a gelatin-based MR elastomer and a formoplast-based MR elastomer within a three-layered beam are evaluated by Demchuk and Kuzmin.[60] Lokander and Stenberg developed some isotropic MR rubber materials by using irregular iron particles.[63,64] They

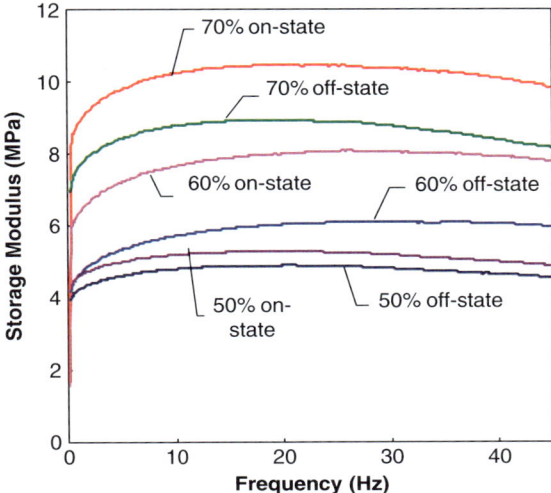

Figure 14.7 Effects of weight percentage of iron particles and frequency on storage modulus.

showed that when these particles have a relatively low critical particle volume concentration, about 38% by volume, a substantial MR effect was obtained. However, these materials have a high zero-field modulus for such high iron particle concentrations; thus the relative modulus increment in response to an applied field is low.

The advantages of MR elastomers over MR fluids include no leakage, no attrition or sedimentation of particles, and possible higher preyield strength. For vibration and noise-control applications, the field-dependent modulus of MREs can be utilised as a variable stiffness.[56,62,65,66] For conventional elastomeric or rubber-like materials, through appropriate formulations, their stiffness and damping can be tailored to specific applications; however, they cannot always be adjusted independently or optimised for all operating conditions. A field-controllable elastomer may overcome these limitations. For example, Ginder et al.[62,66] designed and tested tunable automotive mounts and bushings employing MR elastomers. Except for the additional MR materials and a wire coil, most of the components are similar to their passive counterparts, but the device with MR elastomers can create variable stiffness and damping with an input current that could be used to improve the performance of, for example, vehicle-suspension systems.

MR elastomers can also be used in tuned vibration absorbers (TVAs). Jolly et al.[58] studied an adaptive TVA based on MR elastomers. They reported an 18% change in natural frequency achieved under an applied magnetic field. Ginder et al.[56,66] tested MR elastomers over a broad range of frequencies. They found that the controllable modulus of MREs is effective to vibration frequencies over 1 kHz.

14.3.3 Rheological Behaviour of MR Fluids

Magnetorheological fluids typically consist of micrometre-sized magnetisable solid particles (1–10 μm) suspended in a nonconducting carrying fluid, such as, a mineral or silicone oil. These materials show very distinctive reversible rheological behaviour in the presence of a magnetic field. Without a magnetic field present, the suspensions of magnetisable particles are randomly distributed, and MR fluids behave similar to a Newtonian fluid. However, under the influence of a magnetic field, the particles form chains, columns or more complex structures aligned with the field direction. This causes the MR fluid to exhibit solid-like behaviour with increased shear-yield stress.[2,18,67] A viscoplastic model with a variable shear-yield stress can describe the rheology of many MR materials under steady shearing. The Bingham plastic model, which is commonly used, can be expressed, as follows:

$$\tau(\dot{\gamma}, B) = \tau_y(B) + \mu_p \dot{\gamma} \quad \text{for} \quad \tau > \tau_y$$

$$\dot{\gamma} = 0 \quad \text{for} \quad \tau \leq \tau_y \qquad (14.1)$$

where $\tau(\dot{\gamma}, B)$ is the shear stress, $\tau_y(B)$ is the dynamic yield stress induced by the magnetic flux density B, $\dot{\gamma}$ is the shear rate, and μ_p is the plastic viscosity independent of magnetic-field strength. Figure 14.8 shows the shear stress behaviour of a MR fluid, developed at the University of Nevada, Reno (UNR)[68] with steady shearing deformation under various magnetic fields. It can be seen that the Bingham fluid model can fit the experimental data well.

Not all the experimental data support the Bingham plastic model. Felt et al.[69] studied a monodisperse MR fluid and found that the Bingham model does not accurately describe the shear thinning observed in these systems. In order to incorporate the shear-thinning behaviour of MR/ER fluids, Marksmeier

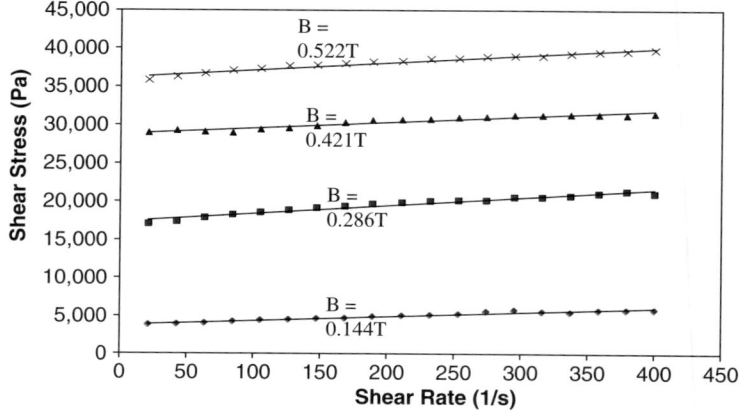

Figure 14.8 A typical MR fluid's shear stress versus shear rate measured at different magnetic fields. The solid lines represent the curve fit of the experimental data using the Bingham plastic model.[68]

et al.,[70] Wang and Gordaninejad,[71,72] and Lee and Wereley[73] suggested using the Herschel–Bulkley model, which states that:

$$\begin{cases} \tau(\dot{\gamma}, B) = \tau_y(B) + k|\dot{\gamma}|^{n-1}\dot{\gamma} & |\tau| > \tau_y \\ \dot{\gamma} = 0 & |\tau| \leq \tau_y \end{cases} \quad (14.2)$$

where k and n are fluid index parameters. The Herschel–Bulkley model assumes a nonlinear postyield behaviour for non-Newtonian fluids.

Li[74] studied the rheological properties of MR fluids under steady shear with a commercial MR parallel-plate rheometer. Figure 14.9 presents a typical result of shear stress versus shear strain rate at various magnetic-field strengths for a commercial MR fluid (LORD MRF-132LD). Figure 14.9[71] shows that at a range of low magnetic-field strength, the MR fluid exhibits Bingham plastic behaviour with a constant plastic viscosity being equal to the zero-field viscosity of Newtonian fluids. However, at higher magnetic-field strengths, the MR fluid exhibits pseudoplastic behaviour with a field-dependent yield stress. Considering the shear thinning effect, the Herschel–Bulkley model is adapted to represent the measured shear stress versus shear strain rate data.

Although most of MR fluid devices operate under oscillatory shear, only a few experiments for oscillating-shear rheology of MR fluids have been presented in the literature. Weiss et al.[75] reported experimental data on the elastic and viscous behaviour for a conventional MR fluid through the use of oscillatory rheometry techniques. They found that in the preyield region (strain

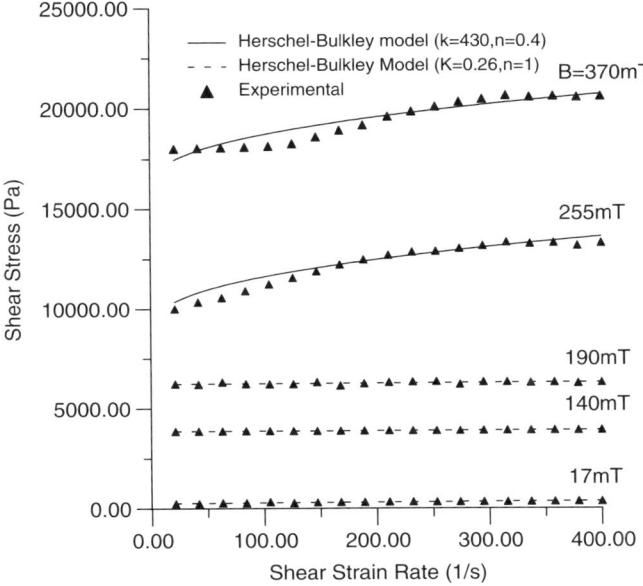

Figure 14.9 Shear stress versus shear strain rate data at various magnetic-field strengths.[71]

amplitudes below 10^{-2}), MR fluids behave similar to a viscoelastic solid, and the storage modulus of MR fluids that was tested reached as high as 4.2 MPa for applied fields of 160 kA/m at high frequency (16 Hz). Li[74] extended Weiss's work in shear mode by using a commercial parallel-plate rheometer. Several effects including strain amplitude, frequency, field strength, particle-volume fraction and temperature on the storage modulus and loss modulus were evaluated. Nakano et al.[76] studied the viscoelasticity of an MR fluid in flow mode by using a newly developed oscillatory pressure-flow-type rheometer. They also found that the dynamic viscoelasticity of MR fluids depends on the fluid-strain amplitude and magnetic-field strength. With an increase in strain amplitude, the viscoelastic plastic property becomes apparent.

Claracq et al.[77] presented an extensive experimental study of the viscoelastic behaviour of MR fluids under small deformations. They found that, when scaled with the magnetic-field strength, a master curve of the complex shear modulus (G^*) as a function of the magnetic-field strength (H), frequency (ω) and particle volume fraction (ϕ) could be established with the following equation (see Figure 14.10):

$$\frac{G^*}{(\phi H)^{1.7 \pm 0.1}} = f\left(\frac{\eta_0 \omega}{(\phi H)^{1.7 \pm 0.1}}\right) = f(\omega \tau) \qquad (14.3)$$

where η_0 is the viscosity of the structured fluid.

A chain model proposed by Klingenberg[78] predicts that the complex shear modulus (G^*) is only a function of the frequency scaled by the square of the applied field strength for a particular suspension at a given concentration at very small strain amplitudes. However, only a few experimental data are available to verify this model.

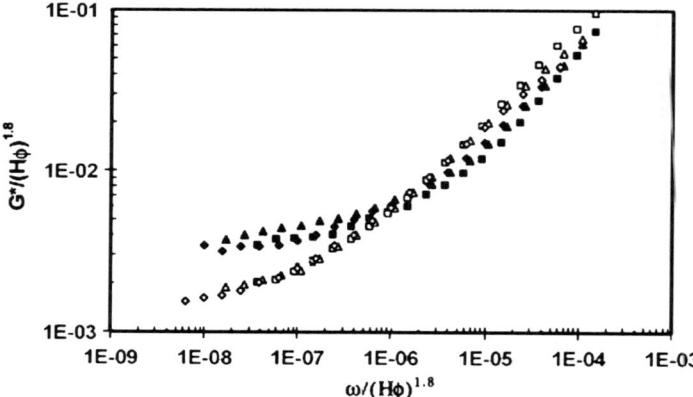

Figure 14.10 Complex shear modulus as a function of frequency (for $\gamma = 10^{-3}$, 5 vol.%) for three different magnetic intensities (G'(solid symbols), G''(hollow symbols)).[77]

14.3.4 Models for Shear-yield Stress

The shear-yield stress is the key parameter that characterises the viscoplastic properties of a MR fluid. Two different shear-yield stress values, the dynamic shear-yield stress and the static shear-yield stress have been commonly reported. Parthasarathy and Kingenberg[12] stated that, "the dynamic yield stress for steady shear flow is defined as the value of the shear stress in the limit of the zero shear rates. The static yield stress is commonly defined as the smallest shear stress required inducing flow or unbounded deformation."

Several researchers reported that the dynamic and the static yield stresses of MR fluids were different based on various measurement methods.[74,79,80] Volkova et al.[81] analysed the existence of the two different yield stresses for two types of magnetic particle suspensions in the presence of a magnetic field. They explained that the dynamic yield stress (referred to as the Bingham yield stress) is associated with the rupture of the aggregate that reforms in the presence of the magnetostatic forces, while the static yield stress (referred to as frictional yield stress) is associated with solid friction of the particles on the plates of the rheometer.

Several theoretical models were proposed to predict the yield stress of MR fluids based on the static microstructure analysis. A simple "point-dipole" model predicts that the yield stress is proportional to $\phi\mu_0 H_0^2$ while under a weak external magnetic field.[57,82,83] However, for a stronger applied magnetic field, magnetic saturation may occur at the polar region of each particle, indicating that the linear "point-dipole" model fails to treat nonlinearity, which is inherent in all magnetic materials. Ginder et al.[82,83] used a finite-element method to predict the inter particle force and yield stress in a single row chain of MR fluids' particles. They extended their work by proposing a "two-zone" model to account for local magnetisation saturation of particles within a contact zone. They obtained an analytical expression for the static yield stress of MR fluids under various field strengths, as follows:

$$\tau_y = \sqrt{6}\phi\mu_0 M_s^{\frac{1}{2}} H_0^{\frac{3}{2}} \tag{14.4}$$

where ϕ is the volume fraction of particles, μ_0 is the permeability of the vacuum, M_s is the particle saturation magnetisation, and H_0 is the applied field. Equation (14.4) shows that the yield stress of a MR fluid increases as a 3/2 power of applied field H_0 at higher field and no further increase in yield stress is possible once the particles are completely saturated. It also suggests that particles comprised of material with higher saturation magnetisation exhibit larger MR effects. From Figure 14.8, one can obtain the yield stress of an MR fluid by extrapolating shear stress data back to zero strain rate based on the Bingham plastic model. A subquadratic dependence on the moderate magnetic flux density is observed, as shown in Figure 14.11.[68] Similar phenomena have been observed in other experiments indicates that field-dependent yield stress follows the power law with an exponent close to 1.5.[35,84,85] The simple "point-dipole" model can almost always overestimates the value of yield stress, when compared to the experimental data.

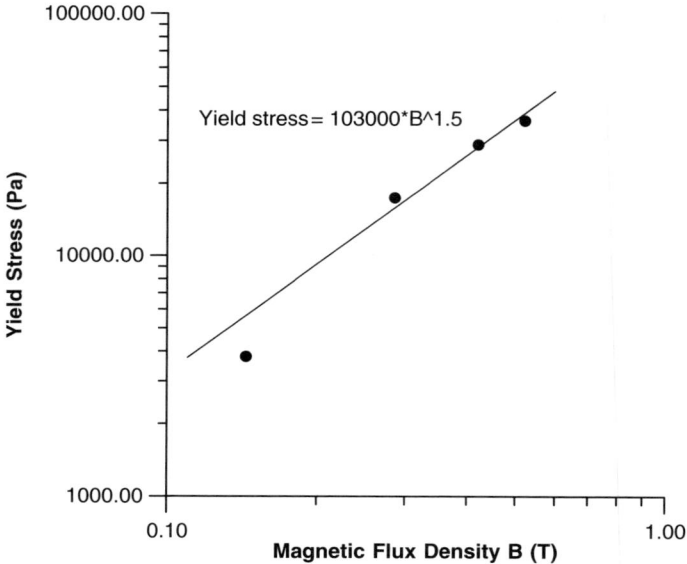

Figure 14.11 The measured yield stress verse applied magnetic flux density B for a UNR MR fluid. The yield stress increase as $B^{1.5}$, as shown by the solid line.[68]

In addition to the above particle-to-particle interaction approach, an alternative model based on the laminar structures of MR fluid-particle aggregations was presented.[86,87] Tang's modified laminar model[86] takes into account the field concentration between particles, as well as the effects of magnetisation saturation of the particles, and presents a reasonably good model for MR static yield stress, when compared to that of experimental data. Other models, however, have considered neither the condition of dynamic deformation of MR structures nor the effects of surfactant interaction with particles.

The static yield stresses of MR fluids predicted by the models are usually taken as a reference for the dynamic yield stress of the Bingham model. Shulman et al.[88] proposed a stationary model to account for the dynamic structure and physical properties of a MR fluid. The model was developed based on a suspension in a magnetic field representing a system of ordered noninteracting ellipsoidal aggregates, which are oriented at some angle to the flow.

Both microscopic and macroscopic models predict that the yield stress is proportional to the volume fraction of particle loading; however, this is only valid at lower volume fractions.[48,85] At higher volume fractions, the yield stress increases faster. For most practical MR fluids which have a high volume fraction ($\phi \geq 0.3$), the relationship of the yield stress and volume fraction is usually represented by a power law $\tau_y \sim \phi^n \, (n>1.0)$.[32] This behaviour can explain how thick aggregates form a stronger structure than that of individual chains resulting in a high yield stress.[21,89]

All proposed theoretical models assumed that no slip occurs between the particles and the MR structure under shear deformations. While in the actual experiments for a static or quasistatic test of MR fluids, particles always slip on the wall surface, if the magnetic interaction between the particles is stronger than that of the friction force between the particles and the walls. This may be a possible explanation as to why the measured yield stress varies from the predicted one.

14.3.5 Field-induced Microstructures

The MR effect results from the interactions and aggregations between particles subjected to magnetic force. The microstructures of MR particle aggregation play an important role in determining the macroscopic properties of MR fluids. The formation of a microstructure and the time scale associated with structure formation is important both in understanding the underlying mechanisms of magnetorheology, and in designing systems that exploit magnetorheology.[90]

Several techniques have been used to study the field-dependent microstructure behaviour of MR fluids. These techniques include, light transmission and scattering techniques on dilute MR fluids or ferrofluids,[91–96] and sound-wave propagation techniques on an MR slurry or suspensions.[97,98] The rheometer technique[99,100] has also been used to investigate the microstructure formation and convolution of MR fluids by monitoring their viscoelastic transition behaviour under various magnetic fields. Cutillas et al.[101] and Flores et al.[94] carried out several experiments to investigate the influence of the static and dynamic structures on the rheology of MR fluids. They observed a column or cylindrical structure formed of MR fluids rather than a chain or lattice structure.

A model accounting for the structure formation time of pairs of particles in the linear magnetic regime is presented as:[83,90]

$$t_c \approx \frac{2\eta_f}{5\mu_0(\beta H)^2} \left[\left(\frac{\pi}{6\phi} \right)^{5/3} - 1 \right] \quad (14.5)$$

where η_f is the viscosity of the suspending fluid and ϕ is the volume fraction of the suspended particles. β is a function of relative permeability of the particles and carrying fluid. With the increase of the field, the structure formation time decreases. Equation (14.5) provides a lower bound on the response time of the MR effect.

The microstructure response of quiescent MR fluids with high iron loading has been measured using magnetoinduction experimental techniques.[90] The measured time scale of microstructure formation subject to an applied magnetic field is within the range of 1–100 ms. The relationship of the time response with the volume fraction, carrier viscosity and magnetic field partially agrees with theoretical models associated with particle aggregation.

14.3.6 Rheometry of MR Fluids

The rheological characteristics of MR fluids are needed in designing, modeling and controlling MR fluid devices. The apparent viscosity depends on the field (*i.e.* magnitude and direction) and the shear rate, making the characterisation of MR fluids difficult. Several experimental setups using different measurement geometries have been investigated to quantify the characteristics of MR fluids.

Most apparatus for rheological measurements of MR fluids are usually realised by modifying conventional rheometers with an additional magnetic circuit that can apply an appropriate magnetic field to the MR-fluid sample. A commercial MR-fluid shear rheometer with a plate-plate configuration, as shown in Figure 14.12, is utilised to study the behaviour of MR fluids. With the plate–plate arrangement, the torque transferred by the MR fluid from the stationary dispenser to the rotor can be measured.

Laun et al.[102] have systematically modified a number of commercial concentric cylinder rheometers with various magnetic field arrangements for MR-fluid measurements. They also designed a MR capillary rheometer for high shear rate measurements. When compared to the concentric cylinder results, they found a satisfactory agreement in the MR fluid data in the overlapping shear-rate range. Ginder et al.[83] have designed and constructed a concentric-cylinder rheometer with ferrous materials to achieve flux densities of over 1 T. Lemaire and Bossis,[103] Li,[74] and Genc and Phule[104] have utilised parallel-plate rheometers to characterise MR fluids. Tang and Conrad[84] have developed a sliding-plate rheometer to study the shear stress-shear strain behaviour of MR fluids at various magnetic flux densities (0.05–0.6 T) in a quasistatic test. A cone-plate rheometer for MR fluid measurement has been utilised by Felt et al.,[69] de Gans et al.[105] and others.[106]

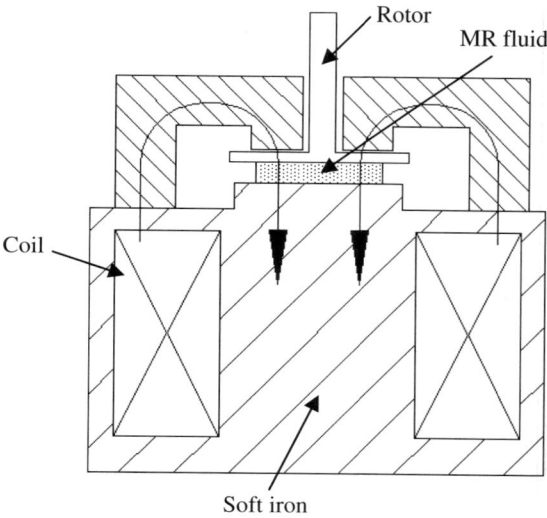

Figure 14.12 Schematic of the MR fluid-shear rheometer.

The rotational viscometers with cone-plate geometry have the advantage of a uniform shear-rate field, but they restrict the orientation of the applied magnetic field and also make it difficult to maintain a constant field across the measured area. Concentric-cylinder rheometers are often preferred to the parallel-plate rheometers, because they can obtain a nearly constant shear rate through the sample for small fluid gaps; while the shear rate in parallel-plate rheometers varies linearly with distance from the axis of the rotation. However, large measurement errors may be caused by nonuniformity of the magnetic field in the rheometer gap and at the fixture edges. The significance of shear-induced heating at high shear rate and shear stress is also reported by Laun et al.[102] for the MR concentric-cylinder rheometers. It is worth mentioning that the MR rheometers discussed above are used to measure the properties of the MR fluids under deformation perpendicular to the field direction. A MR rheometer has been designed and tested by Shorey et al.[107] for the purpose of MR finishing, which can measure the yield stress that depends on the mutual orientation of the magnetic field and the direction of deformation.

Most of the measurements for rheological properties of MR fluids have been performed in Couette flow and low shear rates. However, a MR fluid may undergo Poiseuille flow with an orifice design characteristic in most devices, such as, dampers and pressure-control valves. In general, relatively limited attention has been given to the behaviour of MR suspensions in pressure-driven flows. In addition, conventional rotational rheometry encompasses shear rates only up to a few thousand s^{-1}, while shear rates in most MR fluid devices can reach up to $20\,000\ s^{-1}$ or higher. Therefore, the study of the rheological properties of MR fluids at high shear rates is warranted.

A piston-driven slit-channel rheometer has been developed to estimate the yield stress and dynamic viscoelasticity of MR fluids under a magnetic field by Nakano et al.[76] A slit rheometer utilises an applied uniform magnetic field normal to the slit channel flow direction where forced convection assists in heat dissipation at high shear rates. However, it is difficult to determine the velocity gradient for a non-Newtonian fluid in such a viscometer. This disadvantage can be overcome by careful data processing.[108]

Wang and Gordaninejad[80] have designed and built a piston-driven flow-type rheometer to evaluate the tunable rheological properties of MR fluids in Poiseuille flow. Figure 14.13 shows the piston-driven flow-type MR-fluid rheometer system developed, with a flat channel of rectangular cross section. An electromagnet with a 1200-turn copper coil drives a low-carbon steel induction path that delivers magnetic flux normal to the slit-channel flow. The coil is powered by an operation-amplifier power supply in constant-current mode. The MR fluid is confined in the well-sealed channel cell and is pressurised to flow through the channel between two parallel magnet poles.

Controlling the velocity of the source piston is achieved by connecting the piston-cylinder to an Instron servohydraulic actuator. Two fluid-pressure transducers measure the pressure drop across the test section. The magnetically energised test section has the dimensions of $h = 1.0$ mm height, $w = 10$ mm width and $l = 14$ mm length. A unique design of this flat-channel flow MR

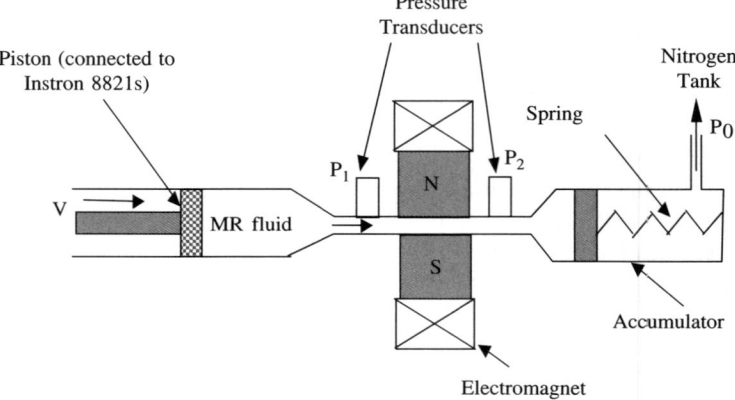

Figure 14.13 Schematic of the MR the flow-mode rheometer.[80]

rheometer is the use of a nitrogen gas high-pressure tank. The tank accumulator functions as a sink for MR fluid when the piston is positively displaced, and as a source, when the piston is negatively displaced making it possible to measure the static yield stress, as well as, the dynamic yield stress of MR fluids. This device can be used to study the apparent viscosity of MR fluids for a wide range of shear strain rates, from 50 s^{-1} to 40 000 s^{-1}.

In order to determine the viscosity, $\eta = \frac{\tau}{\dot{\gamma}}$, one has to relate the shear stress, τ, and shear rate, $\dot{\gamma}$, with measured quantities Δp, pressure drop, and Q, volumetric flow rate. If the dynamic loss is small, the value of τ at the wall is determined by pressure drop across the channel, as follows:

$$\tau_w = \frac{h \Delta p}{2l} \quad \text{for } h << w \quad (14.6)$$

Similar to the Rabinowitch correction used for conventional capillary viscometer,[109] a relationship between shear stress at the wall, τ_w, and shear rate at the wall, $\dot{\gamma}(\tau_w)$, is given as:

$$\dot{\gamma}(\tau_w) = \frac{6Q}{h^2 w}\left[\frac{2n'+1}{3n'}\right] \quad (14.7)$$

where $n' = \frac{d \ln \tau_w}{d \ln Q}$. For a Newtonian fluid $n'=1$, for a non-Newtonian fluid, n' can be determined from the experimental data of $\ln(\tau_w)$ against $\ln(Q/h^2 w)$. For any flow rate, n' is equal to the slop of the curve for that point. A linear regression method can be used to obtain the best fit for $\ln(\tau_w)$ as a function of $\ln(Q/h^2 w)$. By measuring the decline in pressure as a function of flow rate, the apparent viscosity is determined.

Figure 14.14 shows the measured apparent viscosity versus shear strain rate at various magnetic fields for the range of 50 s^{-1} to 40 000 s^{-1}. These results are

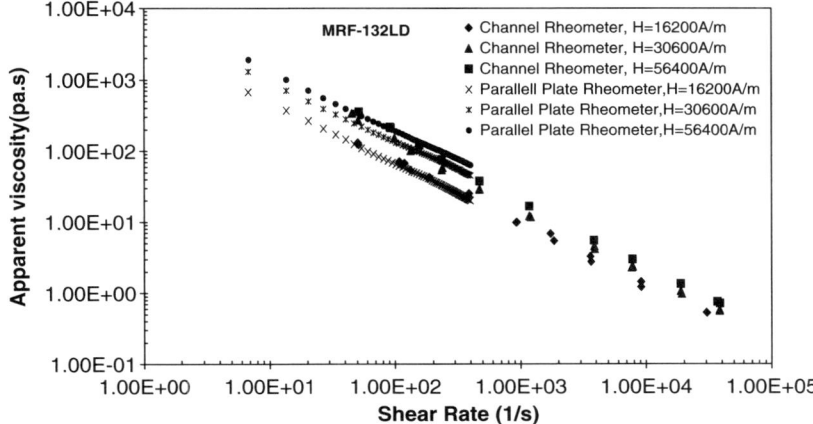

Figure 14.14 Log–log plot of the apparent viscosity of MRF-132LD versus shear rate for various magnetic fields.[80]

compared to those obtained by a commercial parallel-plate MR fluid rheometer.[100] As can be seen from Figure 14.15, the results of both studies agree well in the range of 50 s^{-1} to 300 s^{-1}, where the experimental data of shear rates for these two different rheometers overlap. Thus, the flow-mode rheometer can provide relatively accurate results of the apparent viscosity of MR fluids under various magnetic fields. Furthermore, the channel-flow rheometer extends the shear rate measurement to over 40 000 s^{-1}. Both experimental results show a strong linear relationship on the log–log scale, which implies that there is a power-law dependence $\eta_{app} \propto \dot{\gamma}^{-m}$ of the apparent viscosity η_{app} on the shear rate $\dot{\gamma}$ for MR fluids.

Figure 14.15 presents the experimental results of yield stress dependence on the magnetic flux densities (B_0) for three MR-fluid samples. The dynamic shear-yield stress is obtained by extrapolating shear stress data back to zero-shear strain rate based on the Bingham plastic model. The static shear-yield stress is obtained by evaluating the experimental data of the static pressure drop employing eqn (14.6). The static pressure drop is the minimum pressure that can be induced to a MR fluid to initiate a flow. Note that the dynamic shear-yield stress exceeds the static case over an entire range of magnetic fields for all three samples.

14.3.7 Effects of Surface Roughness

As previously mentioned, the shear-yield stress experimental data traditionally is obtained for MR-fluid flow over a steel surface using a shear rheometer. However, if the material and geometric characteristics of the wall surface change, different shear-yield stress can be obtained. This is because the wall-surface conditions have a significant effect on the field-induced properties of

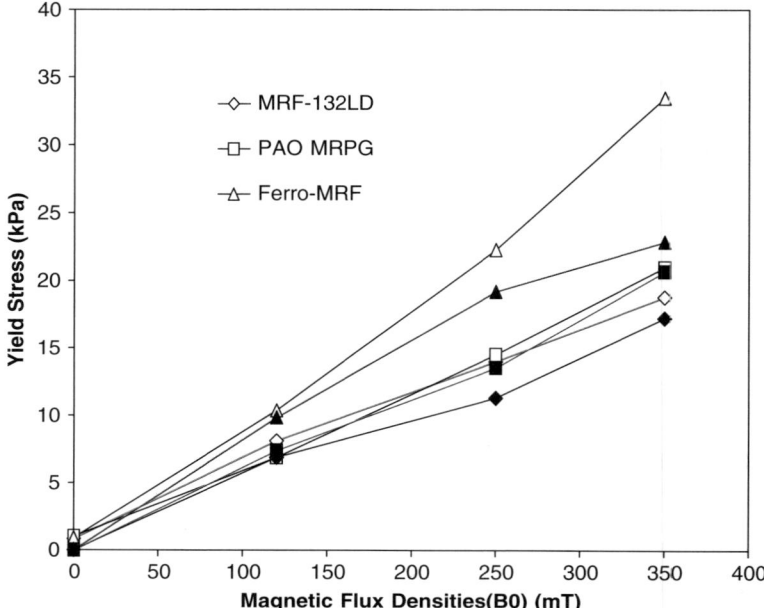

Figure 14.15 Comparison of yield stress versus magnetic field for three MR fluids samples. Hollow symbols are the dynamic yield stress. Solid symbols are the static yield stress.[80]

MR fluids. In this section, the effects of surfaces on the behaviour of MR fluids and how the yield stress is enhanced by surface modifications are reviewed.

Studies on surface effects include macroscopic analysis of MR fluid performance by altering the surface properties on which a MR-fluid flows. In a study to examine different surface effects on the static yield stress of different magnetic colloidal suspensions, the plate roughness and plate material of a parallel-plate rheometer are varied to show the wall effects.[103,110] Two paramagnetic plate surfaces, stainless steel and glass and one ferromagnetic plate surface, iron, are used in these experiments. It was shown that due to the smooth surface of the glass plates, there was nearly zero shear-yield stress at the wall. The stainless steel plates resulted in higher shear-yield stress, due to the roughness of the surface. The roughness of stainless steel and iron plates were identical, however, due to the wall interaction in iron plates, the highest yield stress is observed in measurements with iron plates.[103] Gorodkin et al.[111] experimentally demonstrated that the static yield stress of MR fluids was enhanced by a factor of 2.8 for radially grooved plates as compared to smooth ones. The grooves may create a magnetomechanical barrier that prevents the slip of particles from the wall surface.

It has been shown that wall roughness could significantly enhance the field-induced yield stress, possibly through mechanical entrapment of the particles, at the wall surface. Hence, in practice, improving the wall characteristics may sometimes be more important than improving the MR-fluid shear-yield stress.[84] A compression method was used by Tang et al.[112] to enhance the shear-yield stress of MR fluids by forming a thick microstructure of MR fluids. It is reported that this method could obtain a yield stress as high as 800 kPa under a moderate magnetic field, while the same MR fluid has a yield stress of 80 kPa without compression.

A study on the interaction of MR fluids with various surface topologies subject to various magnetic fields in Poiseuille flow has been conducted by a piston-driven flow-type rheometer, as shown in Figure 14.13.[113] A relation for the pressure loss across the channel in terms of applied magnetic field, surface roughness and flow rate is obtained, *without using the concept of shear yield stress*. Selected typical experimental results are shown in Figures 14.16 and 14.17. Without a magnetic field, the viscous pressure drops are almost identical in the tested range of the volumetric flow rate for all the surface samples, which suggests that the surface roughness does not affect the viscous pressure drop.

However, when the magnetic field is applied, the effect of surface roughness on the pressure drop is apparent. For a given magnetic field as the surface roughness increases, an increase in the pressure drop is observed. The increase in the pressure drop is more effective at high magnitude magnetic fields, such as for 0.4 T.

The dimensionless pressure drop across a channel can be represented by the friction factor that is defined, as follows:

$$c_\mathrm{f} = \frac{\tau_\mathrm{w}}{\frac{1}{2}\rho V_\mathrm{m}^2} \tag{14.8}$$

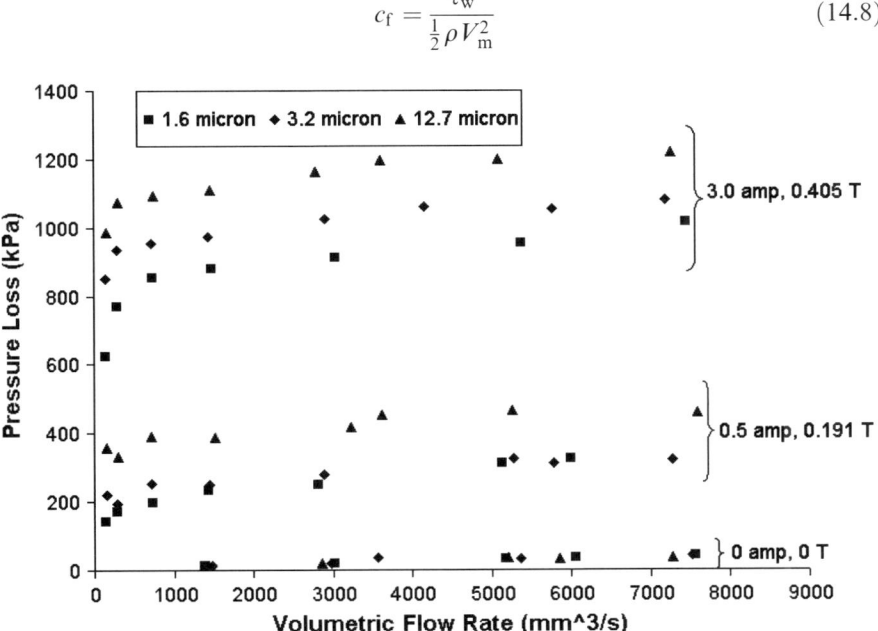

Figure 14.16 Comparison of pressure drop for 1.6, 3.2, and 12.7 μm rough surfaces for various magnetic fields as a function of volumetric flow rate.[113]

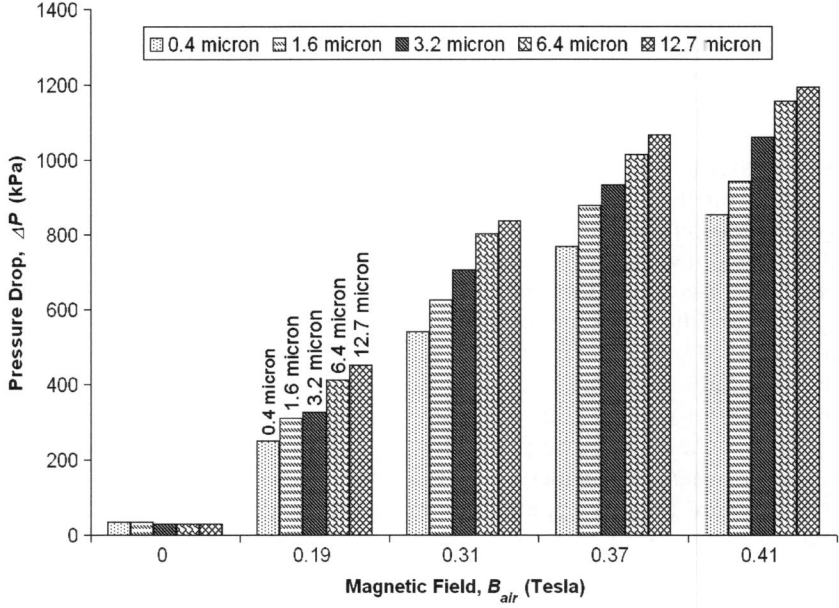

Figure 14.17 The surface effects on the pressure drop of an MR fluid subject to an applied magnetic field under certain volumetric flow rate.[113]

where ρ is the MR fluid density and V_m is the mean MR fluid velocity along the channel ($V_m = Q/A_{ch}$), Q is the volumetric flow rate, A_{ch} is the cross-sectional area of the channel and τ_w is the wall shear stress that is directly related to the pressure drop as:

$$\tau_w = \frac{g_c \Delta P}{2 l_c} \qquad (14.9)$$

where g_c is the channel gap, and l_c is the channel length. To account for the effect of surface on the magnetic field, a *modified* Mason number is defined as:

$$\mathrm{Mn}_{w,\mathrm{mod}} = \frac{8 \eta_f \dot{\gamma}_w}{(\mu_0 \mu_f)^{1-m/\lambda} \beta^2 H_{\mathrm{MR}}^{(2-m/\lambda)} B_{\mathrm{sat}}^{m/\lambda}} \qquad (14.10)$$

where η_f is the MR fluid viscosity, $\dot{\gamma}_w$ is the shear rate of MR-fluid flow at the wall, μ_0 is the vacuum permeability, μ_f is the MR-fluid relative permeability, H_{MR} is the magnetic-field strength inside the MR fluid and β is defined as $\beta = \frac{\mu_p - \mu_f}{\mu_p + 2\mu_f}$, where μ_p is the iron particle relative permeability. It has been found that the friction factor for the given MR fluid in channel-flow mode with different wall surface roughness is collapsed into a single dimensionless curve for various applied magnetic fields. This is presented in Figure 14.18.[113]

The effects of regular grooved surfaces on the friction factor of a MR fluid flowing through a channel under various magnetic fields for different

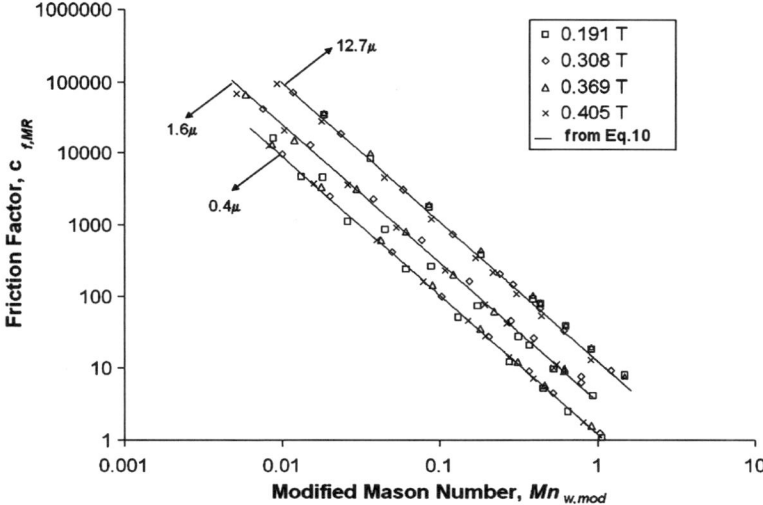

Figure 14.18 The friction factor of MR-fluid flow in a channel as a function of the modified Mason number for various surface roughnesses.[113]

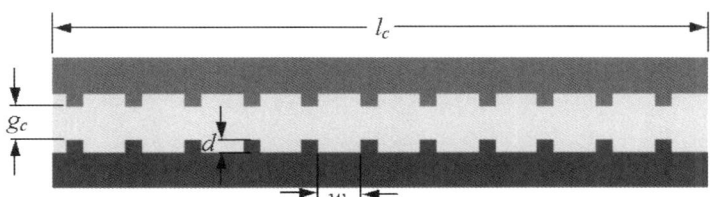

Figure 14.19 Grooved channel cross section and dimensions.[114]

volumetric flow rates are also investigated by using the same method.[114] Nine different groove configurations are tested. Figure 14.19 presents the groove cross section and geometric dimensions. The groove depth, d and width, w are varied to determine the effects of groove depth and width on the pressure drop across the channel. Channel dimensions and groove configurations are listed in Table 14.1 and 14.2, respectively.

Figure 14.20 presents the measured pressured drop for configurations G0, G1, G2, G3 and G6 under different magnetic fields. Again, the surface configurations have less effect on the flow properties in the off-state (zero applied magnetic field). However, when subjected to a magnetic field, the grooved surfaces enhance the MR effects significantly when compared to the smooth surface. The pressure drops increase with increasing depth of the grooves while keeping the width of the grooves constant (when G1, G2 and G3 are compared). The increase in the pressure drop with changing width is relatively small

Table 14.1 Dimensions of the Test Channel.

Dimension	(mm)
Channel length, l_c	14
Channel width, w_c	10
Channel gap, g_c	1

Table 14.2 Dimensions of Different Groove Configurations Tested in the Flow Type Rheometer.

	W (mm)	d (mm)
G1	0.76	0.38
G2	0.76	0.64
G3	0.76	0.76
G4	1.02	0.38
G5	1.02	0.64
G6	1.02	0.76
G7	1.27	0.38
G8	1.27	0.64
G9	1.27	0.76
G0	ungrooved surface	

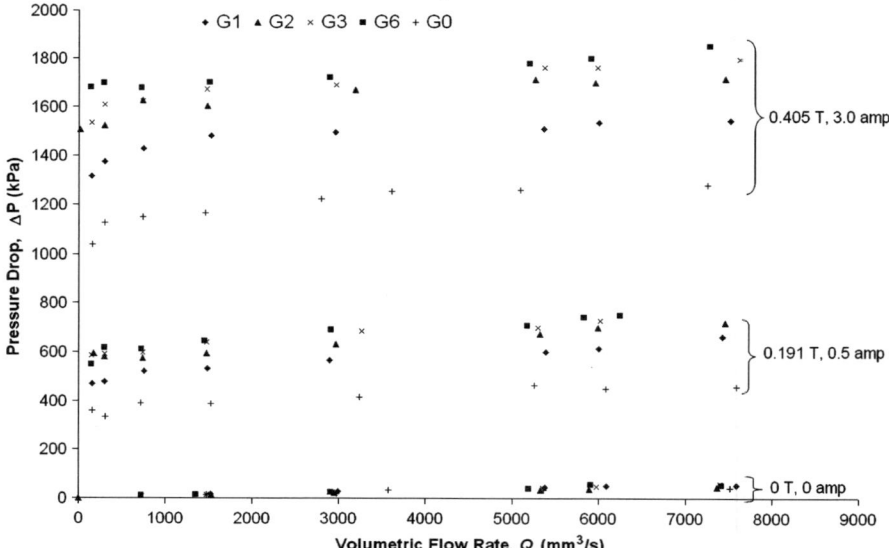

Figure 14.20 Comparison of pressure drop for grooved channel wall configurations G0, G1, G2, G3 and G6 for various magnetic fields.[114]

and negligible when compared to the groove depth effect (when G3 and G6 are compared). As compared to that of an ungrooved regular surface, the enhancement of the MR-fluid strength is between 50–100% for the materials tested. Wall-surface treatment can significantly improve the performance of a MR fluid. The mechanism behind the phenomena observed from the above experiments[114] has not been fully exploited.

14.4 Magnetorheological Fluid Devices

14.4.1 Magnetorheological Fluid Dampers

One of the most promising applications of MR fluids is in dampers/shock absorbers for adaptive shock and vibration control of mechanical systems and structures. MR-fluid dampers generate a controllable force via a change in the fluid's apparent viscosity by activating an electromagnet that energises a valving region; *i.e.* the MR fluid valve. The controllable damping force inherently depends on three basic components: 1) the properties of the MR fluid, 2) the MR-valve configuration, and 3) the design of magnetic path.

The performance of MR fluids falls into two distinct states: (1) the passive state (or off-state) where no magnetic field is present and the fluid behaves as a Newtonian fluid, and (2) the active state (or on-state) where the presence of an external magnetic field alters the rheology of the MR fluid and changes its apparent viscosity as a function of applied magnetic field. The off-state viscosity provides the base damping force to operate in a *fail-safe* mode in the absence of electrical power. In the event of a power supply or control-system failure, the MR-fluid damper can function as a typical passive damper.

A key component of a MR-fluid device is the MR-fluid valve. Generally, there are two basic types of flow geometries that are considered for MR valve design. One is the annular flow geometry where the MR fluid passes through the gap between two concentric cylinders (see Figure 14.21(a)). The other is the radial flow geometry where the gap is formed between two fixed parallel disks (see Figure 14.21(b)). The product of pressure drop across the MR valve and the effective piston area generates the damping force of the MR-fluid damper. The controllable damping force is a function of the applied magnetic flux

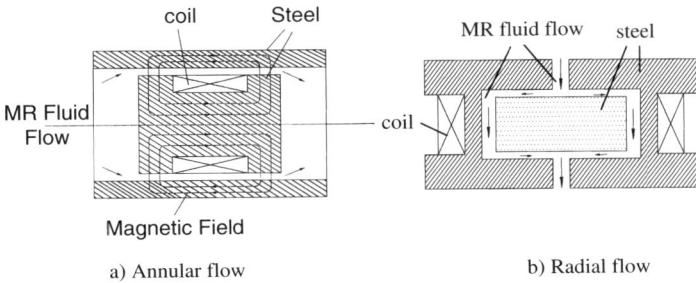

a) Annular flow b) Radial flow

Figure 14.21 Schematic flow paths of two MR-valve configurations.

density inside the MR valve. The following relation can be used to obtain the MR pressure drop induced by the applied magnetic-field strength for both annular flow and radial flow geometries:

$$\Delta P_{MR} = k\tau_y(B)\frac{L}{h} \qquad (14.11)$$

where ΔP_{MR} is the MR pressure drop or controllable pressure drop across the MR valve, k is a coefficient with values between 2 to 3. $\tau_y(B)$ is the yield stress of the MR fluid, which is a function of the applied magnetic flux density B, L is the total length of the MR valve for annular flow, or the radius of the disk-type MR valve for radial-flow geometry, h is the size of the gap.

The key issue for the design of a MR valve is the magnetic path analysis and design. The material and geometric properties of the core materials, wire gages, as well as the number of windings are some of the important parameters for an electromagnet design. The optimum design of the electromagnet for a MR valve can be performed using commercially available software that can provide two and/or three-dimensional electromagnetic finite-element analysis. Estimation of the off-state viscous damping force for the MR valve in a radial-flow configuration is difficult because the minor losses could be significant in this type of geometry.

MR-fluid dampers may be designed in either an internally valved or externally valved configuration. A MR-fluid damper is internally valved if the MR valve is located within the piston of the hydraulic cylinder. An externally valved MR-fluid damper has a bypass MR fluid valve in conjunction with a working piston–cylinder actuator. Compared to internally valved devices, the bypass-type MR-fluid damper offers the following advantages:

(1) Maintenance: The external bypass MR fluid valve has the electromagnetic circuit outside of the hydraulic cylinder that is easily accessible for maintenance.
(2) Flexibility: The bypass MR fluid valve can be fitted into different piston–cylinder configurations to obtain a different semiactive damper system with specific performance capabilities.

Efforts have been undertaken to apply this technology to automotive and aerospace applications, such as semiactive suspension systems,[115–118] and helicopter rotor-damping augmentation.[119,120] A MR-fluid-based active system known as MagneRide™ developed by Delphi Automotive Systems has been installed on the 2002 Cadillac XLS.[31,32]

In addition to land vehicles, a promising and emerging area of application for MR technology is in large structures, such as buildings and bridges.[7,33,121–131] A full-scale, MR seismic damper has been designed and built by Lord Corporation.[7] A bypass-type MR-fluid damper for semiactive control in building structures has been designed and manufactured by the Sanwa Tekki Corporation.[33] Two 30-ton MR-fluid dampers, built by the same company, were installed in the Tokyo National Museum of Emerging Science and Innovation.[121] A modular bypass seismic damper has been designed, built and tested for use in bridges and buildings at the University of Nevada, Reno.[68,122] Another full-scale implementation of MR dampers for bridge applications was achieved in China for the Dongting Lake bridge retrofitted with stay-cable dampers. A total of 312 small-size MR dampers were installed on 156 stayed cables to control cable vibration caused by wind and rain excitation.[121]

14.4.2 Modeling of Magnetorheological-Fluid Dampers

MR-fluid dampers demonstrate highly nonlinear behaviour due to the inherent non-Newtonian behaviour of MR fluids. MR controllable fluids are generally observed to exhibit a strong field-dependent shear modulus and a yield stress that resists the material's flow until shear stress reaches a critical value. Based on the Bingham or Herschel–Bulkley constitutive equation, a quasisteady, field-controllable damper model was developed to predict the damper's performance.[70–73,132–135] However, the quasisteady assumption cannot capture the highly nonlinear force–velocity hysteresis, which is essential for control analysis of MR-fluid dampers.

In order to account for the force–velocity relationship of MR-fluid dampers, several types of models have been proposed. Most of the models are based on the assumption that a MR fluid's preyield viscoelastic properties, which are not accounted for by the Bingham or the Herschel–Bulkley models, could contribute to this highly nonlinear behaviour. Therefore, more parameters associated with the preyield mechanism are added to make the model more accurate. Examples include the augmented six-parameter model proposed by Kamath and Wereley,[136] and the extended Bingham model proposed by Gamota and Filisko.[137] These models can provide a good estimate of MR-fluid damper force–velocity response, however, the model parameters can only be obtained after specific tests, and their effect on the model is difficult to assess. Parametric or phenomenological models have also been attempted to describe the behaviour of MR-fluid dampers. These models are based on the determination of system parameters by curve fitting with experimental data.[138–148]

All of the above parametric and nonparametric models can emulate the dynamic behaviour of the MR damper, however, they cannot provide any physical meanings, and thus they cannot provide valuable tools in damper design and analysis. Meanwhile, the accuracy of these models relies heavily on the experimental data.

Peel et al.[149] and Sims et al.[150,151] extended their quasisteady ER damper model to include dynamic effects accounting for the hysteretic behaviour for an ER long-stroke damper. They developed a lumped-parameter model consisting of a spring, mass, and damper connected in series. The spring stiffness and the mass are related to the fluid bulk modulus and the volume of the working fluid and its density, respectively. Although the model was derived based on the material and geometric properties, some model parameters such as stiffness, yield stress and viscosity still need to be identified using experimental data to improve the accuracy of any numerical prediction.

A dynamic model based on fluid mechanics and the Herschel–Bulkley flow analysis to predict the behaviour of MR/ER fluid dampers has been developed to address this issue.[152,153] The effect of fluid compressibility within the damper is considered by the inclusion of the effective bulk modulus in the proposed model. The fluid-dynamics-based model can accurately capture the dynamic response of a MR-fluid damper over a wide range of operating conditions. Modeling of a MR-fluid damper, based only on the physical parameters of the device as well as the properties of MR fluids, is demonstrated. The schematic of this damper is shown in Figure 14.22. The damper has a through-rod design, *i.e.* the ratio of fluid volume to rod volume in the damper is constant over a stroke. The controllable MR fluid valves are within the piston (internal valve). The mass flow rate continuity for the fluid volume is:

$$\frac{\mathrm{d}}{\mathrm{d}t}(m) = \frac{\mathrm{d}}{\mathrm{d}t}(\rho V) = \rho_{\mathrm{in}} Q_{\mathrm{in}} - \rho_{\mathrm{out}} Q_{\mathrm{out}} \qquad (14.12)$$

where ρ is the density, m is the mass, V is the volume, Q_{in} is the input flow rate, and Q_{out} is the output flow rate. Considering the compressibility of the fluid,

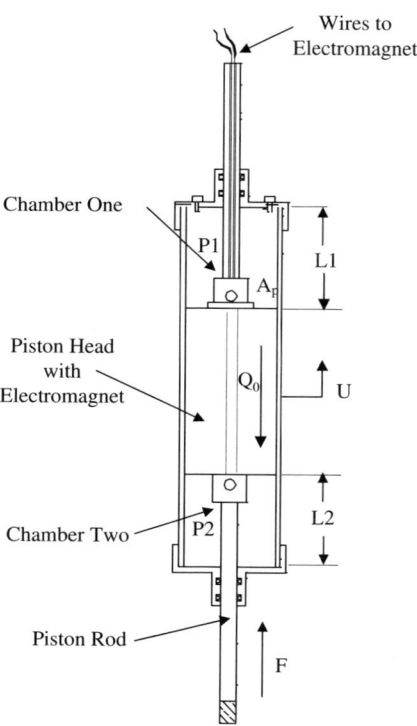

Figure 14.22 Schematic of the prototype a MR-fluid damper.[152]

one has:

$$\frac{d}{dt}(\rho V) = \rho\left(\frac{dV}{dt} + \frac{V}{\beta}\frac{dP}{dt}\right) \quad (14.13)$$

where P is the pressure, and β is the bulk modulus of the fluid. Combining eqns (14.12) and (14.13), and assuming a constant fluid density, one obtains:

$$\frac{dV}{dt} + \frac{V}{\beta}\frac{dP}{dt} = Q_{in} - Q_{out} \quad (14.14)$$

Equation (14.14) presents a mass flow rate continuity equation accounting for the fluid compressibility. For this particular MR-fluid damper Q_{in} is zero, and the flow continuity, eqn (14.13), for chambers one and two are:

$$\begin{cases} \frac{dV_1}{dt} + \frac{V_1}{\beta_1}\frac{dP_1}{dt} = -Q_{out} \\ \frac{dV_2}{dt} + \frac{V_2}{\beta_2}\frac{dP_2}{dt} = Q_{out} \end{cases} \quad (14.15)$$

where

$$V_1 = (L_1 - U)A_p \text{ and } V_2 = (L_2 + U)A_p \quad (14.16)$$

Here U is the piston displacement, A_p is the effective area of the piston, and β_1 and β_2 are the effective bulk moduli of the fluid in chambers one and two, respectively.

The bulk modulus is sensitive to the variations in temperature, pressure, volumetric ratio of fluid, and air content. At room temperature, knowing the volume of air present per unit volume of oil and assuming a perfectly rigid container, the effective bulk modulus is estimated by Stringer,[154] as follows:

$$\frac{1}{\beta} = \frac{1}{\beta_0} + \frac{V_{air}}{V_o}\frac{1}{P} \quad (14.17)$$

where β_0 is the bulk modulus of a pure oil and is about $17 \times 10^8 \text{ N/m}^2$, $\frac{V_{air}}{V_o}$ is the ratio of a volume of air dispersed in a volume of oil (from 0.01% to 1%), and P is the mean value pressure. Here, the chambers mean value pressure is $5.0 \times 10^5 \text{ N/m}^2$. If, for example, 1% of the air is entrapped in the fluid, the effective bulk modulus is $4.86 \times 10^7 \text{ N/m}^2$, while the effective bulk modulus is $1.64 \times 10^9 \text{ N/m}^2$ at 0.01% of air content. Therefore, it is evident that the air content in the fluid can drastically reduce the effective bulk modulus. In practical applications, a small percentage of air is always present in the system, and the effectiveness of the design can significantly be affected, if the design process neglects the variation of compressibility.

It is reasonable to assume that the effective bulk modulus β_1 and β_2 are the same constant. Assuming $\beta_1 = \beta_2 = \beta$, and letting $\Delta P = P_1 - P_2$, eqns (14.15) and (14.16) can be rewritten as:

$$\frac{d\Delta P}{dt} = \beta(A_p \dot{U} - Q_{out})\left[\frac{1}{(L_1 - U)A_p} + \frac{1}{(L_2 + U)A_p}\right] \quad (14.18)$$

where $\dot{U}$ is the piston velocity. Equation (14.18) can be solved if Q_{out} is determined as a function of pressures P_1 and P_2. In this analysis, mass flow rate continuity with the incompressible fluid assumption is considered for the flow of a MR fluid through the channel. Thus, using the Herschel–Bulkley flow analysis, the volume flow rate through MR valves can be expressed as the function of pressure drop, $\Delta P = P_1 - P_2$:

$$Q_m = 0; \quad \left(\tau_y \geq \frac{p'R}{2}\right)$$

$$Q_m = \frac{\left(\frac{p'R}{2} - \tau_y\right)^{\frac{n+1}{n}} \pi R^3}{\left(\frac{p'R}{2}\right)^3 k^{\frac{1}{n}}} \left[\frac{\left(\frac{p'R}{2} - \tau_y\right)^2}{\frac{3n+1}{n}} + \frac{2\tau_y\left(\frac{p'R}{2} - \tau_y\right)}{\frac{2n+1}{n}} + \frac{\tau_y^2}{\frac{n+1}{n}}\right]; \left(\tau_y < \frac{p'R}{2}\right)$$

$$(14.19)$$

where $p' = \frac{P_1 - P_2}{2L_{mrf}}$, and $R = \frac{D_1}{2}$. Assuming incompressible fluid in the piston channel, the total volume flow rate is:

$$Q_{out} = NQ_m \quad (14.20)$$

where N is the number of MR fluid valves. Using eqns (14.18)–(14.20), the pressure drop ΔP can be obtained. The pressure drop ΔP_{vis} generates the viscous damping force. Therefore, the total pressure drop across the piston can be expressed as:

$$F = A_p(\Delta P_{MR} + \Delta P_{vis}) + F_f \text{sgn}(\dot{U}) \quad (14.21)$$

where F_f is the damper's seal friction force. For a Newtonian–Poiseuille flow in a circular channel, the pressure drop ΔP_{vis} is:

$$\Delta P_{vis} = \frac{128\mu Q L_{vis}}{\pi D_2^4} \quad (14.22)$$

where $Q = A_p \dot{U}$.

Figures 14.23 and 14.24 present experimental and theoretical results for force–displacement and force–velocity hysteresis loops with four different electric current inputs of 0.0, 1.0, 1.5, and 2.0 A. The value of bulk modus β in the theoretical analysis is $5.5 \times 10^7 \text{ N/m}^2$ assuming that there is 1.0% of air content in the MR fluid. It can be seen that the high nonlinearity in force–velocity hysteresis loops is mainly due to the compressibility of MR fluids. The proposed fluid-mechanics-based model can provide a simple and direct method to capture this nonlinear behaviour, if the value of the fluid's bulk modulus is known. Both the experimental and theoretical results indicate that in addition to dissipating, MR-fluid dampers store energy. This characteristic may be important in the semiactive control design of these dampers.

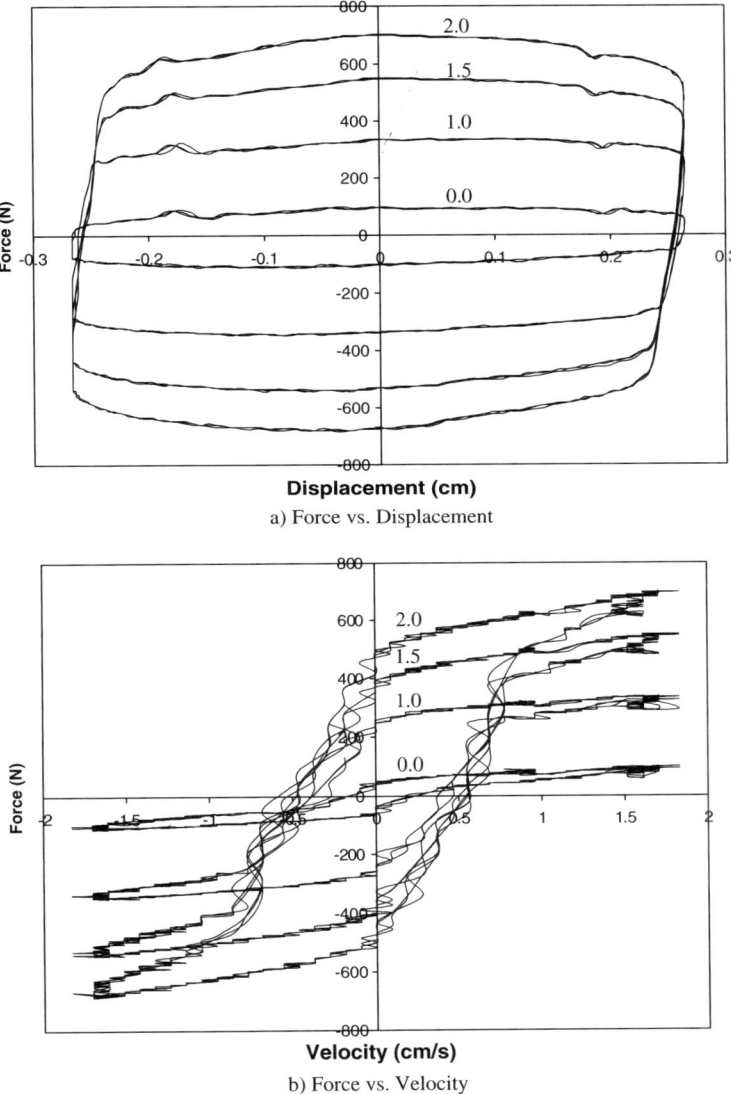

Figure 14.23 Experimental hysteresis loops for UNR MR damper subjected to 0.0 A, 1.0 A, 1.5 A and 2.0 A input currents and harmonic motion at 1.0 Hz and 0.267 cm amplitude.[153]

14.4.3 Effect of Temperature

A MR-fluid shock absorber that emulates the original equipment manufacture (OEM) shock absorber behaviour in its passive mode (off-state) is referred to as a *fail-safe* damper.[115,127,155–162] This means that in the event of a failure in

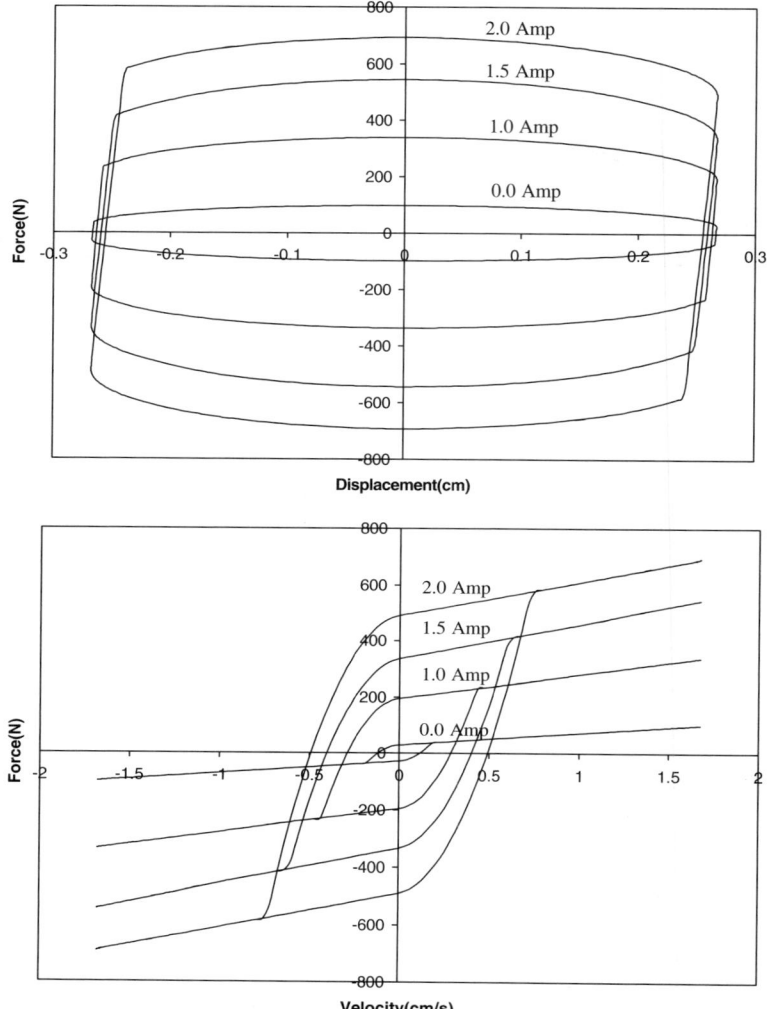

Figure 14.24 Theoretical hysteresis loops for UNR MR damper subjected to 0.0 A, 1.0 A, 1.5 A and 2.0 A input currents and harmonic motion at 1.0 Hz and 0.267 cm amplitude for $\beta = 5.5 \times 10^7 \, \text{N/m}^2$.[153]

magnetic circuit, or control system, or electronics the MR-fluid damper performs as a passive OEM damper; thus a fail-safe MR-fluid damper in its off-state behaves as a viscous damper.

The primary purpose of a fluid viscous damper is the dissipation of energy. The dissipated energy remaining within the mechanical system not only causes unwanted motion, ride discomfort, wear, or ultimately part failure, but also leads to a temperature increase and therefore a force capacity decrease of the system due to a reduction in fluid viscosity. For the case of MR-fluid dampers, this decrease in viscosity (or damping force) may be compensated by an increase in the magnetic filed via electric current input, or an extended surface area of the device that enhances heat transfer from the system.

Limited studies have dealt with heat transfer and temperature effects in field-controllable fluid dampers. Breese and coworkers[163–165] demonstrated constitute-law-based theoretical predictions as well as experimental results on temperature increase of MR-fluid dampers, in which they obtained good agreements between theoretical and experimental results for various size MR fluid devices. Dogruoz et al.[166–168] developed a lumped-system-based theoretical model on determining the temperature increase of automotive size MR-fluid shock absorbers and showed that the reduction in the damping capacity was attenuated by increasing the surface area of the devices. They explored the heat transfer from automotive-size dampers with and without fins experimentally, and compared the experimental results with those of a previously developed model that could predict the temperature changes in any size MR-fluid damper based on the lumped-system approach. Both the temperature and peak force time histories for the unfinned and finned designs were discussed and compared.

The nonlinear constitutive law for the MR-fluid dampers is assumed to be:[163]

$$F(t) = C \left| \frac{dX(t)}{dt} \right|^\alpha \text{sgn}\left(\frac{dX(t)}{dt}\right) \qquad (14.23)$$

where C is the damping coefficient and α is a fractional exponent that accounts for the nonlinearity inherent in a MR-fluid damper. It should be noted that both C and α are functions of input electric current, I, and temperature, Θ. Generally, the range of α is $0 \leq \alpha \leq 1.5$. A value of one represents purely viscous behaviour and a value of zero corresponds to purely rigid plastic behaviour.

Because this is a lumped-system analysis, this quantity represents a sum of all internal energies of the material contained within the damper. The Biot number analysis shows that the temperature gradient within the damper is much less than 5% difference across any distance within the damper, therefore, the rate of change of temperature is assumed to be identical for all the materials within the control volume. The heat balance can be expressed as:

$$\frac{dU(t)}{dt} = \frac{d\Theta(t)}{dt} \sum m\hat{c}_p \qquad (14.24)$$

where $d\Theta(t)/dT$ is the rate of change of the temperature of the system and $\Sigma m\hat{c}_p$ is the summation of the internal energies of the materials contained within the system. For a MR-fluid damper eqn (14.24) becomes,[166]

$$\dot{\Theta}(t) + \lambda(\Theta(t) - \Theta_0) = \frac{C\bar{X}\omega\cos(\omega t)}{\Sigma m\hat{c}_p} |\bar{X}\omega\cos(\omega t)|^\alpha \text{sgn}\left(\frac{dX(t)}{dt}\right) + \eta \qquad (14.25)$$

where

$$\lambda = \frac{hA_S}{m\hat{c}_p} \quad (14.26)$$

and

$$\eta = \frac{I^2 R}{\Sigma m \hat{C}_p} \quad (14.27)$$

Equation (14.25) represents the differential energy balance for the entire lumped system. Due to the nonlinear nature of eqn (14.25), it can only be solved numerically. An exact solution is available if α is either zero or one. For the case of a MR-fluid damper α is a noninteger, therefore, this requires that an approximate solution be developed. For a working period of 400 s, application of a 2-A current causes a temperature increase of 50 °C and 34 °C with the peak force decrease of 38% and 30% compared to that of the room temperature on the unfinned and finned dampers, respectively (Figures 14.25 and 14.26).

The MR-fluid damper's viscous force changes significantly with temperature. For applications such as vehicle-suspension systems the shock absorber is required to operate in a wide range of temperature. Thus, the temperature effects on the performance of MR-fluid damper should be considered in the

Figure 14.25 Finned (left) and unfinned (right) prototype MR-fluid dampers.[166]

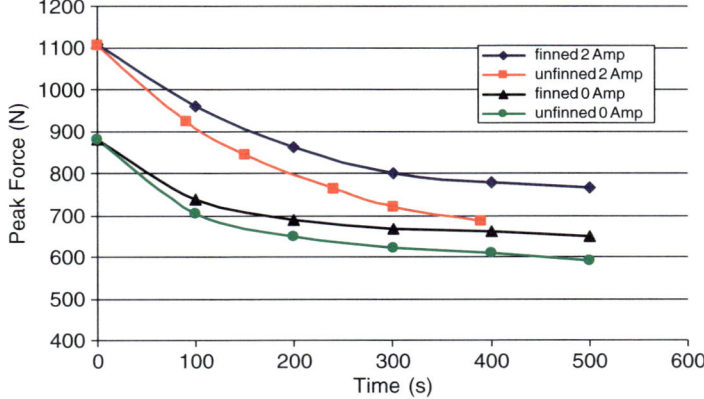

Figure 14.26 Peak force variation of unfinned and finned MRF dampers at a peak velocity of 24π cm/s and currents of 0 and 2 A.[166]

control system design. Liu et al.[169–171] have discussed the MR fluid control system with and without temperature-compensated control. Displacement and acceleration transmissibility and response of the vehicle sprung mass for the quarter car model have been explored at the operating temperature range of a MR-fluid damper. Simulation results under harmonic and off-road excitation have demonstrated that the compensated skyhook control system improves suspension's performance in reducing the sprung-mass displacement and acceleration compared to the uncompensated skyhook control system.[169]

14.4.4 Other Applications

Another application for MR fluids is in MR clutches and rotary brakes for torque transfer. An advantage of MR fluid clutches over conventional (passive) clutches is the continuously variable torque-transferring characteristics. The configurations of most MR clutch designs are usually in two categories: parallel disks and concentric cylinders.[18] The former can be designed with multiplates, which could provide more contact area with the working medium and thus more torque capability in a given volume.

Many MR devices have been developed, for example see Refs. 172–174. A rotary MR brake consisting of two rotors, with a thin gap of MR fluid between them is shown in Figure 14.27.[172] By activating the electromagnet in the device, the rotary motion of the input rotor can be controlled. General Motors,[175,176] Venture Gear Inc.,[177] and Vibratech Inc.[178] have each been issued patents describing controllable rotary devices using MR fluids. Design analysis and experimental evaluation of a MR fluid clutch has also been carried out by other researchers; for example, Lee et al.;[179] and Lampe and Grundmann.[180]

High-force multiplate MR fluid clutches have been designed and developed for limited slip differential (LSD) of land vehicles.[181] Figure 14.28 shows a

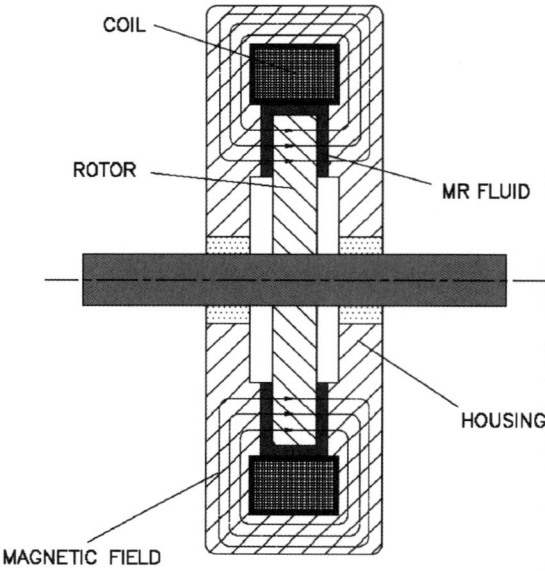

Figure 14.27 Cutaway view of a MR rotary brake.[172]

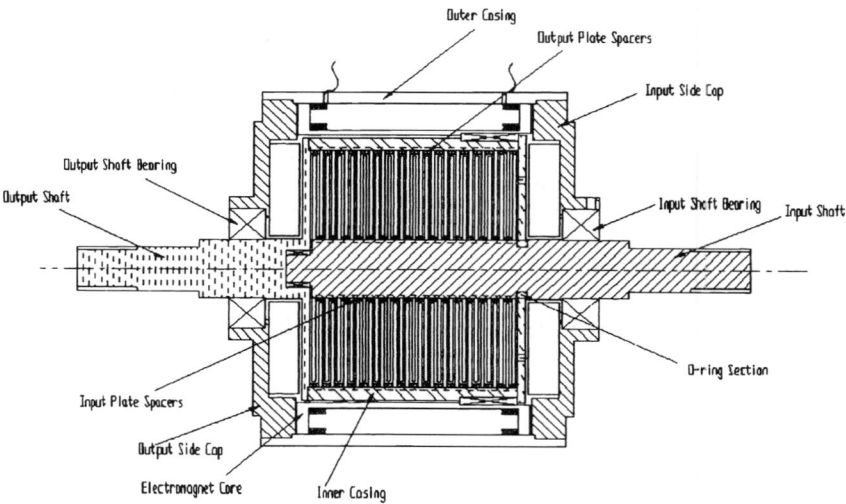

Figure 14.28 Cross-sectional view of multiplate MR fluid limited-slip differential clutch.[181]

cross-sectional view of this MR fluid LSD clutch. The MR fluid LSD clutch consists of 43 plates with 0.5 mm gap between each plate. The MR fluid fills in the gap between the input and output plates. The magnetic circuit of this clutch consists of an electromagnetic coil, which is wound around an electromagnetic core. The MR fluid clutch is activated by a power supply connected to two ends of the electromagnet. The total length of the clutch is 152.4 mm. The 43-plate clutch pack is located inside a 114.3-mm outer diameter inner casting. The torque output of the clutch is shown in Figure 14.29 for different rotational speeds of 30–120 rpm. It is evident that the MR fluid clutch can provide transmitted torque as high as 244 N m for an input current of 3 A, and the output torques almost linearly increase with the input currents. The operating speed of the clutch (for these speeds) does not affect the torque performance.

Since the flow pressure of MR fluid could be controlled by applying a magnetic field, the concept of exploiting MR technology in servovalve application for actuator systems is also proposed.[17,182] The conventional servovalves employ mechanical moving parts such as spools or solenoid to control the pressure and direction of fluid flow, which make them complex, and expensive. It is possible to develop a new simpler servovalve with no need for mechanically driven parts by using MR fluid as the working medium.

MR fluids may also be used in manufacturing and process applications.[183–188] An application is in precision polishing. In this technique, MR fluids function as abrasives, where the material removal over the portion of the workpiece surface is determined by the magnetically controlled hydrodynamic flow of a MR polishing fluid. An advantage of MR fluids over existing technologies is that the polishing tool does not wear, since the recirculated fluid is continuously monitored and maintained. The resulting debris is continuously removed and the heat due to friction is dissipated. The technique requires no dedicated tooling or special setup.[183] Rong et al.[188] proposed an application of MR fluids for flexible fixturing in manufacturing. With the application of a compression technique, very high shear strength of the MR fluid is achieved under a moderate magnetic field, hence the workpiece can be held firmly inside MR fluids when the material is changed into the solid state.

Other possible applications of MR technology are in medical applications, such as the embolisation of blood vessels in cancer treatment,[189,190] real-time controlled damper for human artificial joints,[191,192] exercise equipment for the

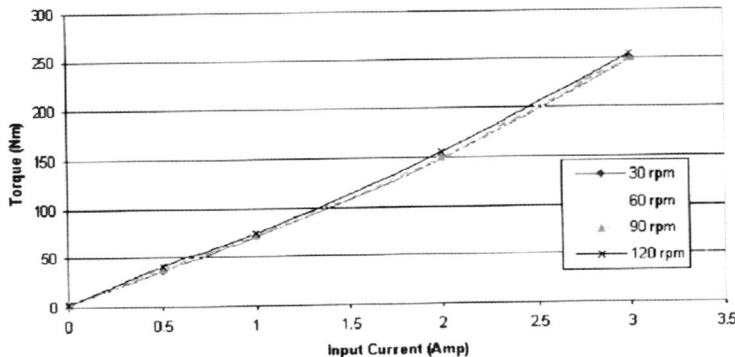

Figure 14.29 Experimental static torque values for different velocities.[181]

rehabilitation of muscles in physical therapy,[174] and haptic display to replicate perceived biological tissue compliance.[193]

14.5 Summary

The ability of MR materials to alter their rheological properties in response to an applied magnetic field has been realised in a variety of applications. The research and development in the areas of science and engineering of magnetorheological (MR) materials and devices have been significantly accelerated in the last 15 years. This chapter presents an overview of some of the key components of recent efforts in this area. The MR effect; compositions of MR materials; measurable macroscopic properties of MR fluids and their relationship with their microstructures; and modeling and performance of certain MR fluid devices, are reviewed. An outline of other MR applications such as clutches, precision finishing and physical therapy devices is also presented.

Although considerable progress has been made in the commercialisation of some MR devices over the last decade, a thorough understanding of the technology underlying both materials and devices are still needed in many practicable designs.

Acknowledgments

The assistance of Dr. Alan Fuchs in providing some of the figures for this manuscript is warmly appreciated.

References

1. A.V. Srinivasan and D.M. Mcfarland, *Smart Structures Analysis and Design*, Chapter 4, Cambridge University Press, UK, 2001, 73–95.
2. P.P. Phule and J.M. Ginder, *MRS Bull.*, 1998, 19–21.
3. K. Minagawa and K. Koyama, *Curr. Org. Chem.*, 2005, **9, No. 16**, 1643–1663.
4. W.I. Kordonsky, *J. Intell. Mater. Syst. Struct.*, 1993, 4, No. 1, p. 65.
5. J.D. Carlson, K.D. Weiss, *Machine Des. Aug.*, **8**, 1994, 61–66.
6. J.D. Carlson, D.N. Catanzarite and K. St. Clair, *Proceedings 5th Int. Conf. on ER Fluids, MR Suspensions and Associated Technology*, W. Bullough, ed., World Scientific, Singapore, 1996, pp. 20–28.
7. J.D. Carlson and B.F. Spencer, *Proceedings of the 3rd International Conference on Motion and Vibration Control*, Chiba, Japan, III, 1996, pp. 35–40.
8. H. Block and J.P. Kelly, *J. Phys. D. Appl. Phys.*, 1988, **21**, 1661–1677.
9. T.C. Jordan and M.T. Shaw, *IEEE Trans. Elect Insul.*, 1989, **24**, 849–879.
10. A.P. Gast and C.F. Zukoski, *Adv. Colloid Interf. Sci.*, 1989, 153–202.

11. C.F. Zukoski, *Ann. Rev. Mater. Sci.*, 1993, **23**, 45–78.
12. M. Parthasarathy and D.J. Klingenberg, *Mater. Sci. Eng. R*, 1996, **17**, 57–103.
13. H. See, *App. Rheol.*, 2001, **11, No. 2**, 70–82.
14. P.J. Rankin, J.M. Ginder and D.J. Klingenberg, *Curr. Opin. Colloid Interf. Sci.*, 1998, **3, No. 4**, 373–381.
15. T. Hao, *Adv. Mater.*, 2001, **13, No. 24**, 1847.
16. T. Hao, *Adv. Colloid Interf. Sci.*, 2002, **97**(1–3), 1–35.
17. O. Ashour, C.A. Rogers and W. Kordonsky, *J. Intell. Mater. Struct.*, 1996, **7**, 123–130.
18. J.M. Ginder, *Encycl. Appl. Phys.*, 1996, **16**, 487–503.
19. P.P. Phule, *Smart Mater. Bull.*, 2001, 7–10.
20. D.J. Kingenberg, *AICHE J.*, 2001, **47, No. 2**, 246–249.
21. G. Bossis, O. Volkova, S. Lacis and A. Meunier, *Ferrofluids. Magnetically Controllable Fluids and their Applications*, Springer-Verlag, Berlin, Germany, 2002, pp. 202–30.
22. J. Rabinow, US Patent Number 2575360, 1951.
23. J. Rabinow, *AIEE Trans.*, 1948, **67**, 1308–1315.
24. J.D. Carlson, *Proceedings 8th Int. Conf. on ER Fluids, MR Suspensions and Associated Technology*, G. Bossis, ed., World Scientific, Singapore, 2002, pp. 63–69.
25. W.M. Winslow, US Patent Number 2,417,850, 1947.
26. W.M. Winslow, *J. Appl. Phys.*, 1949, **20**, 1137–1140.
27. W.I. Kordonsky, *J. Magn. Magn. Mater.*, 1993, **122, No. 1–3**, p. 395.
28. K.D. Weiss, T.G. Duclos, J.D. Carlson, M.J. Chrazn and A.J. Margalia, *SAE Technical Paper Series # 932451*, presented at the 1993 *International Off-Highway & Power plant Congress and Exposition, Milwaukee, Wisconsin*, 1993.
29. K.D. Weiss and T.G. Duclos, *Proc. 4th International Conference on ER Fluids*, ed. R. Tao and G.D. Roy, World Scientific, Singapore, 1994, pp. 43–59.
30. M.R. Jolly, J.W. Bender and J.D. Carlson, *SPIE 5th Annual Int. Symposium on Smart Structures and Materials*, 1998, **3327**, pp. 262–274.
31. http://delphi.com/manufacturers/auto/chsteer/ride/magneride/.
32. J.D. Carlson, *Int. J. Mod. Phys. B*, 2005, **19, No. 7–9**, 1463–1470.
33. H. Fujitani, H. Sodeyama, T. Tomura, T. Hiwatashi, Y. Shiozaki, K. Hata, S. Sunakoda, S. Morishita and S. Soda, *Proceedings of SPIE Conference on Smart Materials and Structures, Damping and Isolation*, ed. G.S. Agnes and K-W. Wang, 2003, **5052**, pp. 265–276.
34. A. Fuchs, M. Xin, F. Gordaninejad, X. Wang. G. Hitchcock, H. Gecol, C. Evrensel and G. Korol, *J. Appl. Polym. Sci.*, 2004, **92, No. 2**, 1176–1182.
35. P.P. Phule and A.D. Jatkar, *Proceedings 6th Int. Conf. on ER Fluids MR Suspensions and Associated Technology*, M. Nakano and K. Koyama, ed. World Scientific, Singapore, 1998, pp. 502–510.
36. P.P. Phule and J.M. Ginder, *Int. J. Mod. Phys. B*, 1999, **13, No. 14–16**, 2019–2027.

37. A.J. Margida, K.D. Weiss and J.D. Carlson, *Proceedings of the 5th International Conference on Electro-Rheological Fluids, Magneto-Rheological Suspensions and Associated Technology*, Sheffield, UK, 1995, pp. 544–550.
38. R.Z. Gong, Z.K. Feng, J.G. Guan and R.Z. Yuan, *J. Mater. Sci. Technol.*, 2001, **17, No. 5**, 511–516.
39. V.I. Kordonsky and S.A. Demchuk, *J. Intell. Mater. Struct.*, 1996, **7**, 522–525.
40. P.P. Phule, *MRS Bull.*, 1998, 23–25.
41. E. Lemaire, A. Meunier and G. Bossis *J. Rheol.*, 1995, **39, No. 5**, 1011–1020.
42. B.J. de Gans, N.J. Duin, D. van den Ende and J. Mellema, *J. Chem. Phys.*, 2000, **113, No. 5**, 2032–2042.
43. C, Kormann, H.M. Laun and H.J. Richter, *Int. J. Mod. Phys. B*, 1996, **10, No. 23–24**, 3167–3172.
44. R.T. Foister, US Patent 5667715, 1997.
45. A. Bombard, M. Alcantara, Knobel and P. Volpe *Int. J. Mod. Phys. B* 2005, **19, 7–9**, 1332–1338.
46. A.M. Trendler and H. Bose, *Int. J. Mod. Phys. B*, 2005, **19, No. 7–9**, 1416–1422.
47. D. Kittipoomwong, D.J. Klingenberg and J.C. Ulicny, *J. Rheol.*, 2005, **49, No. 6**, 1521–1538.
48. P.J. Rankin, A.T. Horvath and D.J. Klingenberg, *Rheol. Acta*, 1999, **5**, 471–477.
49. J.H. Park, M.H. Kwon and O.O. Park, *Kor. J. Chem. Eng.*, 2001, **18, No. 5**, 580–585.
50. J.M. Ginder, D. Larry, Y. Elie and C.D. Lloyd, US Patent 5549837, 1996.
51. A. Fuchs, F. Gordaninejad, D. Blattman and G.H. Hamann, US Patent 6527972, 2003.
52. A. Fuchs, B. Hu, F. Gordaninejad and C. Evrensel, *J. Appl. Polym. Sci.*, 2005, **98, No. 6**, 2402–2413.
53. M.T. Lopez-Lopez, J. de Vicente and F. Gonzalez-Caballero, *Colloid Surf. A-Physicochem. Eng. Aspects*, 2005, **264, No. 1–3**, 75–81.
54. Z.Y. Chen, X. Tang, G.C. Zhang, Y. Yin, W. Ni and Y.R. Zhu, *Proceedings 6th Int. Conf. on ER Fluids MR Suspensions and Associated Technology*, M. Nakano and K. Koyama, ed. World Scientific, Singapore, 1998, pp. 486–493.
55. I.B. Jang, H.B. Kim, J.Y. Lee, J.L. You, H.J. Choi and M.S. Jhon, *J. Appl. Phys.*, 2005, **97, No. 10**, Art. No. 10Q912.
56. J.M. Ginder, M.E. Nicholes, L.D. Elie and J.L. Tardiff, *Proceedings of the SPIE-The International Society for Optical Engineering*, 1999, **3675**, pp 131–138.
57. M.R. Jolly, J.D. Carlson and B.C. Munoz, *Smart Mater. Struct.*, 1996, **5, No. 5**, 607–614.
58. M.R. Jolly, J.D. Carlson, B.C. Munoz and T.A. Bullions, *J. Intell. Mater. Syst., Struct.*, 1996, **7**, 613–622.

59. L.C. Davis, *J. Appl. Phys.*, 1999, **85**, 3348–3351.
60. S.A. Demchuk and V.A. Kuzmin, *J. Eng. Phys. Thermophys.*, 2002, **75**, No. 2, 396–400.
61. G.Y. Zhou, *Smart Mater. Struct.*, 2003, **12, No. 1**, 139–146.
62. J.M. Ginder, M. E. Nicholes and S.M. Clark, *Proceedings of the SPIE-The International Society for Optical Engineering*, 2000, **3985**, 418–425.
63. M. Lokander and B. Stenberg, *Polym. Test.*, 2003, **22**, 245–251.
64. M. Lokander and B. Stenberg, *Polym. Test.*, 2003, **22**, 677–680.
65. J.D. Carlson and M.R. Jolly, *Mechatronics*, 2000, **10, No. 4–5**, 555–569.
66. J.M. Ginder, W.F. Schlotter and M.E. Nicholes, *Proceedings of the SPIE-The International Society for Optical Engineering*, 2001, **4331**, 103–110.
67. J.M. Ginder, MRS Bull., *August*, 1998, 26–29.
68. X. Wang, F. Gordaninejad, K.K. Bangrakulur, G. Hitchcock, A. Fuchs, J. Elkins, C.A. Evrensel, S. Ruan, M. Siino and M. Kerns, *Damping and Isolation, Proc. SPIE Conference on Smart Materials and Structures*, ed. K.-W. Wang, 2004, **5386**, 226–237.
69. D.W. Felt, M. Hagenbuchle, J. Liu and J. Richard, *J. Int. Mater. Sys. Struct.* 1996, **7, No. 5**, 589–593.
70. T.M. Marksmeier, E. Wang, F. Gordaninejad and A. Stipanovic, *J. Intell. Mater. Syst. Struct.*, 1998, **9, No. 9**, 693–703.
71. X. Wang and F. Gordaninejad, *J. Intell. Mater. Syst. Struct.*, 1999, **10, No. 8**, 601–608.
72. X. Wang and F. Gordaninejad, *Proc. SPIE Smart Structure and Materials Conference, Newport Beach, California*, 2000, **3989**, 232–243.
73. D.Y. Lee and N.M. Wereley, *Proceedings of the 7th International Conference on ER Fluids and MR Suspensions*, R. Tao, ed. World Scientific, Singapore, 2000, pp. 579–586.
74. W.H. Li, "Rheology of MR fluids and MR damper dynamic response: experimental and modeling approaches", Ph.D. Dissertation, School of Mechanical and Production Engineering, the Nanyang Technological University, Singapore, 2000.
75. K.D. Weiss, J.D. Carlson and D.A. Nixon, *J. Intell. Mater. Syst. Struct.* 1994, p. 772.
76. M. Nakano, H. Yamamoto and M.R. Jolly, *Int. J. Mod. Phys. B*, 1999, **13, No. 14–16**, 2068–2076.
77. J. Claracq, J. Sarrazin, and J.P. Montfort, *Rheol. Acta*, 2004, **43, No. 1**, 38–49.
78. D.J. Klingenberg, *J. Rheol*, 1992, **37, No. 2**, 199–214.
79. W. Kordonski, S. Gorodkin and N. Zhuravski, *Int. J. Mod. Phys. B*, 2001, **15, No. 6–7**, 1078–1084.
80. X. Wang and F. Gordaninejad, *Rheol. Acta*, 2006, **45, No. 6**, 899.
81. O. Volkova, G. Bossis, M. Guyot, M. Bashtovoi and A. Reks, *J. Rheol.*, 2000, **44, No. 1**, 91–104.
82. J.M. Ginder and L.C. Davis, *Appl. Phys. Lett.*, 1994, **65, No. 26**, 3410–3412.
83. J.M. Ginder, L.C. Davis and L.D. Elie, *Int. J. Mod. Phys. B*, 1996, **10, No. 23–24**, 3293–3303.

84. X. Tang and H. Conrad, *J. Rheol.*, 1995, **40**, 1167–1178.
85. J.H. Park, B.D. Chin and O.O. Park, *J. Colloid Interf. Sci.*, 2001, **240**, 349–454.
86. X.L. Tang and H. Conrad, *J Phys. D. Appl. Phys.* 2000, **33, No. 23**, 3026–3032.
87. R.E. Rosensweig, *J. Rheol.*, 1995, **39**, 179.
88. Z.P. Shulman, V.I. Kordonsky, E.A. Zaltsgendler, I.V. Prokhorov, B.M. Khusid and S.A. Demchuk, *Int. J. Multiphase Flow*, 1986, **12, No. 6**, 935–955.
89. G. Bossis, E. Lemaire, O. Volkova and H. Clercx, *J. Rheol*, 1997, **41, No. 3**, 687–704.
90. M.R. Jolly, J.W. Bender and R.T. Mathers, *Int. J. Mod. Phys. B*, 1999, **13, No. 14–16**, 2036–2043.
91. E. Lemaire, Y. Grasselli and G. Bossis, *J. Phys.* II, 1992, **2, No. 3**, 359–369.
92. Y. Grasselli, G. Bossis and E. Lemaire, *J. Phys., II*, 1994, **4, No. 2**, 253–263.
93. J. Promislow and A.P. Gast, *Langmuir*, 1996, **12, No. 17**, 4095–4102.
94. G.A. Flores, J. Liu, M. Mohebi and N. Jamasbi, *Int. J. Mod. Phys. B*, 1999, **13, No. 14–16**, 2093–2100.
95. S. Cutillas and J. Liu *Int. J. Mod. Phys. B*, 2001, **15, No. 6–7**, 803–810.
96. S. Melle, M.A. Rubio and G.G. Fuller, *Int. J. Mod. Phys. B*, 2001, **15, 6–7**, 758–766.
97. Y. Nahmad-Molinari, C.A. Arancibia-Bulnes and J. C. Ruiz-Suarez, *Phys. Rev. Lett.*, 1999, **82**, 727–730.
98. F. Donado, J.L. Carrillo and M.E. Mendoza, *J. Phys. Condens. Matter*, 2002, **14, No. 9**, 2153–2157.
99. W.H. Li, H. Du, G. Chen and S.H. Yeo, *Mater. Sci. Eng. A-Struct.* 2002, **333, No. 1–2**, 368–376.
100. K. Wollny, J. Lauger and S. Huck, *Appl. Rheol.*, 2002, **12**, 125–131.
101. S. Cutillas, G. Bossis and A. Cebers, *Phys. Rev. E*, 1998, **57, No. 1**, 804–811.
102. H.M. Laun, C. Kormann and N. Willenbacher, *Rheol. Acta*, 1996, **35, No. 5**, 417–432.
103. E. Lemaire and G. Bossis, *J. Phys. D. Appl. Phys.* 1991, **24, No. 8**, 1473–1477.
104. S. Genc and P.P. Phule, *Smart Mater. Struct.*, 2002, **11, No. 1**, 140–146.
105. B.J. de Gans, C. Blom, J. Mellema and A.P. Philips, *Proceedings 6th Int. Conf. on ER Fluids MR Suspensions and Associated Technology*, M. Nakano and K. Koyama, ed. World Scientific, Singapore, 1998, pp. 462–469.
106. Y. Zhu, M. McNeary, N. Breslin and J. Liu, *Int. J. Mod. Phys. B*, 1999, **13, No. 14–16**, 2044–2051.
107. A.B. Shorey, W.I. Kordonski, S.R. Gorodkin, S.D. Jacobs, R.F. Gans, K.M. Kwong and C.H. Farny, *Rev. Sci. Instrum.*, 1999, **70, No. 11**, 4200–4206.

108. M.M. Maiorov, *Magnitnaya Gidrodinamika*, 1980, **16, No. 4**, 11–18.
109. A.A. Zaman, NSF Engineering Resource Center for Particle Science & Technology, University of Florida, Gainesville, Florida, 1998.
110. G. Bossis, P. Khuzir, S. Lacis and O. Volkova, *J. Magn. Magn. Mater.*, 2003, **258**, 456–458.
111. S. Gorodkin, N. Zhuravski and W. Kordonski, *Int. J. Mod. Phys. B*, 2002, **16, No. 17–18**, 2745–2750.
112. X. Tang, X. Zhang and R. Tao, *Int. J. Mod. Phys. B*, 2001, **15, No. 6–7**, 549–556.
113. B.M. Kavicoglu, X. Wang, F. Gordaninejad and G. Hitchcock, *Proceedings of the ASME Aerospace Division: Adaptive Materials and Systems, Aerospace Materials and Structures*, 2004, **69**, pp. 83–88.
114. F. Gordaninejad, B.M. Kavlicoglu and X. Wang, *Int. J. Mod. Phys. B*, 2005, **19, No. 7–9**, 1297–1303.
115. F. Gordaninejad and S.P. Kelso, *J. Intell. Mater. Syst. Struct.*, 2000, **11, No. 5**, 395–406.
116. E.O. Ericksen and F. Gordaninejad, *Int. J. Veh. Des.*, 2003, **33, Nos. 1–3**, 138–152.
117. D. Simon and M. Ahmadian, *J. Vibrat. Acous.-Trans. ASME*, 2001, **123, No. 3**, 365–375.
118. H.S. Lee and S.B. Choi, *J. Intell. Mater. Syst. Struct.*, 2000, **11, No. 1**, 80–87.
119. F. Gandhi, K.W. Wang and L.B. Xia, *Smart Mater. Struct.*, 2001, **10, No. 1**, 96–103.
120. S. Marathe, F. Gandhi and K.W. Wang, *J. Intell. Mater. Syst. Struct.*, 1998, **9, No. 4**, 272–282.
121. J.D. Carlson, *Proceedings of the 3rd World Conference on Structural Control*, ed. F. Casciati, Published by John Wiley & Sons, Ltd, England, 2003, **1**, pp. 227–236.
122. F. Gordaninejad, X. Wang, G. Hitchcock, K. Bangrakulur, A. Fuchs, J. Elkins, C. Evrensel, S. Ruan, M. Siino, M. Kerns, *Proceedings of the Fourth International Workshop on Structural Control*, ed. A. Smith and R. Betti, DEStech Publications Inc., Lancaster, PA, USA, 2004, pp. 140–145.
123. L.M. Jansen and S.J. Dyke, *ASCE J. Eng. Mech.*, 2000, **126, No. 8**, 795–803.
124. F. Gordaninejad, M. Saiidi, B.C. Hansen and F.-K. Chang, *Smart Systems for Bridges, Structures and Highways, Proceedings of the 1998 SPIE Conference on smart materials and structures*, 1998, **3325**, pp. 2–11.
125. F. Gordaninejad, M. Saiidi, B.C. Hansen, E.C. Ericksen and F.-K. Chang, *J. Intell. Mater. Syst. Struct.*, 2002, **13, No. 2–3**, 167–180.
126. B.C. Hansen, F. Gordaninejad, M. Saiidi and F.-K. Chang, *Structural Health Monitoring*, ed. Fu-Kuo Chang, Technomic Publishing Co., Lancaster, PA, USA, 1997, 81–90.
127. Y. Liu, F. Gordaninejad, C. Evrensel, X. Wang and G. Hitchcock, *ASCE J. Struct. Eng.*, 2005, **131, No. 5**, 743–751.

128. G.J. Hiemenz, Y.T. Choi and N.M. Wereley, *Proceedings of SPIE's 7th Int. Symposium on Smart Materials and Structures*, Newport Beach, California, 2000, **3988**, 217–228.
129. Y.L. Xu, W. L. Qu and J.M. Ko, *Earthquake Eng. Struct. Dynam.*, 2000, **29**, 557–575.
130. Y. Ribakov and J. Gluck, *Struct. Des. Tall Build.*, 2002, **11**, 171–195.
131. Y.Q. Ni, Y. Chen, J.M. Ko and D.Q. Cao, *Eng. Struct.*, 2002, **24**, 295–307.
132. Z. Lou, R.D. Ervin and F.E. Filisco, *J. Fluid Mech. Eng. Tans. ASME*, 1994, **116**, 570–576.
133. R.W. Phillips, "Engineering applications of fluids with a variable yield stress," Ph.D. Dissertation, University of California, 1969.
134. H.P. Gavin, R.D. Hanson and F.E. Filisko, *J. Appl. Mech.*, 1996, **63**, 669–675.
135. G. Yang, B.F. Spencer, J.D. Carlson and M.K. Sain, *Eng. Struct.*, 2002, **24, No. 3**, 309–323.
136. G.M. Kamath and N.M. Wereley, *J. Guid., Cont. Dynam.*, 1997, **20**, 1125.
137. D.R. Gamota and F.E. Filisko, *J. Rheol.*, 1991, **35**, 399–425.
138. H.P. Gavin, R.D. Hanson and F.E. Filisko, *J. Appl. Mech.*, 1996, **63**, 676–682.
139. S. B. Choi, S.K. Lee and Y.P. Park, *J. Sound Vibr.*, 2001, **245, No. 2**, 375–383.
140. W.H. Li, G.Z. Yao, G. Chen, S.H. Yeo and F.F. Yap, *Smart Mater. Struct.* 2000, **9, No. 1**, 95–102.
141. B.F. Spencer, S.J. Dyke, M.K. Sain and J.D. Carlson, *J. Eng. Mech.-ASCE*, 1997, **123, No. 3**, 230–238.
142. T. Butz and von O. Stryk, *Z. Angew. Math. Mech.*, 2002, **82, No. 1**, 3–20.
143. R.C. Ehrgott and S.F. Masri, *Smart Mater. Struct.*, 1992, **1**, 275–285.
144. A. Leva and L. Piroddi, *Smart Mater. Struct.* 2002, **11, No. 1**, 79–88.
145. Jin Gang, M.K. Sain, K.D. Pham, B.F. Spencer, Jr. and J.C. Ramallo, *Proceedings of the 2001 American Control Conference*. (Cat. No.01CH37148), 2001, p. 429.
146. S.A. Burton, N. Makris, I. Konstantopoulos and P.J. Antsaklis, *J. Eng. Mech.-ASCE*, 1996, **122, No. 9**, 897–906.
147. C.C. Chang and P. Roschke, *J. Intell. Mater. Syst. Struct.*, 1998, **9, No. 9**, 755–764.
148. K.C. Schurter and P.N. Roschke, *Ninth IEEE International Conference on Fuzzy Systems. FUZZ- IEEE* (Cat. No.00CH37063), 2000, p. 122.
149. D.J. Peel, R. Stanway and W.A. Bullough, *Smart Mater. Struct.*, 1996, **5, No. 5**, 591–606.
150. N.D. Sims, R. Stanway, D.J. Peel, W.A. Bullough and A.R. Johnson, *Smart Mater. Struct.*, 1999, **8, No. 5**, 601–615.
151. N.D. Sims, D.J. Peel, R. Stanway, A.R. Johnson and W.A. Bullough, *J. Sound Vib.*, 2000, **229, No. 2**, 207–227.
152. X. Wang and F. Gordaninejad, *Damping and Isolation, Proceedings of SPIE Conference on Smart Materials and Structures*, ed. D.J. Inman, 2001, **4331**, pp. 82–91.

153. X. Wang and F. Gordaninejad, *J. Appl. Mech.*, 2007, **74, No. 1**, 13.
154. J.D. Stringer, *Hydraulic Systems Analysis: An Introduction*, John Wiley & Sons, New York, NY, 1976.
155. F. Gordaninejad and D.G. Breese, "Magneto-Rheological Fluid Dampers," US Patent No. 6,019,201, 2000.
156. F. Gordaninejad and S.P. Kelso, "Magneto-Rheological Fluid Device," US Patent No. 6,471,018 B1, 2002.
157. F. Gordaninejad and E.O. Ericksen, "Controllable Magneto-Rheological Fluid Damper," US Patent No. 6,510,929 B1, 2003.
158. G. Hitchcock and F. Gordaninejad, "Magneteorheological Fluid Device," US Patent No. 6,823,895, 2004.
159. D.G. Breese and F. Gordaninejad, *Int. J. Veh. Des.*, 2003, **33, Nos. 1–3**, 128–138.
160. G.H. Hitchcock, F. Gordaninejad and X. Wang, *Proceedings of the 3rd World Conference on Structural Control*, ed. F. Casciati, John Wiley & Sons, Ltd, England, 2003, **2**, 121–126.
161. H. Sahin, F. Gordaninejad, X. Wang, Y. Liu, C.A. Evrensel, A. Fuchs, and B. Hu, *Proceedings of the Second ANCRiSST Workshop on Advanced Smart Materials and Smart Structures Technology*, Chung-Bang Yun and B.F. Spenser (eds), Technopress, Korea, 2005, p. 499.
162. U. Dogruer, F. Gordaninejad and C. Evrensel, *Damping and Isolation, Proceedings of SPIE Conference on Smart Materials and Structures*, ed. Kon-Well Wang, 2004, **5386**, pp. 195–203.
163. D.G. Breese, F. Gordaninejad and E.O. Ericksen, *Smart Systems for Bridges, Structures and Highways, Proceedings of SPIE Conference on Smart Materials and Structures*, ed. Chi Liu, 2000, **3988**, pp. 450–457.
164. F. Gordaninejad and D.G. Breese, *J. Intell. Mater. Syst. Struct.*, 1999, **10, No. 8**, 634–645.
165. D.G. Breese and F. Gordaninejad, *Smart Systems for Bridges, Structures and Highways, Proceedings of the 1999 SPIE Conference on smart materials and structures*, 1999, **3671**, pp. 2–10.
166. M.B. Dougruoz, E.L. Wang, F. Gordaninejad and A.J. Stipanovic, *J. Intell. Mater. Sys. Struct.*, 2003, **14, No. 2**.
167. M.B. Dogruoz, F. Gordaninejad, E.L. Wang, A.J. Stipanovich, *Damping and Isolation, Proceedings of SPIE Conference on Smart Materials and Structures*, ed. D.J. Inman, 2001, **4331**, pp. 343–353.
168. M.B. Dogruoz, F. Gordaninejad and L.C. Wang, *Smart Systems for Bridges, Structures and Highways, Proceedings of SPIE Conference on Smart Materials and Structures*, ed. Chi Liu, 2000, **3988**, pp. 84–93.
169. Y. Liu, F. Gordaninejad, C.A. Evrensel, U. Dogruer, M.-S. Yeo, E.S. Karakas and A. Fuchs, *Proceedings SPIE Smart Structures and Materials Conference, Industrial and Commercial Applications*, ed. E.H. Anderson, 2003, **5054**, pp. 332–340.
170. Y. Liu, F. Gordaninejad, C. Evrensel, S. Karakas and U. Dogruer, *Industrial and Commercial Applications of Smart Structures Technologies,*

Proceedings of SPIE Conference on Smart Materials and Structures, ed. J.H. Jacobs, 2004, **5388**, pp. 338–347.
171. Y. Liu, F. Gordaninejad, C.A. Evrensel, E.S. Karakas and U. Dogruer, *Smart Structures and Integrated Systems, Proceedings of SPIE Conference on Smart Materials and Structures*, ed. A.B. Flatau, 2005, **5764**, pp. 360–368.
172. http://www.rheonetic.com.
173. J.D. Carlson, D.F. LeRoy, J.C. Holzheimer, D.R. Prindle and R.H. Marjoram, "Controllable Brake," US Patent Number 5,842,547, 1998.
174. J.D. Carlson, "Portable controllable fluid rehabilitation devices," US Patent Number 5,711,746, 1998.
175. S. Gopalswamy and G.L. Jones, "Magnetorheological transmission clutch," US Patent Number 5,823,309, 1998.
176. S. Gopalswamy, G.L. Jones and S.M. Linzell, "Magnetorheological fluid clutch with minimized reluctance," US Patent Number 5,845,752, 1998.
177. E.A. Bansbach, "Torque transfer apparatus using magnetorheological fluids," US Patent Number 5,823,309, 1998.
178. M. Mokeddem, "Magnetorheological torsional vibration damper," US Patent Number 5,829,319, 1998.
179. U. Lee, D, Kim, N. Hur and D. Jeon, *J. Intell. Mater. Syst. Struct.*, 1999, **10, No. 9**, 701–707.
180. D. Lampe and R. Grundmann, *Proceedings of the 7th International Conference on New Actuators and International Exhibition on Smart Actuators and Drive Systems*, 2000. p. 555.
181. B. Kavlicoglu, F. Gordaninejad, C.A. Evrensel, F. Fuchs and G. Korol, *Proceedings SPIE Smart Structures and Materials Conference, Industrial and Commercial Applications*, ed. E.H. Anderson, 2003, **5054**, pp. 341–349.
182. J.H. Yoo, J. Sirohi and N.M. Wereley, *J. Intell. Mater. Syst. Struct.*, 2005, **16 (11–12)**, 945–953.
183. W.I. Kordonski and D. Golini, *J. Intell. Mater. Syst. Struct.*, 1999, **10, No. 9**, 683–689.
184. S.D. Jacobs, S.R. Arrasmith, I.A. Kozhinova, L.L. Gregg, A.B. Shorey, H.J. Romanofsky, D. Golini, W.I. Kordonski, P. Dumas and S. Hogan, *Am. Ceram. Soc. Bull.*, 1999, **78, No. 12**, 42–48.
185. T. Shimizu, A. Iwabuchi and N. Umehara, *J. Jpn. Soc. Tribol.*, 2002, **47, No. 6**, 436–442.
186. K. Shimada, Y. Akagami, S. Kamiyama, T. Fujita, T. Miyazaki and A. Shibayama, *Proceedings 8th Int. Conf. on ER Fluids, MR Suspensions and Associated Technology*, G. Bossis, ed. World Scientific, Singapore, 2002, pp. 9–15.
187. X. Tang, X. Zhang and R. Tao, *J. Intell. Mater. Syst. Struct.*, 1999, **10, No. 9**, 690–694.
188. Y. Rong, R. Tao and X. Tang, *Int. J. Adv. Manuf. Technol.*, 2000, **16, No. 11**, 822–829.

189. J. Liu, G.A. Flores and R.S. Sheng, *J. Magn. Magn. Mater.*, 2001, **225, No. 1–2**, 209–217.
190. O. Rotariu and N. Strachan, *J. Magn. Magn. Mater.*, 2005, **293, No. 1**, 639–646.
191. J.D. Carlson, W. Matthis and J.R. Toscano, *Proc. SPIE Smart Structure and Materials Conference*, 2001, **4332**, Newport Beach, California, p. 308.
192. J.L. Johansson, D.M. Sherill, P.O. Riley, P. Bonato and P.H. Herr, *Am. J. Phys. Med. Rehab.*, 2005, **84, No. 8**, 563–575.
193. E.P. Scilingo, A. Bicchi, D. De Rossi and A. Scotto, *1st Annual International IEEE-EMBS Special Topic Conference on Microtechnologies in Medicine and Biology. Proceedings* (Cat. No.00EX451), 2000, p. 200.

CHAPTER 15
Metal Hydrides as Intelligent Materials and Artificial Muscles

KWANG J. KIM,[1] GEORGE LLOYD[1] AND MOHSEN SHAHINPOOR[2]

[1] Active Materials and Processing Laboratory, Mechanical Engineering Department, University of Nevada, Reno NV 89557, USA
[2] Mechanical Engineering, College of Engineering, University of Maine, Orono, Maine 04469, USA

15.1 Metal Hydrides in General

Metal hydrides are the binary combination of hydrogen and a metal or metal alloy. They can absorb large amounts of hydrogen via surface chemisorption and subsequent hydriding reactions as illustrated in Figure 15.1. At a given temperature metal hydrides form condensed phases with hydrogen in the presence of hydrogen partial pressure. The useful characteristics of metal hydrides are the large uptake/discharge capacity of hydrogen, safe operation (hydrogen desorption is a highly endothermic process), rapid kinetics, and environmentally benign. Metal hydrides have been traditionally used for hydrogen storage and thermal devices.[1-9]

In most metal hydrides, there are two distinct phases as shown in Figure 15.2 (α and β phases). An isotherm gives the absolute equilibrium absorption or desorption pressure as a function of the hydrogen concentration, H/M (M = metal atom). Initially, hydrogen dissolves within the solid lattice of the metal hydride. Continuous addition of hydrogen results in a sample consisting of the chemisorbed phase. All interstitial hydrogen is chemically bounded in the solid lattice. The endpoints, H/M_α and H/M_β are called the phase limits of the *plateau* region. They are generally not sharply defined. In a dehydriding process (desorption) frequently *hysteresis* is observed, with the dehydriding isotherm lying slightly below the hydriding isotherm. A typical metal hydride is the rare-earth intermetallic $LaNi_5$ (lanthanum-pentanickel). The hydriding/dehydriding

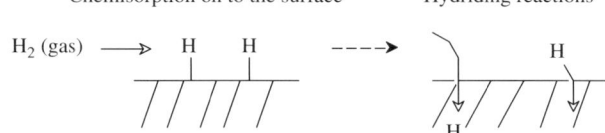

Figure 15.1 Hydrogen absorption by metal hydrides (from Kim et al.[7]).

reaction can be written as,

$$\text{LaNi}_5 + \frac{x}{2}\text{H}_2 \Leftrightarrow \text{LaNi}_5\text{H}_x + \Delta\text{H}_a \tag{15.1}$$

where x and ΔH_a are a nonstoichiometric constant which is about 6–6.7 for this particular compound and the heat of absorption giving off $(-3.1 \times 10^4\,\text{kJ}/\text{kgmole of H}_2$, for LaNi$_5$), respectively. It is usually close to the heat of desorption, ΔH_d ($\Delta\text{H}_a \approx -\Delta\text{H}_d$). The equilibrium behaviour of metal hydrides in the plateau region can be described by van't Hoff plots, according to the following relation,

$$\ln P_{\text{H2}}(\text{atm}) = \frac{\Delta\text{H}_a}{RT} - \frac{\Delta S}{R} \tag{15.2}$$

where R is the molar gas constant, equal to 8.314 kJ/kgmole K, T is the absolute temperature in K, ΔH_a is the heat of absorption in kJ/kgmole of H$_2$, and ΔS is the standard entropy of formation in kJ/kgmole of H$_2$-K. The van't Hoff plots and the static p–H/M–T data available for particular metal hydrides are the usual basis for thermomechanical design. Figure 15.2 also shows the van't Hoff plot for typical AB5-type metal hydrides including LaNi$_5$. Depending upon pressure/temperature requirements and available temperatures, desired hydrides can be selected for use in various actuator systems.

In practical actuation devices, metal hydrides are typically enclosed in a simple cylindrical container (such as shown in Figure 15.3). In an actuator configuration each container communicates with one heat source (driven by an electric input). Although minimising heat and mass transfer resistance is ultimately ideal, the intrinsic hydriding/dehydriding kinetics and possibly external finite-sized heat-transfer time would be the characteristic cycle time.

15.2 Metal-hydride-actuation Principle

The principle of the metal hydride artificial muscle is shown in Figure 15.4. It functions using hydrogen gas pressure from the metal hydride by manipulating the thermoelectric input (or any kind of heating/cooling elements such as heat radiation panels and direct/indirect heat exchangers). The thermoelectric elements are located near the metal hydride to provide appropriate heat sources (either heating or cooling) by simply changing the direction of the electric

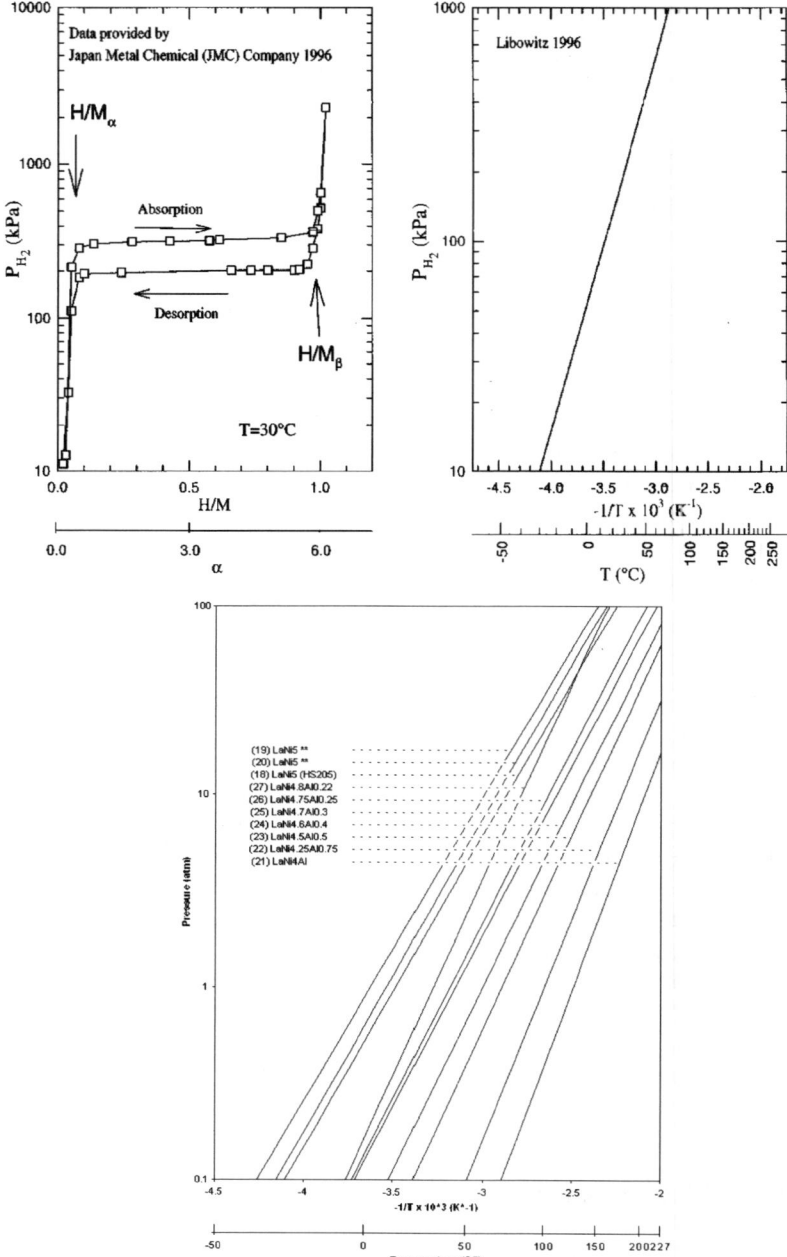

Figure 15.2 (Top) the thermodynamic properties of LaNi$_5$ (from Kim et al.[2]) H$_2$ pressure P_{H2} vs. atom ratio H/M and/or nonstoichiometric coefficient α H$_2$ pressure P_{H2} vs. negative inverse temperature $-1/T$ and/or temperature T Bottom graph of P_{H2} vs. $-1/T$ for a number of AB5-type hydrides.

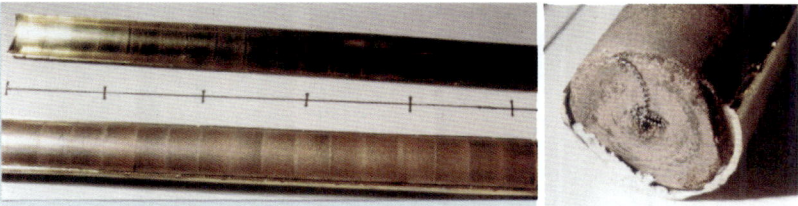
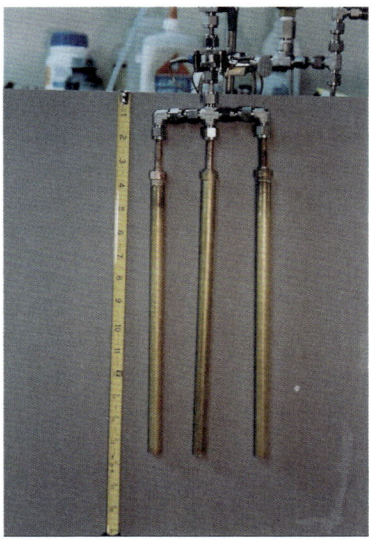

Figure 15.3 View of metal hydride, LaNi$_5$, placed in a simple cylindrical container (top-left, scale is inches.) and a cross-section of the container with metal hydride (top-right). A fabricated reactor module having three reactors is shown (bottom).[2]

current to the element. The expandable inner bladder polymeric material (*i.e.* silicon rubber) that contains the hydrogen gas can construct the functioning part of the metal hydride artificial muscle. The key parameter of the expandable material is the capability to sustain repeated strains of over 100%. The selection of rubber materials is good since they manage these large strains with no plastic strain or creep.

When heat is applied to the metal hydride, hydrogen gas is immediately desorbed from the metal hydride. Thus, the functioning part of the shell expands in the middle causing the two ends to approach each other and produce effective and powerful contraction. This operation causes a strong pulling force between the endpoints of the actuator. The maximum force at a given pressure is obtained when the shell is fully expanded in the middle. This operation is similar to biological muscle contractions such as contraction of the biceps muscles of the human arm. The relationship between pressure and force

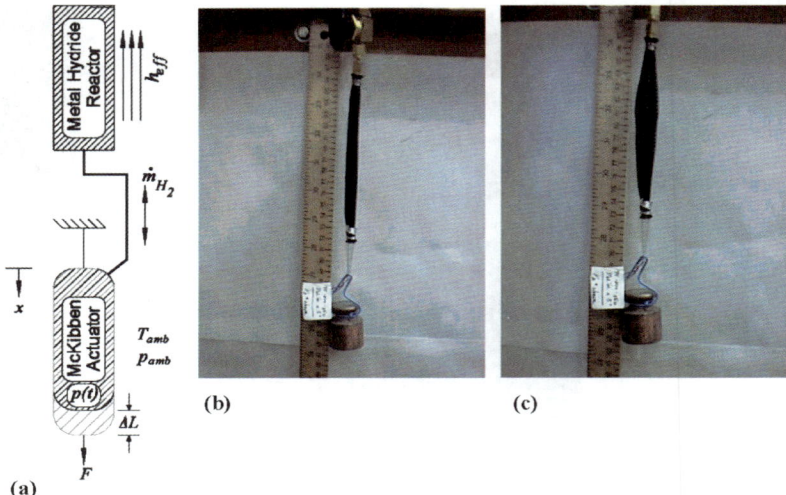

Figure 15.4 Schematic diagram of a metal hydride actuator (a) and a metal hydride actuator in action (before and after actuation, b and c).[15]

is nearly linear. In fact, this allows the distance moved to be set by regulating the H_2 pressure in the system by controlling the heat input to the metal hydride. When metal hydride is cooled, the hydrogen gas is absorbed back into the metal hydride. Therefore, internal pressure decreases and the actuator moves back to its initial position. See Refs. [10–14] for detailed discussion on metal hydride actuators.

15.2.1 Modeling

It is difficult to optimise a thermal metal-hydride reactor. This is because the high latent energy density in comparison with typical thermal resistance promotes the development of strong concentration gradients. This phenomenon complicates the prediction of any thermally operated reactor and generally requires the solution of the coupled heat- and mass-transfer problem. A local thermal equilibrium formulation of the governing heat and mass-transfer equations for one-dimensional radial fluxes suitable for analysing the reactor geometry shown in Figure 15.4 has been derived and discussed previously;[15] we summarise the essential elements of the formulation here. The governing equation for H_2 flow in the hydride matrix is:

$$\left(\frac{K\gamma_{H_2}}{\bar{R}\mu}\right)\frac{T}{r}\frac{\partial}{\partial r}\left[\frac{r}{T}p\frac{\partial p}{\partial r}\right] + \varphi T = \left(\frac{\theta\gamma_{H_2}}{\bar{R}}\right)\left[\frac{\partial p}{\partial T} - \frac{p}{T}\frac{\partial T}{\partial t}\right] \quad (15.3)$$

The energy equation for the hydride matrix is,

$$\frac{k_{\text{eff}}}{r}\frac{\partial}{\partial r}\left(r\frac{\partial T}{\partial r}\right) + \left(\frac{Kc_{v_f}\gamma_{H_2}}{\bar{R}\mu}\right)\frac{1}{r}\frac{\partial}{\partial r}\left(rp\frac{\partial p}{\partial r}\right) + g \\ = (1-\theta)\rho_s c_s \frac{\partial T}{\partial t} + \frac{\theta c_{v_f}\gamma_{H_2}}{\bar{R}}\frac{\partial p}{\partial t} \quad (15.4)$$

The source terms, g and φ, in these equations are given by the following expressions:

$$\varphi = -\frac{1}{2}N_0\left.\frac{\partial x}{\partial t}\right|_r \gamma_{H_2}\zeta\left(\frac{\text{kg}_{H_2}}{\text{s m}^3}\right); \quad g = \frac{1}{2}N_0\left.\frac{\partial x}{\partial t}\right|_r |\Delta H|\zeta\left(\frac{\text{J}}{\text{s m}^3}\right) \quad (15.5)$$

where N_0 is the particle molar mass (1.0187×10^{-14} (kmole$_h$/part$_h$)), ζ is the particle density (1.0106×10^{15} (m^{-3})), and the local reaction rate is an Arrhenius type:

$$\left.\frac{\partial x}{\partial t}\right|_r = Ae^{-E_a/\bar{R}T}\ln\left(\frac{p}{p_{\text{eq}}}\right), \quad \left(\frac{\text{kmole}_{H_2}}{\text{s particle}_h}\right) \quad (15.6)$$

In eqn (15.6), $-E_a$ is the activation energy (30×10^6 (J/kmoleH$_2$)) and A is the reaction prefactor (10^5 (kmoleH/kmole$_h$/s)).

The effluxing hydrogen is assumed to flow without pressure losses to the bladder and to rapidly equilibrate through the thin membrane, so that the pressure at the metal-hydride boundary is in equilibrium with the bladder pressure:

$$p(r_0, t) = \frac{m^b_{H2}(t)\bar{R}T_{\text{amb}}}{V_b(t)\gamma_{H_2}} \quad (15.7)$$

For hydrogen, behaving as an ideal gas in the pressure and temperature ranges of interest, the resulting mass flow rate (calculated at local conditions) is consequently,

$$\dot{m}_{H_2} = v(r_0, t)\rho A_0 \quad (\text{kg}_{H_2}/\text{s}) \quad (15.8)$$

where A_0 is the area associated with the internal artery.

The compression work done by the actuator that is transferred to the mechanical load is given by,

$$W(t) = \int_0^t P(t)\,dt \quad (15.9)$$

The instantaneous efficiency and net efficiency are given by,

$$\dot{\eta} = \frac{\text{mechanical energy output}}{\text{thermal power}}, \quad \eta = \frac{W_{\text{mech}}}{Q_{\text{input}}} \quad (15.10)$$

Figure 15.5 shows computations for the optimised 4 (mm) reactor. The parameter of interest in this pair of simulations is the thermal input, Q_{input} that was set to be 80 W. It is clear that peak power and efficiency occur before peak displacement (which occurs at high compression ratios). The most noteworthy prediction from these simulations is the predicted improvement of time to peak power to less than thirty seconds. Details regarding the modeling of metal-hydride artificial muscle can be found in Refs. [10–14, 16].

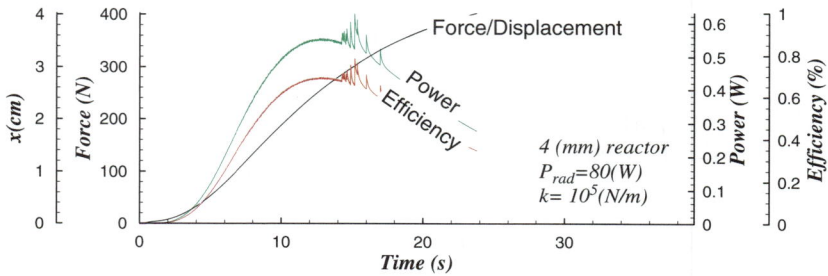

Figure 15.5 Simulation results.[16]

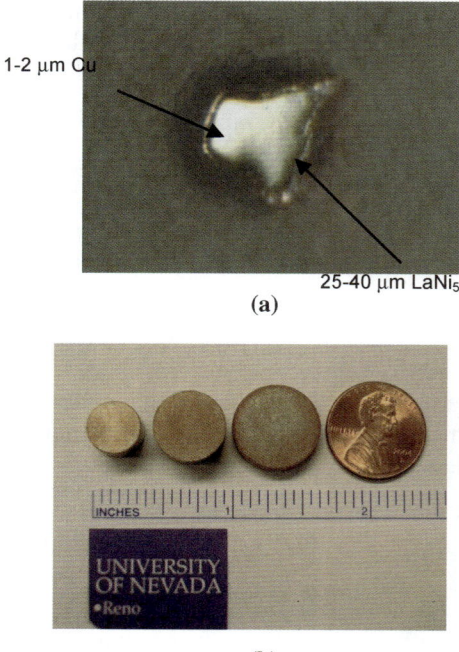

Figure 15.6 (a) A micrograph of the Cu-microencapsulated metal-hydride particle (b) a photograph of porous metal hydride compacts.[16]

15.2.2 Experiments

In most metal hydrides undergoing absorption/desorption cycles, high volumetric strains leads to *decripitation* of metal hydrides into a powder bed of microsized particles. Although metal hydrides themselves have rapid intrinsic kinetics, the poor thermal conductivity of such powder beds ($k_{\text{eff}} \sim 10^{-1}$ W/m K) limits heat-transfer communication with the beds and, therefore, retards the apparent kinetics. In order to obtain reasonably rapid kinetics, reactor fabrication must improve the thermal conductance of the metal hydride reactor while maintaining good hydrogen permeabity. Also, fine hydride particles tend to migrate through the system, potentially limiting long-term reliability. A critical engineering concern is containing these particles while maintaining high heat- and mass-transfer properties. As one of the technical innovations, the "microencapsulation" technique is used to improve the thermal conductivity of

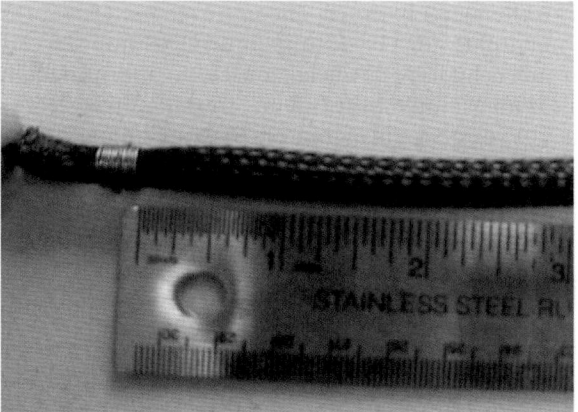

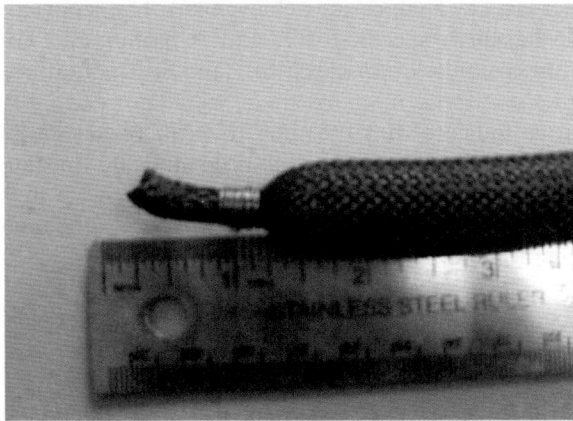

Figure 15.7 The standing position (above) and actuating (below) positions of the functioning part, respectively (this particular design gives off an approximately 20% linear strain).

the metal-hydride material. Metal-hydride particles are sieved to a diameter of 25–45 μm (100 mesh, see Figure 15.6(a)) and then microencapsulated with thin copper using an electroless plating technique. This microencapsulation will allow the decripitated metal-hydride particles to be contained inside a copper shell even after many absorption/desorption cycles. Provided with a driving force, hydrogen readily diffuses through a thin copper layer. When such microencapsulated particles are mixed with binders and compressed into small porous metal-hydride compact pellets, significantly improved thermal conductivity (~ 3–4 W/m K) of the hydride bed is obtained, which provides good heat transfer between the hydride bed and the reactor shell.

The fabricated metal hydride reactor body and its associated functioning part are shown in Figure 15.7. The expandable inner silicon bladder (3/8" diameter) is inserted inside the shell. Upon hydrogen being desorbed from the metal hydride, a constant pressure is maintained, causing a pulling force between the endpoints. When the metal hydride is cooled, the hydrogen is absorbed back to the metal hydride. Therefore, the internal pressure decreases and the functioning part goes back to its initial position. In order to fabricate and attach the metal-hydride artificial muscle to the principal flexors of the elbow joint (this is an example), the full integration of triggering signals, energy sources, power converters, and actuators into an integrated system may be considered. Such integration is currently under development and will be reported in the future.

15.3 Summary

The current development effort is reported on a new generation of artificial muscle using metal-hydride technology suitable for use in industrial and biomedical applications. It should be noted that improved thermal conductivity of metal hydrides is a key engineering achievement. The reported metal-hydride artificial muscle has high specific-force, and is biomimetic, compact, operationally safe, lubricationless, noiseless, fast and soft actuating, and environmentally benign.

The simulations indicate that the time to peak power can realistically be made less than thirty seconds. The simulations indicate that large forces over moderate deflections can be rapidly developed, albeit with high compression ratios and low net efficiencies. Future study will address the muscle model and detailed performance characteristics.

References

1. K.J. Kim, B. Montoya, A. Razani and K.-H. Lee, *Int. J. Hydrogen Ener.*, 2001, **26, No. 6**, 609–613.
2. K.J. Kim, K.T. Feldman, Jr., G. Lloyd and A. Razani, *ASHRAE Trans.*, 1998-Winter Meeting, San Francisco, 1998, 104, Pt. 1, SF-98-18-4.

3. K.J. Kim, G. Lloyd, A. Razani and K.T. Feldman Jr., *Powder Technol. Int. J.*, 1998, **99**, 40–45.
4. K.J. Kim, K.T. Feldman Jr., G. Lloyd and A. Razani, *Int. J. Hydrogen Ener.*, 1998, **23, No. 5**, 355–362.
5. K.J. Kim, K.T. Feldman Jr., G. Lloyd and A. Razani, *Appl. Therm. Eng.*, 1998, **18, No. 12**, 1325–1336.
6. K.J. Kim, K.T. Feldman Jr., G. Lloyd and A. Razani, *Appl. Therm. Eng*, 1997, **17**, 551–560.
7. K.J. Kim, K.T. Feldman Jr. and A. Razani, *Energy*, 1997, **22**, 787–796.
8. G. Lloyd, K.J. Kim, K.T. Feldman Jr. and A. Razani, *AIAA J. Thermophys. Heat Transf.*, 1998, **12, No. 2**, 132–137.
9. G. Lloyd, A. Razani and K.J. Kim, *ASME Transactions: J. Ener. Resour. Technol. (JERT)*, 1998, **20, No. 4**, 304–313.
10. M. Shahinpoor and K. Kim, "Novel Metal Hydride Artificial Muscles", US Patent No. 6,405,532, June 18, 2002.
11. M. Shahinpoor, *Proceedings of the First World Congress On Biomimetics and Artificial Muscle, Biomimetics 2002*, December 9–11, 2002, Albuquerque Convention Center, Albuquerque, New Mexico, USA, 2002.
12. K. Kim, J. Detweiler, G. Lloyd, M. Shahinpoor and A. Razani, *Proceedings of the First World Congress On Biomimetics and Artificial Muscle, Biomimetics 2002*, December 9–11, 2002, Albuquerque Convention Center, Albuquerque, New Mexico, USA, 2002.
13. G.M. Lloyd, K.J. Kim, A. Razani and M. Shahinpoor, *ASME J. Solar Ener. Eng.*, 2003, **125**, 95–100.
14. G. Lloyd, K.J. Kim, A. Razani and M. Shahinpoor, IMECE2004-60449, *Proceedings of ASME-IMECE2004, 2004 ASME International Mechanical Engineering Congress and RD&D Exposition*, November 13–19, 2004, Anaheim Convention center/Hilton, Anaheim, California, 2004.
15. G. Lloyd, K.J. Kim and A. Razani, *Proceedings of Solar '98 Conference*, Albuquerque, New Mexico, 1998, 439–444.
16. G. Lloyd and K.J. Kim, "Smart Metal Hydride Actuators", *Int. J. Hydrogen Ener.*, 2007, **32**, 247–255.

CHAPTER 16

Dielectric Elastomer Actuators as Intelligent Materials for Actuation, Sensing and Generation

GUGGI KOFOD[1] AND ROY KORNBLUH[2]

[1] Advanced Condensed Matter Physics, University of Potsdam Germany
[2] SRI International, Menlo Park, USA

16.1 Introduction

Dielectric elastomers are a type of electroactive polymer based on the deformation of rubbery (elastomeric) insulating (dielectric) polymer films caused primarily by the electrostatic forces acting on the surfaces of these films. Like many electric-field-controlled materials, dielectric elastomers are fast acting and energy efficient. However, dielectric elastomers have demonstrated deformation strains significantly greater than other solid field-controlled materials – greater than 100% in some cases. This performance, combined with the low cost of dielectric polymer materials and simplicity of design and operation has resulted in rapidly increasing interest in dielectric elastomer actuators for actuation as well as for sensing and power generation.

The first experimental accounts of a dielectric deforming due to an electric field were presented in the early part of the 19th century, an effect that was explained by Röntgen in a paper in 1880.[1] Two glasses were used as dielectrics, English flint glass and Thuringian glass, and thickness changes of just a few parts per million were observed. With systematic measurements, the observed deformation was enough to establish a connection between an applied electric field and an expansion, at least for inorganic materials. Later, electrostatically induced pressures acting to compress dielectrics, such as those found in capacitors, became known as "Maxwell stress".

Polymers show a variety of electromechanical responses due to an applied electric field in addition to Maxwell stress. For instance, piezoelectricity and electrostriction are found in some crystalline polymers. In polymers with relatively high moduli of elasticity, such as most crystalline polymers, the strain response due to Maxwell stress is relatively small. While some researchers noted the advantages of an enhanced strain response by using lower-modulus polymers,[2,3] most researchers continued to focus on enhancing the piezoelectric or electrostrictive response of crystalline polymers as an approach to actuation.

In contrast Pelrine et al.[4] noted that by deliberately choosing polymers with relatively low moduli of elasticity, the field-induced strain response due to Maxwell stress can be large. Polymer actuators designed to exploit Maxwell stress in this manner became known as "dielectric elastomers".[5]

The achievement of strains of greater than 100% reported by Pelrine et al.[6] caused a large number of researchers to take note of the potential advantages of what was now known as *dielectric elastomer actuators*.

Since the pioneering work by Pelrine and others at SRI International, many research and development groups have entered the field, further expanding our understanding of this simple technology, and brought forward very interesting new applications. The first commercial uses of this technology are now emerging.

This chapter surveys the basics of dielectric elastomers from a materials and operational perspective. Then we offer a sample of the many actuator, generator and sensor designs that are being developed for a variety of applications. Finally, we discuss technical challenges that must be overcome to allow the widespread use of dielectric elastomers and offer some suggestions on ways to surmount these challenges.

16.2 Actuation Basics

A dielectric elastomer actuator can be described as a rubbery capacitor, where both the dielectric layer and electrodes are highly compliant. When the rubbery capacitor is charged, the positive charges on one electrode and negative charges on the other attract each other, squeezing the material in between them. Since rubbery polymers are relatively incompressible, the dielectric layer expands in area as it decreases in thickness (see Figure 16.1).

The dimensional changes of this configuration are estimated by assuming that the thickness strain, s, is proportional to the stress, p, in the material (which is in principle only true for very small strains),

$$s = Y \cdot p \tag{16.1}$$

in which Y is the Young's modulus. The pressure from the charges on the electrodes, the Maxwell stress, is

$$p = \varepsilon \varepsilon_0 E^2 \tag{16.2}$$

where ε is the relative dielectric constant of the elastomer, ε_0 is the permittivity of free space, and E is the applied electric field (note that this is twice the

Figure 16.1 A rubbery capacitor. When a voltage potential difference is applied between the two electrodes, the pressure from the electrodes on the dielectric elastomer forces it to compress along the electric field, and expand in the perpendicular directions.

Maxwell stress present in capacitors with rigid electrodes). Combining the two equations yields

$$s = \frac{\varepsilon \varepsilon_0 E^2}{Y} \qquad (16.3)$$

This equation describes the behaviour of any material exhibiting only Maxwell's stress, and to a first order only. A more detailed account of how to combine the theory of hyperelasticity with Maxwell's stress is presented in Section 16.5.

Because dielectric elastomers do not rely on any particular crystalline structure, a wide variety of low-modulus polymers can exhibit high actuation response. Silicone rubbers (based on polydimethyl siloxane backbones) and certain commercially available acrylate blends (such as 3M Corporation's VHB series) have generated the most research interest, each having produced strains in excess of 100% and energy densities exceeding other field-induced actuation technologies such as piezoelectric ceramics.[6]

Softer elastomers might have a Young's modulus of about 0.5 MPa and a relative dielectric constant of 3. If they can maintain an applied electric field of 100 MV/m then the resulting thickness strain is roughly 50%, much higher than possible with ceramic piezoelectric materials. Even though the above analysis is based on simplified equations, it nicely illustrates the principle and operation of dielectric elastomer actuators.

Due to the quadratic nature of the Maxwell stress, the greatest actuation happens close to electrical breakdown, therefore the operating voltage is often quite high. Since an unstrained elastomer film for dielectric elastomer actuators might typically be around 50 µm thick, the voltage at breakdown is around 5 kV. Note that the strain is dependent on the field, so if the thickness is lowered to 5 µm, the voltage at breakdown is still about 500 V.

Equation (16.3) highlights the important basics of dielectric elastomer actuators. Since the highest possible strain is sought after, materials and conditions should be found that help to maximise strain.

A high dielectric constant, high electric breakdown field and a low Young's modulus all contribute to an increase in strain. It is difficult to find elastomers that have a favourable combination of these parameters, and often the improvement of one property will result in trading off the other two properties.

Not surprisingly, such tradeoffs are well known to researchers who have developed compact capacitors (*e.g.* Ref. 7).

16.3 Pre-stress Bias

The stress–strain state of the elastomer dielectric can dramatically affect its performance, particularly for the acrylic materials. As an illustration of the importance of prestress for the actuation properties of dielectric elastomer actuators, the reported performance of the circular actuator based on VHB4910 serves as an example. The circular actuator consists of a biaxially prestretched film, fixed to a circular frame (Figure 16.2). Circular electrodes are placed in the centre, such that the central area actuates, while the surrounding passive area provides pre-stress. With low amounts of biaxial prestrain the maximum area strain achieved was just 40%. With high prestrain, the same film achieved an area strain of 158%.[6]

In the following, pre-stress and pre-strain will be used interchangeably, since the correspondence between stress and strain is unique. In fact, during manufacture of actuators usually prestrain is the controlled parameter. Both symmetric and asymmetric biaxial pre-strain configurations are routinely used, and for VHB4910, preparation pre-strains of more than 400% by 400% are common. Pre-strain can also enhance the performance of other polymers, such as silicones, but the amount of pre-strain is not typically as great.

The increase in maximum strain is due to several factors. Perhaps most significantly, the electrical breakdown strength of the polymer can increase significantly. For example, it has been shown that in VHB4905 the electric field at breakdown may increase from just 18 MV/m for unstrained material, to more than 200 MV/m when the material is strained to 500% by 500%.[8] The mechanism behind this increase is still under investigation.

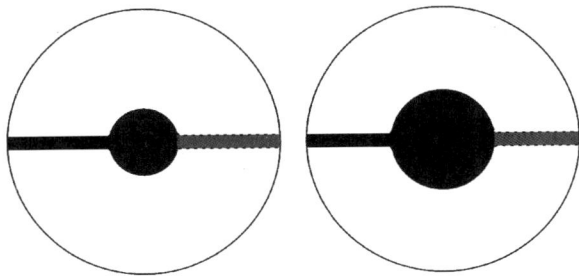

Figure 16.2 Unactuated and actuated states of a circular actuator.

16.4 Compliant Electrodes

Critical to the operation of dielectric elastomers are the compliant electrodes. The elastomers can undergo elongations of more than 100%, and so must the electrodes. Low-modulus electrodes capable of maintaining sufficient conductivity over large strains is a challenge that was not adequately addressed by existing materials. Thus, dielectric elastomer researchers have developed new approaches to making such electrodes.

16.4.1 Percolating Conductive Particle Networks

The most commonly used approach is based on percolating networks of conducting particles in a matrix.[4] The matrix material provides mechanical compliance, while the conducting particles provide conductivity. Usually the conducting particles themselves are stiff, and would be unable to support the high strain required. Ideally, conducting fillers must be chosen that have a low percolation threshold, such that the mechanical properties are mainly determined by the properties of the matrix.

Carbon grease electrodes can be produced from silicone oil and carbon black directly, a procedure is described in ref. [9]. Carbon grease is helpful for quick demonstrations of new actuator designs, however, the grease easily rubs off, and grease-based electrodes have been shown to lower some performance characteristics in comparison to other electrode types.[10]

More robust elastomeric electrodes may be formed following the same approach, mixing of a matrix with a percolating conducting filler [*e.g.* Ref. 9]. This approach has been employed in an automated manufacturing station, which was able to make actuator stacks with several hundreds of layers.[11]

16.4.2 Structured Metal Electrodes

Metals are interesting for electrode applications, since they are typically highly conductive even when they are extremely thin. But metals suffer from two problems. First, their relatively high stiffness imparts mechanical constraints on the expansion of the film. Second, the mechanical properties of metals do not allow them to stretch much. A study of gold electrodes on top of an elastomeric substrate showed that the connectivity and conductivity of the supported metal film could be retained to strains of about 8%.[12] These strain levels are not enough for many applications. Both problems can be surmounted when the metal electrode is patterned in a specific way.

Initial attempts were made through a MEMS approach, in which UV lithography was employed to define thin gold zigzag lines, which could retain conductivity even though the underlying material underwent an 80% linear deformation.[13] A similar approach is behind corrugated electrodes, prepared by direct deposition of a metal film on top of a silicone film with a corrugated surface, prepared by casting of the silicone elastomer in a mold.[14]

Another advantage of metal-film electrodes is that they are potentially self-repairing: when an electric breakdown occurs, it is usually localised to a particular point in the elastomer. Breakdown manifests itself as arcing from one electrode to the other through the film defect. The arc results in sputtering of the metal electrode, which evaporates around the defect. This can be seen as a circle expanding around the defect. When the path from one electrode to the other becomes longer, the voltage at which arcing occurs increases, until the applicable voltage becomes high enough that actuation is restored. Depending upon the size of the defect, most of the efficiency of the actuator is restored.

Corrugated electrodes may be realised by a self-assembly approach, in which the silicone elastomer is simply uniaxially prestrained, before the electrode material is deposited by evaporation.[15]

16.5 Theory and Modeling

We have already presented an overview of the basic theory of operation and the governing analytical equations for dielectric elastomer actuation. A more detailed understanding can aid in the design of improved actuators as well as help guide the development of improved materials. In this section we survey more detailed efforts to model dielectric elastomer actuation. The combination of both large nonlinear elastic properties and the nonlinear response to electrical fields is challenging.

The standard expression for the pressure from two oppositely charged, parallel electrodes, is known as the Maxwell stress, stated above in eqn (16.2). For a more rigorous approach, it is necessary to invoke tensor notation. A more general expression for the stress in an electrostrictive material due to an electric field is[16]

$$\sigma_{ij}^{elec} = \frac{\epsilon_0}{2}(\epsilon_{rel} - a_1)E_iE_j - \frac{\epsilon_0}{2}(\epsilon_{rel} + a_2)E^2\delta_{ij} \tag{16.4}$$

where ϵ_0 is the vacuum permittivity, ϵ_{rel} is the relative permittivity of the unstrained material, E_i is the electric field, and a_1 and a_2 are dielectric parameters. a_1 and a_2 specifically describe the change in permittivity of the material due to deformation, an effect known as *electrostriction*. For a soft, polar elastomer (isotropic) the contributions from electrostriction are negligible, which means that a_1 and a_2 are very close to zero.[17] Further, if the electric field is applied only along the 3 direction, eqn (16.4) reduces to

$$\sigma_{11j}^{elec} = -\tfrac{1}{2}\epsilon_{rel}\epsilon_0 E_3^2 = \sigma_{22}^{elec} \text{ and } \sigma_{33}^{elec} = \tfrac{1}{2}\epsilon_{rel}\epsilon_0 E_3^2 \tag{16.5}$$

The total stress inside the material is a sum of all internal stresses, and the additional stresses acting on the boundary. For the dielectric elastomer this includes just the nonelectrostrictive stress, eqn (16.5), as well as the elastic stress, σ_{ij}^{elast},

$$T_{ij} = \sigma_{ij}^{elec} + \sigma_{ij}^{elast} - p\delta_{ij} \qquad (16.6)$$

To solve eqn (16.6) at static equilibrium, the boundary conditions must be specified as well. As an example, the stress of the electrodes on the dielectric elastomer, T_{33}, is found for the case when the dielectric elastomer is free to move, $T_{11} = T_{22} = 0$. By elimination of p, the total stress on the electrodes becomes

$$T_{33} = \sigma_{33}^{elec} + \sigma_{33}^{elast} - \sigma_{11}^{elec} - \sigma_{11}^{elast} = \epsilon_{rel}\epsilon_0 E^2 + (\sigma_{33}^{elast} - \sigma_{11}^{elast}) \qquad (16.7)$$

At static equilibrium T_{33} must also be zero, by which eqn (16.7) rearranges to

$$\epsilon_{rel}\epsilon_0 E^2 = (\sigma_{11}^{elast} - \sigma_{33}^{elast}) \qquad (16.8)$$

showing that the electrostatic pressure is balanced by the elastic stresses.

This derivation also shows that there are two contributions to the Maxwell stress. For vacuum capacitors, the perpendicular term σ_{11}^{elec} does not exist, and the corresponding Maxwell stress is halved (as noted above). The introduction of σ_{11}^{elec} has been described qualitatively in many ways, the most intuitive being that the like charges within one electrode will try to maximize their internal distance. This maximizing is achieved by increasing the size of the electrode, which is only possible if the electrode is compliant. In this view, the dielectric elastomer acts merely as a flexible vacuum, required to keep the electrodes separate.

There are several other purposes of the elastomer than that of keeping compliant electrodes separate. The elastomer acts as a dielectric, allowing for storage of energy in the electric displacement of the polar groups constituting the material. The elastomer also allows for storage of mechanical energy, which is the mechanism by which it becomes a transducer between electric and mechanic energy. Finally, the elastomer helps to increase the permissible electric field.

There are a very large number of constitutive models that describe the mechanical properties of the elastomer, ranging from the simple Hookean spring (which does not conserve volume), to the hyperelastic constitutive models of Ogden and Yeoh, and beyond.

The simplicity of the neo-Hookean model permits easy evaluation, but unfortunately it only holds well for strains to at most 100%. Simply stated, the elastic stress tensor is proportional to the strain tensor, such that

$$\sigma_{11}^{elast} - \sigma_{33}^{elast} = G\left(\frac{1}{\alpha_3} - \alpha_3^2\right) \qquad (16.9)$$

α_3 is the deformation ratio in the thickness direction, $x_3 = \alpha_3 x_3'$, where x_3' is the initial, and x_3 the actual thickness of the dielectric elastomer. G is the shear modulus, which can be measured directly in uniaxial stress–strain experiments, and is related to the tensile modulus by $E = 3G$.

The electric field depends upon the applied voltage and the actual thickness of the elastomer, $E = V/x_3 = V/\alpha_3 x_3'$, which is introduced in eqns (16.8) and (16.9), and rearranged to

$$\frac{\epsilon_{rel}\epsilon_0 V^2}{x_3'^2 \alpha_3^2} = G(\alpha_3^{-1} - \alpha_3^2) \quad (16.10)$$

a fourth-order polynomial in α_3, which can be solved using standard mathematical software.

Each side of eqn (16.10) is plotted separately in Figure 16.3, left. The lower-lying curve shows the Maxwell stress (left-hand side), while the higher-lying curve shows the elastic stress (right-hand side). The two stresses are balanced in the two points where the curves intersect. A thorough analysis reveals that only the right-most intersection gives rise to a stable minimum.

When the voltage is increased, the elasticity curve remains fixed in the diagram, while the Maxwell-stress curve shifts uniformly upwards, which shifts the intersection point to lower values of the deformation thickness. This development is plotted in Figure 16.3, right. At some voltage (here around 1500 V) the stress becomes higher than can be supported by the elasticity of the material, and the situation becomes unstable, leading to what is commonly known as pull-in. Practically, pull-in can cause undesired electrical breakdown of the elastomer.

The above analysis applies to the situation when the actuator carries no load. The analysis can be expanded to the loaded situation, reflecting the situation for the simplest planar actuator capable of outputting useful actuation. The simplest case is when the width of the actuator is fixed, $x_2 = x_2'$ (the 2-direction is along the width). This constraint makes the length and thickness deformation ratios inversely proportional, $\alpha_1 = \alpha_3^{-1}$. The load is taken to be due to a constant mass m attached to the bottom of the actuator. Then, the tension on the face of the elastomer in the 1-direction is $T_{11} = mg/x_2 x_3 = mg/x_2' x_3' \alpha_3$,

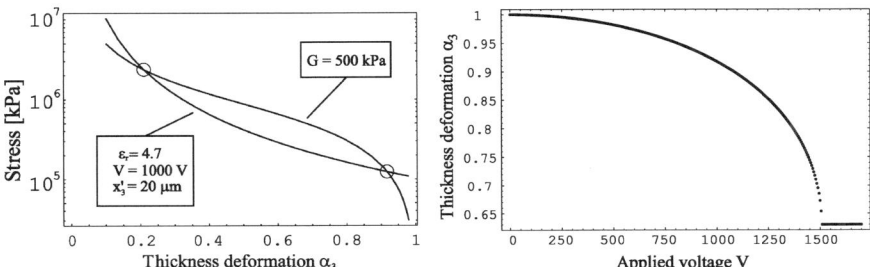

Figure 16.3 Left: plots of each side of equation 10, for the parameters stated. Right: thickness deformation ratio versus voltage, for the same parameters as in the left plot.

where $x_2 x_3$ is the end face area in the 1-direction. Combined with the above, the result is

$$\alpha_1^2 \frac{\epsilon_{\text{rel}} \epsilon_0 V^2}{x_3'^2} + \alpha_1 \frac{mg}{x_2' x_3'} = G(\alpha_1 - \alpha_1^{-2}) \quad (16.11)$$

which can be solved numerically for each set of mass and voltage chosen. The actuation strain is defined as the length of the actuator at a given voltage and mass with respect to the unactuated length for the same mass,

$$\varepsilon^{\text{act}} = \frac{x_1(m, V) - x_1(m, 0)}{x_1(m, 0)} = \frac{\alpha_1(m, V) - \alpha_1(m, 0)}{\alpha_1(m, 0)} \quad (16.12)$$

and is a practical way to present the actuation performance of any linear actuator.

Shown in Figure 16.4 are the deformation ratios and actuation strains calculated for a typical actuator ($G = 100$ kPa, $\epsilon = 2.5$, $x_2' = 100$ mm, $x_3' = 100$ μm). Although the strains referred to the initial unloaded, zero voltage state are high, the achieved actuation strains are not. In this rather simple model (which underestimates the elastic stress because it neglects strain hardening) it is seen that the actuation strain at a given voltage can be optimised by changing the load. A similar model, based on the nonlinear constitutive model proposed by Ogden,[18] and having no free parameters was shown to account fairly well for the actuation strain of a planar, silicone dielectric elastomer actuator.[9,19]

A model to describe the deformation of an annular dielectric elastomer actuator was presented by Yang et al.[20] It was based on Mooney's strain-energy formulation of the stress–strain behaviour, and was shown to be physically viable, but was not experimentally verified. The inflation and variable compliance of an electro-elastic rubber diaphragm was investigated analytically by Goulbourne et al.[21] This paper explored the electroelastic diaphragm mainly for application as a cardiac device; however no experimental results were obtained.

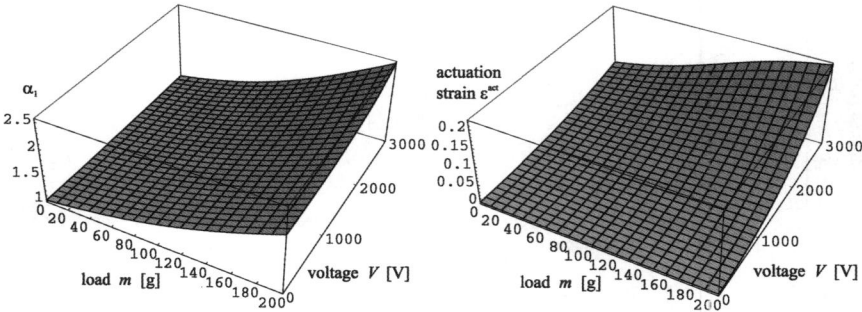

Figure 16.4 Left: deformation ratio of flat-film actuator, based upon Mooney–Rivlin model. Right: actuation strain for same model.

A model for the description of the actuation of a tube was presented by Carpi and de Rossi.[22] The measured actuation strains were in this case quite small, and permitted the use of the neo-Hookean constitutive equations. It was shown how the model, again with no free parameters, accounted well for the experimental actuation curves obtained on a medical-purpose silicone elastomer tube with grease electrodes.

A common standard actuator design for assessing the actuation properties of a given dielectric elastomer material is the expanding planar circle actuator of Figure 16.2 [*e.g.* Ref. 6]. This actuator was modeled by Wissler and Mazza,[23] specifically for the case of VHB4910. Their model was based upon the Yeoh model of hyperelasticity, and they were able to capture some of the viscoelastic (time-dependent) properties of the actuator by turning to a finite-element modeling approach. The models were based upon parameters obtained from experiments, in which the time dependence of the stress and strain was measured as well. It was shown experimentally that it was possible to predict the time-dependent actuation of a circular actuator.

To conclude this section, models that are meant to describe the large strain and actuation strain capabilities of dielectric elastomer actuators must include hyperelastic formulations of the constitutive equations.

Most of our discussion so far has been based on quasistatic models of performance. The often large components of viscoelastic loss of some polymers (particularly acrylics) require that a theoretical description must be based on hyperelastic constitutive equations with a viscoelastic part, in order to properly account for the time-dependent behaviour of actuation. Finally, for a complete analysis of a more complex actuator structure, a finite-element approach is often unavoidable.

16.6 Actuator Design: Geometry and Structure

The basic functional element of a dielectric elastomer actuator is the elastomeric film sandwiched between compliant electrodes as shown in Figure 16.1. This basic element must have both mechanical and electrical connections.

Mechanical connections are particularly critical since the polymer materials are relatively soft and poor design could result in little energy being coupled to the load. The dielectric elastomer material is essentially incompressible. In the functional element, the electrodes uniformly impart an effective compressive stress in the z-direction, which tends to decrease the thickness of the polymer film. The incompressible polymer film expands in area so that the total volume of polymer between the electrodes is conserved. Depending on the loading and constraints, the area expansion can occur equally in both planar directions or only in a single direction. In many applications, motion is desired only in a single planar direction. Undesirable deformation can be restrained by adding anisotropy into the material, or by changing the geometry, such as making the active area much wider than it is long.

Note that it is not always necessary to couple two directions of planar expansion into a single direction of output. For example, in diaphragm actuators that operate across a pressure gradient, such as might be used for a pump, the two directions of planar deformation are implicitly coupled to the output. In this case, the planar deformations displace the diaphragm and thereby impart momentum to the fluid being pumped. In some instances, the output of the actuator is produced by only a change in the geometry of a material or coating, with little or no force applied. In some cases it is desirable to expand the dielectric elastomer uniformly in both planar directions, such as in a solid-state optical aperture.

Note that the pressure that is produced in a dielectric elastomer, as given by eqn (16.2), does not depend on the elasticity of the polymer. Thus, while the dielectric elastomer materials themselves are relatively soft, the pressure or force need not be small.

If the elastomer is properly coupled to the load, then the pressure or forces that can be produced are determined by the strength of the electric field that can be produced in the material.

In the following section we highlight different applications of dielectric elastomer actuators. The various applications selected also highlight a variety of different actuator configurations. The ways in which these actuator configurations deal with the loading issues is discussed.

16.7 Applications

Dielectric elastomers may be expected to find application where they provide either performance or cost benefits. Since dielectric elastomers excel in many performance parameters and are composed of low-cost materials there are a wide range of possible applications.

As a prelude to our discussion of applications, it is useful to compare the performance metrics of dielectric elastomers against other existing actuators and intelligent materials. Table 16.1 provides such a comparison.

The comparison reveals that dielectric elastomers do not excel in all measures of performance but they do have good performance in a number of parameters. Combining this performance with the inherent low material cost and simple elemental structure reveals the following features that can be exploited in applications:

Simplicity: Large strains confer the ability to produce large motions with no bearings or sliding parts for very simple and quiet devices.
Matrices and modularity: Simple fabrication of large numbers of functional elements by patterning different electrode regions on a single substrate can produce large arrays of elements or "intelligent skins".
Biomimetic: The strain, stress, speed of response and mechanical impedance can match natural muscle suggesting use in robotics or prosthetics or orthotics.

Table 16.1 Comparison of candidate artificial muscle actuator technologies and natural muscle (adapted from Ref. [24]).

Actuator type (specific example)	Max. strain (%)	Max. pressure (MPa)	Specific elastic energy density (J/g)	Elastic energy density (J/cm³)	Max. efficiency (%)	Relative speed (full cycle)
Dielectric elastomers[a]						
Acrylic	380	8.2	3.4	3.4	60–80	Medium
Silicone	63	3.0	0.75	0.75	90	Fast
Electrostrictive polymer						
P(VDF-TrFE-CFE)[b]	4.5	45	>0.6	1.0	—	Fast
Graft Elastomer[c]	4	24	0.26	0.48	—	Fast
Electrostatic devices (integrated force array)[d]	50	0.03	0.0015	0.0015	>90	Fast
Electromagnetic (voice coil)[e]	50	0.10	0.003	0.025	>90	Fast
Piezoelectric						
Ceramic (PZT)[f]	0.2	110	0.013	0.10	>90	Fast
Single crystal (PZN-PT)[g]	1.7	131	0.13	1.0	>90	Fast
Polymer (PVDF)[h]	0.1	4.8	0.0013	0.0024	n/a	Fast
Shape-memory alloy (TiNi)[i]	>5	>200	>15	>100	<10	Slow
Thermal (expansion)[j]	1	78	0.15	0.4	<10	Slow
Electrochemomechanical conducting polymer (polyaniline)[k]	10	450	23	23	<1%	Slow

Table 16.1 (*Continued*).

Actuator type (specific example)	Max. strain (%)	Max. pressure (MPa)	Specific elastic energy density (J/g)	Elastic energy density (J/cm^3)	Max. efficiency (%)	Relative speed (full cycle)
Mechanochemical polymer/gels (polyelectrolyte)[l]	>40	0.3	0.06	0.06	30	Slow
Magnetostrictive (terfenol-D, Etrema Products)[m]	0.2	70	0.0027	0.025	60	Fast
Natural muscle (human skeletal)[n]	>40	0.35	0.07	0.07	>35	Medium

[a] Ref. [25]
[b] Ref. [26]
[c] Ref. [27]
[d] Ref. [28]
[e] These values are based on an array of 0.01 m thick voice coils, 50% conductor, 50% permanent magnet, 1 T magnetic field, 2 ohm cm resistivity, and 40 000 W/m^2 power dissipation.
[f] PZT B, at a maximum electric field of 4 V/μm.
[g] Ref. [29]
[h] PVDF, at a maximum electric field of 30 V/μm.
[i] Ref. [30]
[j] Aluminum, with a temperature change of 500°C.
[k] Ref. [31]
[l] Ref. [32]
[m] Terfenol-D Etrema Products.
[n] Ref. [33]

Soft: Low modulus of the materials confers comfort and safety when interacting loosely with humans or other objects.
Reversibility: The technology can "run backwards" efficiently for use as a generator or sensor.
Multifunctionality: The ability to act simultaneously as an actuator and a sensor or sensing structure.

There may be many other features that are important for certain niche applications. For example, the fact that dielectric elastomer actuators can be made without any metallic or highly conductive materials can allow for their use inside of MRI machines.[34]

In the following sections we present examples of actuators and devices that illustrate the range of possible applications and exploit many of these unique capabilities. Some of the applications are based on devices or systems that do not yet commonly exist outside laboratories or are simply not feasible with traditional actuator technologies. Exploiting the unique capabilities of EAPs may make such devices and systems, as well as a multitude of others, a reality.

16.7.1 Artificial Muscles for Biomimetic Robots

A variety of actuator configurations have been explored for use as artificial muscles in robotic applications. Rolled actuators have high force and stroke in a relatively compact package. Because, like muscle, this package is cylindrical, rolled actuators are an obvious choice for artificial muscles. One of the more successful types of rolled actuators is the "spring roll" (see Figure 16.5). This actuator is formed from an acrylic film that is stretched in tension and rolled around an internal spring.

Rolled actuators have been made in a variety of lengths and diameters with maximum stokes of up to 2 cm and forces of up to 33 N. Figure 16.5 shows a rolled dielectric elastomer actuator that weighs approximately 15 g and can

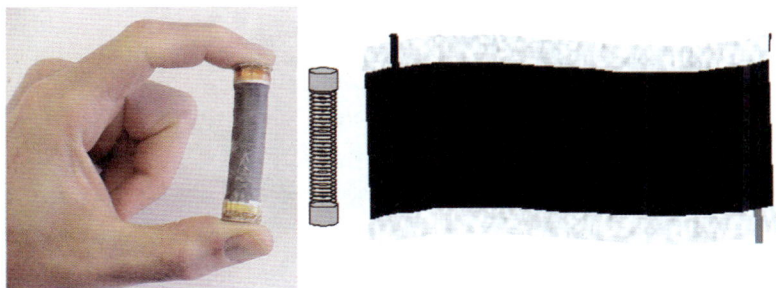

Figure 16.5 Spring roll linear actuator. The structure of the actuator is shown at right (adapted from Refs. 35, 36).

produce peak forces of more than 5 N with maximum displacements of 5 mm (a strain of about 25%).

The maximum performance of dielectric elastomer devices is expected to increase greatly as designs are improved, since these devices are still well below the measured peak performance of the polymer materials.

The main disadvantage of the rolled actuator is that it is somewhat more difficult to fabricate, since it is no longer a flat structure. Additionally, the energy coupling of a long roll is not as good as that of the bow-tie or trench type of actuator, since only one direction of deformation is coupled to the load. The energy coupling of short rolls is better, however.

SRI International developed a biomimetic robot that incorporated spring roll actuators (see Figure 16.6). The robot, named "Flex" is loosely based on the kinematics of a cockroach.[37,38] Flex is approximately 30 cm in length and could move at a maximum speed of 14 cm/s. While this speed and its performance on uneven terrain are not particularly good, it should be noted that Flex is one of the first self-contained robots powered by EAPs. Robots such as Flex serve as test beds for assessment of the advantages of muscle-like actuation. In the future, such robots will exploit the spring-like and viscoelastic behaviour of the artificial muscles to enable robots to run efficiently and reject disturbances due to obstacles or uneven terrain, much as in the dynamics of the musculo-skeletal system of a cockroach help it to run efficiently over uneven terrain without large disturbances of its torso.[39]

Artificial muscles based on rolled actuators have also been explored for use in prosthetic limbs.[40,41] SRI has also developed spring-roll actuators that have separately addressable electrode regions in different quadrants of the roll.[42] These actuators are capable of bending in addition to extension. Such actuators are being explored for a variety of applications including serpentine manipulators and robots.

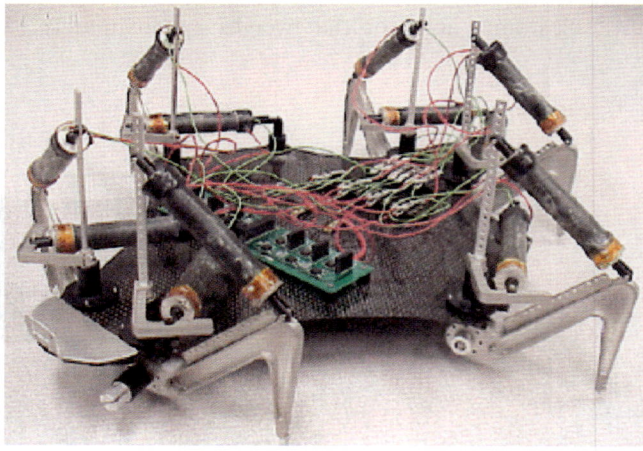

Figure 16.6 FLEX 2 Robot (adapted from Ref. 38).

16.7.2 Linear Actuators for Industrial Applications

While not directly exploiting their muscle-like behaviour, rolled actuators offer a more powerful and energy-efficient alternative to electromagnetic solenoids in applications such as proportional valves or other linear positioners.

An alternative to rolled actuators for industrial positioning applications is the Universal Muscle Actuator™ (UMA, Figure 16.7) developed by Artificial Muscle Inc. [Menlo Park, California, USA, www.artificialmuscle.com]. The UMA is based on two opposing diaphragm actuators attached to a common central platform. The narrow annular area of each diaphragm effectively couples most of the actuation stress to the central platform. UMAs have been made in range of sizes. The number of layers of the diaphragms can vary as well. The counteropposed design offers additional advantages. First, the actuator does not have to operate against an internal spring load (which reduces the energy available for output). Second, the actuator can be electrically driven in a "push-pull" mode in which the stroke response of the actuator is basically linear (in contrast to the more typical quadratic response evident from eqn (16.3)).

16.7.3 Diaphragm Actuators for Pumps and Arrays

Diaphragm actuators (*e.g.* Figure 16.8), consisting of a stretched film layer (or layers) over an opening in a relatively rigid substrate, can form the basis of extremely simple pumps. In this configuration the actuator itself also serves as the pumping membrane. Both directions of planar deformation of the elastomer are therefore coupled to the load (the fluid, in this case). The large strains

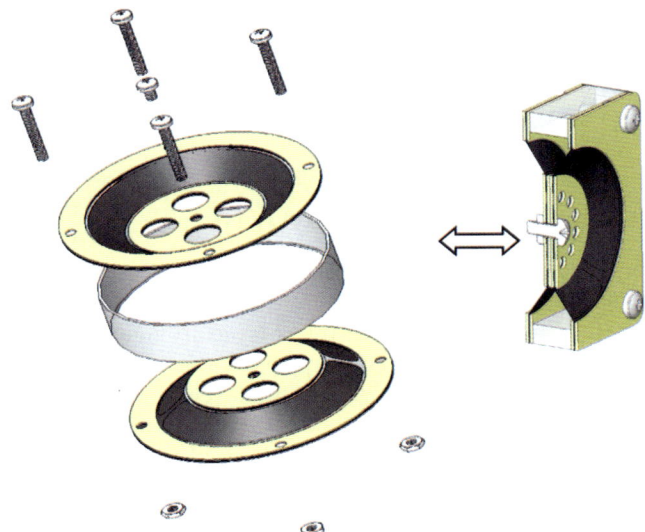

Figure 16.7 The Universal Muscle Actuator™.

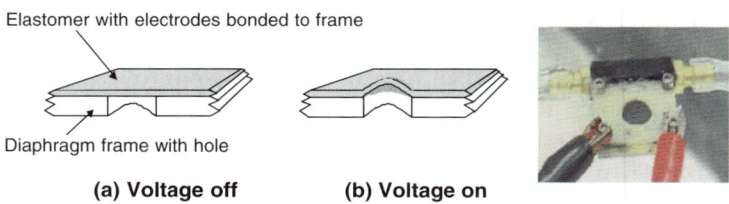

Figure 16.8 Diaphragm actuators for pumps: (a) basic structure, (b) acrylic diaphragm incorporated into a low-profile pump for liquids demonstrated by SRI International (pump diameter is approximately 1 cm).

of dielectric elastomers permit strokes that are significantly larger than those produced with other types of diaphragms, such as piezoelectrically driven diaphragms. At low frequencies, acrylic elastomers can actuate from an essentially flat configuration to one that is hemispherical. Passive flapper valves can be used, simplifying the pump design. The large volume displacements per stroke of pumps based on dielectric elastomers make them tolerant of greater leakage and backflow losses in the valves.

These simple dielectric elastomer pumps may be well suited for use as small implantable pumps, such as for drug-infusion applications that require a compact energy-efficient design and quiet operation. In implantable devices the biocompatibility of materials is important. Dielectric elastomers can be based on materials, such as silicones, that are known to have good biocompatibility. Pumps based on dielectric elastomers have been investigated for a variety of applications including artificial hearts[21] and synthetic jets for flow control over surfaces.[43]

Diaphragm actuators can also be fabricated in arrays using a common substrate of dielectric elastomers. Arrays could be used to make multistage pumps as well as a variety of other devices. Arrays have been used to make low-profile loudspeakers[44] and Braille displays.[45] Figure 16.9 shows several such devices.

16.7.4 Enhanced-thickness Mode Arrays

The actuators discussed so far have relied on the inplane deformation of the elastomer. In some cases it is desirable to use the change in thickness of the film. Since the film thickness of a single layer is typically much less than 1 mm, it would require a large number of layers to produce significant out-of-plane motions. Such an approach was taken in developing a haptic display by Jungmann and Schlaak,[46] where up to 200 layers of stacked silicone films were used. An alternative configuration is the "enhanced-thickness-mode actuator", as shown in Figure 16.10.[47]

In the enhanced-thickness-mode actuator, a relatively thick passive layer is deposited on top of the active dielectric elastomer. This layer thins or thickens as the layer below it expands or contracts. If the layer is soft enough, the

Dielectric Elastomer Actuators as Intelligent Materials

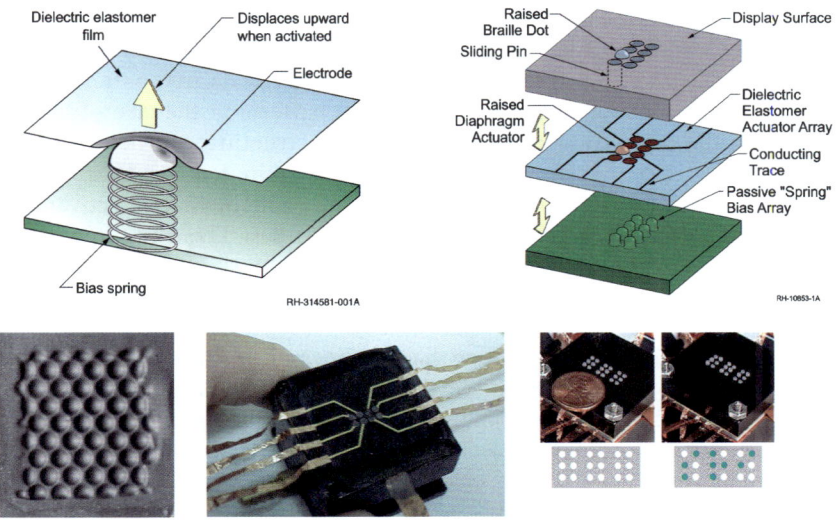

Figure 16.9 Devices based on arrays of diaphragm actuators that use a common elastomer film substrate.

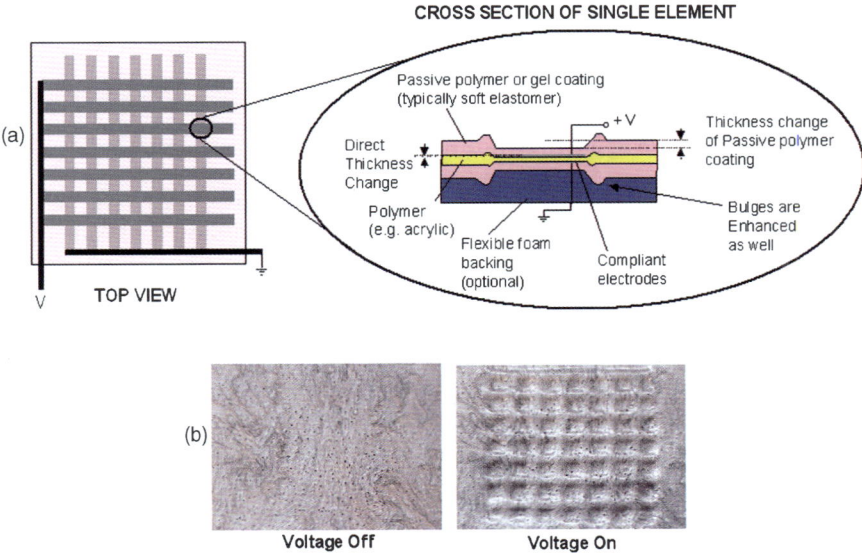

Figure 16.10 Enhanced-thickness-mode actuation: (a) underlying electrode grid pattern and schematic of basic element, (b) resulting surface change in silicone gel on top of acrylic dielectric elastomer, with electrodes in grid pattern (each block of the grid is about 1 cm × 1 cm).

thickness strain is identical to that of the underlying dielectric layer. Since the passive layer is much thicker than the dielectric layer, the absolute change in thickness is large. It is possible to create virtually any desired pattern of bumps and troughs on a single substrate by simply patterning the electrodes on one surface of the dielectric elastomer. The entire structure can be attached to a rigid frame. Alternatively, the structure can be laminated to a flexible foam or gel backing. In this way, a thin flexible skin can be made that can cover large or small areas and contains finely patterned electrodes. This skin could be quite rugged because it is made entirely out of rubbery materials. Although the device gives out-of-plane motion, it can be fabricated with only two-dimensional patterning.

Changes in thickness can change the surface texture or contours and are ideal for displays, integrated microfluidic devices, haptic displays and smart surfaces for controlling fluid or electromagnetic properties. If an enhanced-thickness-mode structure is brought into close proximity with a rigid layer, the bumps and troughs could act as valves or pumping elements in a "lab-on-a-chip" type of microfluidic system. The all-elastomer pad conforms comfortably to the body. This pad can have a massaging or sensory augmentation function.

The soft feel of dielectric elastomer devices is ideal for applications that require contact with the skin. The low modulus affords greater comfort, safety and better mechanical coupling with the soft tissue of the body.

The anticipated low cost and chemical safety of dielectric elastomer materials suggests that many devices could be made disposable, an important feature for many biomedical applications. Unlike materials such as piezoelectrics, it is easy to make very large-area polymer films. Thus it is conceivable that smart skins based on dielectric elastomers could cover the surface of aircraft or watercraft, for example.

16.7.5 Framed Actuator for Optics

A framed actuator consists of a polymer film stretched over a rigid frame. The active region defined by the area of overlapping electrodes on each side of the film can form any part of the area in the frame. A framed actuator can make a good linear actuator if it includes a mechanical coupling. In some cases, no energetic coupling to the environment is necessary. The circular actuator of Figure 16.3 is an example of a framed actuator with no mechanical output. Such actuators can be used for many applications, such as an optical switch, in which an opaque electrode area interrupts a light beam when actuated. The advantage of this approach is the simplicity and therefore low cost of the structure. The apparatus is basically solid state and has just one moving part—the electroded polymer film.

A binary switch of this type can be useful, but in some cases it may be desirable to continuously modulate the amount of light transmitted. This modulation can also be done with a framed actuator. The electrode gradually becomes less opaque as the electroded area increases. Figure 16.11 shows an example of such an actuator.

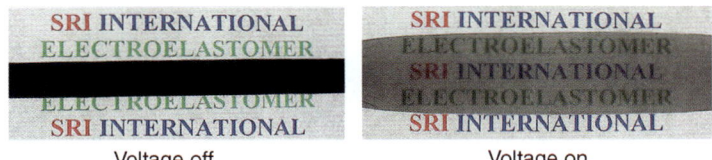

Figure 16.11 Solid-state optical aperture based on a framed actuator. The letters behind the electrode are visible only when the voltage is applied.[35]

In addition to simple switches, such framed actuators could be the basis of diffraction gratings with controllable spacing, Fresnel or other lenses with controllable focal length or a number of other electro-optic devices.

16.7.6 Sensors

We have already noted that dielectric elastomer actuators are also intrinsically position or strain sensors. It has been shown, for example, that measurement of the capacitance can indicate the displacement of a rolled actuator.[14,48] Dielectric elastomer sensors have certain advantages in their own right, even if the actuation function is not included. The large strain capabilities and environmental tolerance of dielectric elastomer materials allow for sensors that are very simple and robust. In sensor mode, it is often not important to maximise the energy density of the sensor materials since relatively small amounts of energy are converted. Thus, the selection of dielectric materials can be based on criteria such as biocompatibility, maximum strain, environmental survivability, and even cost. As sensors, dielectric elastomers can be used in all of the same configurations as actuators, as well as others. Figure 16.12 shows examples of several dielectric elastomer sensors, which can be thin tubes, flat strips, arrays of diaphragms, or large-area sheets. In many applications these sensors can replace bulkier and more costly devices such as potentiometers and encoders. In fibre or ribbon form, a dielectric elastomer sensor can be woven into textiles and provide position feedback from human motion. Such sensors have been shown to be capable of measuring respiration, replacing bulkier pressure-sensing chest bands. The use of such strain sensors has also been explored for measuring the expansion of arteries in animal models. These simple one-piece sensors can be made very small and thin. They can replace saline- or mercury-filled tubes that are commonly used by physiologists to measure the expansion of body parts. The softness and compliance of dielectric elastomers is ideal for interaction with the human body. Dielectric elastomer sensors can also be laminated to structures or skins to provide position information for multi-functional smart materials. Sensors such as the diaphragm array can be used to measure force or pressure as well as motion. Enhanced thickness mode pads can act as sensory skins.

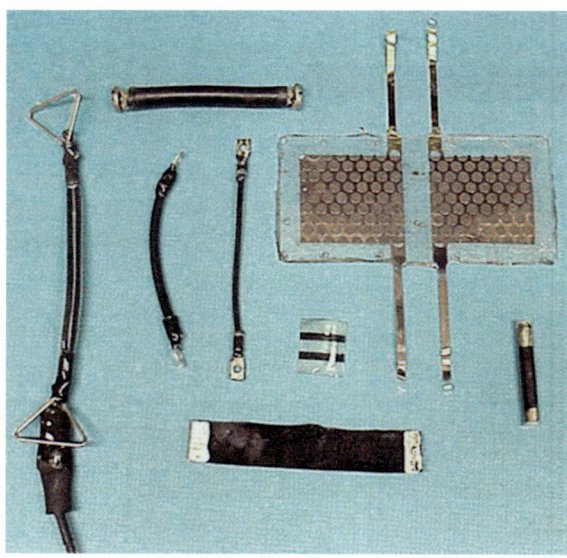

Figure 16.12 Representative dielectric elastomer sensors (adapted from Ref. 25).

16.7.7 Generators

Thus far, we have described actuator and sensor applications of dielectric elastomers. Since the electromechanical phenomena that causes deformation in actuation is reversible, dielectric elastomers also function quite well in generator mode. In generator mode, a voltage is applied across the elastomer and the elastomer is deformed with external work.[49] As the shape of the elastomer changes, the effective capacitance also changes and, with the appropriate electronics, electrical energy can be generated. Just as the maximum energy density of the dielectric elastomer material as an actuator exceeds that of other field-activated materials, the energy density in generator mode is high. An energy density of 0.4 J/g has been demonstrated with the acrylic elastomer. Performance projections based on material properties are even greater.

Many of the design principles and potential advantages of dielectric elastomers that are true in actuator mode also apply to generator-mode operation. In particular, this type of generation is well suited to applications where electrical power must be produced from relatively large motions. The stress–strain match of the dielectric material to natural muscle further suggests a good impedance match to motions produced by natural muscle, such as human activity.

Figure 16.13 shows a device developed by SRI International that can be used to capture the energy produced in walking or running when the heel strikes the ground. This generator, located in the heel of a shoe, effectively couples the compression of the heel to the deformation of an array of multilayer diaphragms. Other configurations are also being explored. The use of dielectric

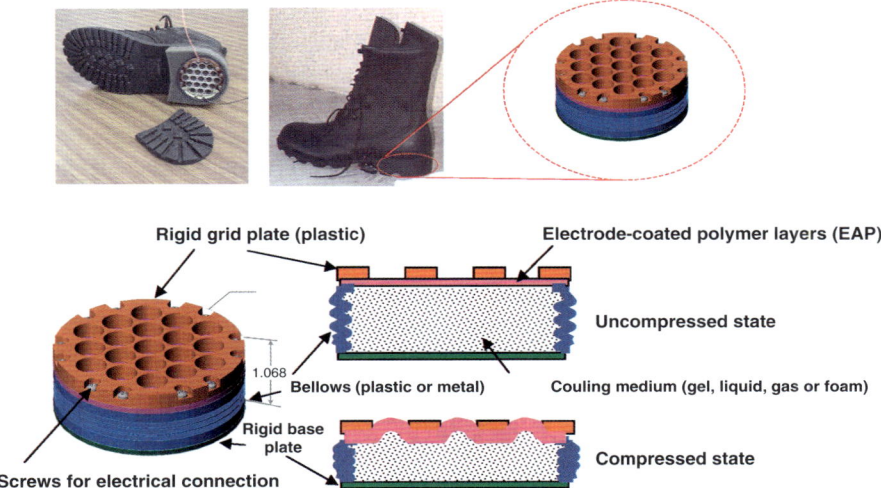

Figure 16.13 Heel-strike generator located in a boot heel and based on the deformation of dielectric elastomers during walking or running.

elastomers is appropriate for a heel-strike generator because of the large deflections produced in the heel with low to moderate pressures. The heel-strike generator has thus far produced a maximum of 0.8 J per cycle. With further development we expect that such devices will be able to generate about 1 W per foot during normal walking. The generator is being developed for the military to help supply the power needs of future soldiers, but also has many commercial applications including cell phone recharging and PDAs or power on-shoe devices such as lights, health and performance monitors, or navigation devices.

16.8 Implementation Challenges for Dielectric Elastomers

Dielectric elastomers have shown exceptional performance on a material and actuator level. However, in order to be a successful actuation technology, they must also make sense at a system level.

One critical system-level issue for dielectric elastomers is the operating voltage. In order to get maximum performance, the operating voltage must be relatively high. At a system level, high voltage has both advantages and disadvantages. Typically a separate voltage conversion unit is required. This unit can increase the overall size, weight and cost of the device and decrease efficiency (particularly if the device is battery-powered). On the other hand, high voltage also implies low current and therefore allows thinner wires and smaller connectors. Further, while high voltage does present a risk of electrical shock, we note that this risk can be mitigated with proper packaging and electrical circuit design. Further, in many cases, the maximum current can be low enough so as not to present a serious safety threat.

Voltage converters can be small compared to the actuator size. Therefore, another option to address the high voltage issue is to use distributed voltage conversion. Each actuator can have a small voltage converter integrated into its packaging. In this approach, the packaged actuator is a low-voltage device that could even operate on battery power.

The overall efficiency of dielectric elastomers can be quite high, as discussed above. However, for cyclic operation, the highest efficiency requires charge recovery since a dielectric elastomer actuator is basically a capacitor. Fortunately, many charge-recovery circuits and amplifiers are already known to operate with piezoelectric devices and could easily be applied to dielectric elastomers as well.

Perhaps the most critical challenge for dielectric elastomers is lifetime. The exceptional performance demonstrated at the material level cannot, in general, be maintained in packaged actuators for the extremely large numbers of cycles demanded of many applications. The typical mode of failure is an electrical breakdown. It is necessary to reduce the maximum permissible operating field in order to achieve long lifetimes. Understanding the conditions for obtaining high output while maintaining an acceptable average lifetime remains a challenge.

There are several ways to improve lifetime. One is to develop improved dielectric materials, as will be discussed below. Another way that has already been mentioned is similar to the approach that has been taken with thin-film capacitors, to allow for self-healing. Such healing or tolerance of an occasional failure could dramatically improve lifetime.

16.9 The Future: Materials Development for New Elastomers

The characteristics of the elastomer dielectric are critical to the performance of this actuator technology. The majority of research has thus far focused on the use of existing commercially available polymers. However, the development of new polymers or polymer formulations specifically optimised for use as dielectric elastomers is beginning to emerge and could allow for dramatic improvements in the technology in years to come. It is therefore fitting that we conclude this chapter with a look to the future of materials development.

As already mentioned, the dielectric elastomer serves several purposes for the dielectric elastomer actuator. Both its electrical and mechanical properties are important, hence there are several approaches to improving actuator performance by varying the properties of the dielectric elastomer. It must be emphasised, however, that it is not likely that one particular dielectric elastomer material will be optimal for all applications. Rather, the large number of possible applications demands that a variety of actuation strains, stresses, response speeds, viscoelastic losses and their combinations are possible.

Typically, low mechanical loss is desired, but for some applications an inherent passive damping or shock absorption is necessary, which can then be facilitated by a manipulation of the viscoelastic loss spectrum. Currently, researchers are attempting to expand the library of materials available for applications, most often basing their approach on the two hitherto most successful elastomers, polydimethyl siloxane (silicone) rubber and polyacrylate rubber.

16.9.1 Improving Elastic Properties

The design of materials for high-strain applications is based on the general nature of the stress–strain behaviour displayed by most hyperelastic elastomers that are able to support strains higher than 500%.

Since the dielectric elastomer is typically found in a prestrained configuration, the Young's modulus (Region I, Figure 16.14) becomes secondary to the modulus found at higher strains. II marks the plateau region, where the elastomer becomes softer, and III marks the strain-hardening region, at which a substantial number of chains in the elastic network reach their full extensibility. The behaviour presented in Figure 16.14 is generally found for all elastomers, with variations in the range of the regions and in the particular slope progression of the curve. An ideal elastomer for prestrained actuator applications would have a very wide plateau region (Region II), and a very late onset of strain-hardening (Region III). For actuator designs that do not incorporate prestrain, the Young's modulus (Region I) becomes overwhelmingly important, and if high strain is sought, the modulus must be low.

The structure of the elastomer network determines its elastic properties, and the actual structure itself depends upon a number of other parameters, some of which can be controlled.[50] The chain length between crosslinks plays a minor role, since entanglements take over the role of crosslinks when the chain length exceeds the entanglement length, which is commonly the case for elastomers useful for dielectric elastomer actuators. To lower the overall stress the number of entanglements must be minimized, which can be achieved by adjusting the reaction conditions. One approach that results in a reduction of the number of

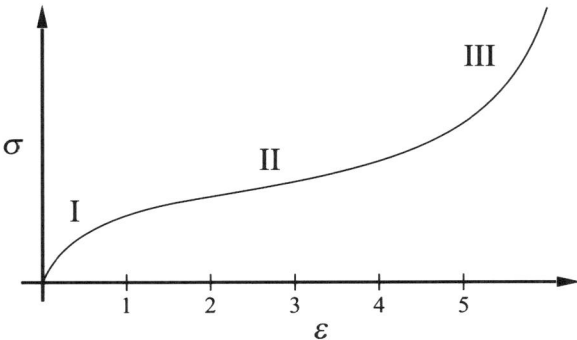

Figure 16.14 Typical hyperelastic stress–strain curve.

entanglements involves adding a considerable (10% to 40% or more) amount of inert solvent during crosslinking. Another approach is to apply a shearing stress during crosslinking; the shearing stress aligns the polymer chains, thus smoothing out the entanglements.[50]

16.9.2 Improving Dielectric Properties

The permittivity of a dielectric elastomer is a frequency dependent, complex value, $\varepsilon^*(\omega) = \varepsilon_R(\omega) - j\varepsilon_I(\omega)$ (the use of a negative sign is conventional). The imaginary part of the permittivity is commonly referred to as dielectric loss, and written in terms of the real part of frequency-dependent conductivity, $\varepsilon_I(\omega) = \sigma_R(\omega)/j\omega$. At low frequencies (mHz to Hz) the permittivity of most rubbers is fairly constant, and is therefore often referred to as the dielectric constant of that material. The dielectric constant for isotropic, polar rubbers is between 2 and 5, for instance, for silicones and polyacrylates. Partly crystalline polyurethane rubbers may reach values of the dielectric constant between 3 and 14.[3] Then, the mechanism of actuation becomes more complex, involving specific electrostriction and also polarisation by charge injection.[51]

Polyurethane served as the base polymer in a recently reported attempt to increase the dielectric constant by adding a high dielectric constant filler, in this case copper phthalocyanine (o-CuPc).[52] With a careful self-assembly approach, the dielectric constant was made to increase to more than 1000, while only doubling the elastic modulus, allowing for an actuation strain of 8% at 12 MV/m. The adding of small, high dielectric constant particles, such as certain ceramics, in an effort to increase the dielectric constant of the composite elastomer has been reported by several groups (e.g. Ref. 53).

As noted, increases in dielectric constant are typically accompanied by an undesirable decrease in breakdown strength and an increase in losses. Thus, while the performance of such polymers at a given voltage or field is good, the overall maximum strains or stresses produced by such materials are typically not as high as the lower dielectric constant materials already in use.

16.9.3 Improving Breakdown Properties

Of the elastomer properties, possibly the electric breakdown strength is the most important, since the Maxwell stress is proportional to the electric field squared (eqn (16.2)). Breakdown manifests itself as a spark between the two electrodes, which can occur anywhere in the film, i.e. in the centre of the actuator electrodes, on the edges or near the electrical leads.

Regardless of the breakdown mechanism, the thickness of the elastomer film must have a high degree of uniformity. An indentation that gives rise to a thickness decrease to one half of the nominal thickness doubles the local electric field, increasing the likelihood of breakdown in that spot. Other obvious reasons for breakdown are defects and impurities, which can be dealt with through an improvement of manufacturing conditions.

If defects are impossible to avoid, a practical solution can be to place two elastomer films back to back. If two defects are placed on top of each other, breakdown is likely to occur. However, if the defect density is sufficiently low, it is unlikely that two defects will be placed on top of each other, thus breakdown is avoided.[14]

As noted, prestrain has been shown to enhance the electric breakdown field tremendously, in VHB 4910. From no prestrain to a biaxial prestrain of 500% by 500%, the breakdown field increased by almost 10 times.[8] Building upon this discovery, it was shown that the actuation properties of VHB 4910 can be improved by soaking the material in a monomer material, which is allowed to diffuse into the VHB 4910 completely, and then crosslinked by increasing the temperature.[48] This approach locks the prestrain, while balancing out the prestress, which can be of great help in some applications.

16.10 Conclusion

The past 15 years have seen the dielectric elastomer actuator concept evolve from an idea to a commercially available product, carried through by major efforts in materials development and discovery, actuator design, modeling and theory and applications studies. While this initial development effort was largely carried out by groups directly connected with the principle inventors of the technology, a large number of researchers and groups from all over the world are now adding their efforts. We can expect even greater growth in the field in the next few years.

The unique capabilities of dielectric elastomers should allow for their use in a wide variety of applications, not just as replacements for existing actuators but as enablers of new devices and systems. Realising this vision will require addressing lifetime issues, manufacturability and systems issues such as electrical requirements. Successful innovation does not require any fundamental improvement in performance. Nonetheless, great advances are still possible. Many groups are undertaking research to develop new materials with drastically improved mechanical and dielectric properties. New actuator designs are constantly being developed. Perhaps most importantly, new applications that exploit the unique capabilities of dielectric elastomers are constantly being explored. With these developments, the goal of providing compact, lightweight and powerful artificial muscles is within reach.

References

1. W. Röntgen, *Ann. Phys.*, 1880, **247**, 771.
2. M. Zhenyi, J.I. Scheinbeim, J.W. Lee and B.A. Newman, *J. Polym. Sci., Part B: Polym. Phys.*, 1994, **32**, 2721.
3. T. Ueda, T. Kasazaki, N. Kunitake, T. Hirai, J. Kyokane and K. Yoshino, *Synth. Met.*, 1997, **85**, 1415.
4. R.E. Pelrine, R.D. Kornbluh and J.P. Joseph, *Sens. Actuators, A*, 1998, **64**, 77.

5. S. Wax and R. Sands, *Proc. SPIE-Int. Soc. Opt. Eng.*, 1999, **3669**, 2.
6. R. Pelrine, R. Kornbluh, Q. Pei and J. Joseph, *Science*, 2000, **287**, 836.
7. G.R. Love, *J. Am. Ceram. Soc.*, 1990, **73**, 323.
8. G. Kofod, R.D. Kornbluh, R. Pelrine and P. Sommer-Larsen, *Proc. SPIE-Int. Soc. Opt. Eng.*, 2001, **4329**, 141.
9. G. Kofod, Ph.D. Thesis, Risø National Laboratory/Danish Technical University, Pitney Bowes Management Services, Denmark A/S, 2001.
10. F. Carpi, P. Chiarelli, A. Mazzoldi and D. de Rossi, *Sens. Actuators, A*, 2003, **107**, 85.
11. H.F. Schlaak, M. Jungmann, M. Matysek and P. Lotz, *Proc. SPIE-Int. Soc. Opt. Eng.*, 2005, **5759**, 121.
12. T. Li, Z. Huang, Z. Suo, S. Lacour and S. Wagner, *Appl. Phys. Lett.*, 2004, **85**, 3435.
13. R.D. Kornbluh, R. Pelrine, J. Joseph, R. Heydt, Q. Pei and S. Chiba, *Proc. SPIE-Int. Soc. Opt. Eng.*, 1999, **3669**, 149.
14. M. Benslimane, P. Gravesen and P. Sommer-Larsen, *Proc. SPIE-Int. Soc. Opt. Eng.*, 2002, **4695**, 150.
15. S.P. Lacour, S. Wagner, Z. Huang and Z. Suo, *Appl. Phys. Lett.*, 2003, **82**, 2404.
16. Yuri M. Shkel and Daniel J. Klingenberg, *J. Appl. Phys.*, 1998, **83**, 7834.
17. T. Yamwong, A.M. Voice and G.R. Davies, *J. Appl. Phys.*, 2002, **91**, 1472.
18. R.W. Ogden, *Proc. R. Soc. London, Ser. A*, 1972, **326**, 565.
19. G. Kofod and P. Sommer-Larsen, *Sens. Actuators, A*, 2005, **122**, 273.
20. E.E. Yang, M. Frecker and E. Mockensturm, *Proc. ASME Int. Mech. Eng. Cong. IMECE*, 2003, 43676.
21. N. Goulbourne, M. Frecker, E. Mockensturm and A. Snyder, *Proc. SPIE-Int. Soc. Opt. Eng.*, 2003, **5051**, 319.
22. F. Carpi and D. de Rossi, *Mater. Sci. Eng., C*, 2004, **24**, 555.
23. A. Wissler and E. Mazza, *Sens. Actuators, A*, 2005, **120**, 184.
24. R. Kornbluh, R. Pelrine, Q. Pei, M. Rosenthal, S. Stanford, N. Bonwit, R. Heydt, H. Prahlad and S. Shastri, in *Electroactive Polymer (EAP) Actuators as Artificial Muscles: Reality, Potential and Challenges*, Y. Bar-Cohen, ed., SPIE Press Bellingham, Washington, 2004, 2nd edn Chapter 16.
25. R. Pelrine, J. Eckerle and S. Chiba, *Proc. 3rd Intl. Symp. Micro Mach. Hum. Sci.*, Nagoya, Japan, 1992.
26. F. Xia, H. Li, C. Huang, M.Y.M. Huang, H. Xu, F. Bauer, Z. Cheng and Q.M. Zhang, *Proc. SPIE-Int. Soc. Opt. Eng.*, 2003, **5051**, 133.
27. J. Su, personal communication.
28. S. Bobbio, M. Kellam, B. Dudley, S. Goodwin-Johansson, S. Jones, J. Jacobson, F. Tranjan and T. DuBois, *Proc. IEEE-MEMS*, 1993, 149.
29. Seung-Eek Park and T.R. Shrout, *J. Appl. Phys.*, 1997, **82**, 1804.
30. I. Hunter, S. Lafontaine, J. Hollerbach and P. Hunter, *Proc. IEEE-MEMS*, 1991, 166.
31. R. Baughman, L. Shacklette, R. Elsenbaumer, E. Pichta and C. Becht, in *Conjugated Polymeric Materials: Opportunities in Electronics, Optoelectronics*

and *Molecular Electronics*, J.L. Bredas and R.R. Chance ed., Kluwer Academic Publishers, The Netherlands, 1990, p. 559.
32. M. Shahinoor, *J. Intel. Mater. Sys. Struct.*, 1995, **6**, 307.
33. I. Hunter and S. Lafontaine, *Proc. IEEE, Solid-State Sensor and Actuator Workshop, 5th Technical Digest*, 1992, 178.
34. J. Vogan, A. Wingert, J. Plante, S. Dubowsky, M. Hafez, D. Kacher and F. Jolesz, *Proc. IEEE Int. Conf. Rob. Autom. ICRA*, 2004, 2498.
35. Q. Pei, R. Pelrine, S. Stanford, R.D. Kornbluh, M.S. Rosenthal, K. Meijer and R.J. Full, *Proc. SPIE-Int. Soc. Opt. Eng.*, 2002, **4698**, 246.
36. R. Kornbluh, R. Pelrine, Q. Pei, R. Heydt, S. Stanford, S. Oh and J. Eckerle, *Proc. SPIE-Int. Soc. Opt. Eng.*, 2002, **4698**, 254.
37. J. Eckerle, S. Stanford, J. Marlow, R. Schmidt, S. Oh, T. Low and S. Shastri, *Proc. SPIE-Int. Soc. Opt. Eng.*, 2001, **4329**, 269.
38. R. Pelrine, R.D. Kornbluh, Q. Pei, S. Stanford, S. Oh, J. Eckerle, R.J. Full, M.S. Rosenthal and K. Meijer, *Proc. SPIE-Int. Soc. Opt. Eng.*, 2002, **4695**, 126.
39. R.J. Full, D.R. Stokes, A.N. Ahn and R.K. Josephson, *J. Exp. Biol.*, 1998, **201**, 997.
40. H.M. Herr and R.D. Kornbluh, *Proc. SPIE-Int. Soc. Opt. Eng.*, 2004, **5385**, 1.
41. R. Kornbluh, J. Bashkin, R. Pelrine, H. Prahlad and S. Chiba, *J. Jpn. Soc. Pol. Proc.*, 2004, **16**, 631.
42. Q. Pei, M.A. Rosenthal, R. Pelrine, S. Stanford and R.D. Kornbluh, *Proc. SPIE-Int. Soc. Opt. Eng.*, 2003, **5051**, 281.
43. A. Pimpin, Y. Suzuki and N. Kasagi, *Proc. 17th IEEE Int. Conf. MEMS*, 2004, 478–481.
44. R. Heydt, R. Pelrine, J. Joseph, J. Eckerle and R. Kornbluh, *J. Acoust. Soc. Am.*, 2000, **107**, 833.
45. R. Heydt and S. Chhokar, *Proc. 23rd Int. Disp. Res. Conf.*, 2003, 7.5.
46. M. Jungmann and F. Schlaak, *Proc. ACTUATOR*, 2002, B7.2.
47. H. Prahlad, R. Pelrine, R.D. Kornbluh, P. von Guggenberg, S. Chhokar, J. Eckerle, M.S. Rosenthal and N. Bonwit, *Proc. SPIE-Int. Soc. Opt. Eng.*, 2005, **5759**, 102.
48. Q. Pei, R. Pelrine, M.S. Rosenthal, S. Stanford, H. Prahlad and R.D. Kornbluh, *Proc. SPIE-Int. Soc. Opt. Eng.*, 2004, **5385**, 41.
49. R. Pelrine, R.D. Kornbluh, Q. Pei, J. Eckerle, P. Jeuck, S. Oh and S. Stanford, *Proc. SPIE-Int. Soc. Opt. Eng.*, 2001, **4329**, 148.
50. A.L. Larsen, P. Sommer-Larsen and O. Hassager, *Proc. SPIE-Int. Soc. Opt. Eng.*, 2004, **5385**, 108.
51. Masashi Watanabe, Hirofusa Shirai and Toshihiro Hirai, *J. Appl. Phys.*, 2001, **90**, 6316.
52. C. Huang, M., Li, Y. and Rabeony, M. Zhang, *Appl. Phys. Lett.*, 2005, **87**, 182901.
53. J. Szabo, J. Hiltz, C. Cameron, R. Underhill, J. Massey, B. White and J. Leidner, *Proc. SPIE-Int. Soc. Opt. Eng.*, 2003, **5051**, 180.

CHAPTER 17
Azobenzene Polymers as Photomechanical and Multifunctional Smart Materials

KEVIN G. YAGER AND CHRISTOPHER J. BARRETT

Department of Chemistry, McGill University, 801 Sherbrooke Street W., Montreal, QC, Canada

17.1 Introduction

Among the many classes of novel advanced materials now being researched worldwide are photofunctional and photoresponsive materials. The advantages of materials that respond to light are numerous: natural light is a plentiful and effectively unlimited energy source, light is correlated to many everyday (and even industrial) activities, light stimulation can be performed remotely and without disturbing intervening materials, and light activation can be highly localised and specific. The intent with photofunctional materials is to generate substances that respond automatically to light in a desired way. As with all "smart" materials, the desired response is programmed into the material architecture, rather than being actively induced after material is prepared. Photofunctional smart materials promise to make a significant impact in modern industry, which relies heavily upon optical processes (e.g. telecommunications and lithography), and in daily life, where photoresponsive materials could be designed to respond appropriately to changing illumination. A variety of photoresponsive molecules have been characterised, including those that photodimerise, such as coumarins[1] and anthracenes;[2] those that allow intramolecular photoinduced bond formation, such as fulgides, spiro-pyrans,[3] and diarylethenes;[4] and those that exhibit photoisomerisation, such as stilbenes,[5] crowded alkenes and azobenzene. Photoinduced molecular motion can be used as a molecular rotor[6,7] or as a molecular device.[8]

This chapter will discuss azobenzene materials. Azobenzene is a small-molecule dye with strong absorption properties and a unique photochemistry.

Azobenzene and its substituted-derivatives (usually collectively referred to simply as "azobenzenes") have a remarkably clean and efficient photochemistry that has been exploited to generate a wide range of functional and responsive materials. This chapter aims to highlight the many ways in which azobenzene (azo) photochemistry can be exploited to generate useful and even "smart" materials. We will provide a broad overview of the numerous systems that have incorporated azobenzene units. These photoresponsive systems can be roughly divided into photoswitching materials (which exhibit two distinct states, with interconversion elicited by light), photoresponsive materials (which exhibit a continuous response to varying light levels), and photodeformable materials (which exhibit mechanical motion/deformation in response to light, with little change in static properties after light exposure). Obviously these categories are not rigidly defined, and are used only to structure the discussion. What all these materials have in common is that significant changes in a material can be elicited with light irradiation, in large part due to the amplification of azo molecular motion through cooperative effects.

17.2 Azobenzenes

Azobenzene is an aromatic molecule where an azo linkage ($-N=N-$) joins two phenyl rings (Figure 17.1). A large class of compounds (usually simply referred to as "azobenzenes" or simply "azos") can be obtained by substituting the aromatic rings with groups. This class of chromophores share numerous spectroscopic and photophysical properties. In particular, the conjugated system gives rise to a strong electronic absorption in the UV and/or visible portions of the spectrum. The exact spectrum can be tailored via the ring-substitution pattern. The azo molecules are also rigid and anisotropic, making

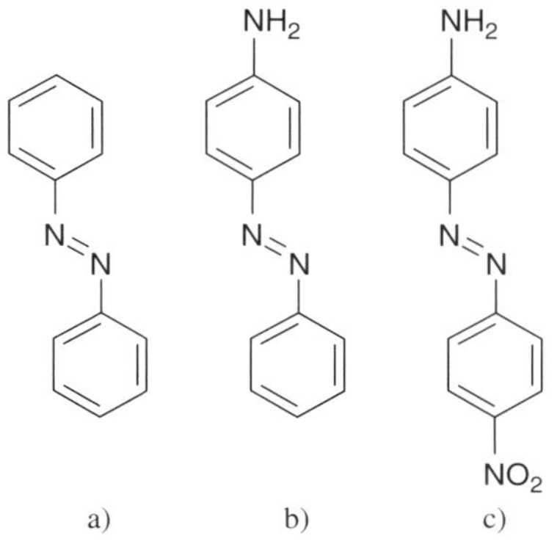

Figure 17.1 Examples of azobenzene chromophores of the (a) azobenzene class, (b) the aminoazobenzene class, and (c) the pseudo-stilbene class. The spectroscopic and photophysical properties change dramatically with changes in ring-substitution pattern.

them ideal liquid crystal mesogens under appropriate conditions. Both smallmolecule and polymeric azobenzenes can exhibit LC phases.[9,10] The most interesting behaviour common to all azos is the efficient and reversible photoisomerisation, which occurs upon absorption of a photon within the absorption band (Figure 17.2). Azobenzenes have two isomeric states: a thermally stable *trans* configuration, and a metastable *cis* form. Under irradiation, *trans*-azobenzenes will be converted to the *cis* form, which will thermally revert to the more stable *trans* on a timescale dictated by the molecule's particular substitution pattern. This exceedingly clean photochemistry gives rise to the numerous remarkable photoswitching and photoresponsive behaviours observed in these systems.

The chromophores can be divided into three spectroscopic classes (see Figure 17.1 for examples), as described by Rau:[11] azobenzene-type molecules, which are similar to the unsubstituted azobenzene; aminoazobenzene-type molecules, which are *ortho-* or *para*-substituted with an electron-donating group; and pseudo-stilbenes, which are substituted at the 4 and 4' positions with an electron-donating and an electron-withdrawing group (such as an amino and a nitro group). The strong absorption spectra give rise to the prominent colours of the compounds: yellow, orange, and red, for the azobenzenes, aminoazobenzenes, and pseudo-stilbenes, respectively. The pseudo-stilbene class (so named due to the similarity with stilbene photochemistry) is especially interesting because the contraposed electron-withdrawing and -donating groups create a highly asymmetric electron distribution within the conjugated system. This leads to a large molecular dipole, and inherent nonlinear optical properties. Also noteworthy is that the *trans* and *cis* absorption spectra of the pseudo-stilbenes generally have significant overlap. Thus, in these systems a single wavelength of illuminating light can induce both the forward (*trans* → *cis*) and the reverse (*cis* → *trans*) photoisomerisation. This leads to a continuous cycling of chromophores between isomeric states, which can be beneficial for many photoresponsive effects. For the other classes of azos, the absorption spectra will not overlap, meaning that two different wavelengths of light can be

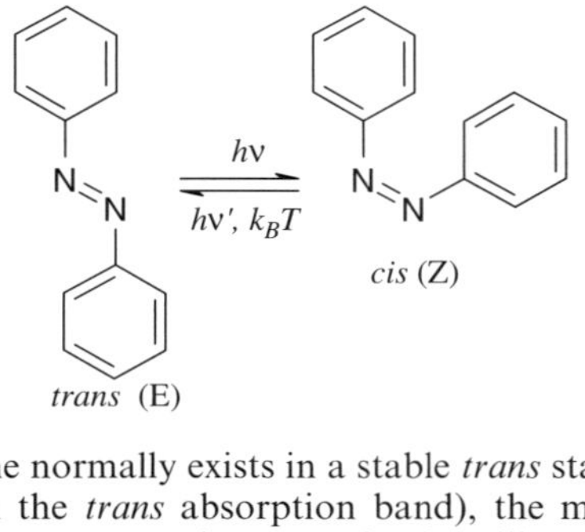

Figure 17.2 Azobenzene normally exists in a stable *trans* state. Upon absorption of a photon (in the *trans* absorption band), the molecule isomerises to the metastable *cis* state. The *cis* molecule will thermally relax back to the *trans* state, or this isomerisation can be induced with irradiation at a wavelength in the *cis* absorption band.

used to switch between two different states, which is ideal for photoswitchable materials.

It should be emphasised, however, that the *cis* state is nearly universally metastable. That is, a *cis*-azobenzene will thermally relax to the *trans* state on a timescale dictated by its ring-substitution pattern and local environment. This can be considered inconvenient from the point of view of generating stable two-state photoswitchable systems. Lifetimes of the *cis* state are typically on the order of hours, minutes, and seconds, for the azobenzenes, aminoazobenzenes, and pseudo-stilbenes, respectively. The energy barrier for thermal isomerisation is on the order of 90 kJ/mol.[12,13] Considerable work has gone into elongating the *cis* lifetime, with the goal of creating truly bistable photoswitchable systems. Bulky ring substituents can be used to hinder the thermal backreaction. For instance, a polyurethane main-chain azo exhibited a lifetime of 4 days (thermal rate-constant of $k = 2.8 \times 10^{-6}$ s^{-1}, at 3 °C),[14] and an azobenzene *para*-substituted with bulky pendants had a lifetime of 60 days ($k < 2 \times 10^{-7}$ s^{-1}, at room temperature).[15] The conformational strain of macrocylic azo compounds can also be used to lock the *cis* state, where lifetimes of 20 days ($k = 5.9 \times 10^{-7}$ s^{-1}),[16] 1 year (half-life 400 days, $k = 2 \times 10^{-8}$ s^{-1}),[17,18] or even 6 years ($k = 4.9 \times 10^{-9}$ s^{-1})[19] were observed. Similarly, using the hydrogen bonding of a peptide segment to generate a cyclic structure, a *cis* lifetime of ~40 days ($k = 2.9 \times 10^{-7}$ s^{-1}) was demonstrated.[20] Of course, one can also generate a system that starts in the *cis* state, and where isomerisation (in either direction) is completely hindered. For instance, attachment to a surface,[21] direct synthesis of ring-like azo molecules,[22] and crystallisation of the *cis* form[23,24] can be used to maintain one state, but such systems are obviously not bistable photoswitches. The azobenzene thermal back-relaxation is generally a first-order kinetic process, although a polymer matrix can lead to a distribution of constrained conformations, and hence anomalously fast decay contributions.[25–28] Similarly, matrix crystallinity tends to increase the decay rate.[29]

17.3 Azobenzene Systems

Azobenzenes are robust moieties, and are amenable to incorporation into a wide variety of materials. The azo chromophore can be doped into a matrix, or covalently attached to a polymer. Both amorphous and liquid-crystalline (LC) systems have been extensively investigated (for examples, see Figure 17.3). Other studies have demonstrated self-assembled monolayers and superlattices,[30] sol-gel silica glasses,[31] and biomaterials.[32–34] The azo group is sufficiently nonreactive that it can be incorporated into numerous synthesis strategies, and has thus been included in crown ethers,[35] cyclodextrins,[36,37] proteins,[38] and three-dimensional polycyclics.[39,40] Thin polymer films are a convenient material matrix for study of azo materials, and realisation of useful photofunctional devices. Although doping the chromophores in a matrix is convenient,[41,42] the resultant films often exhibit instabilities, such as phase separation and microcrystallisation. This occurs due to the mobility of the azo chromophores in the

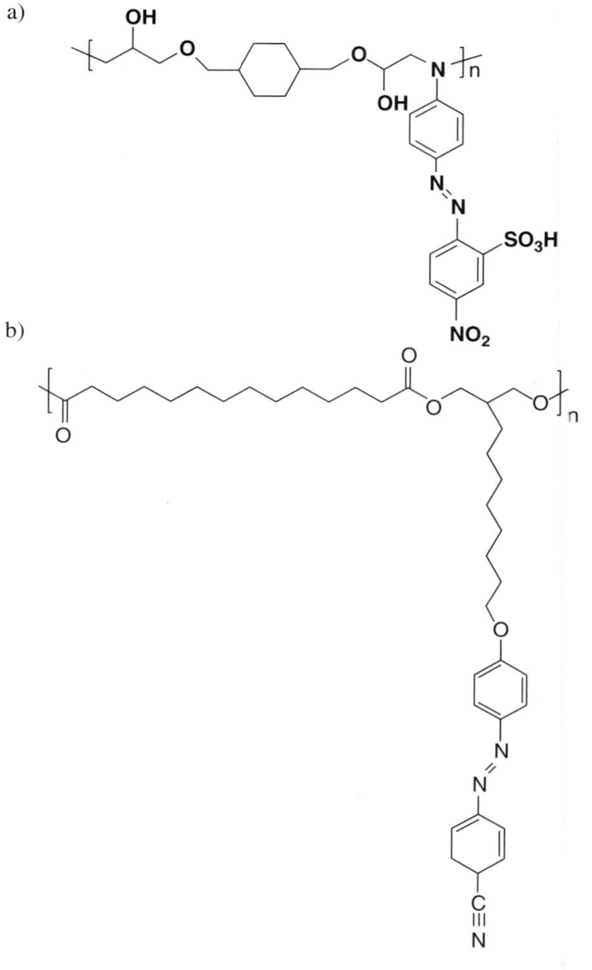

Figure 17.3 Examples of azo polymers: (a) an amorphous side-chain azo polymer; (b) a longer linker between the polymer backbone and the azo unit enables the creation of liquid-crystalline photofunctional polymers.[35,46]

matrix, and the propensity of the dipolar azo units to form aggregates. Higher-quality films are obtained when the azo moiety is covalently bound to the host polymer matrix. These materials combine the stability and processability of polymers with the unusual photoresponsive behaviour of the azo groups. Side-chain and main-chain azo polymers are possible,[43] with common synthesis strategies being divided between polymerising azo-functionalised monomers,[44,45] and postfunctionalising a polymer that has an appropriate pendant (usually a phenyl).[46–48] A wide variety of polymer backbones have been investigated. The most common are the acrylates,[49] methacrylates,[50] and isocyanates,[51] but there are also examples of imides,[52] esters,[53,54] urethanes,[55] ethers,[56] ferrocene,[57] and even conjugated polymers including polydiacetylenes,[58] polyacetylenes,[59] and main-chain azobenzenes.[60,61] A unique strategy that allows for the simplicity of doping while retaining the stability of covalent polymers is to engineer complementary noncovalent attachment of the azo dyes to the polymer backbone. In particular, ionic attachment can lead to (under nominally dry conditions) a homogenous and stable matrix.[62] The use of surfactomesogens (molecules with ionic and liquid-crystalline properties) also enables a simple and programmatic way of generating new materials.[63] It has been demonstrated that azobenzenes can be solubilised by guest–host interactions with cyclodextrin,[64] and it is thus possible that similar strategies could be fruitfully applied to the creation of bulk materials.

Considerable research has also been performed on azobenzene dendrimers[65–67] and molecular glasses (see example in Figure 17.4).[68] These inherently monodisperse materials offer the possibility of high stability, excellent sample homogeneity (crucial for high-quality optical films), and excellent spatial control (with regard to lithography, for instance) without sacrificing the useful features of amorphous linear polymers. The synthetic control offered with such systems allows one to carefully tune solubility, aggregation, thermal stability, and crystallinity.[69–71] The unique structure of dendrimers can be used to exploit azobenzene's photochemistry.[72–74] For instance, the dendrimer structure can act as an antenna, with light-harvesting groups at the periphery, making energy available, via intramolecular energy transfer, to the dendrimer core.[75,76] Thus a dendrimer with an azo core could be photoisomerised using a wavelength outside of its native absorption band. The dendrimer architecture can also be used to amplify the molecular motion of azo isomerisation. For instance, a dendrimer with three azobenzene arms exhibited different physical properties for all the various isomerisation combinations, and the isomers could be separated by thin-layer chromatography on this basis.[77]

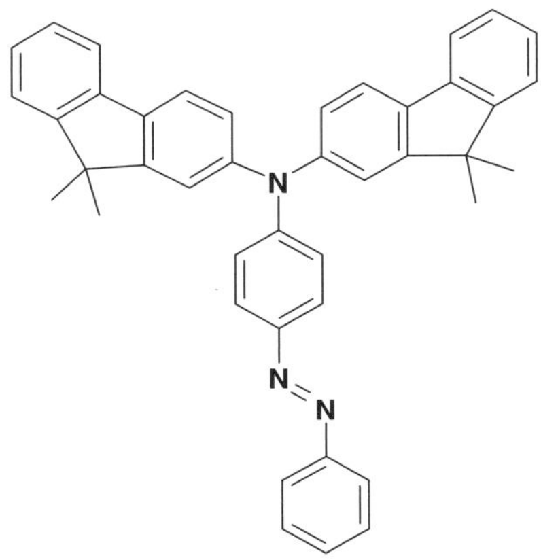

Figure 17.4 Example of an azobenzene molecular glass. This amorphous material exhibits the photophysical and photomechanical phenomena characteristic of the azos.[183]

Thin films of azo material are typically prepared with spin-coating, where polymer solution is dropped onto a rotating substrate. This technique is fast and simple, and generally yields high-quality films that are homogeneous over a wide area. Films can also be prepared via solvent evaporation, the Langmuir–Blodgett technique,[78–81] or self-assembled monolayers.[82] Recently, a technique has emerged that offers the possibility of creating thin films of controlled internal structure in a simple and robust way. This layer-by-layer electrostatic self-assembly technique, pioneered by Decher,[83] involves immersing a charged substrate into an aqueous bath of oppositely charged polymer (polyelectrolyte). The electrostatic adsorption in general overcompensates the surface charge, making it possible to adsorb a layer of another polyelectrolyte. With repeated and alternating immersion in polyanion and polycation baths, one can build up an arbitrary number of layers. These polyelectrolyte multilayers (PEMs) are simple to produce, use benign (all-aqueous) chemistry, and are inherently tunable.[84–86] For instance, by adjusting the ionic strength[87–90] or pH[91–94] of the assembly solutions, the polyelectrolyte chain conformation is modified, hence the final film architecture is controlled. This film-preparation technique has many advantages. The adsorption is quasithermodynamic, which makes the films stable against dewetting and pinhole defects that can arise in spin-cast films. The technique is not limited to flat surfaces: any geometry that can be immersed in (or exposed to) the assembly solution is suitable (coating of capillary walls,[95] colloids,[96,97] and living cells[98] has been demonstrated, for instance). The all-aqueous technique is also readily biocompatible. Perhaps most importantly, the technique is amenable to the incorporation of a wide variety of secondary functionalities. Any molecular unit that has a charged unit (or that can be covalently attached to a charged polymer) can be included. Many groups have demonstrated the incorporation of azobenzene units into polyelectrolyte multilayers.[99–103] Copolymers of azo groups and ionic groups have been investigated,[104,105] however, the inherent insolubility of the azo moiety in water typically limits the extent of loading. Azo-ionomers[106,107] and polymers where the charge is attached to the azo unit[108,109] can alleviate these problems. The azo chromophore can also be generate by postfunctionalising a PEM thin film.[110] It is important to note that the azo-containing PEMs have demonstrated many of the unique photoinduced changes associated with azobenzene, including induced birefringence,[111,112] and surface mass transport.[108] Typically the extent of the photophysical change is somewhat diminished in these systems, due to the effective crosslinks that the ionic attachment points create.

17.4 Photoswitchable Azo Materials

The azobenzene molecular photomotion has been exploited to generate photoswitchable materials, where two distinct states or phases (with notably different material properties) can be generated by appropriate irradiation. The azobenzene unit is rigid and anisotropic, and thereby exhibits liquid-crystalline

(LC) phases in many systems. The ordering of the LC phase can, however, be reversibly switched with light, since the *cis* form of azobenzene is a poor LC mesogen. Irradiation of an LC *trans*-azo sample will disrupt the order and induce a phase transition from the ordered LC state to the isotropic phase. This effect enables fast isothermal control of LC phase transitions,[113–116] even when the azo chromophore is incorporated only to a small extent.[117] This all-optical material response is obviously attractive for a variety of applications, especially for display devices, optical memories,[118] and electro-optics.[119]

Examples of photoswitchable phase changes,[120] phase separation[121] (or reversal of phase separation[122]), solubility changes,[123,124] and crystallization[125] have been found. These suggest a highly promising route towards novel functional materials: the incorporation of photophysical effects into self-assembling systems. The inherent amplification of molecular order to macroscopic material properties can be coupled with molecular-scale photoswitching. For instance, in amphiphilic polypeptide systems, self-assembled micelles were stable in the dark, but could be disaggregated with light irradiation.[126] This construct can act as a transmembrane structure, where the reversible formation and disruption of the aggregated enabled photoswitchable ion transport.[127] In another example, cyclic peptide rings connected by a *trans*-azo unit would hydrogen-bond with their neighbours, forming extended chains. The *cis*-azo analogue, formed upon irradiation, participates in intramolecular hydrogen bonding, forming discrete units and thereby disrupting the higher-order network.[20,128] A system of hydrogen-bonding azobenzene rosettes was found to spontaneously organise into columns, and these columns to assemble into fibres. Upon UV irradiation, this extended ordering was disrupted,[129] which converted a solid organogel into a fluid. An azobenzene surfactant in aqueous solution also showed reversible fluidity, with irradiation destroying self-assembled order, thereby transforming the solid gel into a fluid.[130] Similarly, large changes in viscosity can be elicited by irradiating a solution of azo polyacrylate associated with the protein bovine serum albumin.[131] In a liquid-crystal system, light could be used to induce a glass-to-LC phase transition.[132] A wide variety of applications (such as microfluidics) is possible for functional materials that change phase upon light stimulus.

Azo-containing self-assembled structures in solution can also be controlled with light. Azo block-copolymers can be used to create photoresponsive micelles[133–137] and vesicles.[138] Since illumination can be used to disrupt vesicle encapsulation, this has been suggested as a pulsatile drug-delivery system.[139] The change in azo dipole moment during isomerisation plays a critical role in determining the difference between the aggregation in the two states, and can be optimised to produce a highly efficient photofunctional vesicle system.[140] The use of azo photoisomerisation to disrupt self-assembled systems may be particularly valuable when coupled with biological systems. With biomaterials, one can exploit the powerful and efficient biochemistry of natural systems, yet impose the control of photoactivation. The azobenzene unit in particular has been applied to photobiological experiments with considerable success.[33] Similar to the case of liquid-crystals, order–disorder transitions can be photoinduced in biopolymers. Azo-modified polypeptides may undergo transitions from ordered chiral helices to disordered solutions,[141–143] or even undergo reversible α-helix to β-sheet conversions.[144] In many cases catalytic activity can be regulated due to the presence of the azo group. A cylcodextrin with a histidine and azobenzene pendant was normally inactive because the *trans* azo would bind inside the cyclodextrin pocket, whereas the photogenerated *cis* version liberated the catalytic site.[145] The activity of papain[146,147] and the catalytic efficiency of lysozyme[148] were, similarly, modulated by photo-induced disruption of protein structure. Instead of modifying the protein structure itself, one can embed the protein in a photofunctional matrix[147,149,150] or azo derivatives can be used as small-molecule inhibitors.[151] Azobenzene can also be coupled with DNA in novel ways. In one system, the duplex formation of an azo-incorporating DNA sequence could be reversibly switched,[152] since the *trans* azobenzene intercalates between base pairs, stabilising the binding of the two strands, whereas the *cis* azobenzene disrupts the duplex.[153] The incorporation of an azobenzene unit into the promoter region of an otherwise natural DNA sequence allowed photocontrol of gene expression,[154] since the polymerase enzyme has different interaction strengths with the *trans* and *cis* azo isomers. The ability to create biomaterials whose biological function is activated on demand with light is of interest for fundamental biological studies, and, possibly, for biomedical implants.

17.5 Photoresponsive Azo Materials

17.5.1 Photo-orientation

Many azo materials have been generated that exhibit a continuous response to light stimulus. Perhaps the most well-known photoresponsive effect in azo materials is the photoinduced birefringence that can be generated.[155,156] The root of this effect is the photo-orientation of the anisotropic azo chromophores, upon exposure to a polarised light source. This molecular photo-orientation occurs in a statistical fashion (see mechanism in Figure 17.5). An azobenzene chromophore will preferentially absorb light polarised along its transition dipole axis (long axis of the molecule), whereas the probability of absorption of light polarised perpendicular to the dipole is vanishing (the absorption probability varies as the square of the cosine of the angle between the dipole and the light polarisation). Thus, azo molecules oriented along the polarisation direction will tend to absorb, and reorient randomly, whereas those oriented against the polarisation will not absorb and will remain fixed. For any given initial angular distribution of chromophores, there will thus be a depletion of chromophores oriented along the polarisation direction, with a concomitant increase in the population of chromophores oriented perpendicular to the incident light's electric-field vector. This statistical alignment (orientation hole burning) leads to birefringence (anisotropy of refractive index) and dichroism (anisotropy of absorption spectrum) that grows with the duration of the

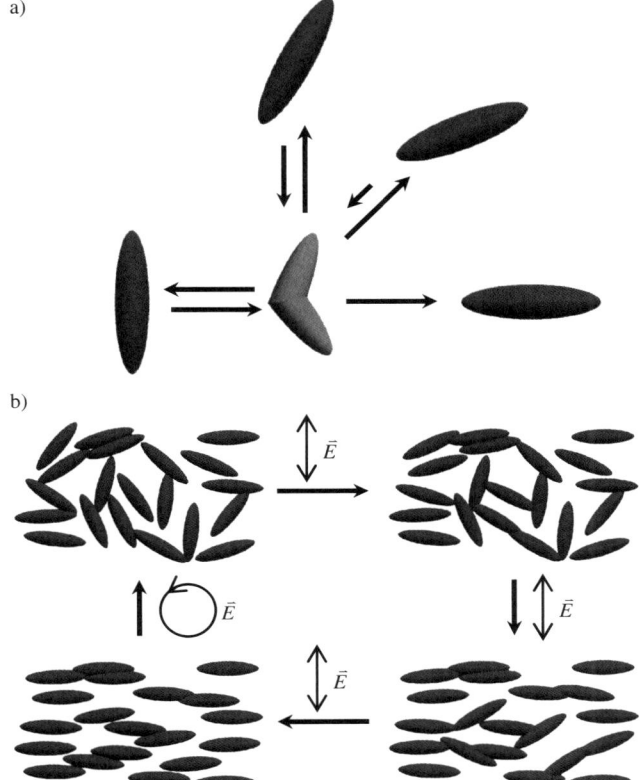

Figure 17.5 The azobenzene molecule can be photo-oriented with polarised light. (a) If azo molecules are irradiated with polarised light (vertical electric field in the diagram), the probability of absorption, and hence random motion, depends on the angle of the chromophore. As a result, chromophores oriented perpendicular to the incident light polarisation (horizontal in the diagram) cannot isomerise and will remain fixed. (b) This leads, over time, to a statistical accumulation of chromophores oriented in the perpendicular direction. Thus, polarised irradiation leads to an oriented sample. Circularly polarised light can restore isotropy.

polarised illumination. It is also reversible, given that circularly polarised light (or unpolarised light) will restore a random, isotropic distribution of chromophores. Importantly, the response is proportional to the incident light flux (irradiation time and power), and can be tuned to a desired value. The ability to generate stable, localised birefringence differences of variable extent has been suggested for applications ranging from waveplates[157] and polarisation filters,[158] to data storage and optical switching.[159] The dynamic and reversible nature of this optical effect makes all azo samples remarkable photoresponsive materials. This persistent photoresponse can, in some cases, be undesired, since the material's optical properties will continually change during optical measurements.

The photo-orientation has been exploited in liquid-crystal systems to positive effect. The cooperative motion of liquid crystals allows azo orientation to be transferred to a bulk LC sample. Thus, even at low concentration, azo chromophores doped into an LC phase can be used to align the director using polarised light.[160,161] Similarly, "command surfaces" can be created, where the alignment in a bulk LC phase can be controlled by photoirradiation of a film of azos tethered to a surface in contact with the phase.[118,162,163] Using proper irradiation, one can force the LC phase to adopt an inplane order (director parallel to the surface), homeotropic order (directory perpendicular to the surface), tilted or even biaxial orientation.[164] The changes are fast, efficient, and fully reversible, suggesting uses as optical displays, memories,[118] switches,[119] *etc.*

17.5.2 Surface Properties

The ability to modulate the surface properties of materials with light is of interest for lithography, "smart" coatings (that respond to the ambient light level), microfluidic devices, and many others. Photoregulation of the surface energy, as measured by the contact angle, has been demonstrated in azo monolayers[165,166] and polymers.[29] Fluorinated polymers can be especially useful in tailoring the photoresponse,[167] which has been used to demonstrate photopatterning of wettability.[168] A monolayer of azo-modified calixarene exhibited photoresponse proportional to light intensity. A gradient in light intensity was used to create a gradient in surface energy sufficient to move a macroscopic oil droplet.[169] Applications to microfluidics are obvious. Recently an azobenzene copolymer assembled into a polyelectrolyte multilayer showed a modest 2° change in contact angle with UV light irradiation. However, when the same copolymer was assembled onto a patterned substrate, the change in contact angle upon irradiation was enhanced[170] to 70°. That surface roughness plays a role in contact angle is well established, and shows that many systems can be optimised to give rise to a large change in surface properties.

17.6 Photodeformable Azo Materials

17.6.1 Surface Mass Transport

One of the most remarkable photoresponsive effects in the azo-polymers systems is the all-optical surface patterning that occurs when a film is exposed to a light-intensity gradient (detailed reviews of this field are available[171–174]). In 1995, it was discovered[175,176] that thin films of the azo-polymer pdr1A (Figure 17.6) exhibited spontaneous surface deformations that reproduced the intensity and polarisation gradient of any incident light field. This surface patterning occurs at low laser power, and is not the result of destructive

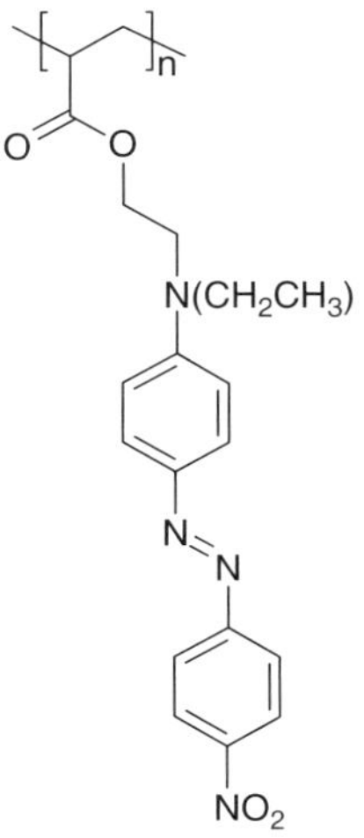

Figure 17.6 Chemical structure of poly(disperse red 1 acrylate), or pdr1A, a polymer that was found to exhibit a remarkable surface mass-transport phenomenon.

ablation. Importantly, the original film thickness and flat topography can be recovered by heating the polymer film above its glass-to-rubber transition temperature, T_g. During this patterning process, polymer material is being moved over nanometre to micron length scales, at temperatures well below the material's T_g. The process requires the presence of azo chromophores (but occurs in a wide range of azo polymers), and moreover, requires the cycling of these chromophores between *trans* and *cis* isomeric states. A typical experiment for demonstrating surface patterning involves exposing the film to the interference pattern caused by intersecting two coherent laser beams. The sinusoidal variation in light intensity (or polarisation, depending on what polarisation combination is used in the incident beams) is encoded into the material as a sinusoidal surface topography, *i.e.* a surface-relief grating (SRG). Figure 17.7 shows an AFM image of a typical SRG, where an initially flat film now exhibits surface features hundreds of nanometers in height. Other experiments have created localised Gaussian "dents" using a focused laser spot.[177] Considering the low laser power sufficient to induce the effect (a few mW/cm^2), it is surprising that it is able to generate such large-scale material motion. This single-step patterning can be used to generate a topographical master quickly and cleanly.

The mechanism of the surface patterning is not entirely understood. Thermal modeling indicates that for typical inscription parameters thermal effects (both thermal gradients and bulk photoheating) can be neglected.[178] Asymmetric diffusion models fit in nicely with the statistical photo-orientation known to occur in the azos,[179,180] yet are difficult to reconcile with the high molecular weight azo photomotion observations (where entire polymer chains are clearly migrating). Mean-field theories[181,182] do not fit the observed phase behaviour,

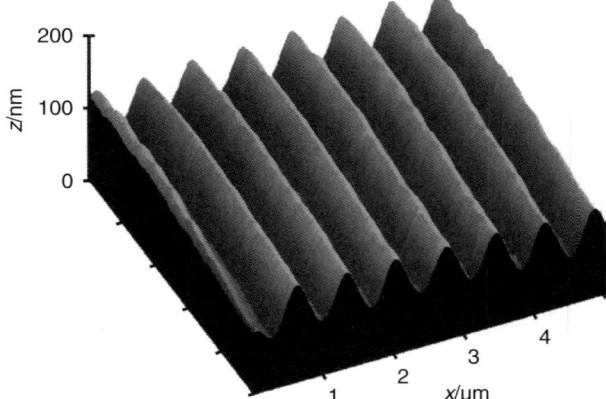

Figure 17.7 AFM image of a sinusoidal surface relief grating, inscribed in an azo polymer film (of pdr1A) by irradiating with a sinusoidal interference pattern. The surface deforms in response to the incident light field gradient, with material motion over hundreds of nanometres possible.

and models examining the interaction between the electric field of the incident light and the polarised material[171,183–185] appear to predict force densities that are much too small.[186] An assumption of a gradient in pressure in the material[187,188] only partially accounts for the polarisation dependence of the patterning. In reality it is likely that some combination of these effects is giving rise to the observed efficient mass transport.

A wide variety of azo-containing materials have been found to exhibit the surface-patterning phenomenon. The process is most stable and efficient in polymer systems of intermediate molecular weight. Monomeric azo systems, or azos doped into a polymer matrix, do not exhibit the effect. Conversely, very high molecular weight polymers (or crosslinked systems) do not have sufficient motion and freedom to exhibit the effect.[187] These general trends notwithstanding, there are noteworthy counterexamples. High molecular weight polypeptides[189] (MW $\sim 10^5$) and azo-cellulose polymers[190,191] (MW $\sim 10^7$) were found to produce SRGs. At the other end of the spectrum, much progress has been made with molecular glasses.[192–194] These amorphous monodisperse systems, in fact, give rise to superior photoresponsive patterning effects, when compared to similar polymeric systems.[195] Molecular glasses allow fine control of architecture, and hence the properties of functional materials.

With regard to lithography, one is limited by the inherent diffraction limit of optics, and by material resolution. The diffraction limit can be overcome using near-field optical patterning. Near-field patterning on azo materials has been established by irradiation sub-diffraction limit objects in proximity to an azo material[196–199] (which can even be used as a means of imaging the near field of nanostructures[200]). The use of a SNOM tip for near-field patterning of azo materials is also possible.[201,202] The use of molecular glasses, which are small and monodisperse, allows one to overcome some resolution limitations of conventional polymer systems. The surface deformation response of azo systems is not useful merely for patterning, however. Clearly this photoresponse could be useful for a wide range of uses beyond simple patterning, since it allows nanoscale motion to be induced when desired, using light.

17.6.2 Photomechanical Effects

The azo molecular photomotion gives rise not only to the nanoscale surface patterning just described, but also to larger-scale photomechanical motions. It was shown that a macroscopic, freestanding azo-polymer LC film could be caused to bend and unbend in response to polarised light irradiation.[203,204] This effect occurs because of photocontraction of the free surface, with correspondingly less contraction deeper in the film (where, due to absorption, the light intensity decreases). The direction of bending can be controlled with the polarisation direction (since photo-orientation occurs), which allows directional control of the effect.[205] A detailed ellipsometry study on thin films of amorphous polymers showed that the films expanded (on the order of 4%) when irradiated uniformly with light.[206] The expansion has two contributions: a large irreversible viscoelastic expansion of the material, and a smaller, reversible elastic expansion that occurs only when the illuminating beam is turned on. It is interesting to note that irradiation produces a contraction in LC materials yet an expansion in amorphous materials. This is confirmed by an experiment involving irradiation of azo films floating on a water surface, where contraction in the direction of polarised light was seen for the LC materials, whereas expansion was seen for amorphous materials.[207] Similar photomechanical responses have been observed in photoresponsive gels containing other molecular units.[208–210]

The ability to control expansion and contraction with light could be used to create photoactuators. The photomechanical effect can be optimised,[211] and has been used to, for instance, induce bending of a coated microcantilever.[212] In related experiments, it has been shown that azo-colloids can be permanently photodeformed into ellipsoids.[213,214] This photomechanical deformation can be used to create novel photofunctional materials. A photonic crystal of azobenzene colloids was prepared, and the regular colloidal array could be anisotropically deformed with polarised light irradiation.[215] This ability to tune the photonic bandgap, and to do so anisotropically, creates new opportunities for photonic materials. Azo LC infiltrated into a photonic crystal[216] and an azo-containing multilayer defect[217] can also be used to control the light-interaction properties of this class of materials.

17.7 Conclusion

The azobenzene chromophore is a unique molecular switch, exhibiting a clean and reversible photoisomerisation that induces molecular motion. This motion can be exploited as a switch, and amplified so that larger-scale material properties are switched or altered in response to light. Thus, azo materials offer the possibility to generate responsive materials. Light control allows for both materials that respond to ambient light levels, useful in a variety of settings. Light is also an ideal triggering mechanism, since it can be localised (in time and space), selective, nondamaging, and allows for remote activate and remote delivery of energy to a system. Thus, for sensing, actuation, and transport, photofunctional materials are promising. Azo materials have demonstrated a wide variety of switching behaviour, from altering optical properties, to altering surface energy, to even eliciting material phase changes. Azobenzene is considered a promising route for next-generation materials because of its ease of incorporation, and efficient photochemistry.

References

1. S.R. Trenor, A.R. Shultz, B.J. Love and T.E. Long, *Chem. Rev.*, 2004, **104**, 3059.
2. H.D. Becker, *Chem. Rev.*, 1993, **93**, 145.
3. N. Tamai and H. Miyasaka, *Chem. Rev.*, 2000, **100**, 1875.
4. M. Irie, *Chem. Rev.*, 2000, **100**, 1685.
5. A. Momotake and T. Arai, *J. Photochem. Photobiol. C: Photochem. Rev.*, 2004, **5**, 1.
6. N. Koumura, E.M. Geertsema, A. Meetsma and B.L. Feringa, *J. Am. Chem. Soc.*, 2000, **122**, 12005.
7. B.L. Feringa, R.A. van Delden, N. Koumura and E.M. Geertsema, *Chem. Rev.*, 2000, **100**, 1789.
8. V. Balzani, G. Bergamini, P. Ceroni, A. Credi and M. Venturi, Photochemically Controlled Molecular Devices and Machines, In *Intelligent Materials*, ed. M. Shahinpoor and H.J. Schneider, Royal Society of Chemistry, Cambridge UK, 2007.
9. S. Kwolek, P. Morgan and J. Schaefgen, *Encyclopedia of Polymer Science and Engineering*, John-Wiley, New York, 1985.
10. G. Möhlmann and C. van der Vorst, *Side Chain Liquid Crystal Polymers*, Plenum and Hall, Glasgow, 1989.
11. H. Rau, Photoisomerization of Azobenzenes, In *Photochemistry and Photophysics*, ed. J. Rebek, CRC Press, Boca Raton, FL, 1990.
12. E.V. Brown and G.R. Granneman, *J. Am. Chem. Soc.*, 1975, **97**, 621.
13. P. Haberfield, P.M. Block and M.S. Lux, *J. Am. Chem. Soc.*, 1975, **97**, 5804.
14. L. Lamarre and C.S.P. Sung, *Macromol.*, 1983, **16**, 1729.
15. Y. Shirota, K. Moriwaki, S. Yoshikawa, T. Ujike and H. Nakano, *J. Mater. Chem.*, 1998, **8**, 2579.
16. Y. Norikane, K. Kitamoto and N. Tamaoki, *J. Org. Chem.*, 2003, **68**, 8291.

17. H. Rau and D. Roettger, *Mol. Cryst. Liq. Cryst. Sci. Tech. A*, 1994, **246**, 143.
18. D. Rottger and H. Rau, *J. Photochem. Photobiol., A*, 1996, **101**, 205.
19. S.A. Nagamani, Y. Norikane and N. Tamaoki, *J. Org. Chem.*, 2005, **70**, 9304.
20. M.S. Vollmer, T.D. Clark, C. Steinem and M.R. Ghadiri, *Angew. Chem., Int. Ed.*, 1999, **38**, 1598.
21. B.K. Kerzhner, V.I. Kofanov and T.L. Vrubel, *Zh. Obsh. Khim.*, 1983, **53**, 2303.
22. U. Funke and H.F. Gruetzmacher, *Tetrahedron*, 1987, **43**, 3787.
23. G.S. Hartley, *Nature (London, UK)*, 1937, **140**, 281.
24. G.S. Hartley, *J. Chem. Soc., Abst.*, 1938, 633.
25. W.J. Priest and M.M. Sifain, J. Polym. Sci., *Part A: Polym. Chem.*, 1971, **9**, 3161.
26. C.S. Paik and H. Morawetz, *Macromol.*, 1972, **5**, 171.
27. C. Barrett, A. Natansohn and P. Rochon, *Macromol.*, 1994, **27**, 4781.
28. C. Barrett, A. Natansohn and P. Rochon, *Chem. Mater.*, 1995, **7**, 899.
29. N. Sarkar, A. Sarkar and S. Sivaram, *J. Appl. Polym. Sci.*, 2001, **81**, 2923.
30. S. Yitzchaik and T.J. Marks, *Acc. Chem. Res.*, 1996, **29**, 197.
31. D. Levy and L. Esquivias, *Adv. Mater. (Weinheim, Ger.)*, 1995, **7**, 120.
32. M. Sisido, Y. Ishikawa, K. Itoh and S. Tazuke, *Macromol.*, 1991, **24**, 3993.
33. I. Willner and S. Rubin, *Angew. Chem., Int. Ed. Engl.*, 1996, **35**, 367.
34. B. Gallot, M. Fafiotte, A. Fissi and O. Pieroni, *Macromol. Rapid Commun.*, 1996, **17**, 493.
35. S. Shinkai, T. Minami, Y. Kusano and O. Manabe, *J. Am. Chem. Soc.*, 1983, **105**, 1851.
36. J.H. Jung, C. Takehisa, Y. Sakata and T. Kaneda, *Chem. Lett.*, 1996, 147.
37. H. Yamamura, H. Kawai, T. Yotsuya, T. Higuchi, Y. Butsugan, S. Araki, M. Kawai and K. Fujita, *Chem. Lett.*, 1996, 799.
38. A.K. Singh, J. Das and N. Majumdar, *J. Am. Chem. Soc.*, 1996, **118**, 6185.
39. S.H. Chen, J.C. Mastrangelo, H. Shi, A. Bashir-Hashemi, J. Li and N. Gelber, *Macromol.*, 1995, **28**, 7775.
40. S.H. Chen, J.C. Mastrangelo, H. Shi, T.N. Blanton and A. Bashir-Hashemi, *Macromol.*, 1997, **30**, 93.
41. R. Birabassov, N. Landraud, T.V. Galstyan, A. Ritcey, C.G. Bazuin and T. Rahem, *Appl. Opt.*, 1998, **37**, 8264.
42. F. Lagugne-Labarthet, T. Buffeteau and C. Sourisseau, *J. Phys. Chem. B.*, 1998, **102**, 2654.
43. N.K. Viswanathan, D.Y. Kim, S. Bian, J. Williams, W. Liu, L. Li, L. Samuelson, J. Kumar and S.K. Tripathy, *J. Mater. Chem.*, 1999, **9**, 1941.
44. A. Natansohn, P. Rochon, J. Gosselin and S. Xie, *Macromol.*, 1992, **25**, 2268.
45. M.S. Ho, C. Barrett, J. Paterson, M. Esteghamatian, A. Natansohn and P. Rochon, *Macromol.*, 1996, **29**, 4613.

46. X. Wang, S. Balasubramanian, L. Li, X. Jiang, D.J. Sandman, M.F. Rubner, J. Kumar and S.K. Tripathy, *Macromol. Rapid Commun.*, 1997, **18**, 451.
47. X. Wang, J.-I. Chen, S. Marturunkakul, L. Li, J. Kumar and S.K. Tripathy, *Chem. Mater.*, 1997, **9**, 45.
48. X. Wang, J. Kumar, S.K. Tripathy, L. Li, J.-I. Chen and S. Marturunkakul, *Macromol.*, 1997, **30**, 219.
49. S. Morino, A. Kaiho and K. Ichimura, *Appl. Phys. Lett.*, 1998, **73**, 1317.
50. A. Altomare, F. Ciardelli, B. Gallot, M. Mader, R. Solaro and N. Tirelli, *J. Polym. Sci., Part A: Polym. Chem.*, 2001, **39**, 2957.
51. N. Tsutsumi, S. Yoshizaki, W. Sakai and T. Kiyotsukuri, *MCLC S&T Section B: Nonlin. Opt.*, 1996, **15**, 387.
52. F. Agolini and F.P. Gay, *Macromol.*, 1970, **3**, 349.
53. K. Anderle, R. Birenheide, M. Eich and J.H. Wendorff, *Makromol. Chem., Rapid Commun.*, 1989, **10**, 477.
54. S. Hvilsted, F. Andruzzi, C. Kulinna, H.W. Siesler and P.S. Ramanujam, *Macromol.*, 1995, **28**, 2172.
55. J. Furukawa, S. Takamori and S. Yamashita, *Angew. Makromol. Chem.*, 1967, **1**, 92.
56. M.C. Bignozzi, S.A. Angeloni, M. Laus, L. Incicco, O. Francescangeli, D. Wolff, G. Galli and E. Chiellini, *Polym. J. (Tokyo)*, 1999, **31**, 913.
57. X.-H. Liu, D.W. Bruce and I. Manners, *Chem. Commun. (Cambridge, UK)*, 1997, 289.
58. M. Sukwattanasinitt, X. Wang, L. Li, X. Jiang, J. Kumar, S.K. Tripathy and D.J. Sandman, *Chem. Mater.*, 1998, **10**, 27.
59. M. Teraguchi and T. Masuda, *Macromolecules*, 2000, **33**, 240.
60. A. Izumi, M. Teraguchi, R. Nomura and T. Masuda, *J. Polym. Sci., Part A: Polym. Chem.*, 2000, **38**, 1057.
61. A. Izumi, M. Teraguchi, R. Nomura and T. Masuda, *Macromolecules*, 2000, **33**, 5347.
62. A. Priimagi, S. Cattaneo, R.H.A. Ras, S. Valkama, O. Ikkala and M. Kauranen, *Chem. Mater.*, 2005, **17**, 5798.
63. C.M. Tibirna and C.G. Bazuin, *J. Polym. Sci., Part B: Polym. Phys.*, 2005, **43**, 3421.
64. S.A. Haque, J.S. Park, M. Srinivasarao and J.R. Durrant, *Adv. Mater. (Weinheim, Ger.)*, 2004, **16**, 1177.
65. A.W. Bosman, H.M. Janssen and E.W. Meijer, *Chem. Rev.*, 1999, **99**, 1665.
66. H.B. Mekelburger, K. Rissanen and F. Voegtle, *Chem. Ber.*, 1993, **126**, 1161.
67. D.M. Junge and D.V. McGrath, *Chem. Commun. (Cambridge, UK)*, 1997, 857.
68. V.A. Mallia and N. Tamaoki, *J. Mater. Chem.*, 2003, **13**, 219.
69. K. Naito and A. Miura, *J. Phys. Chem.*, 1993, **97**, 6240.
70. H. Ma, S. Liu, J. Luo, S. Suresh, L. Liu, S.H. Kang, M. Haller, T. Sassa, L.R. Dalton and A.K.-Y. Jen, *Adv. Funct. Mater.*, 2002, **12**, 565.

71. V.E. Campbell, I. In, D.J. McGee, N. Woodward, A. Caruso and P. Gopalan, *Macromolecules*, 2006, **39**, 957.
72. O. Villavicencio and D.V. McGrath, Azobenzene-containing Dendrimers, In *Advances in Dendritic Macromolecules*, ed. G.R. Newkome, JAI Press, Oxford, 2002.
73. A. Momotake and T. Arai, *Polymer*, 2004, **45**, 5369.
74. A. Momotake and T. Arai, *J. Photochem. Photobiol. C.*, 2004, **5**, 1.
75. D.-L. Jiang and T. Aida, *Nature (London, UK)*, 1997, **388**, 454.
76. T. Aida, D.-L. Jiang, E. Yashima and Y. Okamoto, *Thin Solid Films*, 1998, **331**, 254.
77. D.M. Junge and D.V. McGrath, *J. Am. Chem. Soc.*, 1999, **121**, 4912.
78. T. Seki, M. Sakuragi, Y. Kawanishi, T. Tamaki, R. Fukuda, K. Ichimura and Y. Suzuki, *Langmuir*, 1993, **9**, 211.
79. G. Jianhua, L. Hua, L. Lingyun, L. Bingjie, C. Yiwen and L. Zuhong, *Supramol. Sci.*, 1998, **5**, 675.
80. J. Razna, P. Hodge, D. West and S. Kucharski, *J. Mater. Chem.*, 1999, **9**, 1693.
81. J.R. Silva, F.F. Dall'Agnol, O.N. Oliveira Jr. and J.A. Giacometti, *Polymer*, 2002, **43**, 3753.
82. S.D. Evans, S.R. Johnson, H. Ringsdorf, L.M. Williams and H. Wolf, *Langmuir*, 1998, **14**, 6436.
83. G. Decher, *Science (Washington, DC, USA)*, 1997, **277**, 1232.
84. W. Knoll, *Curr. Opin. Colloid Interf. Sci.*, 1996, **1**, 137.
85. G. Decher, M. Eckle, J. Schmitt and B. Struth, *Curr. Opin. Colloid Interf. Sci.*, 1998, **3**, 32.
86. P.T. Hammond, *Curr. Opin. Colloid Interf. Sci.*, 1999, **4**, 430.
87. G.B. Sukhorukov, J. Schmitt and G. Decher, *Ber. Bunsen-Gesell.*, 1996, **100**, 948.
88. M. Lösche, J. Schmitt, G. Decher, W.G. Bouwman and K. Kjaer, *Macromolecules*, 1998, **31**, 8893.
89. M.R. Linford, M. Auch and H. Möhwald, *J. Am. Chem. Soc.*, 1998, **120**, 178.
90. R. Steitz, V. Leiner, R. Siebrecht and R.v. Klitzing, *Colloids Surf. A*, 2000, **163**, 63.
91. S.S. Shiratori and M.F. Rubner, *Macromolecules*, 2000, **33**, 4213.
92. A.J. Chung and M.F. Rubner, *Langmuir*, 2002, **18**, 1176.
93. T.C. Wang, M.F. Rubner and R.E. Cohen, *Langmuir*, 2002, **18**, 3370.
94. S.E. Burke and C.J. Barrett, *Langmuir*, 2003, **19**, 3297.
95. O. Mermut and C.J. Barrett, *Analyst (Cambridge, UK)*, 2001, **126**, 1861.
96. G.B. Sukhorukov, E. Donath, H. Lichtenfeld, E. Knippel, M. Knippel, A. Budde and H. Möhwald, *Colloids Surf., A*, 1998, **137**, 253.
97. F. Caruso, *Adv. Mater. (Weinheim, Ger.)*, 2001, **13**, 11.
98. A. Diaspro, D. Silvano, S. Krol, O. Cavalleri and A. Gliozzi, *Langmuir*, 2002, **18**, 5047.
99. R.C. Advincula, E. Fells and M.-K. Park, *Chem. Mater.*, 2001, **13**, 2870.

100. S. Balasubramanian, X. Wang, H.C. Wang, K. Yang, J. Kumar, S.K. Tripathy and L. Li, *Chem. Mater.*, 1998, **10**, 1554.
101. F. Saremi and B. Tieke, *Adv. Mater. (Weinheim, Ger.)*, 1998, **10**, 389.
102. K.E. Van Cott, M. Guzy, P. Neyman, C. Brands, J.R. Heflin, H.W. Gibson and R.M. Davis, *Angew. Chem., Int. Ed.*, 2002, **41**, 3236.
103. O. Mermut and C.J. Barrett, *J. Phys. Chem. B*, 2003, **107**, 2525.
104. L. Wu, X. Tuo, H. Cheng, Z. Chen and X. Wang, *Macromolecules*, 2001, **34**, 8005.
105. I. Suzuki, K. Sato, M. Koga, Q. Chen and J.-i. Anzai, *Mater. Sci. Eng., C*, 2003, **23**, 579.
106. J.-D. Hong, B.-D. Jung, C.H. Kim and K. Kim, *Macromolecules*, 2000, **33**, 7905.
107. B.-D. Jung, J.-D. Hong, A. Voigt, S. Leporatti, L. Dähne, E. Donath and H. Möhwald, *Colloids Surf., A*, 2002, **198–200**, 483.
108. X. Wang, S. Balasubramanian, J. Kumar, S.K. Tripathy and L. Li, *Chem. Mater.*, 1998, **10**, 1546.
109. H. Wang, Y. He, X. Tuo and X. Wang, *Macromolecules*, 2004, **37**, 135.
110. S.-H. Lee, S. Balasubramanian, D.Y. Kim, N.K. Viswanathan, S. Bian, J. Kumar and S.K. Tripathy, *Macromolecules*, 2000, **33**, 6534.
111. M.-K. Park and R.C. Advincula, *Langmuir*, 2002, **18**, 4532.
112. J. Ishikawa, A. Baba, F. Kaneko, K. Shinbo, K. Kato and R.C. Advincula, *Colloids Surf., A*, 2002, **198–200**, 917.
113. T. Ikeda, S. Horiuchi, D.B. Karanjit, S. Kurihara and S. Tazuke, *Macromolecules*, 1990, **23**, 42.
114. T. Ikeda and O. Tsutsumi, *Science (Washington, DC, USA)*, 1995, **268**, 1873.
115. T. Hayashi, H. Kawakami, Y. Doke, A. Tsuchida, Y. Onogi and M. Yamamoto, *Eur. Polym. J.*, 1995, **31**, 23.
116. T. Kato, N. Hirota, A. Fujishima and J.M.J. Frechet, *J. Polym. Sci., Part A: Polym. Chem.*, 1996, **34**, 57.
117. M. Eich and J. Wendorff, *J. Opt. Soc. Am. B*, 1990, **7**, 1428.
118. W.M. Gibbons, P.J. Shannon, S.-T. Sun and B.J. Swetlin, *Nature (London, UK)*, 1991, **351**, 49.
119. Y.-Y. Luk and N.L. Abbott, *Science (Washington, DC, USA)*, 2003, **301**, 623.
120. K. Aoki, M. Nakagawa and K. Ichimura, *J. Am. Chem. Soc.*, 2000, **122**, 10997.
121. S. Kadota, K. Aoki, S. Nagano and T. Seki, *J. Am. Chem. Soc.*, 2005, **127**, 8266.
122. J.J. Effing and J.C.T. Kwak, *Angew. Chem., Int. Ed. Engl.*, 1995, **34**, 88.
123. H. Yamamoto, A. Nishida, T. Takimoto and A. Nagai, *J. Polym. Sci., Part A: Polym. Chem.*, 1990, **28**, 67.
124. K. Arai and Y. Kawabata, *Macromolecules Rapid Commun.*, 1995, **16**, 875.
125. T.D. Ebralidze and A.N. Mumladze, *Appl. Opt.*, 1990, **29**, 446.
126. M. Higuchi, N. Minoura and T. Kinoshita, *Chem. Lett.*, 1994, 227.

127. M. Higuchi, N. Minoura and T. Kinoshita, *Macromolecules*, 1995, **28**, 4981.
128. C. Steinem, A. Janshoff, M.S. Vollmer and M.R. Ghadiri, *Langmuir*, 1999, **15**, 3956.
129. S. Yagai, T. Nakajima, K. Kishikawa, S. Kohmoto, T. Karatsu and A. Kitamura, *J. Am. Chem. Soc.*, 2005, **127**, 11134.
130. H. Sakai, Y. Orihara, H. Kodashima, A. Matsumura, T. Ohkubo, K. Tsuchiya and M. Abe, *J. Am. Chem. Soc.*, 2005, **127**, 13454.
131. G. Pouliquen and C. Tribet, *Macromolecules*, 2005, **39**, 373.
132. P. Camorani and M.P. Fontana, *Phys. Rev. E*, 2006, **73**, 011703.
133. G. Wang, X. Tong and Y. Zhao, *Macromolecules*, 2004, **37**, 8911.
134. P. Ravi, S.L. Sin, L.H. Gan, Y.Y. Gan, K.C. Tam, X.L. Xia and X. Hu, *Polymer*, 2005, **46**, 137.
135. S.L. Sin, L.H. Gan, X. Hu, K.C. Tam and Y.Y. Gan, *Macromolecules*, 2005, **38**, 3943.
136. E. Yoshida and M. Ohta, *Colloid Polym. Sci.*, 2005, **283**, 872.
137. E. Yoshida and M. Ohta, *Colloid Polym. Sci.*, 2005, **283**, 521.
138. H. Sakai, A. Matsumura, T. Saji and M. Abe, *Stud. Surf. Sci. Catal.*, 2001, **132**, 505.
139. X.-M. Liu, B. Yang, Y.-L. Wang and J.-Y. Wang, *Chem. Mater.*, 2005, **17**, 2792.
140. X. Tong, G. Wang, A. Soldera and Y. Zhao, *J. Phys. Chem. B.*, 2005, **109**, 20281.
141. G. Montagnoli, O. Pieroni and S. Suzuki, *Polym. Photochem.*, 1983, **3**, 279.
142. H. Yamamoto and A. Nishida, *Polym. Int.*, 1991, **24**, 145.
143. A. Fissi, O. Pieroni, E. Balestreri and C. Amato, *Macromolecules*, 1996, **29**, 4680.
144. A. Fissi, O. Pieroni and F. Ciardelli, *Biopolymers*, 1987, **26**, 1993.
145. W.-S. Lee and A. Ueno, *Macromol. Rapid Commun.*, 2001, **22**, 448.
146. I. Willner, S. Rubin and A. Riklin, *J. Am. Chem. Soc.*, 1991, **113**, 3321.
147. I. Willner and S. Rubin, *React. Polym.*, 1993, **21**, 177.
148. T. Inada, T. Terabayashi, Y. Yamaguchi, K. Kato and K. Kikuchi, *J. Photochem. Photobiol. A*, 2005, **175**, 100.
149. I. Willner, S. Rubin, R. Shatzmiller and T. Zor, *J. Am. Chem. Soc.*, 1993, **115**, 8690.
150. I. Willner, S. Rubin and T. Zor, *J. Am. Chem. Soc.*, 1991, **113**, 4013.
151. K. Komori, K. Yatagai and T. Tatsuma, *J. Biotechnol.*, 2004, **108**, 11.
152. H. Asanuma, X. Liang, T. Yoshida and M. Komiyama, *ChemBioChem*, 2001, **2**, 39.
153. X. Liang, H. Asanuma, H. Kashida, A. Takasu, T. Sakamoto, G. Kawai and M. Komiyama, *J. Am. Chem. Soc.*, 2003, **125**, 16408.
154. M. Liu, H. Asanuma and M. Komiyama, *J. Am. Chem. Soc.*, 2005, **128**, 1009.
155. K. Ichimura, *Chem. Rev.*, 2000, **100**, 1847.
156. Y.L. Yu and T. Ikeda, *J. Photochem. Photobiol., C*, 2004, **5**, 247.

157. Y. Shi, W.H. Steier, L. Yu, M. Chen and L.R. Dalton, *Appl. Phys. Lett.*, 1991, **59**, 2935.
158. A. Natansohn and P. Rochon, *Adv. Mater. (Weinheim, Ger.)*, 1999, **11**, 1387.
159. A. Shishido, O. Tsutsumi, A. Kanazawa, T. Shiono, T. Ikeda and N. Tamai, *J. Am. Chem. Soc.*, 1997, **119**, 7791.
160. S.T. Sun, W.M. Gibbons and P.J. Shannon, *Liq. Cryst.*, 1992, **12**, 869.
161. K. Anderle, R. Birenheide, M.J.A. Werner and J.H. Wendorff, *Liq. Cryst.*, 1991, **9**, 691.
162. K. Ichimura, Y. Hayashi, H. Akiyama, T. Ikeda and N. Ishizuki, *Appl. Phys. Lett.*, 1993, **63**, 449.
163. A.G. Chen and D.J. Brady, *Appl. Phys. Lett.*, 1993, **62**, 2920.
164. O. Yaroschuk, T. Sergan, J. Lindau, S.N. Lee, J. Kelly and L.-C. Chien, *J. Chem. Phys.*, 2001, **114**, 5330.
165. L.M. Siewierski, W.J. Brittain, S. Petrash and M.D. Foster, *Langmuir*, 1996, **12**, 5838.
166. N. Delorme, J.-F. Bardeau, A. Bulou and F. Poncin-Epaillard, *Langmuir*, 2005, **21**, 12278.
167. C.L. Feng, Y.J. Zhang, J. Jin, Y.L. Song, L.Y. Xie, G.R. Qu, L. Jiang and D.B. Zhu, *Langmuir*, 2001, **17**, 4593.
168. G. Moller, M. Harke, H. Motschmann and D. Prescher, *Langmuir*, 1998, **14**, 4955.
169. K. Ichimura, S.-K. Oh and M. Nakagawa, *Science (Washington, DC, USA)*, 2000, **288**, 1624.
170. W.H. Jiang, G.J. Wang, Y.N. He, X.G. Wang, Y.L. An, Y.L. Song and L. Jiang, *Chem. Commun. (Cambridge, UK)*, 2005, 3550.
171. N.K. Viswanathan, S. Balasubramanian, L. Li, S.K. Tripathy and J. Kumar, *Jpn. J. Appl. Phys., Part 1*, 1999, **38**, 5928.
172. J.A. Delaire and K. Nakatani, *Chem. Rev.*, 2000, **100**, 1817.
173. K.G. Yager and C.J. Barrett, *Curr. Opin. Solid State Mater. Sci.*, 2001, **5**, 487.
174. A. Natansohn and P. Rochon, *Chem. Rev.*, 2002, **102**, 4139.
175. P. Rochon, E. Batalla and A. Natansohn, *Appl. Phys. Lett.*, 1995, **66**, 136.
176. D.Y. Kim, S.K. Tripathy, L. Li and J. Kumar, *Appl. Phys. Lett.*, 1995, **66**, 1166.
177. S. Bian, L. Li, J. Kumar, D.Y. Kim, J. Williams and S.K. Tripathy, *Appl. Phys. Lett.*, 1998, **73**, 1817.
178. K.G. Yager and C.J. Barrett, *J. Chem. Phys.*, 2004, **120**, 1089.
179. P. Lefin, C. Fiorini and J.-M. Nunzi, *Opt. Mater.*, 1998, **9**, 323.
180. P. Lefin, C. Fiorini and J.M. Nunzi, *Pure Appl. Opt.*, 1998, **7**, 71.
181. T.G. Pedersen and P.M. Johansen, *Phys. Rev. Lett.*, 1997, **79**, 2470.
182. T.G. Pedersen, P.M. Johansen, N.C.R. Holme, P.S. Ramanujam and S. Hvilsted, *Phys. Rev. Lett.*, 1998, **80**, 89.
183. J. Kumar, L. Li, X.L. Jiang, D.Y. Kim, T.S. Lee and S. Tripathy, *Appl. Phys. Lett.*, 1998, **72**, 2096.

184. S.P. Bian, W. Liu, J. Williams, L. Samuelson, J. Kumar and S. Tripathy, *Chem. Mater.*, 2000, **12**, 1585.
185. O. Baldus and S.J. Zilker, *Appl. Phys. B-Lasers Opt.*, 2001, **72**, 425.
186. M. Saphiannikova, T.M. Geue, O. Henneberg, K. Morawetz and U. Pietsch, *J. Chem. Phys.*, 2004, **120**, 4039.
187. C.J. Barrett, A.L. Natansohn and P.L. Rochon, *J. Phys. Chem.*, 1996, **100**, 8836.
188. C.J. Barrett, P.L. Rochon and A.L. Natansohn, *J. Chem. Phys.*, 1998, **109**, 1505.
189. S.Z. Yang, L. Li, A.L. Cholli, J. Kumar and S.K. Tripathy, *Biomacromolecules*, 2003, **4**, 366.
190. S.Z. Yang, L. Li, A.L. Cholli, J. Kumar and S.K. Tripathy, *J. Macromol. Sci., Pure Appl. Chem.*, 2001, **38**, 1345.
191. S. Yang, M.M. Jacob, L. Li, K. Yang, A.L. Cholli, J. Kumar and S.K. Tripathy, *Polym. News*, 2002, **27**, 368.
192. H. Nakano, T. Takahashi, T. Kadota and Y. Shirota, *Adv. Mater. (Weinheim, Ger.)*, 2002, **14**, 1157.
193. M.-J. Kim, E.-M. Seo, D. Vak and D.-Y. Kim, *Chem. Mater.*, 2003, **15**, 4021.
194. E. Ishow, B. Lebon, Y. He, X. Wang, L. Bouteiller, L. Galmiche and K. Nakatani, *Chem. Mater.*, 2006, **18**, 1261.
195. H. Ando, T. Takahashi, H. Nakano and Y. Shirota, *Chem. Lett.*, 2003, **32**, 710.
196. M. Hasegawa, T. Ikawa, M. Tsuchimori, O. Watanabe and Y. Kawata, *Macromolecules*, 2001, **34**, 7471.
197. T. Ikawa, T. Mitsuoka, M. Hasegawa, M. Tsuchimori, O. Watanabe, Y. Kawata, C. Egami, O. Sugihara and N. Okamoto, *J. Phys. Chem. B*, 2000, **104**, 9055.
198. T. Ikawa, T. Mitsuoka, M. Hasegawa, M. Tsuchimori, O. Watanabe and Y. Kawata, *Phys. Rev. B: Condens. Matter*, 2001, **64**.
199. O. Watanabe, T. Ikawa, M. Hasegawa, M. Tsuchimori, Y. Kawata, C. Egami, O. Sugihara and N. Okamoto, *Mol. Cryst. Liq. Cryst.*, 2000, **345**, 629.
200. C. Hubert, A. Rumyantseva, G. Lerondel, J. Grand, S. Kostcheev, L. Billot, A. Vial, R. Bachelot, P. Royer, S.H. Chang, S.K. Gray, G.P. Wiederrecht and G.C. Schatz, *Nano Lett.*, 2005, **5**, 615.
201. P.S. Ramanujam, N.C.R. Holme, M. Pedersen and S. Hvilsted, *J. Photochem. Photobiol., A*, 2001, **145**, 49.
202. B. Stiller, T. Geue, K. Morawetz and M. Saphiannikova, *J. Microsc.*, 2005, **219**, 109.
203. Y. Yu, M. Nakano and T. Ikeda, *Nature (London, UK)*, 2003, **425**, 145.
204. T. Ikeda, M. Nakano, Y. Yu, O. Tsutsumi and A. Kanazawa, *Adv. Mater. (Weinheim, Ger.)*, 2003, **15**, 201.
205. Y.L. Yu, M. Nakano, T. Maeda, M. Kondo and T. Ikeda, *Mol. Cryst. Liq. Cryst.*, 2005, **436**, 1235.
206. O.. Tanchak and C.J. Barrett, *Macromolecules*, 2005, **38**, 10566.

207. D. Bublitz, M. Helgert, B. Fleck, L. Wenke, S. Hvilsted and P.S. Ramanujam, *Appl. Phys. B: Lasers Opt.*, 2000, **70**, 863.
208. A. Suzuki and T. Tanaka, *Nature*, 1990, **346**, 345.
209. T. Seki, J.Y. Kojima and K. Ichimura, *J. Phys. Chem. B*, 1999, **103**, 10338.
210. L. Frkanec, M. Jokic, J. Makarevic, K. Wolsperger and M. Zinic, *J. Am. Chem. Soc.*, 2002, **124**, 9716.
211. N. Tabiryan, S. Serak, X.M. Dai and T. Bunning, *Opt. Exp.*, 2005, **13**, 7442.
212. H.F. Ji, Y. Feng, X.H. Xu, V. Purushotham, T. Thundat and G.M. Brown, *Chem. Commun. (Cambridge, UK)*, 2004, 2532.
213. Y.B. Li, Y.N. He, X.L. Tong and X.G. Wang, *J. Am. Chem. Soc.*, 2005, **127**, 2402.
214. Y. Li, Y. He, X. Tong and X. Wang, *Langmuir*, 2006, **22**, 2288.
215. Y. Li, X. Tong, Y. He and X. Wang, *J. Am. Chem. Soc.*, 2006, **128**, 2220.
216. S. Kubo, Z.Z. Gu, K. Takahashi, A. Fujishima, H. Segawa and O. Sato, *Chem. Mater.*, 2005, **17**, 2298.
217. F. Fleischhaker, A.C. Arsenault, V. Kitaev, F.C. Peiris, G. von Freymann, I. Manners, R. Zentel and G.A. Ozin, *J. Am. Chem. Soc.*, 2005.

CHAPTER 18

Intelligent Chitosan-based Hydrogels as Multifunctional Materials

ARTHUR F.T. MAK[1] AND SHAN SUN[2]

[1] Department of Health Technology and Informatics, Hong Kong Polytechnic University, Hong Kong, China
[2] Department of Bioengineering, University of Illinois at Chicago, Chicago IL 60607, USA

18.1 Introduction

Chitosan is one of the few naturally occurring intelligent materials. Different from the other natural polysaccharides such as cellulose, alginic acid, agarose and pectin, which are neutral or acidic, chitosan can be highly basic due to a large amount of amino groups on its backbone. Protonation and deprotonation of these groups determine the amount of ions on the residuals and influence static electric repulsion and intramolecular or intermolecular hydrogen bonds. This causes molecular chains either to compact or become repulsive, and the dimensions to undergo dramatic changes in response to external stimuli including pH, ions, thermal and electrical stimulations in solution; therefore chitosan becomes a member of the so-called "smart polymers".

Many efforts in modification of chitosan and preparation of their hybrids have been made to improve their sensitivities in response to the environmental changes. The presence of amino and hydroxyl groups allows chemical modifications of chitosan, such as acylation, N-phthaloylation, tosylation, alkylation, Schiff base formation, reductive alkylation, O-carboxymethylation, N-carboxyalkylation, silylation *etc.*[1,2] Chitosan-based materials are easily made under mild conditions into multiple scales, from macro-sized scaffold to nanoscale particles, and can be tailored into almost any types of forms including membrane, fibres, gels, beads, capsules, sponges and scaffolds.[3] The other distinctive advantages include their availability, biocompatibility, biodegradability, nontoxicity, antimicrobial properties, heavy metal ions chelation, gel forming properties, ease of chemical modification and high affinity to proteins.

Intelligent features render the function of chitosan more selective and more versatile, especially for the biomedical applications such as drug-delivery systems, artificial muscles and new array biosensors. For decades, chitosan has been widely used in industry, environment, food and health care.[4]

This chapter is focused on chitosan-based hydrogels as intelligent materials. Besides the chemical, physical and biological characteristics of chitosan, we will discuss the intelligent properties of various chitosan-based materials and the associated underlying mechanisms, review the applications and future directions based on the stimuli-responsive properties of chitosan and their derivatives.

18.2 Characteristics of Chitosan

18.2.1 Physical and Chemical Properties of Chitosan

Chitosan is a natural product derived from deacetylation of chitin, and chitin is the second most abundant natural polysaccharide found in the exoskeleton of shellfish like shrimps or crabs. Chitosan is a linear polysaccharide composed of randomly distributed β-(1-4)-linked D-glucosamine (deacetylated unit) and N-acetyl-D-glucosamine (acetylated unit), their molecular structures are shown in Scheme 18.1. To obtain chitosan, chitin is subjected to N-decetylation by treatment with a 40–45% NaOH solution, followed by purification procedures. Because of protonatable amino groups, chitosan is a cationic polymer that can be formed into gels with polyanions. Chitosan solution is viscous and electrically conductive, and thus is a polyelectrolyte, like many other biological molecules such as many polypeptides and DNA.

Molecular weight and degree of N-deacetylation are two most important indices of chitosan. Molecular weight can be determined by means of viscometry. According to the Mark–Houwink equation, the intrinsic viscosity is expressed as

$$[\eta] = KM^{\alpha} = 1.81 \times 10^{-3} M^{0.93}$$

where the constants K and α have been determined in 0.1 M acetic acid and 0.2 M sodium chloride solution,[5] as shown above.

The degree of N-deacetylation is the ratio of glucosamine to N-acetyl glucosamine on chitosan. This ratio reflects the amount of amino groups, which determine the chitosan's solubility and activity as a functional material, usually in the range of 60–100%.[6] It is also an important index to reflect the stimuli-responsive ability of chitosan-based materials. Chitosan molecules have a rod-like shape or coiled shape at low degree of deacetylation due to the low

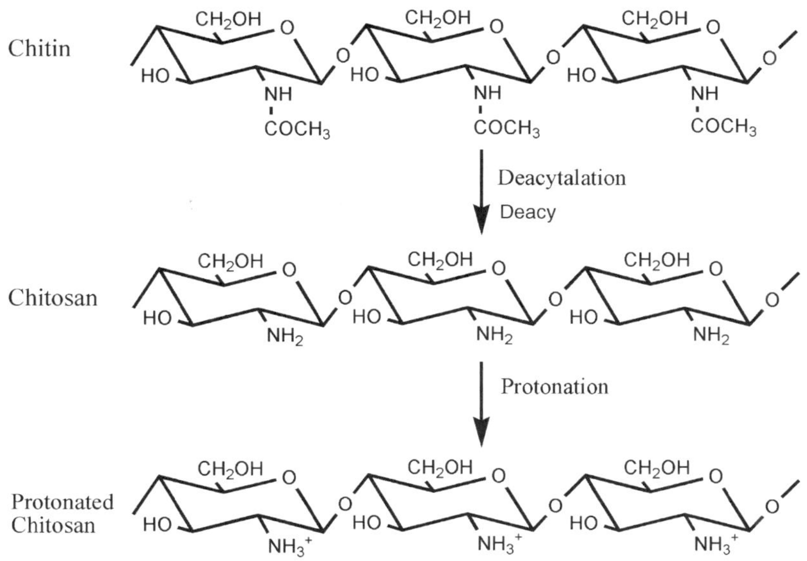

Scheme 18.1 Molecular structures of chitin, chitosan, and protonated chitosan.

charge density in polymer chain, and have an extended and flexible chain at high degree of deacetylation.[7]

18.2.2 Biological Properties of Chitosan

Chitosan has very good biocompatability, noncytoxity and antibacterial activity.[8] Chitosan can degrade to oligosaccharides of variable length in the presence of lysozyme and other enzymes in vivo. It has been proven to be useful in promoting tissue growth in tissue repair and accelerating wound healing and bone regeneration.[9,10] Chitosan can absorb fats and cholesterol intensively and initiates clotting of red blood cells due to the positive charges, allowing it to bind to negatively charged surfaces such as hair and skin to make it a useful ingredient in nutritional and cosmetic products.

18.2.3 Solvent and Solubility

Solvents significantly influence the solubility and stimuli-responsive properties of chitosan. Chitosan is soluble in acids such as acetic acid, formic acid, HCl, glutamic acid and lactic acid, yielding viscous solution, which can be processed into various forms. A chemically or physically crosslinked chitosan may remain as whole but swell in the aqueous solution. Chitosan is positively charged in acidic to neutral solution due to that its amino group has a pKa value in the range of 6.3–7.0.[11] The degree of deacetylation, substitute groups and composite polymers of chitosan, pH, ions and ionic strength may all influence the charges on residuals. As a result, the expansion of macromolecules can be very different. The accumulation of such differences at the level of single molecules gives rise to dramatic change in shape or volume of the materials, which is responsible for the stimuli-responsive sensitivity.

18.3 Intelligent Properties

18.3.1 pH Sensitivity

Cationic chitosan exhibits pH sensitive swelling properties. The degree of swelling, usually defined as the ratio of weight uptake in wet state to the weight in dry state, may increase in acidic and decrease in alkali media, and abrupt change is likely to occur at low pH value of 1.0–3.0. Most chitosan derivatives and composites have similar properties. For example, the swelling ratio of chitosan-blended polyethylene glycol (PEG) hydrogel increased about 6 folds as pH 2.0 changed to pH 1.0 (Figure 18.1). The pH-dependent swelling is reversible as shown in Figure 18.2, but the process is diffusion dependent, as are most pH-sensitive hydrogels.[12] The rate of stimuli-response is relatively slow as compared to the other types of intelligent materials.[13]

As mentioned above, the pH sensitivity of chitosan-based materials is mainly attributed to protonation and deprotonation of amino groups in various media. The IR spectrum of hydrogel swollen in acidic solution has confirmed the formation of $-NH_3^+$ from protonation $-NH_2$.[14] These charged groups cause electrostatic repulsion between adjacent ionised residual groups and dissociation of hydrogen bonds, leading to chain expansion and eventually increase of

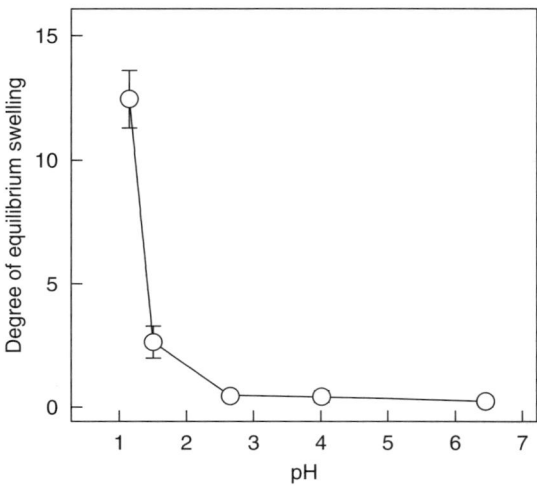

Figure 18.1 Equilibrium degree of swelling of chitosan/PEG copolymer in different pH solutions.

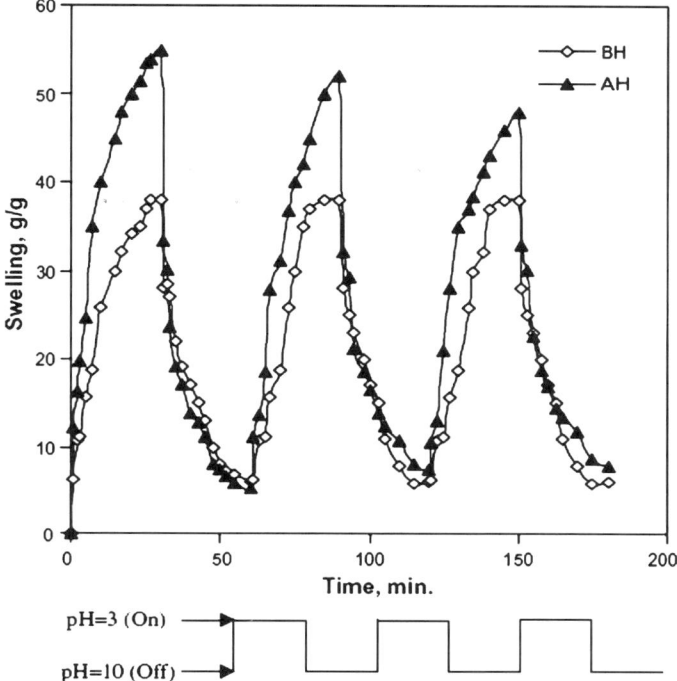

Figure 18.2 On-off switching behaviour as reversible swelling (pH 3) and deswelling (pH 10) of the pH-responsive hydrogel, chitosan-g-poly(AA-co-AAm). AH: after hydrolysis, bH: before hydrolysis. AA-AAm weight ratio is 1.0.

the water uptake. In neutral and alkaline solution, $-NH_3^+$ are deprotonated to noncharged $-NH_2$. Such swelling at low pH is mainly driven by solvent diffusion instead of the chain relaxation.[15]

The pH-dependent degree of swelling can be influenced by crosslinking or substituting groups on chitosan, accompanied by a change in sensitivity to external stimuli. For example, to reduce the solubility of chitosan in aqueous solvents or to increase resistance to chitosan degradation, linear chitosan chains are used to be crosslinked with bifunctional reagent glutaric dialdehyde (GA), ethylene glycol diglycidyl ether (EGDE), or multivalent counterions of sodium tripolyphosphate.[16] Confined by the intra- or intermolecular crosslinkage, the degree of swelling may be decreased significantly depending on the extent of the crosslinking, or the gel can even become no longer responsive to the pH change as high crosslinking regent in the gel.

Multiple maxima peaks of swelling are likely to appear as function of pH with modified chitosan or composites of chitosan. For example, chitosan-g-poly(AA-co-AAm) hydrogel can reach peak swelling at pH 3, pH 6 and pH 8.[17] This is because both amine (chitosan backbone) and carboxylate (PAA chains) are present on the polymer chains. Chitosan is a weak base with an intrinsic pKa of 6.5, and PAA contains carboxylic groups that become ionised at pH values above its pKa of 4.7. Either protonated $-NH_3^+$ or deprotonated $-COO^-$ groups increase the charge density on the polymer, causing an enhancement of the osmotic pressure inside the gel particles because of the $-NH_3^+ - -NH_3^+$ or $-COO^- - -COO^-$ electrostatic repulsion. This osmotic pressure difference between the internal and external solution of the network is balanced by the swelling of the gel.[18]

18.3.2 Ionic Strength Sensitivity

Because ionic strength determines osmotic pressure and the electrostatic repulsion of the adjacent molecular chains, it can influence the swelling behaviours of polyelectrolyte hydrogels. A higher degree of swelling is likely in lower ionic strength solution. Chitosan-based hydrogels exhibit various swelling capacity in different salt solutions with the same molar concentrations. In a constant pH solution, when hydrogel is treated alternatively with NaCl and $CaCl_2$ solutions with equal molarity, the swelling of the hydrogel showed a cyclic response: lower in $CaCl_2$ solution than that of in NaCl, as shown in Figure 18.3. While at a pH value below 3, the ionic strength produced by the acid itself becomes so large that the additional salts have only small effects on swelling. These swelling

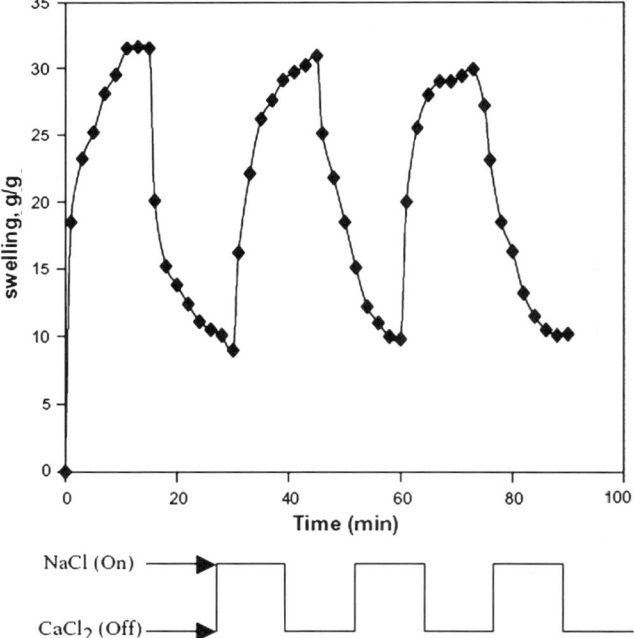

Figure 18.3 Reversible cation exchange ability of the chitosan-g-poly(AA-co-AAm) hydrogel. AA-AAm weight ratio is 1.0.

changes are due to ionic strength along with osmotic pressure, which are determined by the valency difference of salts.[12]

18.3.3 Organic Effectors Sensitivity

Expansion of a chitosan covalently bonded with anthryl unit gel can be triggered by aromatic effectors. It is known that aromatic substrates may lead to dimensional changes especially when positively charged nitrogen atoms are present on chitosan or on these aromatic effectors, which can generate large noncovalent cation-π interactions. Lomadze and Schneider[19] studied systematically the stacking effect of various aromatic effectors on swelling behaviour and corresponding kinetics of the gel film, showing the volume expansions could be quite different due to the structure of the aromatic groups present in pH 5 solution. For example, 45% and 66% of swelling were attributed to the presence of imidazole and benzimidazole, respectively. In contrast, relatively less change in the degree of swelling occurred in the presence of toluenesulfonic acid, pyrazole and pyrimidine amino acids (Scheme 18.2).

18.3.4 Electrosensitivity

Like many other polyelectrolyte hydrogels, chitosan-based materials can undergo reversible change in deformation induced by electrical stimuli in aqueous solution.[20] The deformation response consists of complicated "electrochemomechanical" behaviours,[21,22] including mechanical effects (pressure and diffusive drag), chemical effects (concentration, chemical potential, osmotic pressure) and electrical effects (electrical field intensity, electrical potential). Some experiments and theoretical analyses have been documented for the explanation of the deformation of polyelectrolyte gels under an electric stimulation.[23–25] The deformation is generally regarded to be caused by a pH and

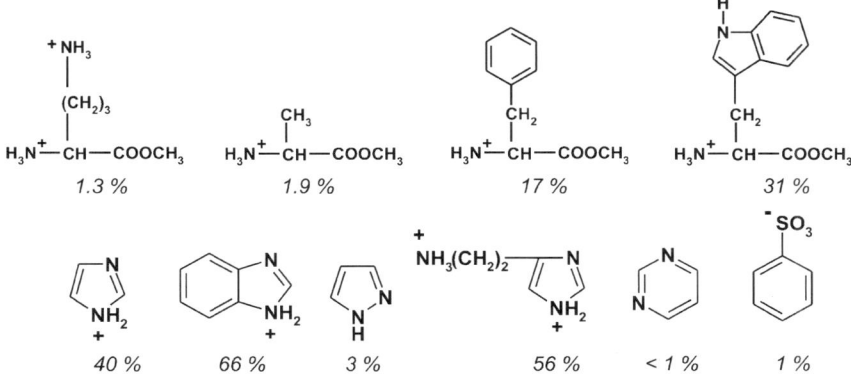

Scheme 18.2 Volume expansions% with different organic effectors (at 0.1 M concentration, pH 5.0); net effects: pH and salt effects deducted.

ionic gradient induced by the applied electric field, as a result, expansion occurred on the side facing the cathode, while contraction occurred on the other side facing the anode, so the gel could bend to the anode and produce a concomitant force. Shiga and Kurauchi[26] measured the different pH of the bath medium at two point sites separately in the half-cells with anode and cathode. Gülch and coworkers[27] measured the Donnan potential by inserting a KCl-filled glass microelectrode in to the gel. Zhou and coworkers[28] simulated the steady-state bending under an external electric field of a chitosan-based hydrogel by using a developed multiphasic theory, and found that with an increase of the applied potential, the gradient of the fixed charge concentration along the thickness of the sample also increased, accompanied by a greater pressure gradient. For a given pressure gradient, water flows into the sample from the half-cell with the cathode, resulting in relatively more water present in the section adjunct to the half-cell with the cathode; A converse situation happened in the section adjunct to the half-cell with the anode, thus generating a deformation gradient of bending.

Maintaining adequate mechanical properties, the rate of deformation could be improved by adjusting a number of extrinsic factors listed above. An index curvature was used here to express the extent of bending. For example, a deformed curvature of 2 cm^{-1} indicates that the deformed segment exhibits a curved shape with a radius of 0.5 cm. The closer to zero a curvature value is, the flatter the sample becomes. As shown in Figure 18.4(a), the bending curvature of a Chitosan/PEG stripe was proportional to the intensity of the applied electric potential, which functioned as a driven force. Figure 18.4(b) showed that a smaller sample diameter could facilitate bending in amplitude and speed, reflecting that ion diffusion played a key role in the process. Figure 18.4(c) illustrated that faster and more bending could be achieved by manipulating the pH value of the solution. These observations were interpreted in terms of fixed charge density and osmotic swelling, which depended on the equilibrium states in different pH and ionic environment, and the electrochemical reactions under the electric field. In the very low pH bath, the gels swelled much more compared to those in the medium with higher pH. Although the number of charged groups could increase in low pH, their density still decreases because the network expanded to a highly swollen state. As discussed earlier, lower charge density could accompany lower liquid pressure and strain. In addition, the reversibility of this phenomenon was examined by changing the direction of the applied potential cycle. The number of such alternate bendings to a prescribed extent within a given time period[20] was affected by the ionic strength and degree of crosslinkage as shown in Figure 18.4(d).

Besides the electrochemomechanical behaviours, chitosan also exhibits electrorheological response (ER) when its particle suspension is exposed to an electric field.[29,30] The ER behaviour produces a rapid and reversible change in suspension viscosity and structure in response to the electric field, which is regarded due to the polarisation of its branched polar amino groups.[31] Chitosan derivatives such as chitosan adipate and dihydroxylpropyl chitosan can also exhibit ER properties under various applied electric field strengths.[32,33]

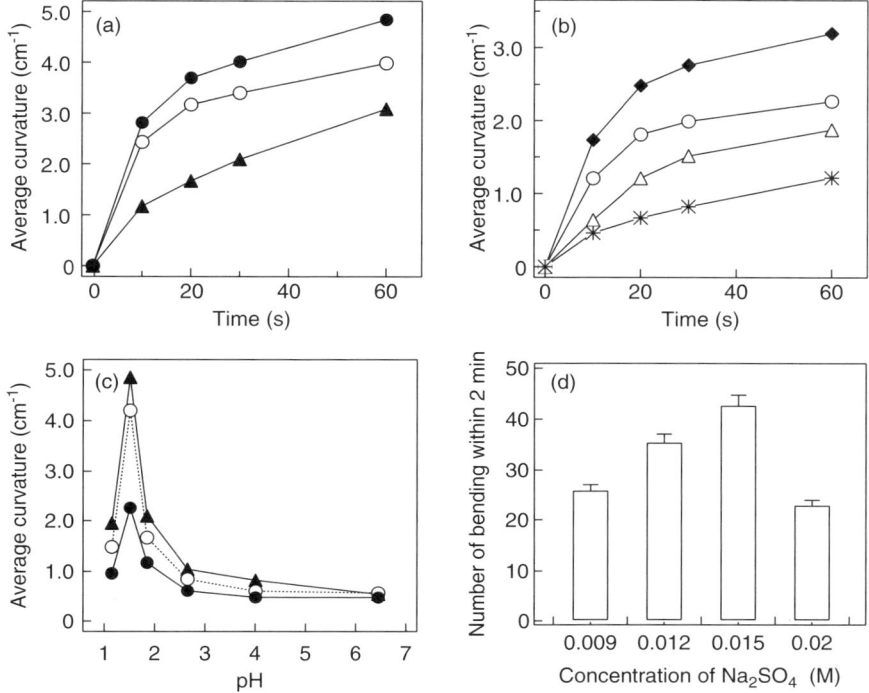

Figure 18.4 Bending of chitosan/PEG copolymer in response to electrical stimuli. Here are the influence of various factors on bending curvature: (a) effect of voltages (7) 10 V, (○) 15 V, (●) 20 V; (b) effect of the diameter of fibrous specimen in the medium of 0.05% HCl and 0.015 M Na_2SO_4 ($V = 10$ V), (◆) 0.325 mm, (○) 0.453 mm, (Δ) 0.763 mm, (*)1.045 mm; (c) effect of pH in 0.015 M Na_2SO_4 solution ($V = 10$ V) at (●) 10 s, (○) 20 s, (7) 30 s; (d) in presence of different concentrations of Na_2SO_4, the number of bending to a prescribed extent within 2 min, which was used as an index of reversibility of chitosan/PEG hydrogel.

However, no study has been reported using the ER properties of chitosan for biomedical applications yet, while the potential application in drug-delivery system has been tested with some biodegradable polymers based on their ER behaviours.[34]

18.3.5 Thermosensitivity

Chitosan alone has not been reported for its thermal sensitivity. Modified chitosan such as hydroxybutyl chitosan (HBC) exhibits sol-gel transition depending on the ambient temperature, and falls into temperature-responsive materials.[35] Once the temperature at or above the lower critical solution temperature (LCST) of the polymer solution, HBC solutions can form gel and can be reversed to a sol state as the temperature decreases below its LCST. The kinetics of gel formation is mainly dependent on the temperature difference between the LCST of the solution and the environmental temperature. Factors such as chitosan molecular weight, total degree of hydroxybutyl substitution, and polymer solution concentration can affect LCST. Generally the gelation temperatures are in a range of 21–26 °C depending on different gel formulations.

18.4 Chitosan-based Intelligent Materials

18.4.1 pH-Responsive Hydrogels

pH-sensitive gels are of general interest for their potential biomedical applications, such as artificial muscles or switches, biochemical separation systems, and controlled-release systems. Unmodified chitosan exhibits weak mechanical strength, uncontrollable porosity and degradability particularly in low pH aqueous solution. The use of pure chitosan formulations in oral administration is also limited due to their fast dissolution in the stomach and their limited capacity for controlling the release of drugs.[36–38] Table 18.1 listed various pH-sensitive chitosan-based polymers and some of their characteristics.[39–42]

So far, many efforts have been made on modification of chitosan and preparation of chitosan blended with other polymers. For examples, N-acylation of chitosan with various fatty acid (C_6–C_{16}) chlorides increased its hydrophobic character and made important changes in its structural features. Chitosan acylated with a short chain length (C_6) possessed similar properties but exhibited significant swelling. Acylation with longer side chains (C_8–C_{16}) resulted in a higher degree of order and crushing strength but lower swelling.[36] Freeze-dried chitosan/polyvinyl pyrrolidone (PVP) has been used to control the porosity of the blend.[37] By graft copolymerisation of specifically d,l-lactic acid (LA) or glycolic acid (GA) onto chitosan, the crystallinity and swelling behaviour of the synthesised chitosan hydrogels were affected significantly.[43]

18.4.2 Thermoresponsive and Dual Stimuli-responsive Polymers

As mentioned above, chitosan may possess thermal sensitivity after chemically conjugating with hydroxybutyl groups to the hydroxyl and amino reactive sites of chitosan.[35] However, in most cases, chitosan is combined with thermoresponsive polymers such as poly(N-isopropylacrylamide) (PNIPAAm), poly(N-acryloylglycine) or glycerol phosphate, to form semi- or full interpenetrating networks.[44–46] Thermoresponsive hydrogels are known to have phase-separation properties in aqueous solution when the temperature is increased above a critical solution temperature. For example, PNIPAAm hydrogels exhibit phase transition in water around 31–34 °C, which is close to the body temperature.[47] For applications requiring mainly thermosensitivity, chitosan is used primarily as a supporting component rather than the responsive element

Table 18.1 pH-stimuli-responsive intelligent chitosan-based materials.

Responsive polymer	Characteristics and application	Reference
Chitosan	Drug-delivery system	39
Chitosan grafted D,L lactic acid and/or glycolic acid	Modify crystallinity and swelling behaviours	43
Polyacrylonitrile grafted chitosan	Several sharp swelling in pH 2–13	12
Crosslinked chitosan and poly(vinyl alcohol) (PVA)	Improve mechanical properties	5
Citration/chitosan	Induce physical crosslink	40
Chitosan/alginate	More prolonged drug release and less erosion than chitosan alone	38
Chitosan and polyacrylamide (PAAm) – semi-interpenetrating networks (semi-IPNs)	Improve mechanical properties, maximum swelling appears at pH4	44–45
Chitosan-poly(ethylene oxide) (PEO)	Hydrogen bonds	41
Poly[N-vinyl-2-pyrrolidone-polyethylene glycol diacrylate]-chitosan	Control porosity for oral delivery	37
Chitosan–polyvinyl pyrrolidone	Antibiotic delivery in an acidic environment	42
Chitosan-poly(ethylene glycol–co-propylene glycol)	Dual stimuli-responsive artificial muscle	55
Chitosan/N-acryloylglycine (NAGly)	Dual stimuli-responsive	45
Chitosan-glycerol phosphate-water	Dual stimuli-responsive	46

due to its biological and chemical properties mentioned before. For application requiring pH/temperature dual stimuli-responsiveness, candidate polymers would need to show both intelligent properties. Such a demand is mainly motivated by the need of vehicles for the controlled release of active compounds under conditions determined at the site of action, at which the dual stimuli-responsive hydrogels are expected to be triggered for phase transition and swelling by the dual controlled factors more precisely. In the system of PNIPAAm and cross-linked chitosan, the onset of phase transition as a function of temperature is likely limited by the degree of crosslinking and the quantity of chitosan.[48]

18.4.3 Magnetic Chitosan Microsphere

Similar to the role in thermal-responsive polymers, chitosan can be used as a supportive structure for magnetic carriers rather than a functional intelligent material in the magnetic-responsive compound. Popular applications include bioaffinity chromatography, wastewater treatment, immobilisation of the enzymes, separation for purification and immunoassay.[49–52] Magnetic inorganic materials like Fe_3O_4 possess high resistance to mechanical, thermal solvent affects, but they do not offer much selective binding due to limited functional groups. Thus, various polymers with functional groups that can be tailored for specific binding are used as carriers for the magnetic materials, and chitosan is one candidate.

The important aspect of the magnetic chitosan particles is the good magnetic quality remained after preparation, in addition to the uniform size distribution by controlling the stirring rate, molecular weight and Fe_3O_4/chitosan ratio.[53] A recent attempt has been made to use these particles to induce hyperthermia in biological tissues for tumour therapy.[54] Hyperthermia is a therapeutic procedure to raise the body temperature of a local region affected by cancer to 42–46 °C. Chitosan particles are able to be loaded with drugs and release them during hyperthermia to enhance the therapeutic effects. Results of the use of such incorporated chitosan seemed promising, particularly with regard to the low cytotoxicity and the high heating rate. These findings supported the use of the chitosan microspheres as a promising magnetic support for biomedical applications with good economical aspects.

18.4.4 Electrical Responsive Polymers

As mentioned above, a chitosan blended with polyethylene glycol (CS/PEG) hydrogel strip/fibre could bend to the anode in aqueous solution with one of the two electrodes on each of its sides. For a given material composition, many factors such as electrical potential, pH, ionic strength of the bath medium, geometric size of the specimen can significantly influence the rate of bending and its reverse. Within the ranges of concentrations of crosslinking agent, pH and ionic strength examined by Sun *et al.*,[55] the experimental results suggested that 0.015 M Na_2SO_4 and 0.05% HCl solution are likely to offer an optimum conditions with regard to the mechanical properties and the bending responsiveness of the CS/PEG subjected to a given electric field.

18.5 Biomedical Applications

As multifunctional materials, chitosan and its derivative polymers can be used in numerous applications, including phase, affinity precipitations, bioactive surfaces, permeation switches, bioreactors, medical diagnostics, and drug-delivery systems.[56] The major field of application based on their stimuli-responsive properties is the biomedical engineering as discussed below.

18.5.1 Drug-delivery and Drug-release Systems

Chitosan-based materials have been extensively investigated as carriers in controlled drug-delivery and -release systems, and various drug formulations in neutral or charged, proteins or genes have been induced into the system, which is in a format of matrix, microparticle or nanoparticle, capsule, fibre or film. Their potential therapies include dental cavity, cancer, mucosal vaccination and gastrointestinal diseases.[57–59]

Like other drug-release systems consisting of polyelectrolyte hydrogels, drug release from chitosan-based particulate systems depends upon the extent of crosslinking, morphology, size and density of the carrier system, physicochemical properties of the drug, and local environment such as pH, polarity and presence of enzymes in the dissolution media. The release of drug from chitosan particulate systems involves four different mechanisms:

(a) Diffusion-controlled systems, in which the embedded drug molecules diffuse through the matrix or from a reservoir embraced by the polymers.
(b) Swelling-controlled systems, in which swelling at low pH can lead to rapid drug release, while shrinking at neutral or alkaline medium retards release.
(c) squeezing systems, in which drug is pushed out by a change in osmotic pressure. Similarly, the above-mentioned thermoresponsive polymers can also squeeze drugs out with sol to gel phase transition in response to temperature change.
(d) erosion/degradation controlled systems,[60] in which drug is released by the breakdown of polymer backbone, side chains, or cleavage of covalent bonds between drugs and polymers. Hybrid devices involving more than one mechanism also exist.

Conventional drug-delivery systems are limited by a discrepancy between the desired and the actual drug-release profile. The introduction of the stimuli-responsive polymers offers new ways to reduce such discrepancy, allowing an appropriate amount of an active agent to be made available at the desired time to a specific body site for therapeutic action.[61] Variations in physiological pH at different body sites like gastrointestinal tract, vagina and blood vessels, and the local pH change due to the emergence of some specific substrates provide the on-site stimuli for the pH-responsive drug-delivery systems.

Besides pH-controlled systems, electrically modulated drug the delivery with chitosan gel as matrix has been carefully studied.[62] Particularly for potential application in transdermal drug delivery, chitosan was acetylated with acetic anhydride to form robust gels, with the degree of acetylation varying from 80% before treatment to about 30% afterwards. Electrical stimuli may cause gel formation or collapse, and gel mass loss increased with increasing applied current and decreased with the degree of acetylation of chitosan, with greater drug release at higher applied current. This chemomechanical process also involved several competition factors such as solvent polarity, pH and ionic strength, electro-osmotic solvent flow, and additional contribution of drug polarity especially for anodic and cathodic drugs. As an example, the charge repulsion between the drugs and chitosan vehicles could account for the different permeation rates under an electric field.[63]

18.5.2 Injectable Gels for Tissue Engineering

Apart from drug-release and drug-delivery systems, thermogelling properties of chitosan hybrids have been investigated for tissue-engineering purposes. These chitosan hybrids are sols at room temperature. Once exposed to body temperature, the gel solution can gel rapidly and reversibly *in situ* within a short period of time. A potential application is to use such chitosan as an injectable delivery of a biologically active scaffold. A chitosan–glycerol phosphate–water system is one such systems.[46] Recently, thermosensitive hydroxybutyl chitosan (HBC) seeded with human mesenchymal stem cells (hMSC) was investigated as a biologically relevant reconstruction of the degenerated disk.[35] Conjugation of hydroxybutyl groups to chitosan makes the polymer more water soluble and thermally responsive. Depending on the formulation of the HBC gels, gelation temperatures could range from 13.0 to 34.6 °C. Minimal cytotoxicity of these polymers in MSC and disk cell cultures were noted up to a concentration of 5 wt%, and cell proliferation without a loss in metabolic activity and extracellular matrix production showed HBC as a promising candidate in tissue engineering.

18.5.3 Artificial Actuators and Muscles

In biomedical applications, artificial actuators are referred to those smart-material-based systems that could undergo significant shape or volume changes upon stimulation.[20,64,65] An ideal for robotic actuation would be an analogue of a human muscle – a contractile material able to be driven by chemical or electrical signals. This is the ultimate goal for the artificial muscles. With the development in polymeric materials and advanced instrumentation, research in artificial actuators has been extended to a wide range of biomedical applications such as prosthetic devices,[66] drug-delivery systems,[67] biosensors,[68] chemical memories and circulatory-assist devices.[25] Smart hydrogels are potential candidates for such a system. The response rate and extent, as well as the concomitant force produced afterwards are mainly determined by the structural design of the polymeric composite. Polyelectrolyte hydrogels have attracted much interest for their reversible deformation in response to external stimulus, especially electrical simulation. Chitosan/poly(ethylene glycol) composite hydrogels prepared were shown to possess such properties.[55] Although the present artificial muscle cannot fully match the real muscle in terms of range of motion, strength to weight ratio and speed of response, research on the materials that have inherent muscle-like properties has been gathering momentum.

Compared to the other polymers including ionic polymers, conductive polymers and electrostrictive polymers used in artificial actuators, chitosan-based hydrogels were shown to be relatively similar in physical constitution as soft tissues, but exhibited relatively slower response and smaller actuating force, with maximum stresses in the range of 0.1–0.3 MPa.[69] It is commonly agreed that among other needs, it is necessary to improve the response rate and the load-carrying capacity of such materials before practical applications could be envisaged.

18.6 Conclusions

Chitosan-based materials have been attracting extensive attention for their potential applications as stimuli-responsive materials. Besides their qualifications for biomaterials such as naturally produced, biocompatible, biodegradable, structurally similar to some biologically polysaccharides including heparin, chondroitin sulfate, and hyaluronic acid, their intelligent properties will make chitosan-based materials promising candidates for effective drug-delivery systems, injectable tissue scaffolds, and artificial actuators. In the future, chemical modification of chitosan and derivatives are still important to get the desired physicochemical properties. Novel functional groups or compounds and nanoscale features are expected to be added to improve and optimise their sensitivity in respond to physical, chemical and biological stimuli, and more in vivo studies need to be carried out. We anticipate that more chitosan-based intelligent hydrogels will become commercially available for biomedical and industrial applications in the near future.

References

1. K. Kurita, *Prog. Polym. Sci.*, 2001, **26**, 1921.
2. M. Minoru, S. Hiroyuki and S. Yoshihiro, *Trends Glycosci. Glycotechnol*, 2002, **14**, 205.
3. B. Krajewska, *Sep. Purif. Technol.*, 2005, **41**, 305.
4. M.N.V. Ravi Kumar, *React. Funct. Polym.*, 2000, **46**, 1.
5. T. Wang, M. Turhan and S. Gunasekaran, *Polym Int*, 2004, **53**, 911.
6. L. Raymond, F.G. Morin and R.H. Marchessault, *Carbohydr. Res.*, 1993, **243**, 331.
7. N. Errington, S.E. Harding, K.M. Varum and L. Illum, *Int. J. Biol. Macromol.*, 1993, **15**, 113.
8. B.K. Choi, K.Y. Kim, Y.J. Yoo, S.J. Oh, J.H. Choi and C.Y. Kim, *Int. J. Antimicrob. Agents*, 2001, **18**, 553.
9. R.A.A. Muzzarelli, *Cell Mol. Life Sci.*, 1997, **53**, 131.
10. E. Ruel-Gariepy, A. Chenite, C. Chaput, S. Guirguis and J.C. Leroux, *Int. J. Pharm.*, 2000, **203**, 89.
11. M. Beppu and C. Santana, *Mater. Res.*, 2002, **5**, 47.
12. G.R. Mahdavinia, A. Pourjavadi, H. Hosseinzadeh and M.J. Zohuriaan, *Eur. Polym. J.*, 2004, **40**, 1399.
13. Y. Qiu and K. Park, *Adv. Drug Deliv. Rev.*, 2001, **53**, 321.
14. K.-D. Yao, T. Peng, H.-B. Feng and Y.-Y. He, *J. Polym. Sci. A: Polym. Chem.*, 1994, **32**, 1213.

15. F.M. Goycoolea, A. Heras, I. Aranaz, G. Galed, M.E. Fernandez-Valle and W. Arguelles-Monal, *Macromol. Biosci.*, 2003, **3**, 612.
16. F.-L. Mi, S.-S. Shyu, S.-T. Lee and T.-B. Wong, *J. Polym. Sci. B: Polym. Phys.*, 1999, **37**, 1551.
17. G.R. Mahdavina, M.J. Zohuriaan-Mehr and A. Pourjavadi, *Polym. Adv. Technol.*, 2004, **15**, 173.
18. Per M. Claesson and B.W. Ninhami, *Langmuir*, 1992, **8**, 1406.
19. N. Lomadze and H.-J. Schneider, *Tetrahedron*, 2005, **61**, 8694.
20. S. Sun and A.F.T. Mak, *J. Polym. Sci. B: Polym. Phys.*, 2001, **39**, 236.
21. E.M. Gutman, *Mechanochemistry of Solid Surfaces*, World Scientific Publishing Co. Pte. Ltd., Singapore, 1994.
22. E.M. Gutman, *Mechanochemistry of Materials*, Cambridge International Science Publishing, Cambridge, UK, 1998.
23. I.C. Kwon, Y.H. Bae and S.W. Kim, *Polym. Sci. Part B: Polym. Phys.*, 1994, **32**, 1085.
24. T.F. Otero, F.J. Huerta, S.A. Cheng, D. Alonso and S. Villanueva, *Electroactive Polymer Actuators and Devices*, ed. Y. Bar-Cohen, Newport Beach, California, 2000, SPIE **3987**, 148.
25. R. Dagni, *Chem. Eng. News*, 1997, 26.
26. T. Shiga and T. Kurauchi, *J. Appl. Polym. Sci.*, 1990, **39**, 2305.
27. R.W. Gülch, J. Holdenried, A. Weible, E. Tübingen, T. Wallmersperger and B. Kröplin, in *Polymer Actuators and Devices*, ed. Y. Bar-Cohen, Newport Beach, California, 2000, SPIE **3987**, 193.
28. X. Zhou, Y.C. Hon, S. Sun and A.F.T. Mak, *Smart Mater. Struct.*, 2002, **11**, 459.
29. J.H. Sung, W.H. Jang, H.J. Choi and M.S. Jhon, *Polymer*, 2005, **46**, 12359.
30. J.H. Sung, H.J. Choi and M.S. Jhon, *Mater. Chem. Phys.*, 2002, **77**, 778.
31. U.S. Choi, *Coll. Surf. A:Physicochem. Eng. Aspects*, 1999, **157**, 193.
32. U.S. Choi and Y.S. Park, *J. Ind. Eng. Chem.*, 2001, **7**, 281.
33. S. Wu, F. Zeng and J. Shen, *J. Appl. Polym. Sci.*, 1998, **6**, 20777.
34. J.L. Davies, I.S. Blagbrough and J.N. Staniforth, *Chem. Commun.*, 1998, 2175.
35. J.M. Dang, D.N. Sun, Y. Shin-Ya, A.N. Sieber, J.P. Kostuik and K.W. Leong, *Biomaterials*, 2006, **27**, 406.
36. C.L. Tien, M. Lacroix, P. Ispas-Szabo and M.A. Mateescu, *J. Contr. Release*, 2003, **93**, 1.
37. M.V. Risbud, A.A. Hardikar, S.V. Bhat and R.R. Bhonde, *J. Contr. Release*, 2000, **68**, 23.
38. C. Tapia, E. Costa, J. Sapag-Hagar, F. Valenzuela and C. Basualto, *Drug. Devel. Industr. Pharm.*, 2002, **28**, 217.
39. P. Giunchedi, C. Juliano, E. Gavini, M. Cossu and M. Sorrenti, *Eur. J. Pharm. Biopharm.*, 2002, **53**, 233.
40. X.Z. Shu, K.J. Zhu and W. Song, *Int. J. Pharm.*, 2002, **233**, 217.
41. V.R. Patel and M.M. Amiji, *Pharm. Res.*, 1996, **13**, 588.
42. M.V. Risbud and S.V. Bhat, *J. Mater. Sci. Mater. Med.*, 2001, **12**, 75.
43. X. Qu, A. Wirse'n and A.-C. Albertsson, *Polymer*, 2000, **41**, 4589.

44. M.Z. Wang, J.C. Qiang, Y. Fang, D.D. Hu, Y.L. Cui and X.G. Fu, *J. Polym. Sci. Part A: Polym. Chem.*, 2000, **38**, 474.
45. I.M. El-Sherbiny, R.J. Lins, E.M. Abdel-Bary and D.R.K. Harding, *Eur. Polym. J.*, 2005, **41**, 2584.
46. A. Chenite, C. Chaput, D. Wang, C. Combes, M.D. Buschmann, C.D. Hoemann, J.C. Leroux, B.L. Atkinson, F. Binette and A. Selmani, *Biomaterials*, 2000, **21**, 2155.
47. C.-L. Lin, W.-Y. Chiu and C.-F. Lee, *J. Colloid Interf. Sci.*, 2005, **290**, 397.
48. L. Verestiuc, C. Ivanov, E. Barbu and J. Tsibouklis, *Int. J. Pharm.*, 2004, **269**, 185.
49. A. Kondo and H. Fujuda, *J. Ferment. Bioeng.*, 1997, **84**, 337.
50. R.J. Ansell and K. Mosbach, *Analyst*, 1998, **123**, 1611.
51. T. Yoshimoto, T. Mihama, K. Takahashi, Y. Saito and Y. Inada, *Biotechnol. Lett.*, 1987, **8**, 877.
52. D.-S. Jiang, S.-Y. Long and J. Huang, H-Y. Xiao and J.-Y Zhou, *Biochem Eng. J.*, 2005, **25**, 15.
53. E.B. Denkbas, E. Kilicay, C. Birlikseven and E. Öztürk, *React. Funct. Polym.*, 2002, **50**, 225.
54. J.-H. Park, K.-H. Im, S.-H. Lee, D.-H. Kim, D.-Y. Lee, D.-K. Lee, K.-M. Kim and K.-N. Kim, *J Magn. Magn. Mater.*, 2005, **293**, 328.
55. S. Sun, Y.W. Wong, K.-D. Yao and A.F.T. Mak, *J. Appl. Polym. Sci.*, 2000, **76**, 542.
56. N.V. Majeti and R. Kumar, *React. Funct. Polym.*, 2000, **46**, 1.
57. M. Prabaharan and J.F. Mano, *Drug Deliv.*, 2005, **12**, 41.
58. K.C. Gupta and M.N.V.R. Kumar, *J. Macromol. Sci., Polym. Rev. Macromol. Chem. Phys*, 2000, **C40**, 273.
59. O. Felt, P. Buri and R. Gurny, *Drug. Devel. Industr. Pharm.*, 1998, **24**, 979.
60. S.A. Agnihotri, N.N. Mallikarjuna and T.M. Aminabhavi, *J. Contr. Release*, 2004, **100**, 5.
61. P. Gupta, K. Vermani and S. Garg, *Drug Discov. Today*, 2002, **7**, 569.
62. S. Ramanathan and L.H. Block, *J. Contr. Rel.*, 2001, **70**, 109.
63. J.Y. Fang, K.C. Sung, J.J. Wang, C.C. Chu and K.T. Chen, *J. Pharm. Pharmacol.*, 2002, **54**, 1329.
64. M. Shahinpoor and K.J. Kim, *Smart Mater. Struct.*, 2001, **10**, 819.
65. M. Solari, V. Recagno, A. Roseo, S. Sacone and R. Tacchino, *J. Intell. Mater. Syst. Struct.*, 1993, **4**, 157.
66. R.H. Baughaman, C.X. Cui and A.A. Zakhidov, *Science*, 1999, **284**, 1340.
67. L.M. Low, S. Seetharaman, K.Q. He and M. Madou, *Sens. Actuators B.*, 2000, **67**, 149.
68. J. Dumont and G. Fortier, *Biotechnol. Bioeng.*, 1996, **49**, 544.
69. Y. Bar-Cohen, *Electroactive Polymers as Artificial Muscles – Capabilities, Potentials and Challenges*, Handbook on Biomimetics, Y. Osada (Chief Editor), Section 11, in Chapter 8, NTS Inc., Tokyo, Japan, 2000.

CHAPTER 19

Polymer-Protein Complexation and its Application as ATP-driven Gel Machine

RYUZO KAWAMURA,[1] AKIRA KAKUGO,[1] YOSHIHITO OSADA[1] AND JIAN PING GONG[1,2]

[1] Graduate School of Science, Hokkaido University, Sapporo 060-0810, Japan
[2] SORST, JST, Sapporo 060-0810, Japan

19.1 Introduction

There are two basic differences between the motion in a man-made machine and in a biological motor. One is in their principles: The motion of a man-made machine, which is constructed from hard and dry materials such as metals, ceramics or plastics, is realised by the relative displacement of the macroscopic constituent parts of the machine. In contrast to this, the motion of a living organism, which consists of soft and wet protein and tissues, is caused by a molecular deformation that is integrated to a macroscopic level through its hierarchical structure.[1–3] The other is in their energy sources: The man-made machine is fueled by electrical or thermal energy with an efficiency of around 30%, but a biological motor is driven by direct conversion of chemical energy with an efficiency as high as 80–90%.[4] In order to create biomimetic systems, polymer gels have been employed due to their reversible size and shape change, thereby realising the motion by integrating the deformation on a molecular level. The contractile collagen fibre is the earliest man-made example of gel actuators by Katchalsky and coworkers.[5,6] Later, Osada and coworkers[7–13] and others developed several gel machines based on synthetic polymers. Recently, various types of peptide-based gels as well as hybrid gels have been developed.[14–18] The specific properties of these gels might be beneficial to constructing soft actuators based on the concept of molecular machines.[19,20]

Polypeptides from living organisms are attractive materials because of their high functionality, such as enzyme activity, ability to self-organise, or selective recognition. Some groups have realised devices by depositing or arranging polypeptides on substrates. For example, thin films composed of bacteriorhodopsin[20] have been successfully used as photoelectric devices, and thin films of glucose oxidase as glucose-response biosensors.[21] Actin-based metallic nanowires and actuators were constructed recently.[22–25]

Furthermore, various kinds of motor proteins, actin myosin[26–28] or microtubule kinesin,[29] as well as other proteins that generate rotational motion like F1-ATPase,[30] also have been used as micromotors.[31–33] These devices using polypeptides would find applications in the medical or pharmaceutical fields. However, most of these devices have been exploited only in one aspect of biological function, partly because polypeptides were immobilised or arranged two-dimensionally on the surface of the substrate. By using the chemical crosslinking technique, polypeptides can have a highly ordered structure as three-dimensional gels.

Actins and myosins are major components of muscle proteins and play an important role in dynamic motion of creatures that is caused by the molecular deformation using the chemical energy released by hydrolysis of ATP. Recently, we have found that F-actins can be self-assembled into large bundles in the presence polycations through polymer-complex formation. These polycation–actin complexes, several tens of times the length of native actin filaments, move along a chemically crosslinked myosin fibrous gel with a velocity as high as that of native actins, by coupling to ATP hydrolysis.[22–24,34] This result indicates that muscle proteins can be tailored into a desired shape and size without sacrificing their bioactivities, and can be used as biomaterials.

19.2 Actin Gel Formed from Polymer–Actin Complexes[34]

Since the isoelectric point of actins is pH 4.7, F-actins are negatively charged in neutral buffer. Therefore, they have been assumed to form complexes with cationic polymers through electrostatic interaction. Figure 19.1(a)–(d) show some examples of fluorescence microscope images of polymer–actin complexes obtained by mixing F-actins with poly-L-Lysine(p-Lys) (Figure 19.1(a)) and x, y-ionene polymers which have the structure of $-[-(CH_2)_x-N^+Br^-(CH_3)_2-(CH_2)_y-N^+Br^-(CH_3)_2-]_n-$, (Figure 19.1(b)–(d)) for 120 min. The morphology of the complexes depends on the chemical structure of the polycations.[34] One can see that large filamentous, stranded and branched complexes of 20–30 μm size are formed in the presence of p-Lys, 3,3- 6,6- 6,12-ionene polymers and their morphological nature, both size and shape, are remarkably different from that of native F-actin (Figure 19.1(e)). The relationship between the morphology of polycation–actin complex and the concentrations of components is discussed in Section 19.7.

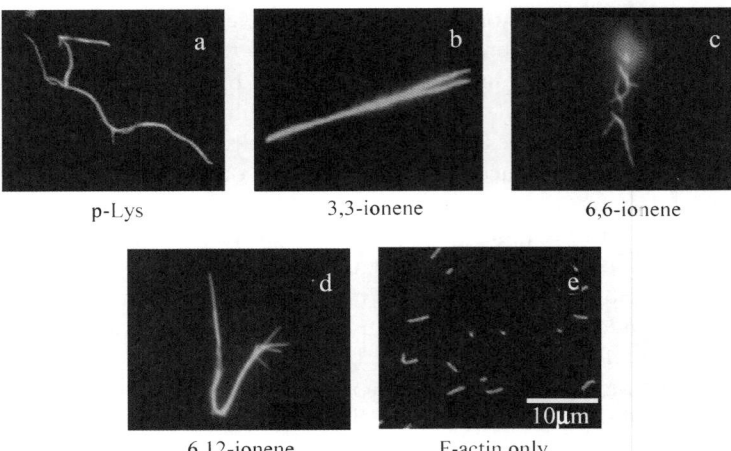

Figure 19.1 Fluorescence microscope images of polymer–actin complexes formed by mixing F-actin and various cationic polymers at room temperature. (a) p-Lys, (b) 3,3-ionene, (c) 6,6-ionene, (d) 6,12-ionene, (e) F-actin only. The molar ratio of ammonium cation of polymer to monomeric actin was kept constant at 30:1 for x,y-ionene polymers and 100:1 for p-Lys, which corresponds to weight ratios of [3,3-ionene]/[actin]) 0.41 g/g, [6,6-ionene]/[actin]) 0.61 g/g, [6,12-ionene]/[actin]) 0.81 g/g, [p-Lys]/[actin]) 0.35 g/g. Actin concentration was 0.001 mg/mL.[34]

The number-average length of the fluorescence image of F-actins is 2.14 μm with a standard deviation of 0.11 μm (average over 784 samples) in the F-buffer that is known to cause a transformation to fibrous polymerised state of actin (F-actin) from globular monomer actin (G-actin). However, polymer–actin complexes grow with time and reach as large as 5–20 μm within 1 or 2 h, which is about 2–10 times larger than that of native F-actin. p-Lys gives out a large complex, whereas 3,3-ionene polymer gives the smallest complexes. These results imply that hydrophobicity and charge density of the ionene polymers are important in complex formation. The effect of electrostatic interactions is proved, since no complexation was observed with negative controls such as p-Glu, DNA and neutral polymer, such as PEG. The average lengths of polymer–actin complexes are shown in Figure 19.2(a) and (b).

Concerning the analysis of the lateral structure of the polymer–actin complexes, transmission electron microscopy (TEM) accompanied with a negative staining technique was employed. Although the samples should be exposed in high vacuum and a dry state in this technique, the higher resolution offers advantage compared to analysing from the images of fluorescent microscopes concerning such a precise structure.

The TEM measurements show that the polymer–actin complexes are bundles that consist of 3–20 filaments.[24] The average width of the p-Lys–actin complex is 21.0 nm with a standard deviation of 2.6 nm. Comparing with the native F-actins that have an average width of 12.1 nm with a standard deviation of

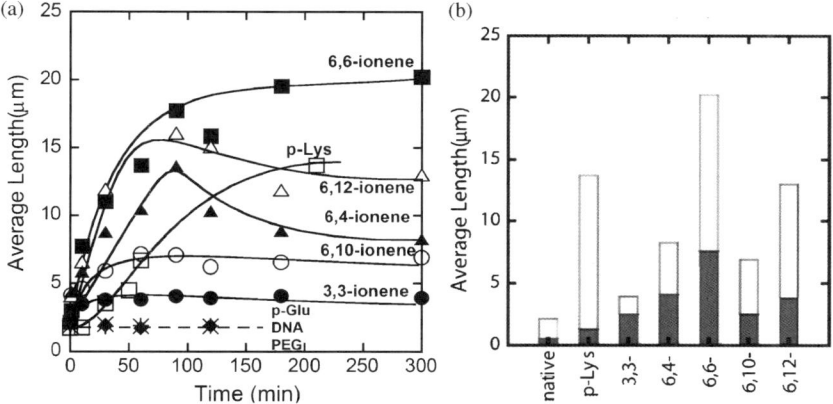

Figure 19.2 (a) Time courses of polymer–actin complexes growth. (●): 3,3-ionene-actin complexes, (▲): 6,4-ionene-actin complexes, (■): 6,6-ionene-actin complexes, (○): 6,10-ionene-actin complexes, (△): 6,12-ionene-actin complexes, (□): p-Lys–actin complexes, (◆): p-Glu-actin complexes, (×): [DNA-actin complexes, (+): PEG-actin complexes. (b) Average length of polymer–actin complexes observed from fluorescence microscope images (white columns) and from transmission electron microscope (TEM) images (shade columns) at 210–300 min. The molar ratio of ammonium cation to monomeric actin was 30:1 for x,y-ionene polymers and 100:1 for p-Lys. The corresponding weight ratios were as follows: [3,3-ionene]/[actin] = 0.41 g/g, [6,4-ionene]/[actin] = 0.54 g/g, [6,6-ionene]/[actin] = 0.61 g/g, [6,10-ionene]/[actin] = 0.74 g/g, [6,12-ionene]/[actin] = 0.81 g/g, [p-Lys]/[actin] = 0.35 g/g, [p-Glu]/[actin] = 0.36 g/g, [DNA]/[actin] = 0.77 g/g, [PEG]/[actin] = 0.10 g/g. Actin concentration: 0.001 mg/mL.[34]

1.2 nm, p-Lys–actin complexes are only slightly thicker than that of the native F-actin with almost the same width scattering. 3,3-ionene complexes also showed a very thin and homogeneous wire-like morphology, showing an average width of 16.1 nm with a standard deviation of 1.7 nm.

In the case of x,y-ionene polymers, they form bundles of thin filaments with F-actin above certain critical concentrations that are dependent on each polymer. The effect of x,y- of ionene polymers was observed in the width of complexes. Actin- 6,4-, 6,6-, 6,10-, and 6,12-ionene complexes have an average width of 79.0, 59.3, 38.7 and 66.1 nm with a standard deviation of 60, 29, 21, and 27 nm, respectively (Figure 19.3). From the large scattering in the width of actin-6,4-, 6,6-, 6,10-, and 6,12-ionene complexes, the morphology of complexes seems to have randomness quantitatively. Indeed, a ring-shaped complex (nanoring) is observed in a 6,6-ionene-actin complex occasionally.[34]

19.3 Polymorphism of Actin Complexes[35]

In order to investigate the polymorphism, the complexation of F-actin with poly-N-[3-(dimethylamino)propyl]-acrylamide(PDMAPAA-Q), which has positive

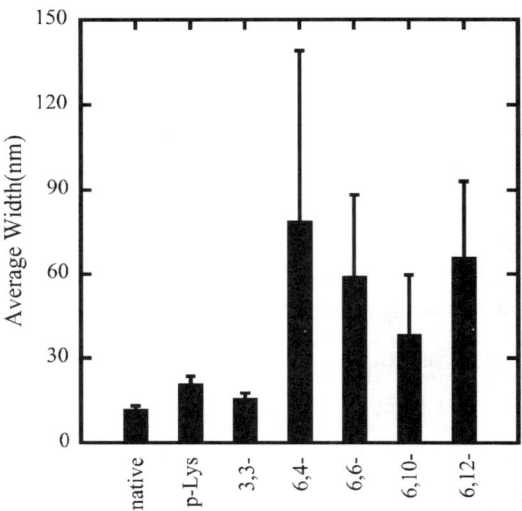

Figure 19.3 Average width of polymer–actin complexes obtained at 240 min by TEM images. Error bar means standard deviation.[34]

charges on its side chain, is further investigated systematically. The polymer–actin complexes exhibit a rich polymorphism in a wide range of the polycation concentrations (C_P) and KCl salt concentrations (C_S), as elucidated by the fluorescent analysis and TEM analysis, which show micro- and nanoscale images, respectively. There are five characteristic phases in the C_P–C_S phase diagram (Figure 19.4). Figure 19.5 shows the fluorescent images and TEM images for the polymer–actin complexes in the five phases. In phase I, F-actin does not grow. In phase II, we observe the coexistence of native F-actin and polymer–actin complexes. In the coexistence phase (II), the fraction of native F-actins increases with the increase of C_S. The borderline between phases I and II shifts to a higher C_P with the increase of C_S. This can be explained by the screening effect of salts on the electrostatic interaction between F-actins and polycations. As C_P and C_S increase, the F-actins form complexes with polycations and exhibit various structures. The polymer–actin complexes evolve from the crosslinked structure dominant phase (III), to the branched-structure dominant phase (IV), and then to the parallel-bundle dominant phase (V) as shown by the TEM images in Figure 19.5.

The fluorescent images in Figure 19.5 show that the polymer–actin complexes of the crosslinked structure dominant phase (III) are in a compact globule state, whereas those of the parallel-bundle-dominant phase are in the extended state (V). The globule size of 15–20 μm is attributed to the persistence length of F-actin, which is assumed as ca. 10 μm. The morphological change of polymer–actin complexes from the compact globule state to the extended-bundle state is due to the increase of bending rigidity that increases with the thickness D of actin bundles, varying as D.[36,37]

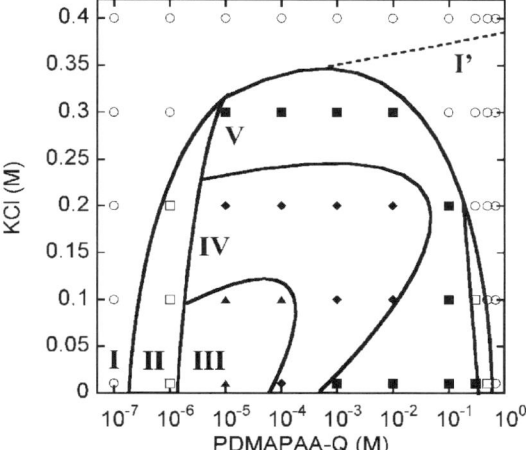

Figure 19.4 Phase diagram for the morphology of polymer–actin complexes, in which the phase behavior is summarised as a function of PDMAPAA-Q concentration C_P and KCl concentration C_S for a constant actin concentration C_A) 0.01 mg/mL (2.32×10^{-7} M). ○, F-actins; □, coexistence (polymer–actin complex and native F-actins); ▲, crosslinked structure dominant phase; ◆, branched structure dominant phase; and ■, parallel-bundle dominant phase. The dotted line shows the possible borderline between the native F-actin and F-actin with charge inversion.[35]

19.4 Oriented Myosin Gel Formed under Shear Flow[22,23]

The chemically crosslinked myosin gel with its oriented filament array can be obtained by reacting the scallop myosin at pH 7.0 using transglutaminase (TG) under shear-flow-induced stretching. The oriented myosin gel is semitransparent, showing a swelling degree of ca. 100, and a Young's modulus of 190 Pa in the oriented direction, which is more than two times larger than that of the myosin gel prepared without stretching. The orientation of myosin fibres in the gels was analysed by scanning electron microscopy (SEM) and atomic force microscopy (AFM). From the images, distinct bundles of regularly oriented filaments ca. 1.5 µm in diameter were observed, indicating that the rod-like myosin molecules are self-organised with orientation to form a hierarchical structure. The molecular orientation within the filaments was confirmed by the strong IR dichroism of carbonyl absorption at 1600 cm^{-1}, which could not be observed in the absence of stretching. The chemically crosslinked myosin gel shows an ATPase activity as high as that of native myosins in the presence of 0.5 w/w native actin. Myosin gels crosslinked by other crosslinking agents, such as glutaraldehyde and 1-ethyl-3-(3dimethyl aminoprolyl)carbodiimide, also showed an ATPase activity, though not as high as those crosslinked by TG. The motion of these myosin gels will be discussed in the next section.

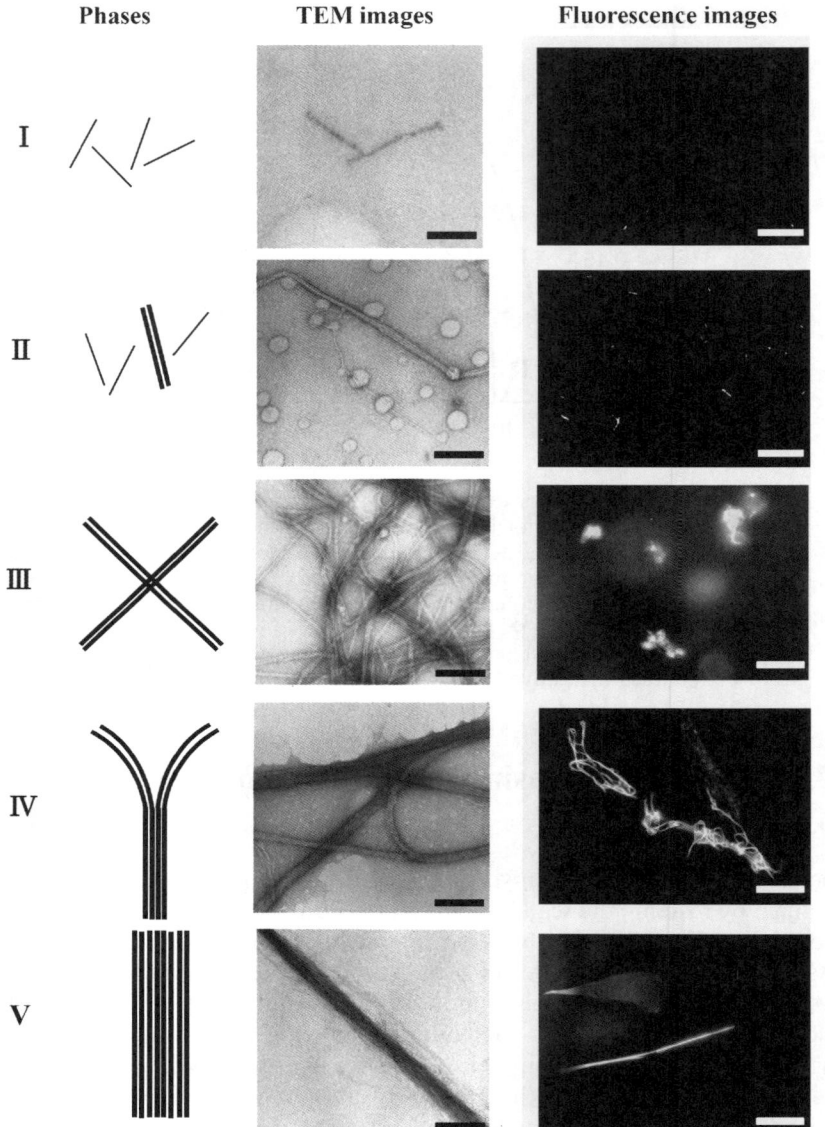

Figure 19.5 Typical morphologies of polymer–actin complexes in phases I, II, III, IV, and V as observed by TEM images and fluorescence images. Scale bars present 200 nm for TEM images and 25 μm for fluorescence images.[35]

19.5 Motility Assay of F-actin on Oriented Myosin Gel[23]

F-actins showed a preferential motion along the axis of oriented myosin gel as elucidated by the degree of anisotropy (DA), which is defined as the ratio of the

Table 19.1 Motility assay of F-actin on oriented myosin gel.

Sample	DA	Mean velocity (±SD)	n
Nonoriented	1.1	0.69 ± 0.24 µm/s	66
Oriented	1.7	0.83 ± 0.30 µm/s	91

square-root average velocity in the fibre direction to that perpendicular to the fibre direction,

$$\mathrm{DA} = \frac{\bar{x}}{\bar{y}} \tag{19.1}$$

where,

$$\bar{x} = \sqrt{\sum_{i=1}^{N} \frac{x_i^2}{N}}, \bar{y} = \sqrt{\sum_{i=1}^{N} \frac{y_i^2}{N}} \tag{19.2}$$

The DA measured on the nonoriented myosin gel was 1.1 (average over 66 samples), and that on the oriented myosin gel was 1.7 (average over 91 samples). The mean velocity on the nonoriented myosin gel was 0.69 µm/s with a standard deviation of 0.24 µm/s, while that on the oriented myosin gel was 0.83 µm/s with a standard deviation of 0.30 µm/s. Thus, F-actin filaments prefer to move along the axis of the oriented myosin gel with an increased velocity Table 19.1.

19.6 Motility Assay of Polymer–Actin Complex Gel

The p-Lys–actin complex can be crosslinked to obtain a stable structure. The complex gel crosslinked with TG also shows a high motility on the oriented myosin gel in spite of its large dimension. The p-Lys–actin complex gels move preferentially along the axis of the oriented myosin gel almost without path deviations. Figure 19.6 shows the velocity distributions of p-Lys–actin complex gels on the oriented myosin gel (closed circles) as a function of the individual filament size. The p-Lys–actin complex gels, about four times larger than that native F-actin, move with an average velocity of 1µm/s, almost the same as that of native F-actins on the oriented myosin gel (opened circles). Some of the p-Lys–actin complex gels move as fast as 2.0 µm/s. In addition, the DA of the p-Lys–actin complex gels was 2.2 (average over 38 samples) on the oriented myosin gel, which was higher than that of the native F-actin (DA = 1.7), indicating an enhanced directional preference along the axis of the oriented myosin gel. Thus, despite its increased mass, the p-Lys–actin complex gels, several tens to hundred times the volume of the native F-actins, move on the covalently crosslinked myosin gel, with an increased velocity. This is rather surprising, since the interaction between the myosin gel and the actin gel can only occur at the two-dimensional interface and, due to crosslinking, a

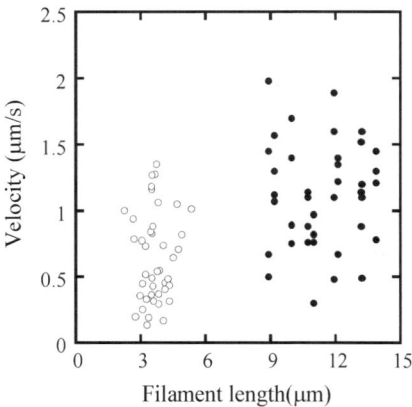

Figure 19.6 Individual velocities of native F-actins (opened circle) and p-Lys–actin gels (closed circle) of various lengths on the oriented myosin gel.[22]

considerable number of actin and myosin molecules are not involved in the sliding motion.[22] These results indicate that actins can be tailored into the desired shape and size without sacrificing their bioactivities by using complex formation with synthetic polymers.

19.7 Polarity of the Actin in Complexes[24]

On a glass surface where myosin molecules are simply immobilised, F-actin is known to show motility. We found that all these polymer–actin complex gels, including those from p-Lys, PDMAPAA-Q, and x,y-ionene bromide polymers ($x = 3$ or 6; $y = 3$, 4 or 10), also exhibit motility and the velocity of the complexes were comparable to that of native F-actin ($0.77 \pm 0.32\,\mu$m/s).[24]

The polarity of polymer–actin complexes, which is considered to be essential in the sliding motion of polymer–actin complex gels, has not been completely clarified yet. This question is important for understanding the cooperative motion by the actin assembly. An arrowhead-like pattern can be observed under transmission electron microscopy (TEM) when heavy meromyosin (HMM) was decorated on native F-actin (Figure 19.7(a)), indicating that F-actin has a well-defined polarity by self-organisation. The pointed end of the arrowhead and the opposite end of the arrowhead are called the p-end and b-end, respectively.

To evaluate the polarity of actin complexes, we also attempted to decorate the actin complexes with HMM. Figure 19.7(a) shows some examples of TEM images of HMM-decorated PDMAPAA-Q–actin complexes and 6,4-ionene-actin complexes. Different from native F-actin, which is a single strand, the polymer–actin complexes are bundles that consist of 3–20 filaments. Arrowhead structures within a filament of the bundle pointed in the same direction,

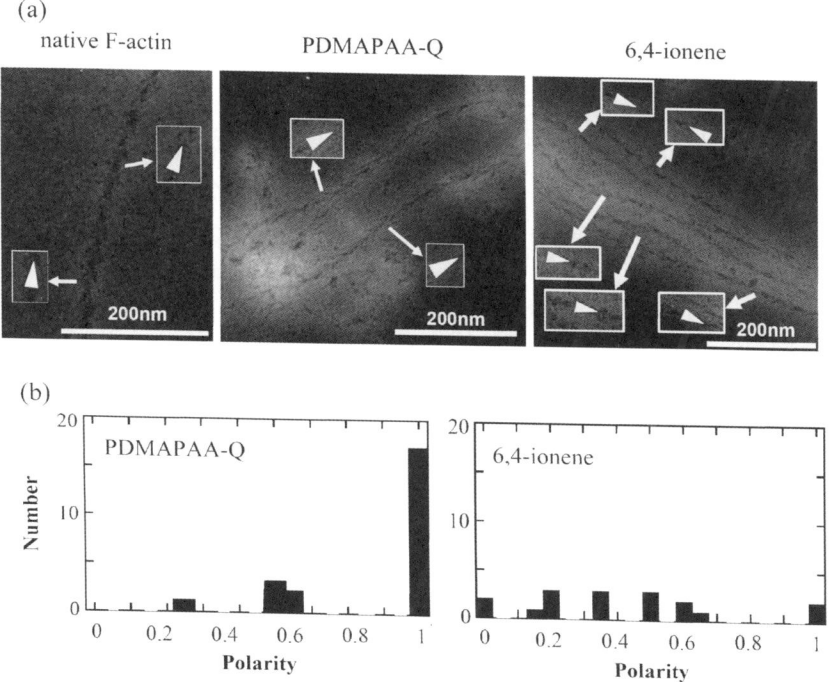

Figure 19.7 (a) Polarity of polymer–actin complexes decorated with HMM. White arrows indicate the direction of arrowhead structures of decorated filaments. (b) Histograms of complex polarity distributions.[24]

although some defects are observed occasionally. However, arrowhead directions of filaments within a bundle are not completely the same. The complex polarity P defined by eqn (19.3) was estimated,

$$P = |n_1 - n_2|/|n_1 + n_2| \qquad (19.3)$$

Here, n_1 and n_2 are the number of filaments pointed in the opposite directions.

The average polarity of the actin complex is shown in Table 19.2. As shown in Table 19.2, polarity depends on the chemical structure of polycations, and PDMAPAA-Q–actin complexes show the highest polarity as 0.89 (average over 23 samples), while 6,4-actins show the lowest value as 0.42 (average over 17 samples). Thus, PDMAPAA-Q prefers to form actin filament bundles having a unipolarity.

Among all these polymer–actin complexes, the PDMAPAA-Q–actin complex, which has the highest polarity, also shows the highest motility with a velocity of 1.3 μm/s. Dendritic complexes, which are occasionally observed when F-actin is mixed with 6,4-ionene, did not exhibit a translational motion but instead migrate around their barycentric position. When the velocity at 3.3 s is plotted against the polarity, a linear relationship is observed, i.e. the velocity

Table 19.2 Polarities of polymer–actin complexes.

Polymer	Polarity
PDMAPAA-Q	0.89 (n = 23)
p-Lys	0.76 (n = 21)
3,3-Ionene	0.50 (n = 22)
6,4-Ionene	0.42 (n = 17)
6,10-Ionene	0.49 (n = 22)

n is the number of samples for the average.[24]

of the complexes is proportional to the polarity. These results suggest that the polarity of the polymer–actin complex is essential in producing a high motility. It is known that a native F-actin always moves toward one direction without moving back to the opposite direction. The direction of the F-actin motion is associated with its polarity.[38] For the polymer–actin bundles that are assembly of F-actin, the whole polarity is determined by the polar direction of F-actin. If the F-actin filaments assemble in an antiparallel way, the whole polarity is cancelled and leads to no sliding motion. This explains why the correlation between polarity and velocity is observed. However, the difference between the velocity of native F-actin and that of the polymer–actin complex gels with a polarity close to 1 (p-Lys–actin and PDMAPAA-Q–actin) cannot be explained only in terms of polarity. The higher velocity observed for the complex gels should, therefore, be attributed to two possible factors: (1) the arrowhead structure that depends on the structures of polycations and (2) the bundle formation. If the change of the arrowhead structure is attributed to the change of the helix pitch of an actin filament, an elongation of arrowhead pattern will cause the extensions of the helix structure of the actin filament. In consequence, the elasticity of the actin filament might increase and cause the effective dynamic interaction with myosin, which favours the motion. On the other hand, by forming an actin bundle, the bending fluctuation is eliminated, and this also leads to an effective integration of driving forces from each myosin molecules.

To elucidate the randomness of the motion, we also investigated the trajectory of the motion.[24] From this characterisation, it is found that all polymer–actin complexes show more translational motion than that of native F-actin. Because polymer–actin complexes are bundles formed from actin filaments, they are less flexible than native F-actin. The less random motion of polymer–actin complexes is attributed to the reduced structural flexibility.

19.8 Conclusions

Biopolymers including F-actin have a lot of potential as functional materials in biotechnology and biomaterials science. It has been revealed that actin and myosin vary their functionality depending on the orientation within the complexes. In the case of myosin, it has been found to have a self-assembly ability. On the other hand, actin has shown an ability to form a functional complex with polycations by electrostatic interaction. Moreover, the cross-linked actin complex gels also have shown motility on the myosin gels. The motility of actin-complex gels depends on the structure that is determined by the conditions such as the concentrations or the structures of the polycations. This may also be applicable for the other functional biopolymers. To develop more functional biomaterials hereafter, how to orientate the functional domain or the polarity may become the most significant issue. The basic analysis and technical development to form and stabilise the biopolymers by crosslinking may contribute to promote the possibility of the biopolymers as functional materials in biotechnology and biomaterials science.

References

1. H.E. Huxley, The mechanism of muscular contraction, *Science*, 1969, **164**, 1356–65.
2. A.F. Huxley, *Reflections on Muscle*, Liverpool University Press, Liverpool, 1980.
3. G.H. Pollack, *Cells, Gels and the Engines of Life*, Ebner and Sons Publishers, Seattle, WA, 2001.
4. K. Kitamura, M. Tokunaga, A.H. Iwane and T. Yanagida, A single myosin head moves along an actin filament with regular steps of 5.3 nanometres, *Nature*, 1999, **397**, 129–34.
5. I.Z. Steinberg, A. Oplatka and A. Katchalsky, Mechanochemical engines, *Nature*, 1966, **210**, 568–71.
6. M.V. Sussman and A. Katchalsky, Mechanochemical turbine: A new power cycle, *Science*, 1970, **167**, 45–7.
7. Y. Osada, H. Okuzaki and H. Hori, A polymer gel with electrically driven motility, *Nature*, 1992, **355**, 242–4.
8. T. Mitsumata, K. Ikeda, J.P. Gong and Y. Osada, Solvent-driven chemical motor, *Appl. Phys. Lett.*, 1998, **73**, 2366–8.
9. Y. Osada and A. Matsuda, Shape memory in hydrogels, *Nature*, 1995, **376**, 219.
10. Y. Osada and S.B. Ross-Murphy, Intelligent gels, *Sci. Am.*, 1993, **268**, 82–7.
11. E.W.H. Jager, E. Smela and O. Inganas, Microfabricating conjugated polymer actuators, *Science*, 2000, **290**, 1540–6.
12. D.J. Beebe, J.S. Moore, J.M. Bauer, Q. Yu, R.H. Liu, C. Devadoss and B.H. Jo, Functional hydrogel structures for autonomous flow control inside microfluidic channels, *Nature*, 2000, **404**, 588–90.
13. S.R. Quake and A. Scherer, From micro- to nanofabrication with soft materials, *Science*, 2000, **290**, 1536.

14. L.A. Haines, K. Rajagopal, B. Ozbas, D.A. Salick, D.J. Pochan and J.P. Schneider, Light-activated hydrogel formation via the triggered folding and self-assembly of a designed peptide, *J. Am. Chem. Soc.*, 2005, **127**, 17025–17029.
15. K. Rajagopal, B. Ozbas, D.J. Pochan and J.P. Schneider, *Biopolymers*, 2005, **80**, 487–487, ibid 2005, **80**, 594–594.
16. A.P. Nowak, V. Breedveld and L. Pakstis, et al., *Nature*, 2002, **417**, 424–428.
17. M.J. Zohuriaan-Mehr, Z. Motazedi, K. Kabiri and A. Ershad-Langroudi, *J Macromol Sci.-Pure Appl. Chem.*, 2005, **A42**, 1655–1666.
18. V. Balzani and A. Credi and M. Venturi, *Molecular Devices and Machines*, Wiley-VCH, Weinheim, 2003.
19. J.F. Stoddart, Molecular machines, *Acc. Chem. Res.*, 2001, **34**, 410–11.
20. T. Koyama, N. Yamaguchi and T. Miyasaka, *Science*, 1994, **265**, 762–765.
21. B.A. Gregg and A. Heller, *Anal Chem.*, 1990, **62**, 258–263.
22. A. Kakugo, S. Sugimoto, J.P. Gong and Y. Osada, Gel machines constructed from chemically cross-linked actins and myosins, *Adv. Mater.*, 2002, **14**(16), 1124–6.
23. A. Kakugo, S. Sugimoto, K. Shikinaka, J.P. Gong and Y. Osada, Characteristics of chemically cross-linked myosin gels, *J. Biomater. Sci. Polym. Ed.*, 2005, **16**(2), 203–18.
24. A. Kakugo, K. Shikinaka, N. Takekawa, S. Sugimoto, Y. Osada and J.P. Gong, Polarity and motility of large polymer–actin complexes, *Biomacromolecules*, 2005, **6**, 845–49.
25. F. Patolsky, Y. Weizmann and I. Willner, Actin-based metallic nanowires as bio-nanotransporters, *Nat. Mater.*, 2004, **3**, 692–695.
26. T.Q.P. Uyeda, H.M. Warrick, S.J. Kron and J.A. Spudich, *Nature*, 1991, **352**, 307–311.
27. H.E. Huxley, *J. Biol. Chem.*, 1990, **265**, 8347–8350.
28. S.J. Kron and J.A. Spudich, *Proc. Natl. Acad. Sci. USA*, 1986, **83**, 6272–6276.
29. R.D. Vale, T.S. Reese and M.P. Sheetz, *Cell*, 1985, **42**, 39–50.
30. H. Noji, R. Yasuda, M. Yoshida and K. Kinosita, Jr., *Nature*, 1997, **386**, 299–302.
31. M. Muratsugu, S. Kurosawa and N. Kamo, *J. Colloid Interf. Sci.*, 1991, **147**, 378–386.
32. T. Miyasaka, K. Koyama and I. Itoh, *Science*, 1992, **255**, 342–344.
33. H. Suzuki, K. Oiwa, A. Yamada, H. Sakakibara, H. Nakayama and S. Mashiko, *Jpn. J. Appl. Phys.*, 1995, **43**, 3937–3941.
34. A. Kakugo, K. Shikinaka, K. Matsumoto and J.P. Gong, Y. Osada. Growth of large polymer–actin complexes, *Bioconjugate Chem.*, 2003, **14**(6), 1185–90.
35. H.J. Kwon, A. Kakugo, K. Shikinaka, Y. Osada and J.P. Gong, Morphology of Actin Assemblies in Response to Polycation and Salts, *Biomacromolecules*, 2005, **6**, 3005–3009.

36. R.K. Meyer and U. Aebi, Bundling of actin filaments by alpha-actinin depends on its molecular length, *J. Cell Biol.*, 1990, **110**, 2013–2024.
37. L.D. Landau and E.M. Lifshitz, *Theory of Elasticity*, Pergamon, Oxford, UK, 1986.
38. G.H. Pollack, *Cells, Gels and the Engines of Life*, Ebner and Sons Publishers, Seattle, WA, 2001.

CHAPTER 20

Intelligent Composite Materials Having Capabilities of Sensing, Health Monitoring, Actuation, Self-Repair and Multifunctionality

HIROSHI ASANUMA

Dept. of Electronics and Mechanical Engineering, Chiba University, 1-33, Yayoicho, Inege-ku, Chiba-shi, Chiba 263-8522, Japan

20.1 Introduction

Conventional composite materials are generally known as high-performance structural materials. The typical one is fibre-reinforced plastic (FRP) such as glass-fibre-reinforced plastic (GFRP) and carbon-fibre-reinforced plastic (CFRP). Some FRPs have also been laminated with metal or alloy sheets known as fibre-metal laminates (FMLs). Aiming at higher-temperature applications, fibre-reinforced metals (FRMs) have been developed. FRPs, FMLs and FRMs provide the opportunity to incorporate sophisticated phases such as optical fibre sensors, shape-memory alloys, piezoelectric ceramic fibres for sensing, health monitoring, actuation, and other capabilities.

An important purpose of developing intelligent composite materials is simplification of mechanical systems just by using active/intelligent material systems or composite materials as shown in Figure 20.1. They may have sensing, health monitoring, actuation, self-repair, and/or other capabilities. Most of the conventional complicated mechanical systems need joints, lubrication, heavy actuators, delicate sensors, and so on. But the active/intelligent material systems or composite materials may be able to eliminate them just by their intelligently designed microstructures.

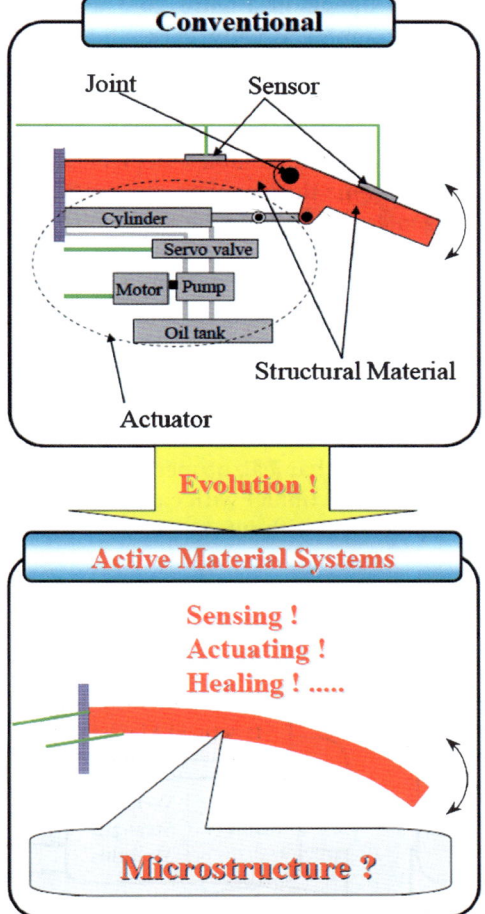

Figure 20.1 Simplification of conventional complicated mechanical systems by using intelligent composite materials.

In this chapter, some intelligence designed in composite materials from an engineering point of view, where simplicity is more important than complexity for safety and reliability, will be introduced. In addition, a new category of composite materials having liquid phases for self-repair and other capabilities will be introduced.

20.2 A New Route to Develop Intelligent Composite Materials

The main route to develop intelligent composite materials is embedding sophisticated functional materials in structural composite materials, as already explained above. This is shown in Figure 20.2 as Type-I route, which is normally taken; many examples already exist and are well known.[1-5] The sophisticated functional materials are usually fragile, heavy and/or expensive and have some negative effects on mechanical properties, and sometimes react with matrix materials during embedding processes. So it is not easy to fabricate highly reliable and commercially available intelligent composite materials by this route. In order to overcome this problem, the author proposed the Type-II route to obtain intelligent/active composite materials without using sophisticated functional materials.

The idea is summarised in Figure 20.2. There exist a couple of competitive structural materials that normally compete with each other because of their similar high mechanical properties such as high strength and high modulus, or high specific strength and high specific modulus. They tend to have other properties that are different from each other or are opposite. So if they are combined together to make a composite material, the similar or common property, normally a high mechanical property, can be maintained and the other dissimilar properties may conflict with each other, which will successfully generate functional properties without using any sophisticated functional materials.

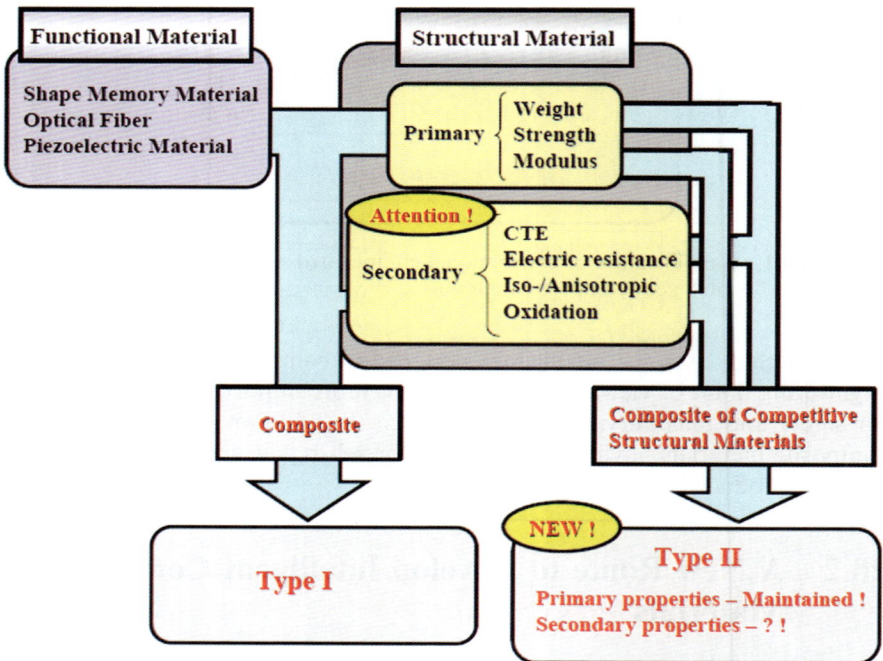

Figure 20.2 Main routes to develop intelligent composite materials.

20.3 Composite Materials Fabricated by the New Route

Two typical examples made through the Type-II route, that is, CFRP/aluminium laminate[6,7] and titanium fibre/aluminium composite material are introduced in this section.

In the case of the CFRP/Al active laminate as shown in Figure 20.3, the main and common properties, that is, low weight and high strength are shown in the overlapped region of the ovals, and the other properties such as thermal expansion and electrical resistance are shown in the other parts. Aluminium has a high and an isotropic CTE (coefficient of thermal expansion) and very low resistivity, but CFRP has anisotropic CTE, that is, very low in the fibre direction and very high in the transverse direction, and relatively high resistivity. So, if they are laminated as shown in Figure 20.4 with an electrical insulation layer by hot pressing, a unique and useful material system can be obtained. This simple laminate is one of the most useful types because of its unidirectional actuation. The laminate does not bend in the transverse direction because the CTEs of the CFRP and aluminium layers are close to each other, but surprisingly bends in the fibre direction due to the large difference in CTEs in this direction. The CFRP layer works as a heater by application of a voltage between its ends, so the curvature of the laminate can be controlled by the applied voltage.

If this material system is regarded as a living body, the aluminium, having a high CTE, works as muscle, and the CFRP, having a nearly zero CTE, especially the carbon fibre itself having a negative CTE, works as bone. In addition, the carbon fibre works as a blood vessel to supply energy to the material system for actuation because it efficiently generates electrical resistant heat, and it also works as a nerve because it can sense temperature change as its electrical resistivity changes.

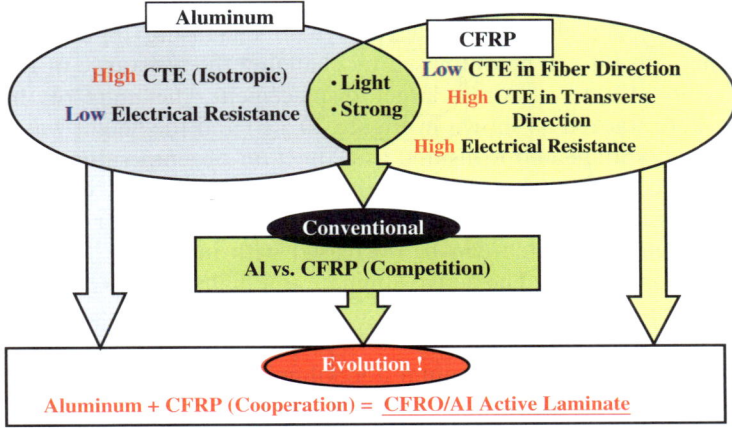

Figure 20.3 The CFRP/Al active laminate as a typical Type-II composite material.

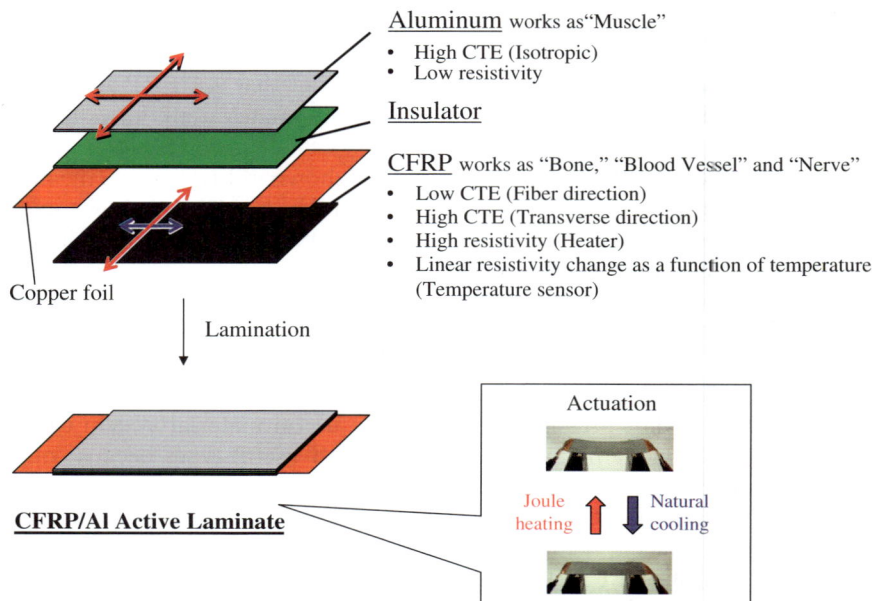

Figure 20.4 Fabrication of the CFRP/Al active laminate and its unidirectional actuation by electrical resistance heating and natural cooling.

Taking advantage of the above-mentioned deformation characteristics, four types of demonstrators as shown in Figure 20.5, that is, a hatch, stack, coil and lift type, were fabricated and demonstrated. In the case of the hatch type shown in Figure 20.5(a), a part of the body was cut out and this part was actuated by applying a voltage, where the initial current slightly increased with electrical resistance decrease caused by a temperature increase. This electrical-resistance change can be used for monitoring its temperature and control of its curvature. Its actuation speed drastically increases with increasing input power. The voltage and current can be optimised by changing the resistance of the CFRP layer. As an idea to obtain a larger displacement, the units of the active laminate were stacked as shown in Figure 20.5(b) and the height was changed by applying a voltage and measured as a function of temperature. The height can be changed almost linearly as a function of temperature, which is a very useful behaviour for this application. A coil-type demonstrator was also made and its diameter and pitch can be decreased with increasing temperature as shown in Figure 20.5(c). The active laminate is a high strength active material, so it does not need to generate force against something, but it can also do so as shown in Figure 20.5(d).

The second example, the titanium fibre/aluminium system, is another interesting and useful example and is summarised in Figure 20.6. The basic concept is common to that of the CFRP/Al active laminates. The material system is very simple but it can generate many useful functions such as heating,

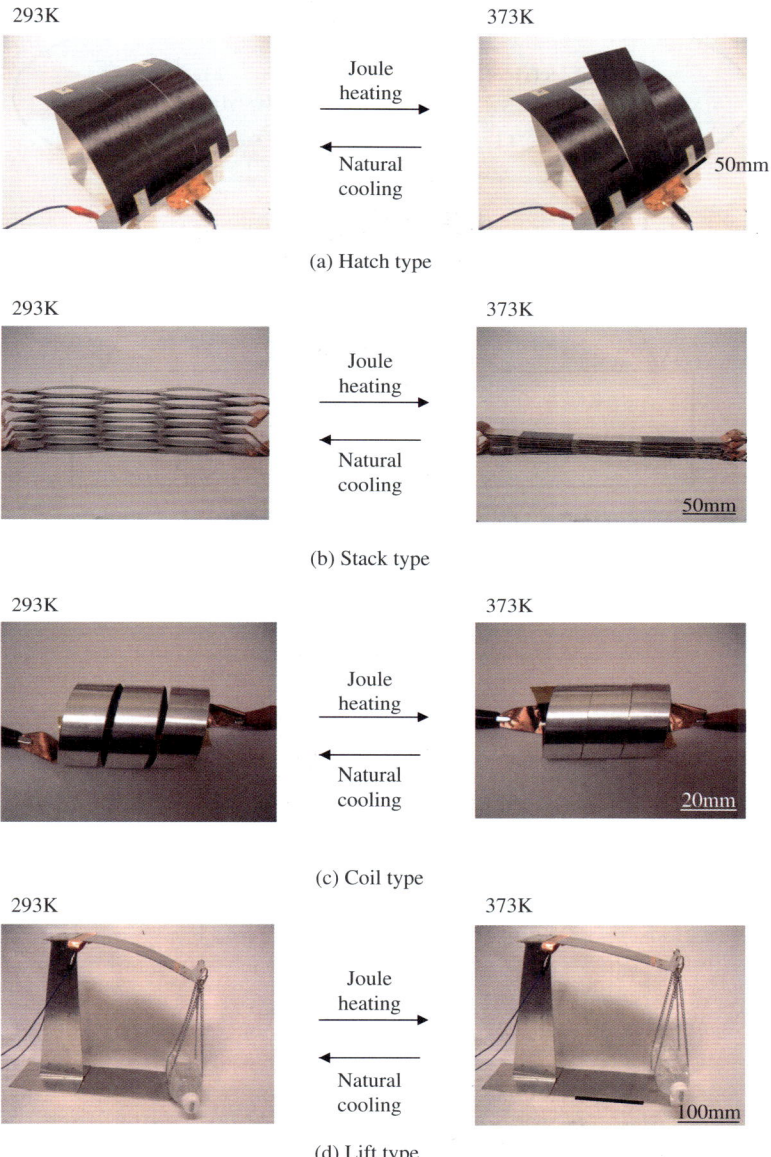

Figure 20.5 Four types of demonstrators made by using the CFRP/Al active laminate.

actuation, temperature sensing, deformation sensing, and so on, as shown in Figure 20.7, and also a self-repair function explained later.

The advantages of the above-introduced active material systems compared to bimetals are directional actuation due to anisotropy of thermal deformation due to the directionality of the continuous fibres and low weight because of

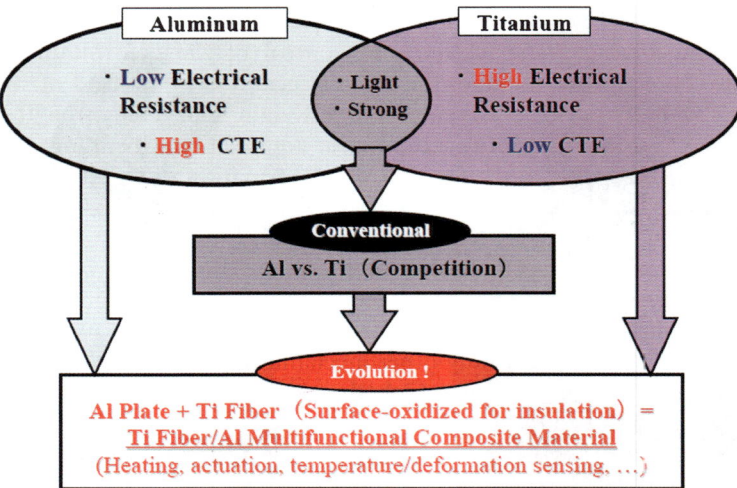

Figure 20.6 The Ti fibre/Al multifunctional composite material made through the Type-II route.

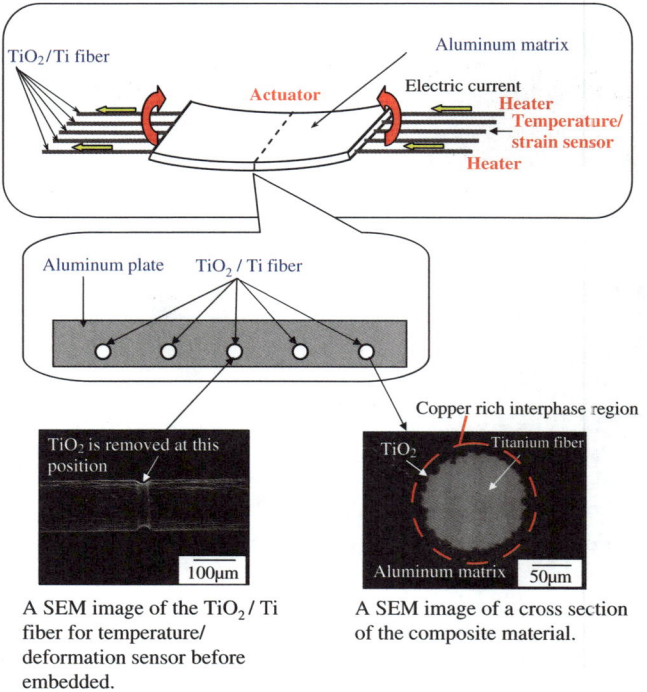

Figure 20.7 The Ti fibre/Al multifunctional composite material as a typical Type-II product and its microstructure.

using only high specific strength materials. By taking advantage of the unique deformation characteristic, design and control of the deformation become available. This positive usage of thermal deformation in the field of structural materials will be able to develop a lightweight and high-strength planar actuator. A simple and useful actuation is unidirectional bending of a lightweight structural material panel. Thermal deformation of structural materials is regarded as a negative phenomenon in general, but it is positively used in these cases. The newly developed active material systems will be able to replace a part of conventional mechanical systems because of simplification and weight reduction by removal of complicated and heavy actuation systems, and making them free from lubrication and wear problems by removal of joints, so it will be used for aerospace and many other mechanical applications. In addition, if the other functions, for example, high electrical and thermal conductivities of aluminium are taken into consideration, more applications will be found.

20.4 A New Category of Composite Materials Having Liquid Phases for Self-repair and Other Capabilities

Composite materials having liquid phases for self-repair and other capabilities can be classified as a new category and are very attractive. In most cases, liquid phases are formed in matrices by melting lower melting point solids embedded in them or encapsulated in spheres or hollow tubes to be released at their breakages caused by some trigger, to react with something else for strengthening or penetrate into defects to be repaired. Polymeric materials, polymer matrix composite materials and concretes have been targeted for realisation of these functions.[8–12]

The author has been challenging realisation of a new type of metal-matrix composite material having a liquid phase.[13] The concept and the mechanical properties are summarised in Figure 20.8 and explained below.

The metal-matrix composite material has a fibre/matrix interphase as shown in Figure 20.8(a) which melts or softens at the temperature T_I, which is lower than the matrix melting temperature T_M. The levels of the tensile strengths of the fibre-reinforced metal (FRM) without the interphase and the unreinforced matrix material are shown in this figure by σ_C and σ_M, respectively, as shown in Figures 20.8(b) and (c). As shown in Figure 20.8(b), the interphase FRM maintains its tensile strength σ_I from the point A (at room temperature) to the point B (just below T_I) and it drastically reduces its tensile strength at T_I down to the point C (at T_I) as low as the level of the unreinforced matrix σ_M because the interphase melts, and then it follows this level up to the point D. When the interphase FRM is cooled down from the point D to the point E (just below T_I), it recovers its tensile strength up to the point F because the interphase solidifies, and then it can go back to the original point G (= A). If the interphase and the matrix causes reaction and the chemical composition of

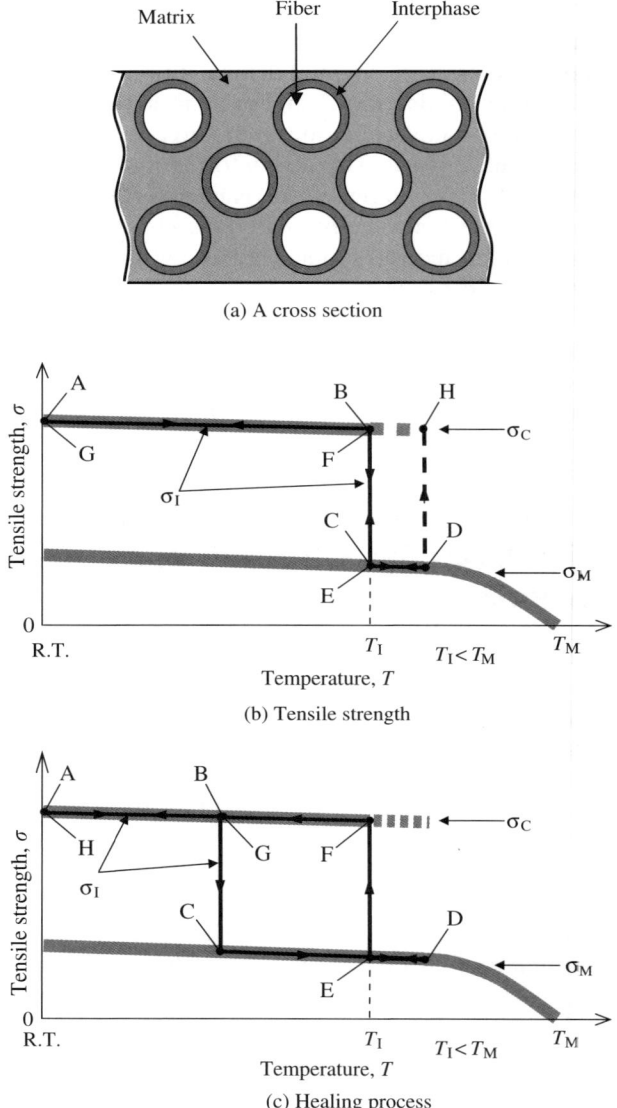

Figure 20.8 The interphase FRM proposed by the author and its self-repair process.

the interphase changes, its tensile strength might start to increase from the point D up to the point H, and then it will go back to the point B and A on cooling.

Taking advantage of the above-mentioned mechanical behaviour, a self-repair process can be considered as shown in Figure 20.8(c). As shown in this figure, if the fibre/matrix debonding takes place at the point B, the tensile strength of the interphase FRM reduces down to the point C as low as the level of the unreinforced matrix σ_M. But, if this damaged composite material is heated to the point D, and then, cooled to the point E (just below T_I), the debonded fibre/matrix interface can rebond and its tensile strength can go up to the point F, then it can go back to the point G (= B) before debonding and then H (= A) when it is cooled down. This cyclic self-repair process is an advantage of the interphase FRM.

A couple of the interphase FRMs were experimentally made by the author,[14] that is, aluminium matrix and stainless-steel fibre were used, and Al-Zn, Al-Cu and Al-Si eutectic alloys were selected to form the low melting temperature interphase. The fabrication technique is the key to realise the special material system easily. This type of composite material cannot be fabricated by normal processes, because if these low melting temperature alloys are coated on the fibres first, and then the coated fibres are embedded in the matrix, the interphase forming alloys melt first during the embedding process and disappear. So the author developed an innovative process called interphase forming/bonding process (IF/B process) [13]. The characteristic of the IF/B process can be briefly explained using Figure 20.9. The matrix plates and the insert material used for bonding of them react with each other and produce a liquid alloy, a part of which remains in the clearances between the reinforcement fibres and the U-grooves previously made for alignment of the fibres, and the rest part is squeezed out of the composite material. Using this technique, bonding of the matrix plates and coating of the low melting temperature alloy on the fibres can be realised at the same time easily.

The mechanical properties of this type of composite materials are unique and useful. Up to just below the melting temperature of the interphase, which is

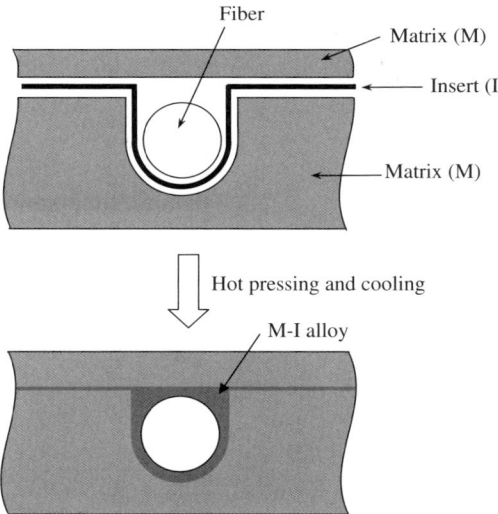

Figure 20.9 The interphase forming/bonding process (IF/B process) proposed by the author.

high enough for service, the tensile strength of the composite material is as high as the reinforced state. When it is heated up to or higher than the melting temperature, the interphase melts and the tensile strength drastically reduces down to the unreinforced matrix level. Using this state, self-repair or plastic forming can be done. After this stage, its tensile strength can recover up to the reinforced level by a temperature decrease. In some cases, such as using Al-Zn alloy, mutual diffusion between molten interphase and solid matrix occurs and the melting temperature of the interphase increases. This phenomenon is useful for recovery of the tensile strength at temperatures higher than the original melting temperature of the interphase.

The Ti/Al multifunctional composite material is also made by the IF/B process, so it has the function of self-repair. The IF/B process enables fabrication of many other composite materials, in particular it is suitable for embedding fragile functional materials such as optical fibre[15] and piezoelectric ceramic fibre,[16,17] so these fabricated composite materials also have capabilities of self-repair.

A self-repair capability for the CFRP/Al active laminate by forming a liquid phase during actuation was also developed.[18] As schematically shown in Figure 20.10, an active laminate having a small amount of solid self-repair phase in contact with a part of the bonding plane was made, and the solid phase was melted in the composite material by the heating capability of the active laminate and advantageously pushed into its delaminated part by the internal pressure induced by its actuation. This type of active laminate having

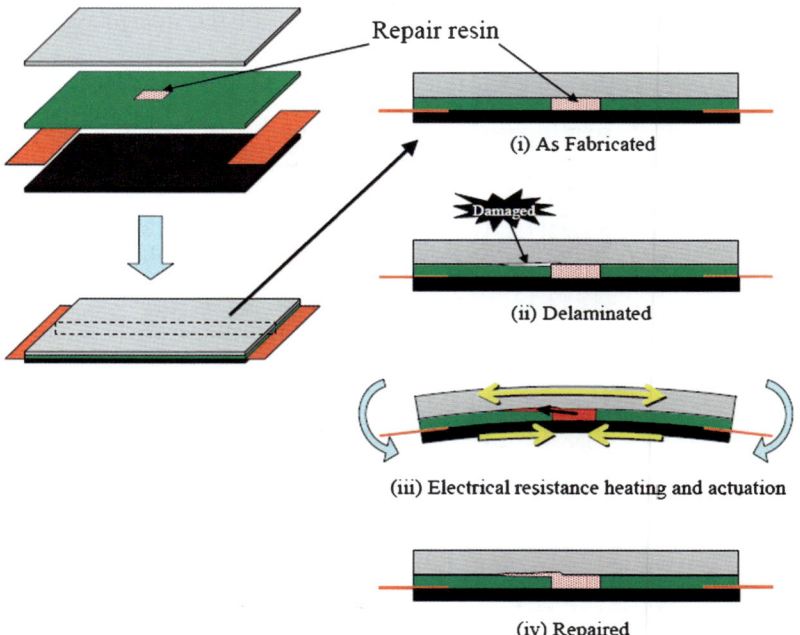

Figure 20.10 The self-repair mechanism developed for the CFRP/Al active laminate.

multifunctionality including autonomous self-repair capability during service is a new and efficient intelligent structural material system.

20.5 Summary and Outlook

In this chapter, a couple of intelligent composite materials fabricated by using high-performance structural materials without using sophisticated functional materials and by simple and innovative routes or processes to be reliable and low cost were mainly introduced. Using these composite materials, some conventional complicated mechanical systems will be converted into dynamic and multifunctional/intelligent material systems capable of sensing, monitoring, actuating, self-repair, and so on, in the near future.

Toward the future, the intelligent composite materials will evolve in a variety of ways to by taking wide areas of science and technology in. In Figure 20.11, an example of a magnified microstructure that will generate a variety of useful functions is shown. The material consists of a couple of phases and paths. The phases will respond to the stimuli or react with them given through the paths depending on the types of the phases. Each phase has its own physical, chemical, mechanical and electrical properties, and so on, to be responsive to a variety of stimuli given through the paths by electric current, light, heat, chemical, fluid, gas, and so on, differently from the other phases. One of the phases would play the role of bone, the other one or two phases would play the

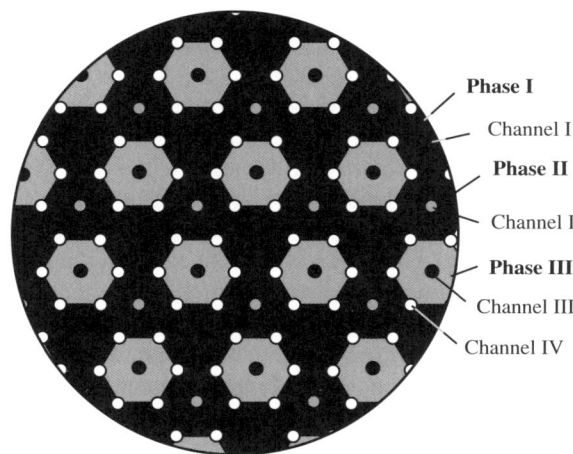

Figure 20.11 A cross section of an intelligent composite material for future developments.

role of muscle, and the channels would play the role of blood vessel and nerve, like a living body. This kind of intelligent composite materials will bring us a bright future.

References

1. M.V. Gandhi and B.S. Thompson, *Smart Materials and Structures*, Chapman & Hall, London, 1992.
2. M. Schwartz (Editor-in-Chief), *Encyclopedia of Smart Materials*, John Wiley & Sons, Inc., New York, Vols 1 and 2, 2002.
3. D.D.L. Chung, *Composite Materials: Science and Applications*, Springer, London, 2002.
4. H. Asanuma, *JOM*, 2000, October, 21–25.
5. H. Asanuma, *Active Composites, Metal and Ceramic Matrix Composites*, 2004, ed. B. Cantor, F. Dunne and I. Stone, Institute of Physics Publishing, Bristol and Philadelphia, 367–382.
6. H. Asanuma, O. Haga, N. Naito and T. Tsuchiya, *Proc. Annual Meeting of the Japan Society for Composite Materials*, 1996, 19–20, (in Japanese).
7. H. Asanuma, O. Haga and M. Imori, *JSME Int. J. Series A*, 2006, **49**(1), 32–37.
8. S. Shimamura, *Adv. Mater.*, Kogyo Chosakai Press, 1982, 209–210, (in Japanese).
9. C.M. Dry, Self-repairing, reinforced matrix materials, United States Patent 5,561,173, October 1, 1996 (Filed: June 6, 1995).
10. C. Dry, *Proc. 2nd Intl. Conf. on Advances in Composites*, 1996, 127–133.
11. M. Zako and N. Takano, *J. Intell. Mater. Syst. Struct.*, 1999, **10**, 836–841.
12. S.R. White, N.R. Sottos, P.H. Geubelle, J.S. Moore, M.R. Kessler, S.R. Sriram, E.N. Brown and S. Viswanathan, *Nature*, 2001, **409**, 794–797.
13. H. Asanuma, M. Hirohashi, M. Kase and T. Kikuchi, *Proc. 1st Japan Int. SAMPE Symp.*, 1989, 979–984.
14. H. Asanuma, M. Hirohashi, T. Kikuchi and H. Kitagawa, *Proc. 6th Japan-U. S. Conf. on Composite Materials*, 1992, 434–442.
15. H. Asanuma, M. Hirohashi, H. Takase and T. Miyazaki , *Proc. 1st Intl. Conf. on Processing Materials for Properties*, 1993, 983–986.
16. H. Asanuma and H. Sato, A functional composite material embedded with a piezoelectric ceramic fibre having a metal core, patent pending in Japan, application No.2005-59552, 2005.
17. D. Askari, H. Asanuma, M.N. Gasemi-Nejhad, *Proc. 2005 ASME International Mechanical Engineering Congress and Exposition*, 2005, Paper No. IMECE2005-79049, 2005.
18. H. Asanuma, O. Haga, M. Imori, S. Akutsu and M. Komori, *Proc. Mechanical Engineering Congress*, 2005 Japan, 2005, **6**, 191–192.

CHAPTER 21

Overview of Liquid-crystal Elastomers, Magnetic Shape-memory Materials, Fullerenes, Carbon Nanotubes, Nonionic Smart Polymers and Electrorheological Fluids as Other Intelligent and Multifunctional Materials

M. SHAHINPOOR[1,]* AND H.-J. SCHNEIDER[2]

[1] Department of Mechanical Engineering, Biomedical Engineering Center, College of Engineering, University of Maine, Orono, Maine, 04469, USA
[2] FR Organische Chemie der Universität des Saarlandes, D 66041 Saarbrücken, Germany

21.1 Liquid-crystal Elastomers as Multifunctional Materials

Liquid-crystal elastomers were first discussed by Finkelmann et al.[1] These materials can be used as an actuator by inducing nematic–isotropic phase transition due to, e.g., a temperature increase that causes them to shrink, as described in a thorough review on these intelligent multifunctional materials.[2]

---

*Formerly Artificial Muscle Research Institute, School of Engineering and School of Medicine, University of New Mexico, Albuquerque, NM 87131, USA

LCEs have been made electroactive by creating composite materials that consist of monodomain nematic liquid-crystal elastomers and a conductive phase such as graphite or conducting polymers that are distributed within their network structure.[3,4] The actuation mechanism of these materials involves a phase transition between nematic and isotropic phases over a period of less than a second. The reverse process is slower, taking about 10 s, and it requires cooling for expansion of the elastomer to its original size.

The mechanical properties of LCE materials can be controlled and optimised by effective selection of the liquid-crystalline phase, density of crosslinking, flexibility of the polymer backbone, coupling between the backbone and liquid-crystal group, and the coupling between the liquid-crystal group and the external stimuli.

It was Nobel laureate Pierre de Gennes and coworkers Hebert and Kant who proposed the use of liquid crystals as artificial muscles[5] and pointed out that they could be used as multifunctional materials and particularly as biomimetic artificial muscles.[6] Later, Finkelmann, Brand, et al.[7-18] made significant contributions towards the multifunctionality of liquid-crystal elastomers. It was shown how blends of liquid-crystal elastomers and a conductive phase such as graphite could lead to smart materials exhibiting electrically controllable biomimetics artificial muscles. Liquid single-crystal elastomers (LSCE) are liquid-crystalline phase structures with preferred macroscopic molecular orientation in their molecular network. At the liquid crystalline to isotropic phase transformation temperature, when the liquid-crystalline network becomes isotropic like a conventional rubber, the dimensions of the network change. Due to the anisotropy in this phase of the molecular structure of liquid-crystal elastomers, the networks shorten in the direction of the optical axis or director axis. De Gennes and coworkers[19,20] have also alluded to the fact that such nematic networks present a potential as artificial muscles just like the ionic polymer–metal nanocomposite (IPMNC) discussed in other chapters of this book. For nematic LSCE synthesised from nematic LC side-chain polymers, the change of this length is normally more than 100%. The velocity of this process is determined by the heat conductivity of the network and not by material-transport processes. If the sample thickness is limited, calculations indicate that the response time of these LSCE is similar to that of natural muscles and is mainly determined by the relaxation behaviour of the polymer chains or network strands.

The magnitude of changes in the dimensions of nematic LSCE with temperature is determined by the coupling between the nematic state of order and the conformation of the polymer main chains. In this regard the conformation of nematic side-chain polymers is only little affected. For nematic main-chain polymers, where the mesogenic units are incorporated into the polymer main chain, large conformational changes should occur and be directly expressed in the thermoelastic behaviour.

Electrical Joule heating of a nematic LSCE that consists of a combination of LC side-chain and main-chain polymers and a conductive phase can be blended into them. These materials can be synthesised by a hydrosilylation reaction of the monofunctional side-chain mesogen (Figure 21.1) and the bifunctional liquid-crystalline main-chain polyether with poly(methylhydrogensiloxane).

Figure 21.1 depicts how the nematic coelastomer in a liquid single-crystalline coelastomer, a weakly crosslinked network is mechanically loaded. Under load, the hydrosilylation reaction is completed. The LSCE Mon-57/43 has a weight ratio between the mesogenic units in the main chain to those in the side chains of 57/43 and a nematic to isotropic phase transition temperature of $T_{n,i} = 104\,^\circ\text{C}$. Macroscopically disordered LCEs result when there is no mechanical load. The LCE Poly-77/23 exhibits the nematic to isotropic phase transition temperature at $T_{n,i} = 104\,^\circ\text{C}$.

In a thermal bath environment LSCE samples exhibit a nematic–isotropic phase transition, which is accompanied by contraction as shown in Figure 21.2.

The nematic state introduces a network anisotropy, which is directly reflected in the thermal expansion behaviour of the elastomers. In Figure 21.2 the thermoelastic behaviour is shown, where the length L of the sample, normalised to the length L_{iso} in the isotropic state, is depicted as a function of the temperature.

These measurements were performed with different constant loads as indicated in Figure 21.2. The thermal expansion behaviour is reversible and only a narrow hysteresis is observed at $T_{n,i}$.

The temperature-dependent elongation exceeds the elongation of the previously investigated LSCEs, which are synthesised from the pure main-chain polymer. As was mentioned before, it was proposed that electrically controllable LSCE could be made by blending them with a conductor phase such as graphite by physical loading as shown in Figure 21.3.

Here, the idea was to repeat the same type of experiment as depicted above in Figure 21.2 in which the elastomer shrinks upon heating in a liquid thermal bath, by Joule heating a LSCE with embedded graphite fibres inside the elastomer. In this endeavour highly conducting graphite fibres were embedded inside a sample of LSCE to induce the phase transition by internal Joule heating.

In summary, coelastomers containing network strands comprising LC side- and main-chain polymers exhibit exceptional thermoelastic behaviour. These systems might be suitable as model systems for artificial muscles. They show that the liquid-crystalline to isotropic phase transition can cause conformational changes of macromolecules, which are sufficient to obtain a large mechanical response of polymer networks.

21.2 Magnetic Shape-memory (MSM) Materials

Magnetically controlled shape-memory (MSM) materials provide a new way to produce motion and force,[21–23] as was demonstrated with a Ni-Mn-Ga alloy as early as 1996. A new mechanism was suggested based on the magnetic-field-induced reorientation of the twin structure of a magnetic shape-memory alloy. Effectively, magnetic control of the shape-memory effect would lead to a much

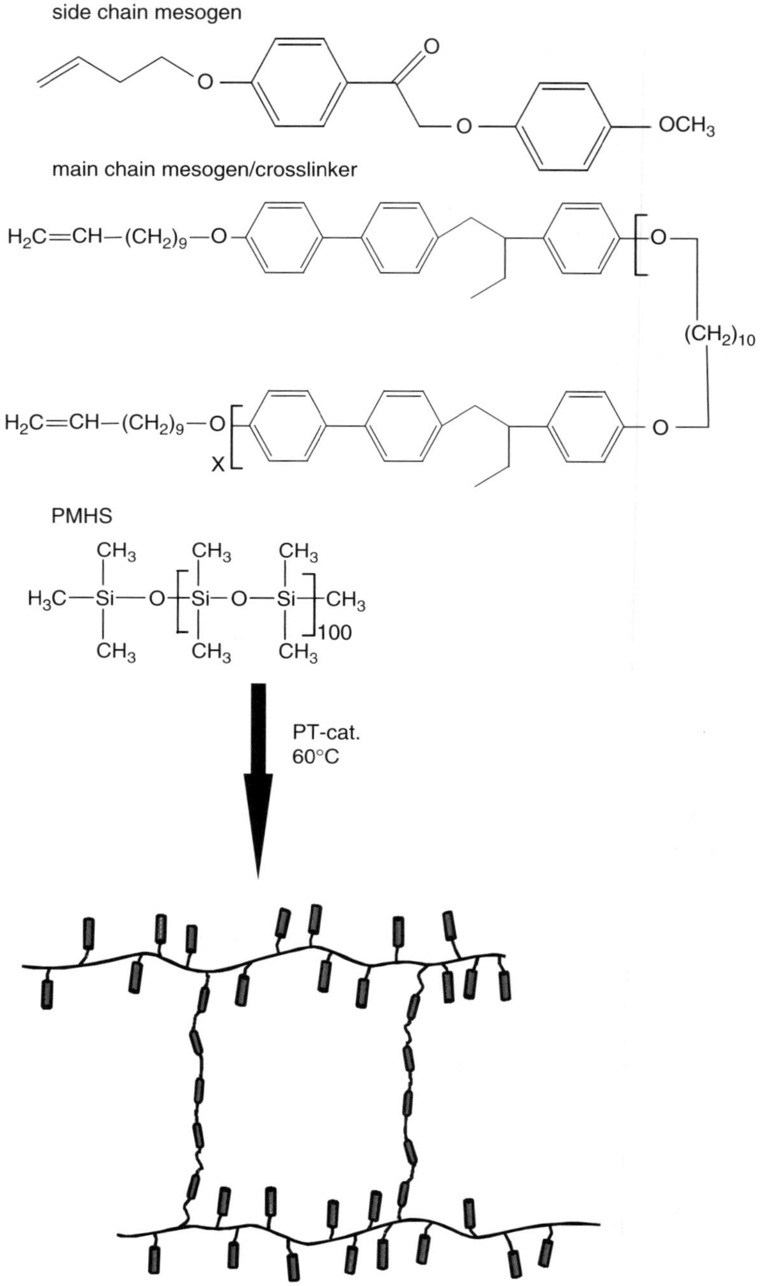

Figure 21.1 Synthesis of a nematic coelastomer.

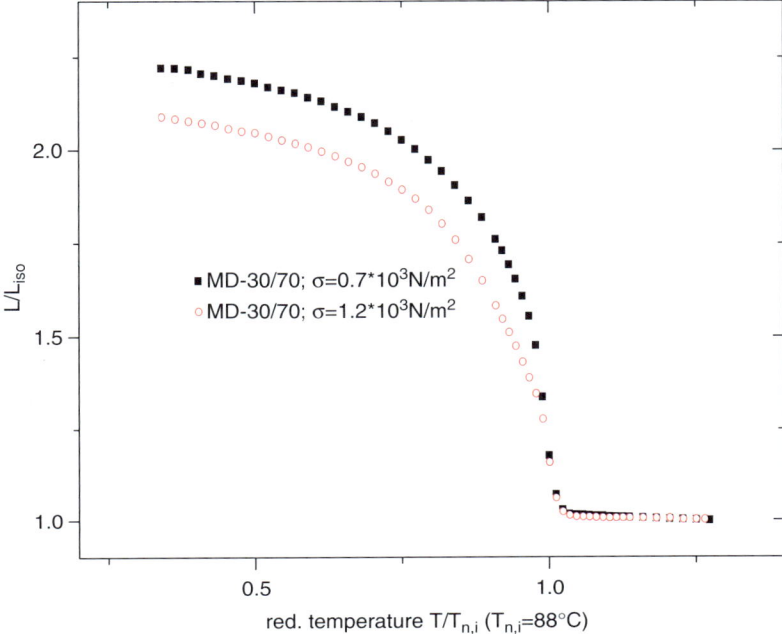

Figure 21.2 Contraction of liquid-crystal elastomer in a temperature bath.

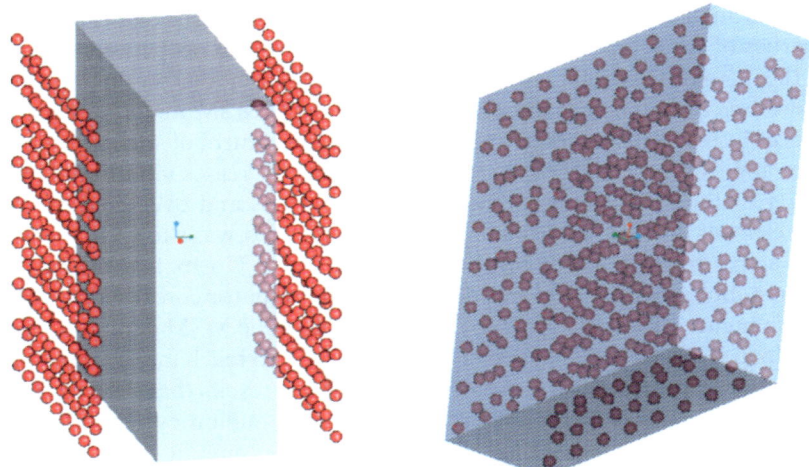

Figure 21.3 Basic loading configuration of graphite powder into liquid-crystal elastomer films.

more rapid response of the actuator than thermal control. Magnetic field controls the reorientation of the twin variants in an analogous way as the twin variants are controlled by the stress in classical shape-memory alloys. The magnetic shape-memory effect has demonstrated that certain shape-memory materials that are also ferromagnetic can show very large dimensional changes (6–10%) under application of a magnetic field. These strains occur within the low-temperature (martensitic) phase. Ferromagnetic shape-memory (FSM) materials are a new class of active materials, which combine the properties of ferromagnetism with those of a diffusionless, reversible martensitic transformation. These materials, such as shape-memory alloy Ni_2MnGa, which has a cubic Heusler structure in the high-temperature austenitic phase and undergoes a cubic-to-tetragonal martensitic transformation, clearly exhibit magnetic shape-memory effects. The ferromagnetic shape-memory (FSM) effect refers to either the reversible field-induced austenite to martensite transformation, or the rearrangement of martensitic variants by an applied magnetic field leading to an overall change of shape. Typically, contraction or expansion in the order of 6% to 10% is observed in these materials. When they are subjected to one externally applied magnetic field, the twins in a favourable orientation grow at the expense of the others. This results in a shape change of the magnetic shape-memory alloy. The theoretical maximum limiting the achievable strain is the value of the local transformation strain (6.6% in Ni_2MnGa).

21.2.1 MSM Alloy Actuators

Motion of the MSM actuators is fast, with a rise time of less than 0.2 ms. Flux generation and the inertia of the moving mass rather than the MSM mechanism of the material limit the velocity. This makes it possible for an actuator to operate at high frequencies and large strokes. The electromechanical hysteresis properties of the actuator were also studied. Hysteresis exists between the strain and the actuator's input current as well as between the strain and stress. Hysteresis is caused by the internal properties of the MSM material. Hysteresis produces losses in the material and complicates the control of some positioning-system applications. On the other hand, hysteresis increases vibration damping capacity of the MSM material and avoids vibrations and overshooting of the MSM element in rapid shape changes of the element. It was shown that with the MSM actuator large forces can be generated.[21,22,23] Up to 1 kN forces were measured, but generally the force depends on the actuator construction and can be much higher. It was also shown that the strain of the MSM actuator depends on the load it has to work against. The optimal load to reach maximal magnetic-field-induced strain is about 1–1.5 MPa. Fatigue test results that showed that the stroke of the actuator does not decrease after 200 million cycles of the alternating magnetic field reveal that MSM actuators can operate for long times at high frequencies without significant fatigue of the actuating element.

21.2.2 Sensing and Multifunctionality Properties of MSM Materials

Sensing and voltage-generation properties of MSM materials have been demonstrated,[24,25] Nonstoichiometric Ni-Mn-Ga in 5M martensite is the most utilised MSM material so far. The material also has a reverse effect: the change in shape of the material with nonzero magnetisation alters the magnetic field in which it is placed. Due to its properties, the material has two promising applications: motion-generating actuators and sensors. In the above works[24,25] both sensors and actuators are studied with interest focused on the modeling and operational parameters of the applications. New MSM applications have been built and tested as linear motors and speed sensors, as described in the above references. The speed sensor can also be used as a voltage generator.

21.3 Fullerenes and Carbon Nanotubes as Multifunctional Intelligent Materials

Carbon nanotubes (CNTs) are tubular structures[26] with a diameter between 0.4 nm and 3 nm; they are formed entirely by sp^2-hybridised carbon atoms, a feature they share with other carbon allotropes such as graphite or with fullerenes (Figure 21.4) Fullerenes, which are nearly spherical (also called buckyballs) accommodate 60 up to 540 carbon atoms ("C_{60}", "C_{540}", *etc.*) within pentagonal, hexagonal or heptagonal rings that enforce nonplanarity of the surface.

The unique electronic, optical and chemical properties of nanotubes have already led to more than 15 000 papers. Several reviews focus on the chemistry of these intriguing nanostructures, including chemical modification and association with different guest compounds.[27] Supramolecular complex formation by mostly dispersive interactions with the π-orbitals can occur either on the outside of the tubes (exobinding), which strongly self-assemble to bundles, or inside the cavities (endobinding). CNTs can noncovalently bind drugs and even proteins and nucleic acids; they are therefore promising candidates for new drug-delivery systems.[28] Also, very large compounds such as the metallofullerene such as $Gd_2@C_{92}$ can be encapsulated within a CNT.[29] Other recent reviews can be consulted for electronic[30,31] NT devices, encompassing rectifiers, FETs, conducting wires, sensors and displays.

Carbon nanotubes were obtained in 1991 from the soot of *e.g.* graphite flames,[32] but were postulated already in 1976,[33] and even patented in 1987.[34] The source of CNTs is the same as that of fullerenes, where the major problem is the separation of buckyballs of different size. As it is beyond the scope of this chapter to discuss all potential applications of fullerene-related systems we will only mention a few. Many donor–acceptor combinations of electron-rich fullerenes with, *e.g.*, porphyrins hold promise as photovoltaic devices, *e.g.* for light harvesting with solar cells.[35]

Fullerenes, as much as related nanotube fibres are most suitable for mechanical reinforcement,[36] also in the form of so-called buckypaper. These lightweight thin sheets also exhibit unusually high heat- and electric-current-carrying capacity, and may perhaps be used for illumination of computer screens. Buckyball thin films are accessible, *e.g.*, by chemical vapour phase decomposition (CVD).[37] Often, double-walled or multiwalled carbon nanotubes (DWNTs,

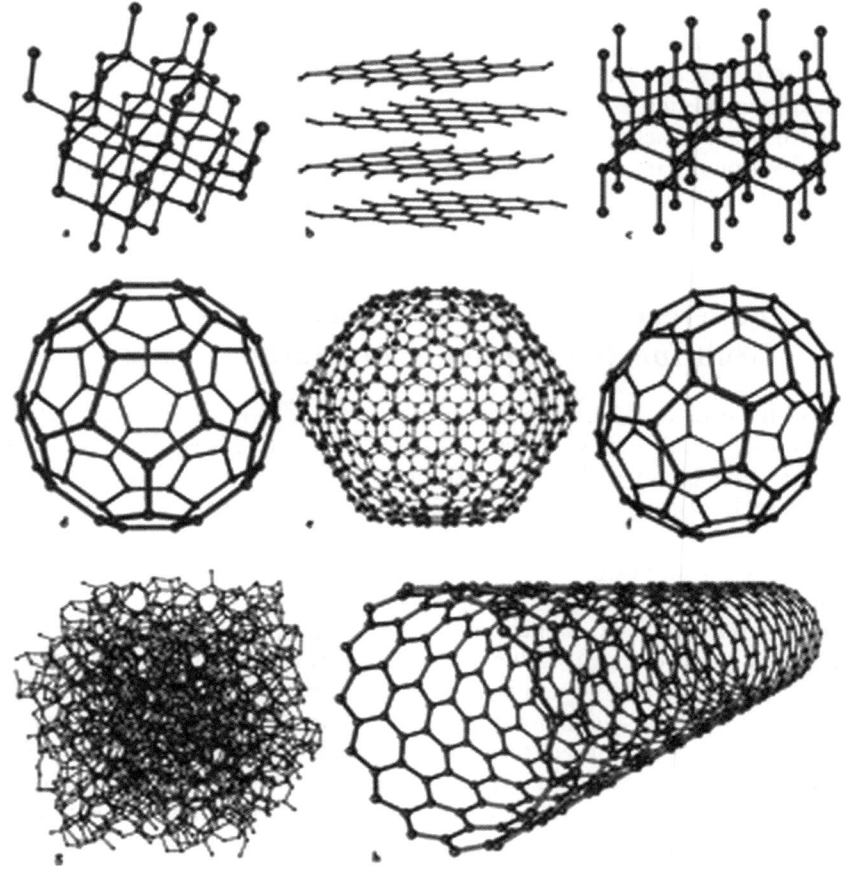

Figure 21.4 Eight allotropes of carbon from top and left to right: diamond, graphite, lonsdaleite, C_{60}, C_{540}, C_{70}, amorphous carbon and a carbon nanotube.

MWNTs) instead of single-walled tubes (SWNTs) are the basis of such assemblies.[38] A bistable device switching between an OFF and an ON state, having ON-OFF current ratios larger than 10^4 and a write-once, read-many times memory has been made with single layers made from fullerene C_{60} mixed with polystyrene, sandwiched between two electrodes.[39] Fullerenol-doped PUE films show electrostriction effects, measured as an induced voltage appearing on the surface of a film by pressing and stretching, similar to piezoelectric material.[40]

Generally, all mechanical deformations of CNTs result in stretching and compression of bonds, as well as of π–π and sigma–π coupling, which in turn will change all the electronic properties, and may, *e.g.*, induce voltage. Inversely, electrical input, in particular injection of electrons, can be used for deformation of such materials.[41] Using, for example, a conducting AFM tip electrons can be injected into the nanotubes, which then expand by a measurable, although quite small degree.[42] Electromechanical actuators based on sheets of SWNC carbon nanotubes also rely on quantum chemical-based expansion due to electrochemical double-layer charging, and are able to generate higher stresses than natural muscles.[43]

The use of CNTs in actuators is illustrated by a most recently described system, which allows mechanical motions driven by, *e.g.*, hydrogen as "fuel". (Figure 21.5).[44] Here, the actuator movements result from charge injection into carbon nanotubes, which are placed with high density on the large surface of a catalyst-containing sheet electrode; the platinum catalyst secures the charge transfer from oxygen in the water. Buckypaper from nanotubes as well as blends with other polymers[45] or carbon nanofibres[46] can be used for electrochemical stimulation of actuators, taking advantage of a higher elastic modulus and strength of such polymers.

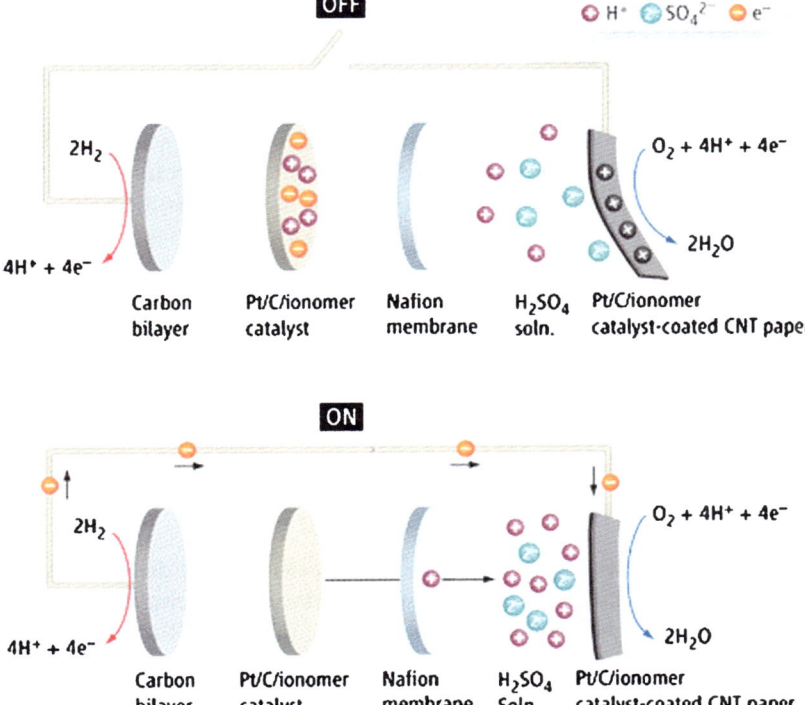

Figure 21.5 A fuel-cell muscle. (Top) The nanotube film expands as a result of the charging, causing it to bend as it expands relative to the platinum/carbon/ionomer layer (Bottom) The counterelectrode is a carbon bilayer-platinum catalyst-Nafion membrane electrode assembly shown in "exploded view". Reproduced with permission from J.D. Maddon, *Science*, 2006, **311**, 1559–1560.

21.4 Nonionic Polymer Gels/EAPs

Nonionic dielectric polymers have not been considered as adequate for electroactive actuator materials because of their poor reaction to the electric field. As electroactive polymeric materials, the polyelectrolytes and conductive polymers have been investigated intensively, since they can show large deformation under low electric field in aqueous media or in the presence of water as an additive. It has been shown that nonionic polymeric materials can be used as electrically active materials with large deformation.[47–52] The observed electrically induced deformation phenomena are "contraction and relaxation", "bending by solvent drag in the gel", "crawling deformation", and "electrotaxis" amoeba-like creep deformation. The materials exhibiting such phenomena are highly swollen dielectric gels through plasticised polymers to nonsolvent-type elastomers. Characteristics of the actuations are particularly large deformation or huge strain under much smaller energy dissipation compared to the conventional polyelectrolyte or conductive polymer actuators. Applications of the materials for pumping, valves, artificial eye pupils have been discussed.[47–52]

Nonionic polymer gels containing a dielectric solvent can be made to swell under a DC electric field with significant strain. Bending and crawling nonionic polymeric artificial muscles have been made using a poly(vinyl alcohol) gel with dimethyl sulfoxide.[47–52] A $10 \times 3 \times 2$ mm^3 actuator gel was subjected to an electrical field and exhibited bending at angles that were greater than 90° in a time of 60 ms. This phenomenon is attributed to charge injection into the gel and a flow of solvated charges that induce an asymmetric pressure distribution in the gel. Another nonionic gel is poly(vinyl chloride) (PVC), which is generally inactive when subjected to electric fields. However, if PVC is plasticised with dioctyl phthalate (DOP), a typical plasticiser, it can maintain its shape and behave as an elastic nonionic gel.

21.5 Electrorheological (ER) Fluids as Multifunctional Smart Materials

Electrorheological (ER) fluids are suspensions consisting of dielectric particles of size 0.1–100 μm and dielectric base fluids. Since the dielectric constant of suspension particles differs from the dielectric constant of the base fluid, the external electric field polarises the particles. These polarised particles interact and form chain-like or even lattice-like organised structures. Simultaneously the rheological properties of the suspension change effectively, *e.g.* the effective viscosity increases dramatically. ER-suspensions also have a magnetic analogue consisting of ferromagnetic particles and the base liquid. The response time of electrorheological fluids is of the order of 1–10 ms, which in principle enables the use of these liquids in such applications as electrically controlled clutches, valves and active damping devices. There are a number of publications on the applications of ER fluids to robotics and intelligent materials and composites.[53–57] It should be noted that when an electric current is applied to ER fluids they behave as a Bingham plastic in the sense that one will need a finite applied stress to initiate shearing.

The unique properties of ER fluids make them ideal materials to transfer energy and control motion electronically. There are two types of ER fluids, heterogeneous and homogeneous. Potential applications can be found in mechanical devices that involve motion transmission, such as vibration dampers, shock absorbers, brakes, clutches, valves, and speed controllers.

One of the exciting products that has recently been developed is the smart tactile display for visually impaired or blind persons.[55,56,57] Previously unavailable due to high manufacturing costs, this full-page graphical display will enable visually impaired users to have interactive access. ER fluids are basically semiactive variable-impedance materials that can be utilised for enhancing human biomechanical performance. They can thus cushion the effects of falls, limit or avoid fractures and support continued mobility despite injury. The variable-impedance materials include electrorheological (ER), magnetorheological (MR), and shear-thickening (ST) fluids. These fluids change their resistance to deformation in response to electric, magnetic, or shear fields, respectively.

21.5.1 Other Applications of ER Fluids

The energy absorbing nature of ERF is useful in commercial products ranging from shock absorbers to protective garments. Each of these products relies on a combination of the energy-absorption mechanism of a specific field responsive fluid and a matched "transmission geometry" to translate the change in material properties at the molecular scale into a usable macroscale change in device properties.

Applications of ER fluids have historically relied on transmitting torque by shearing fluid between two or more counterrotating disks/shells or on pushing activated ER fluid through an orifice. Mechanical clutches and dampers are classic examples of applications for ER fluids. Other applications are in haptic devices working with an electrorheological fluid as well as actuators making use of electrorheological fluids.

References

1. H. Finkelmann, H.-J. Kock and G. Rehage, *Makromol. Chem., Rapid Commun.*, 1981, **2**, 317.
2. H.R. Brand and H. Finkelmann, *Handbook of Liquid Crystals, Vol. 3: High Molecular Weight Liquid Crystals*, D. Demus, J. Goodby, G.W. Gray, H.-W. Spiess and V. Vill ed., Wiley-VCH, Weinheim, 1998, p. 277.

3. M. Shahinpoor, *Proceedings of SPIE 7th International Symposium on Smart Structures and Materials*, Newport Beach, California, SPIE Smart Materials and Structures Publication No. SPIE 3987-27, 2000, pp. 187–192.
4. Finkelmann, H. and M. Shahinpoor, *Proceeding of SPIE 9th Annual International Symposium on Smart Structures and Materials*, San Diego, California, SPIE Publication No. 4695-53 March, 2002.
5. P.G. de Gennes, M. Hebert and R. Kant, *Macromol. Symp.*, 1997, **113**, 39.
6. M. Hebert, R. Kant and P.G. de Gennes, *J. Phys.-I, France*, 1997, **7**, 909–918.
7. J. Küpfer and H. Finkelmann, *Macromol. Chem., Rapid Commun.*, 1991, **12**, 717.
8. H. Finkelmann and H.R. Brand, Trends., *Polym. Sci.*, 1994, **2**, 222.
9. J. Küpfer and H. Finkelmann, *Makromol. Chem.*, 1994, **195**, 1353.
10. H.R. Brandt and K. Kawasaki, *Macromol. Rapid Commun.*, 1994, **15**, 251.
11. H.R. Brandt and H. Pleiner, *Physica*, 1994, **A208**, 359.
12. W. Kaufhold, H. Finkelmann and H.R. Brand, *Makromol. Chemie*, 1991, **192**, 2555.
13. S. Disch, H. Finkelmann, H. Ringsdorf and P. Schumacher, Macromolecules, 1995, **28**, 2424.
14. J. Küpfer, E. Nishikawa and H. Finkelmann, *Polym. Adv. Technol.*, 1994, **5**, 110.
15. E. Nishikawa, Smektische Einkristall-Elastomere, PhD Thesis, Albert-Ludwigs-Universität Freiburg im Breisgau, 1997.
16. G.H.F. Bergmann, H. Finkelmann, V. Percec and M. Zhao, Macromol., *Rapid. Commun.*, 1997, **18**, 353.
17. G.H. F. Bergmann, Flüssigkristalline Hauptketten elastomere: Synthese, Characterisierung und Untersuchungen zu mechanishen, thermischen und Orientierungseigenschaften, PhD thesis, Albert-Ludwigs-Universität Freiburg im Breisgau, 1998.
18. G. Bergmann and H. Finkelmann, Macromol. Chem., *Rapid Commun.*, 1997, **18**, 353–359.
19. P.G. de Gennes, M. Hebert and R. Kant, *Macromol. Symp.*, 1997, **113**, 39.
20. M. Hebert, R. Kant and P.G. de Gennes, *J. Phys.-I, France*, 1997, **7**, 909–918.
21. K. Ullakko, *J. Mater. Eng. Perf.*, 1996, **5**, 405–409.
22. R.C. O'Handley, K. Ullakko, J.K. Huang and C. Kantner, *Appl. Phys. Lett.*, 1996, **69**, 1966–1971.
23. R.C. O'Handley, *J. Appl. Phys.*, 1998, **83**, 3263–3271.
24. J. Tellinen, I. Suorsa, A. Jääskeläinen, I. Aaltio and K. Ullakko, *Proceedings of the 8th international conference ACTUATOR 2002*, Bremen, Germany, june 2002, 10–12.
25. Ilkka Suorsa, "Performance and Modeling of Magnetic Shape Memory Actuators and Sensors", Ph.D Dissertation, Helsinki University of Technology Espoo, Finland, May 2005.

26. For other tubular structures, mostly based on self-assembly of cyclophanes, peptides, cyclodextrins etc, which also have potential as intelligent materials see D.T. Bong, T.D. Clark, J.R. Granja and M.R. Ghadiri, *Angew. Chem. Int., Ed. Engl.*, 2001, **40**, 988.
27. (a) D.A. Britz and A.N. Khlobystov *Chem. Soc. Rev.*, 2006, **35**, 637–659; A.N. Khlobystov, D.A. Britz and G.A.D. Briggs, *Acc. Chem. Res.*, 2005, **38**, 901; (b) P. Avouris, *Acc. Chem. Res.*, 2002, **35**, 1026; D. Tasis, N. Tagmatarchis, A. Bianco and M. Prato, *Chem. Rev.*, 2006, **106**, 1105.
28. Multifunctional nanotube biological transporters: N.W.S. Kam, M. O'Connell, J.A. Wisdom and H.J. Dai, *Proc. Natl. Acad. Sci. USA*, 2005, **102**, 11600; see also P. Kohli and C.R. Martin, *Current Pharm. Biotech.*, 2005, **6**, 35; C.R. Martin and P. Kohli, *Nat. Rev. Drug Discov.*, 2003, **2**, 29; see also G.R. Dieckmann, A.B. Dalton, P.A. Johnson, J. Razal, J. Chen, G.M. Giordano, E. Munoz, I.H. Musselman, R.H. Baughman and R.K. Draper, *J. Am. Chem. Soc.*, 2003, **125**, 1770.
29. K. Suenaga, R. Taniguchi, T. Shimada, T. Okazaki, H. Shinohara and S. Iijima, *Nano Lett.*, 2003, **3**, 1395.
30. M.P. Anantram and F. Leonard, *Rep. Prog. Phys.*, 2006, **69**, 507.
31. R.H. Baughman, A.A. Zakhidov and W. de Heer, *Science*, 2002, **97**, 787.
32. S. Iijima, *Nature*, 1991, **354**, 56; characterization: S. Iijima and T. Ichihashi, *Nature*, 1993, **363**, 603.
33. A. Oberlin, M. Endo and T. Koyama, *J. Cryst. Growth*, 1976, **32**, 335.
34. H.G. Tennent, Carbon fibrils, method for producing the same and compositions containing same, ed. US Patent, 4,663,230, 1987.
35. See, *e.g.*, (a) G. Adamopoulos, T. Heiser, U. Giovanella, S. Ould-Saad, K.I. van de Wetering, C. Brochon, I. Zorba, K.M. Paraskevopoulos and G. Hadziioannou, *Thin Solid Films*, 2006, **511**, 371; (b) H. Imahori, M. Kimura, K. Hosomizu, T. Sato, T.K. Ahn, S.K. Kim, D. Kim, Y. Nishimura, I. Yamazaki, Y. Araki, O. Ito and S. Fukuzumi, *Chem. Eur. J.*, 2004, **10**, 5111; (c) H. Imahori and S. Fukuzumi, *Adv. Funct. Mater.*, 2004, **14**, 525; (d) H. Imahori, K. Mitamura, Y. Shibano, T. Umeyama, Y. Matano, K. Yoshida, S. Isoda, Y. Araki and O. Ito, *J. Phys. Chem. B*, 2006, **110**, 11399; (e) F.S. Meng, J.L. Hua, K.C. Chen, H. Tian, L. Zuppiroli and F. Nuesch, *J. Mater. Chem.*, 2005, **15**, 979; (f) T. Konishi, A. Ikeda and S. Shinkai, *Tetrahedron*, 2005, **61**, 4881; (g) J.F. Nierengarten, U. Hahn, T.M.F. Duarte, F. Cardinali, N. Solladie, M.E. Walther, A. Van Dorsselaer, H. Herschbach, E. Leize, A.M. Albrecht-Gary, A. Trabolsi and M. Elhabiri, *Compt. Rend. Chim.*, 2006, **9**, 1022; (h) D.M. Guldi, A. Rahman, V. Sgobba and C. Ehli, *Chem. Soc. Rev.*, 2006, **35**, 471.
36. J.N. Coleman, U. Khan, W.J. Blau and Y.K. Gun'ko, *Carbon*, 2006, **44**, 1624; see also, *e.g.*, M. Zhang, S.L. Fang, A.A. Zakhidov, S.B. Lee, A.E. Aliev, C.D. Williams, K.R. Atkinson and R.H. Baughman, *Science*, 2005, **309**, 1215.
37. Leading references – (a) I.P. Kang, M.J. Schulz, J.H. Kim, V. Shanov and D.L. Shi, *Smart Mater. Struct.*, 2006, **15**, 737; (b) Y.A. Kim, H. Muramatsu, T. Hayashi, M. Endo, M. Terrones and M.S. Dresselhaus, *Chem.*

Vapor Depos., 2006, **12**, 327; (c) S. Roth and R.H. Baughman, *Curr. Appl. Phys.*, 2002, **2**, 311; (d) Y. Yeo-Heung, A. Miskin, P. Kang, S. Jain, S. Narasimhadevara, D. Hurd, V. Shinde, M.J. Schulz, V. Shanov, P. He, F.J. Boerio, D.L. Shi and S. Srivinas, *J. Intell. Mater. Syst. Struct.*, 2006, **17**, 191; (e) Y.H. Yun, V. Shanov, M.J. Schulz, S. Narasimhadevara, S. Subramaniam, D. Hurd and F.J. Boerio, *Smart Mater. Struct.*, 2005, **14**, 1526.
38. E. Flahaut, R. Bacsa, A. Peigney and C. Laurent, *Chem. Commun.*, 2003, 1442; J.Q. Wei, L.J. Ci, B. Jiang, Y.H. Li, X.F. Zhang, H.W. Zhu, C.L. Xu and D.H. Wu, *J. Mater. Chem.*, 2003, **13**, 1340.
39. H.S. Majumdar, J.K. Baral, R. Osterbacka, O. Ikkala and H. Stubb, *Org. Electron.*, 2005, **6**, 188.
40. Y. Nakama, J. Kyokane, K. Tokugi, T. Ueda and K. Yoshino, *Synth. Met.*, 2003, **135**, 749.
41. E. Munoz, A.B. Dalton, S. Collins, M. Kozlov, J. Razal, J.N. Coleman, B.G. Kim, V.H. Ebron, M. Selvidge, J.P. Ferraris and R.H. Baughman, *Adv. Eng. Mater.*, 2004, **6**, 801.
42. S. Roth and R.H. Baughman, *Curr. Appl. Phys.*, 2002, **2**, 311; see also G.M. Spinks, G.G. Wallace, L.S. Fifield, L.R. Dalton, A. Mazzoldi, D. Rossi, I.I. Khayrullin and R.H. Baughman, *Adv. Mater.*, 2002, **14**, 1728; J.N. Barisci, G. Spinks, G. Wallace, J. Madden and R.H. Baughman, *Smart Mater. Struct.*, 2003, **12**, 549.
43. R.H. Baughman, C.X. Cui, A.A. Zakhidov, Z. Iqbal, J.N. Barisci, G.M. Spinks, G.G. Wallace, A. Mazzoldi, D. De Rossi, A.G. Rinzler, O. Jaschinski, S. Roth and M. Kertesz, *Science*, 1999, **284**, 1340.
44. V.H. Ebron, Z.W. Yang, D.J. Seyer, M.E. Kozlov, J.Y. Oh, H. Xie, J. Razal, L.J. Hall, J.P. Ferraris, A.G. MacDiarmid and R.H. Baughman, *Science*, 2006, **311**, 1580; see also J.D. Maddon, *Science*, 2006, 311, 1559.
45. Y.H. Yun, V. Shanov, M.J. Schulz, S. Narasimhadevara, S. Subramaniam, D. Hurd and F.J. Boerio, *Smart Mater. Struct.*, 2005, **14**, 1526.
46. Y. Yeo-Heung, A. Miskin, P. Kang, S. Jain, S. Narasimhadevara, D. Hurd, V. Shinde, M.J. Schulz, V. Shanov, P. He, F.J. Boerio, D.L. Shi and S. Srivinas, *J. Intell. Mater. Syst. Struct.*, 2006, **17**, 191.
47. T. Hirai, *Intelligent and Smart Textiles*, ed. X. Tao, Woodhead Pub., Cambridge, UK, 2002.
48. T. Hirai, H. Nemoto, M. Hirai and S. Hayashi, *J. Appl. Polym. Sci.*, 1994, **53**, 79–85.
49. M. Hirai, T. Hirai, A. Sukumoda, H. Nemoto, Y. Amemiya, K. Kobayashi and T. Ueki, *J. Chem. Soc., Faraday Trans.*, 1995, **91**, 473–478.
50. J. Zheng, M. Watanabe, H. Shirai and T. Hirai, *Chem. Lett.*, 2000, 500–501.
51. Md.Z. Uddin, M. Yamaguchi, M. Watanabe, H. Shirai and T. Hirai, *Chem. Lett.*, 2001, 360–361.
52. Md.Z. Uddin, M. Yamaguchi, M. Watanabe, H. Shirai and T. Hirai, *J. Robot. Mechatron.*, 2002, **14**(2), 118–123.

53. M.V. Gandhi, B.S. Thompson and S.B. Choi, *J. Compos. Mater.*, 1989, **23**, 1232–1255.
54. M.V. Gandhi, B.S. Thompson, S.B. Choi and S. Shakair, *ASME J. Mech., Transmiss. Automat. Des.*, 1989, **111**, 328–336.
55. J. Furusha and M.S. Sakaguchi, *Int. J. Mod. Phys., B*, 1999, **13**, 2051–2059.
56. J.P. Huang, M. Karttunen, K.W. Yu and L. Dong, *Phys. Rev. E*, 2003, **67**, 021403.
57. R.S. Dwyer-Joyce, W.A. Bullough and S. Lingard, *Int. J. Mod. Phys. B*, 1996, **10**, 3181–3189.

CHAPTER 22

Overview on Biogenic and Bioinspired Intelligent Materials – from DNA-based Devices to Biochips and Drug-delivery Systems

H.-J. SCHNEIDER

FR Organische Chemie der Universität des Saarlandes, D 66041, Saarbrücken, Germany

22.1 Introduction

In view of their complexity and multifunctional properties biological materials offer an intriguing entrance for new technologies, which are just on the verge of development after chemists have learned to manipulate them into desired fashions. Nature achieves most complex functions of life on the basis of single, yet often very large molecules, which therefore are a highly promising starting point for nanoscale intelligent devices. These materials can be used as building elements by themselves, or, more often, they are combined with synthetic materials. Highly appealing biogenic systems aiming at mechanical motions including molecular machines have been already conceptually devised, but still rely on elaborate physical methods for checking their operation. On the other hand, sensor devices in the form of biochips, where a signal from an analyte is transduced into a highly sensitive output, have made an enormous impact already, in particular in the biomedical field. It has been predicted that the market size for biochips alone will grow from $2.0 billion in 2004 to about $5 billion in 2009.[1] In this chapter only a few developments can be illustrated, which should highlight bioinspired possibilities for smart materials and at least provide leading references.

22.2 Biological Materials: Nucleic Acids as an Example

The most traditional examples of using *proteins* as signal-responsive materials are enzyme-based sensors, in particular electrodes that function similarly to ion-sensitive electrodes (ISEs). The most impressive examples of using biological materials, however, are those based on nucleic acids.[2] The controllable sequence of nucleobases in DNA allows assembly of new materials in a predetermined fashion, using association with other molecules that bind selectively to particular nucleobases.[3] Aptamer-linked smart materials that are responsive to a cooperative combination of two analytes, can be used as logic gates with chemical inputs.[4] Furthermore, double-stranded DNA lends itself very much to applications in nanotechnology: it features a nanoscale size with a diameter of about 2 nm, a rather well-defined and controllable length of typically 50 nm, depending on the biological source. Most importantly, and different from most proteins, DNA exhibits a relatively uniform structure of relatively high conformational stability, with repeating units of about 3.5 nm length in the form of nucleobase triplets. Well-defined nanowires with about 50 nm diameter could, *e.g.*, be prepared by binding of silver ions to ds-DNA, followed by reduction of the ions to elementary silver and further aggregation of the metal.[5] The synthesis of DNA polymers by chemical and enzymatic methods, mainly by solid-state, templating and amplification processes,[6] and with the polymerase chain reaction[7] has been greatly advanced, so that any desired sequence and combination with other organic structures[8] no longer presents significant problems.

The molecular recognition of the base sequence is the basis of sensors, *e.g.* of so-called biochips, and may one day serve in molecular computers.[9] Structural changes of DNA can be stimulated by several interaction mechanisms with synthetic compounds or with enzymes: intercalation of aromatic effectors, *e.g.*, lead to reversible prolongation of the polymer; nucleases produce shorter chains; helicases unwind the double strands yielding several flexible single strands. Such reactions have until now barely been developed into functional smart materials. Relatively stable, three-dimensional materials on the basis of ds-DNA, which can be handled not only in solutions, have been made accessible on the basis of branched DNA, yielding well-ordered self-assembled DNA lattices.[2,10] Such scaffolds can be used to bring, *e.g.*, protein molecules in proximity or to assemble nanoelectronic components in two- or three-dimensional arrays.[2,10] Molecular gears, walkers, gauges and translation devices have been designed on this basis, which are activated by small molecules, proteins or also other DNA molecules. Molecular motors with rotation are feasible based on the transition of right-to-left handed DNA ("B" and "Z") helices.[11] DNA molecules themselves as "fuel" are used in molecular machines by kinetic control of DNA hybridisation with complementary oligonucleotides, using DNA catalysts for promoting the hybridisation.[2a,12]

22.3 Biosensors and Biochips

Immobilised proteins can in sensor electrodes serve as intelligent materials if coupled to a suitable signal transduction.[13] Glucoseoxidase immobilised in a dialysis membrane serves to monitor glucose concentration, *e.g.*, in blood with an oxygen electrode by the decrease of oxygen concentration. Fluorescence-based glucose sensing offers many advantages; it can be based on fluorescence resonance energy transfer (FRET) between a fluorescent donor and an acceptor either within a protein that undergoes glucose-induced conformation changes, or with external fluorophores.[14] Semiconductor nanocrystals as fluorescent probes have compared with conventional fluorophores, show a narrower, tunable emission spectrum and can be more stable.[15] Semiconductor quantum dots as fluorescent probes have the advantage of unique size-dependent optical and electronic properties.[16] Entire cells can be used as smart materials, *e.g.* neurochips bear neurocells that under electric stimulation respond to certain neurologically active compounds. As transduction methods one can employ, besides electric potentials, fluorescence, chemiluminescence, SPR (surface plasmon resonance), refraction changes, or mass-sensitive balances. It also should be mentioned that photochromic materials, which hold promise for applications such as optical data storage, sunlight conversion, and light-driven molecular pumps,[17] can also be based on natural material such as bacterial rhodopsin.[17b]

Hybrid materials containing proteins such as calmodulin and hydrogels can produce mechanical motions as result of conformational changes of the protein, induced, *e.g.*, by calcium ions.[18] Aqueous cavities created in the gel matrix of a peptide/protein microarray hydrogel composed of glycosylated amino acetate were found to be a suitable semiwet reaction medium for enzymes, whereas the hydrophobic domains of the gel fibre are useful as a unique site for monitoring the reaction.[19] Chips with gel-immobilised nucleic acids, proteins, or cells can be used to control chemical and enzymatic reactions with the immobilised compounds or samples bound to them.

The most important biosensors rely on nucleic acids as signal-giving hosts.[20] On a typical DNA chip of, *e.g.*, 1.3×1.3 cm surface – most often glass – about 70 000 different oligonucleotides are placed as microscopic spots by photolithographic techniques. On plasmidchips a fluorescent plasmidic DNA is complexed by organic reagents; if the plasmid DNA is taken up by certain cell components in the analyte this can be detected on a corresponding fluorescence array. The same principle is used for adenochips, except that adenoviruses instead of plasmids are used as vectors. Similarly, one uses proteins and peptides immobilised on a chip surface for detection of noncovalently bound analytes in the surrounding matrix. For immunoassays one uses antigen ligands covalently attached to the surface or also enzyme-labelled antibodies or antigens. Related are flowthrough chips, which besides, *e.g.*, immobilised oligonucleotides, has pores, through which the analyte is pumped; nonsequence complementary material is not bound and then washed out. After this, the remaining bound material is treated with a chemiluminescence enzyme, making the corresponding pores in the array visible. Crosslinked DNA-gold-nanoparticles can be used, *e.g.*, for highly sensitive protein detection.[21]

Protein microchips containing immobilised antibodies, antigens, and enzymes have been designed. Stacking interactions with oligonucleotides extend the possibilities of such microchips for analysis of nucleic-acid sequences. This opens ways to identify bacteria and viruses, to detect toxins and to study translocations in the human genome.[22]

Supramolecular polymers based on quadruple hydrogen-bonding ureido-pyrimidinone (UPy) moieties, obtained by simply mixing UPy-functionalised polymers with UPy-modified biomolecules, yield bioactive materials. Formation of single giant cells at the interface between bioactive material and life tissue can be triggered *in vivo*.[23] Antibodies, nucleic acids, antigens, heat-sensitive and fragile bacterial, animal cells, and whole protozoa, have been encapsulated in silica, metal-oxide, organosiloxane and hybrid sol-gel polymers. Such "living ceramics" allow applications as optical and electrochemical sensors, diagnostic devices, catalysts, and even artificial organs.[24]

22.4 Intelligent Bionanoparticles

Recent technologies have made it possible to conjugate biomolecules with nanoparticles, which leads to highly promising applications particularly in biomedical fields,[25,26] based also on biohybrid carbon nanotubes.[27] Figure 22.1 illustrates the possible mechanisms for binding, *e.g.*, proteins or nucleic acids to nanoparticles, which often are gold[28] or silver colloids stabilised and activated by, *e.g.*, citrate reduction. Due to the negative charge on the colloids electrostatic interactions can be used for binding positively charged proteins, also the particularly strong noncovalent binding combination of biotin and streptavidin or immonoglobulin; the most stable and often used methods involve covalent but reversible reaction of gold with thiols. Furthermore, the known interaction between metals and micro-organisms is envisaged to produce bionanoparticles from, *e.g.*, silver nanospheres and yeast or even live fungi.[25]

External electrical or optical signals can switch properties of such bionanoparticles. For example, the hydrogen bonds between diaminopyridine and flavin derivatives on the surface of a Au nanoparticle are much enhanced by electrochemical reduction[29] (Figure 22.2). Hydrogen bonding of diaminopyridine-decanethiolate, attached to a gold surface, with alternatively a ferrocene-terminated uracil and a nonelectroactive uracil was used to add and to remove electroactive functionality in a mesoscale assembly with tunable current–voltage properties. STM tips are used to explore the local current–voltage properties.[30]

Electrostatic association of CdS particles with DNA was used for the formation of a semiconductor nanoparticle DNA wire by association of cationic surfactants at the water/air interface and subsequent compression with the Langmuir–Blodgett technique, providing a densely packed DNA in the layer. Addition of positively charged CdS particles capped with thiocholine

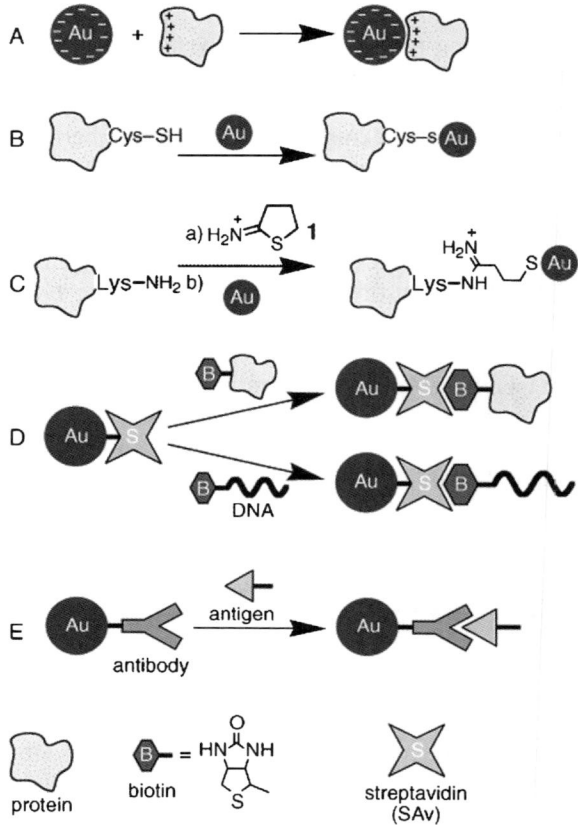

Figure 22.1 Formation of the biomolecule–nanoparticle (NP) hybrids: A) NP–protein conjugation by electrostatic interactions. B), C) conjugation adsorption of NPs on natural (B) and synthetic (C) thiol groups of the protein. D), E) conjugation by use of bioaffinity interactions upon streptavidin–biotin binding (D), and antibody–antigen associations (E). Reproduced with permission from E. Katz and I. Willner, *Angew. Chem. Int. Ed. Engl.*, 2004, **43**, 6042.

resulted in aggregation with the DNA. After transfer of the layer to a solid support they performed as a nanosemiconductor.[31]

Remote electronic control over DNA hybridisation is possible by inductive coupling of a radio-frequency magnetic field to a metal nanocrystal, which is covalently linked to DNA. Inductive coupling to the nanocrystal increases the local temperature of the bound DNA, thereby inducing denaturation while leaving surrounding molecules relatively unaffected.[32]

Figure 22.3 shows how by an optical signal the RF-triggered folding and unfolding of the DNA strand is fully reversible. Electromagnetic field-induced excitation of biocompatible superparamagnetic nanoparticles holds promise in,

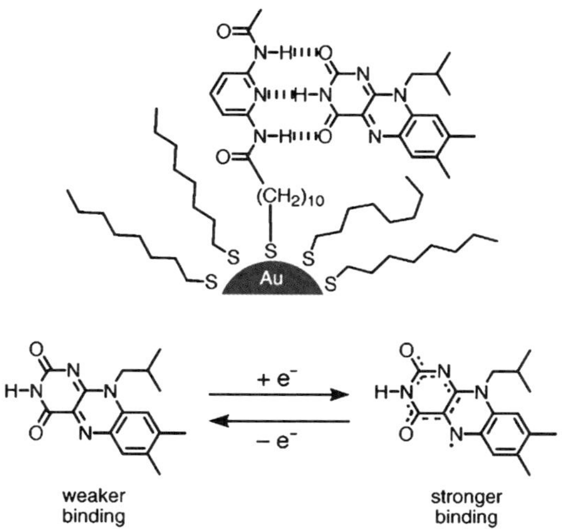

Figure 22.2 Electrochemically controlled recognition on hydrogen-bonded pyridinediamide functionalised on a gold nanoparticle, carrying also alkyl chains. Reproduced with permission from E. Katz and I. Willner, *Angew. Chem. Int. Ed. Engl.*, 2004, **43**, 6042 (after A.K. Boal, V.M. Rotello, *J. Am. Chem. Soc.*, 1999, **121**, 4914).

e.g., tumour treatment, which by the use of nano-Au–conjugates with nucleic acids or peptides could provide for higher selectivity.[33]

22.5 Nanobiosensors

Promising new developments of biosensors make use of the large spectral shifts that accompany changes of size and binding modes of biomolecular metallo-nanoparticles upon interaction with selected analytes. This relates to the known use of staining biological objects by metal particles. Surface plasmon resonance (SPR) and FT-IR spectroscopy can be used to characterise the attachment of DNA or of peptides to gold surfaces with the aid of the thiol linker method.[34]

For example, metal nanoparticles were functionalised with two kinds of nucleic acids, which were complementary to two segments of the DNA to be analysed. Hybridisation with the analysed DNA leads to the nanoparticle aggregation, which can be detected by redshifted interparticle plasmon absorbance of the aggregates.[35] Hybrid materials composed of a single-stranded DNA, a gold nanoparticle, and a fluorophore is highly quenched by the nanoparticle through a distance-dependent process. The fluorescence of this hybrid molecule increases by a factor of up to several thousands as it binds to a complementary ssDNA.[36]

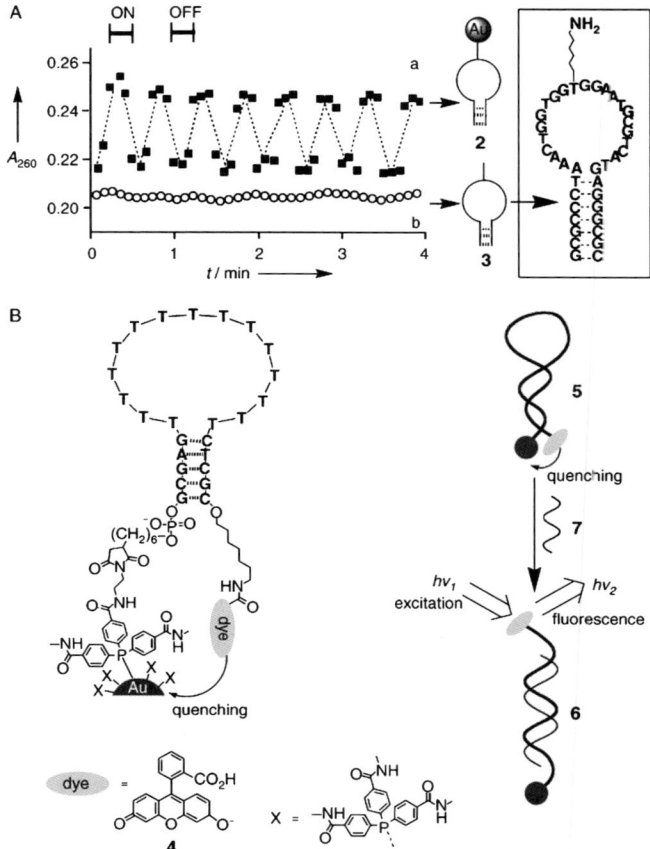

Figure 22.3 Switching DNA conformations on a nanoparticle by a radiofrequency signal. A) The reversible change of the Au NP–DNA molecular beacon conjugate **2** between denaturated and hybridised conformations upon switching "ON" and "OFF", respectively, the radiofrequency electromagnetic field. The DNA hairpin molecule **3**, which lacks the Au nanoparticle, is not affected by the electromagnetic field. B) Fluorescence emission controlled by the transformation from the hairpin conformation **5** to the extended state **6** of a DNA chain upon hybridisation with the complementary oligonucleotide **7**. Reproduced with permission from A. Jordan, R. Scholz, P. Wust, H.F. Wehling and R. Felix, *J. Magn. Magn. Mater.*, 1999, **201**, 413.

22.6 Drug-delivery and Related Systems

Many devices on the basis of smart polymers have been designed for the controlled release of drugs.[37] The engineering of nano- and microcarriers and their subsequent functionalisation provides smart materials in the form of micro- and nanoparticles, also liposomes, which are able to encapsulate drug

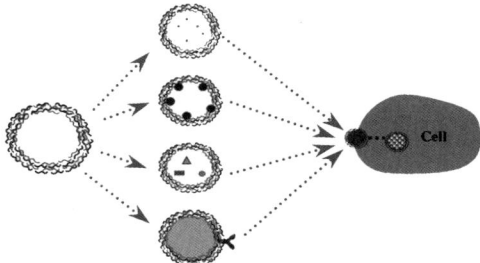

Figure 22.4 Multifunctional nanocapsules targeting a cell.; for explanation see text. Reproduced with permission of the publisher from G.B. Sukhorukov, A.L. Rogach, B. Zebli, T. Liedl, A.G. Skirtach, K. Kohler, A.A. Antipov, N. Gaponik, A.S. Susha, M. Winterhalter and W.J. Parak, *Small*, 2005, **1**, 194.

molecules; polymeric vesicles, generated by self-organisation or interfacial polymerisation, serve the same purpose. Figure 22.4 illustrates how hollow particles are loaded inside with effector molecules, for visualisation often with dyes, or for activation by an outside field with ferromagnetic particles, *etc.* The target cell may be recognised by proper functionalisation of the particle outside, *e.g.* by placing peptides, antibodies *etc* on the outer shell.[37f,38,] With microspheres biodegradation can also be modulated by external biocompounds.[39] Membranes that close drug-filled microreservoirs can degrade and open at time intervals controlled by molecular mass, thickness or composition of the membrane, thus allowing release of the drug in pulsatile fashion over extended time periods.[40]

Controlled release not only of drugs but also of fragrances, flavours, vitamins *etc.*, *e.g.* in food, is most often based on volume changes or/and phase transition triggered by signals, such as changes of temperature, of external chemical or biological compounds; corresponding chemomechanical polymers as well as voltage-activated systems are discussed in other chapters of the present monograph. Recent progress in microfabrication allows implementation of such functions into tiny microchips, which contain on one silicon chip of, *e.g.*, 17 mm by 17 mm by 310 μm 34, possibly over 1000 reservoirs.[41] Such reservoirs containing an active substance to be released are covered with tiny membranes as anodes, which are dissolved by applying an electric potential between the anode and a cathode (Figure 22.5).[42]

Many micro- or nanoscale delivery devices relying on an *electrical signal* input have been described. In corresponding capsules electrophoresis of charged drugs towards an oppositely charged electrode can liberate the entrapped drug.[43] Conductive polypyrrole/acrylic acid/carbon black composite materials showed a fast and reversible electroactuation with low potentials, and could be used in a microfabricated pump.[44] Hydrogels composed of acrylamide and a carboxylic acids, and doped with a polypyrrole/carbon black composite are also electroactive; the presence of the maleic acid increases the osmotic

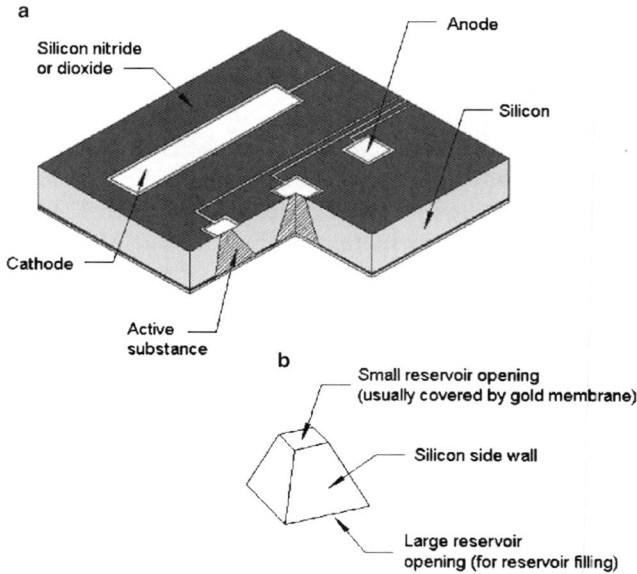

Figure 22.5 A voltage-triggered delivery device on a microchip; for explanation see text. With permission from J.T. Santini, M.J. Cima and R. Langer, *Nature*, 1999, **397**, 335.

pressure arising at the hydrogel/electrolyte interface, thus increasing the bending response of the material.[45]

A special electrically triggered delivery system, called *electroporation*,[46] uses short high-voltage pulses to overcome the barrier of membranes. If the external electric field surpasses the capacity of the cell membrane a membrane breakdown occurs. Molecules inside the cells, encompassing ions, drugs, dyes, tracers, antibodies, peptides, oligonucleotides or nucleic acids for gene transfer can then diffuse out or driven out electrophoretically through the destabilised membrane. In related *iontophoresis* methods[47] a heterogeneous cation-exchange membrane is used as an electrically sensitive and efficient rate-limiting barrier on surfaces of implants.[48]

Electrochemically triggered arrays of polymeric valves on a set of drug reservoirs have been developed, in which the valves are bilayer structures, made as a flap hinged on one side to a valve seat, and consisting of thin films of evaporated gold and electrochemically deposited polypyrrole.[49] Metal or polymeric valves, made, *e.g.*, from polyaniline and poly(2-hydroxy-ethylmethacrylate)-poly(N-vinylpyrrolidinone) blends, can be actuated under the control of additional sensors. The thin valves consist of nonporous layers that can be dissolved or disintegrated by water electrolysis. Reversible polymeric valves were prepared from a blend of redox polymer and hydrogel that swells and shrinks by applying a suitable bias or through a specific chemical reactions.[50]

Electrothermal control of drug delivery is possible with microchip devices containing an array of individually sealed and actuated reservoirs, each capped by a thin metal membrane. Passage of a threshold level of electric current through the membrane causes it to disintegrate, thereby exposing the reservoir contents.[51]

For drug release by acidity differences *pH-sensitive* polymers,[52] in particular chemomechanical hydrogels are an obvious choice, as described in another chapter of the present book. A pH-dependent micelle formation can be used on the basis of self-assembled amphiphilic copolymers (Figure 22.6). Rapid release can be triggered by amine protonation at lowering the pH, inducing micellar dissociation and therefore liberation of a hydrophobic drug.[53]

Acid-catalysed polyanhydride hydrolysis is an interesting way to liberate drugs or proteins as a function of time and pH, using a laminated device with polyanhydrides as isolating layers. Pulsatile protein release from such a device can vary from 30 to 165 h, depending on the selected polyanhydride type and isolating layer thickness.[54]

Enzyme-controlled drug delivery has also been reported.[55] Capsules of a selective membrane permeability with proteins placed within the interior make it possible to generate micro-enzyme reactors, where the enzyme is protected from both proteases and dilution effects. The capsule walls can be triggered to control the enzymatic reactions within or outside the capsules. Rhythmic drug delivery with a hydrogel was made possible by feedback between the swelling state of a gel membrane and the enzymatic conversion of glucose, which generates hydrogen ions. In this way for example the gonadotropin-releasing hormone could be released in short, repetitive pulses.[56]

Temperature-sensitive gels are the basis of several thermally responsive delivery systems.[57]

Temperature-sensitive gel beads in the range of 0.1–2 mm diameter were, *e.g.*, obtained with poly(N-isopropylacrylamide) and alginate, with crosslinking in

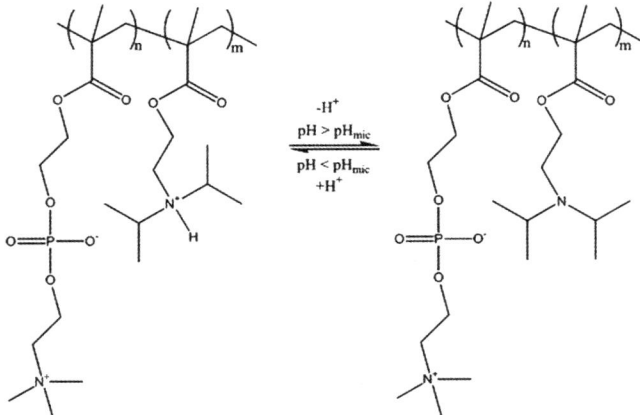

Figure 22.6   pH-dependent micelle formation with amphiphilic copolymers. Reproduced with permission of the publisher from C. Giacomelli, L. Le Men, R. Borsali, J. Lai-Kee-Him, A. Brisson, S.P. Armes and A.L. Lewis, *Biomacromolecules*, 2006, **7**, 817.

the presence of inorganic ferrofluid filling particles, such as $BaTiO_3$, TiO_2, and Ni.[58] Temperature-controlled micelle formation is another alternative.[59]

Ultrasound offers another attractive way to control delivery from smart gels. Low-frequency ultrasound increases release rates from swollen matrices by a factor of 30–500, in contrast to the unswollen matrices, where the release rates increased only by a factor of 2–3. If unswollen matrices contain drug particles that absorb ultrasound on the surface of the matrices this causes ultrasound attenuation. Swollen matrices contain a compact arrangement of fluid pockets separated from each other by thin membranes containing a solution of the dissolved drug, therefore ultrasound penetrates and causes tearing of these membranes. In consequence, the fluid pockets interconnect, and drug molecules diffuse out.[60]

Light-induced delivery systems are based on photochemically sensitive materials, which swell or shrink or decompose upon irradiation.[61] Microcapsules with a fluorescence dye, gold and silver particles in a multilayer absorb light, which can be used to liberate, by near infrared laser light, antitumour agents in cancer tissue.[62]

Magnetic nanoparticles (MNP) can be used to bind drugs; magnetic fields outside the body are focused on specific targets *in vivo*, capturing the particle complex and resulting in enhanced delivery to a target site.[37,63]

Cobalt ferrite magnetic nanoparticles containing effector molecules have been coated with an amorphous silica shell. Figure 22.7 shows how the movement of doped cells under an external magnetic field is monitored with a "magnetic motor".[64] Alternatively, a magnetic ferrogel from poly(vinyl alcohol) and Fe_3O_4 magnetic particles allows drug accumulation around the gel by an external magnetic field, with instant drug release when the magnetic field is switched off.[65] Similarly, the antitumour agent *cis*-platin was encapsulated together with

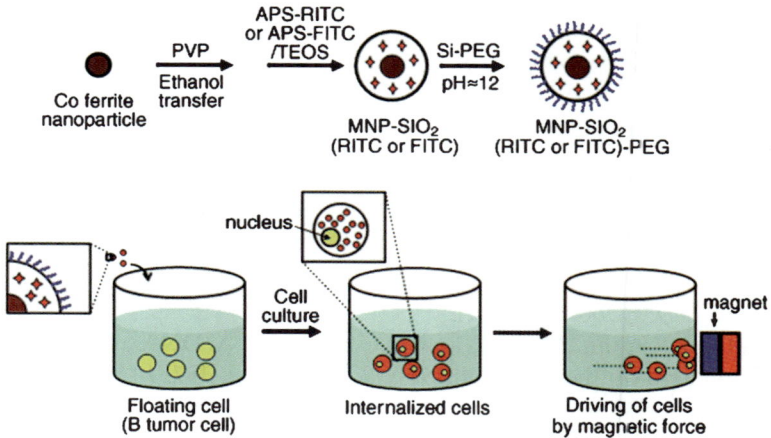

Figure 22.7 Ferrite nanoparticles moved by an external magnetic field. Reproduced with permission of the publisher T.J. Yoon, J.S. Kim, B.G. Kim, K.N. Yu, M.H. Cho and J.K. Lee, *Angew. Chem.-Int. Ed.*, 2005, **44**, 1068.

magnetite, yielding a magnetically triggered system.[66] Magnetic silica-iron oxide composite nanotubes can also contribute to drug delivery as well as to biochemical separations.[67]

References

1. 2005 The Worldwide Biochips & Equipments Market; Fuji-Keizai USA, Inc., June 2005. www.researchandmarkets.com; see also www.mrgco.com/TOC.WorldwideBiochip.html
2. (a) N.C. Seeman, *Nature*, 2003, **421**, 427; (b) N.C. Seeman, *Trends Biochem. Sci.*, 2005, **30**, 119; (c) N.C. Seeman, *Chem. Biol.* 2003, **10**, 1151; N.C Seeman, *Angew. Chem. Int. Edn Engl.* 1998, **37**, 3220; recent review: D. Ho, D. Garcia and C.M. Ho, *J. Nanosci. Nanotech.*, 2006, **6**, 875; see also M. Shahinpoor, *Proc. SPIE 1996 North Amer. Confer. Smart Struct. Materials*, 1996, San Diego, California, vol. 2716, paper no. 33.
3. K. Tanaka and M. Shionoya, *Chem. Lett.*, 2006, **35**, 694.
4. J.W. Liu and Y. Lu, *Adv. Mater.*, 2006, **18**, 1667.
5. E. Braun, Y. Eichen, U. Sivan, G. Ben-Yoseph, *Nature* 1998, **391**, 775, see also F. Patolsky, Y. Weizmann and I. Willner, *Nature Mater.*, 2004, **3**, 692, and references cited therein.
6. X.Y. Li and D.R. Liu, *Angew. Chem. Int. Ed. Engl.* 2004, **43**, 4848; M.H. Caruthers, *Science* 1985, **230**, 281.
7. M. Smith, *Angew. Chem. Int. Ed. Engl.*, 1994, **33**, 1214.
8. See, *e.g.*, G. von Kiedrowski, L.H. Eckardt, K. Naumann, W.M. Pankau, M. Reimold and M. Rein, *Pure and Applied Chemistry*, 2003, **75**, 609; L.H. Eckardt, K. Naumann, W.M. Pankau, M. Rein, M. Schweitzer, N. Windhab and G. von Kiedrowski, *Nature*, 2002, **420**, 286.
9. Leading references see: E. Winfree, *J. Biol. Mol. Struct. Dynam. Conversat.*, 2000, **2**, 263; C. Mao, T. LaBean, J.H. Reif and N.C. Seeman, *Nature*, 2000, **407**, 493–496; R.J. Sha, X.P. Zhang, S.P. Liao, P.E. Constantinou, B.Q. Ding, T. Wang, A.V. Garibotti, H. Zhong, L.B. Israel, X. Wang, G. Wu, B. Chakraborty, J.H. Chen, Y.W. Zhang, H. Yan, Z.Y. Shen, W.Q. Shen, P. Sa-Ardyen, J. Kopatsch, J.W. Zheng, P.S. Lukeman, W.B. Sherman, C.D. Mao, N. Jonosk and N.C. Seeman, in *Structural DNA Nanotechnology: Molecular Construction and Computation*, in: Lecture notes in computer science, 2005, **3699**, Springer, New York.
10. P. Tosch, C. Wälti, A.P.J. Middelberg and A.G. Davie, *J. Physics: Conference Series*, 2007, **61**, 1241.
11. C. Mao, W. Sun, Z. Shen, and N.C. Seeman, A DNA nanomechanical device based on the B–Z transition, *Nature*, 1999, **397**, 144; N.C. Seeman, H. Yan, X.P. Zhang and Z.Y. Shen, *Biophys. J.*, 2002, **82**, 341A; H. Yan, X.P. Zhang, Z.Y. Shen and N.C. Seeman, *Nature*, 2002, **415**, 62.
12. A.J. Turberfield, J.C. Mitchell, B. Yurke, A.P. Mills, M.I. Blakey and F.C. Simmel, *Phys. Rev. Lett.*, 2003, 90; B. Yurke, A.J. Turberfield, A.P. Mills, F.C. Simmel and J.L. Neumann, *Nature*, 2000, **406**, 605.

13. J.M. Kauffmann and G.G. Guilbault, *Methods Biochem. Anal.*, 1992, **36**, 63; G.G. Guilbault, *Methods Enzymol.*, 1988, **137**, 14.
14. J.C. Pickup, F. Hussain, N.D. Evans, O.J. Rolinski and D.J.S. Birch, *Biosens. Bioelectron.*, 2005, **20**, 2555.
15. M. Bruchez, M. Moronne, P. Gin, S. Weiss and A.P. Alivisatos, *Science*, 1998, **281**, 2013.
16. Z.B. Lin, X.G. Su, Y. Mu and Q.H. Jin, *J. Nanosci. Nanotech.*, 2004, **4**, 641.
17. (a) *Photochromism: Molecules and Systems*, H. Dürr, H. Bouas-Laurent, ed., Elsevier Amsterdam, 2nd revised edn, 2003; (b) N. Hampp, C. Bräuchle, *ibid.* p.954.
18. J.D. Ehrick, S.K. Deo, T.W. Browning, L.G. Bachas, M.J. Madou and S. Daunert, *Nature Mater.*, 2005, **4**, 298.
19. S. Kiyonaka, K. Sada, I. Yoshimura, S. Shinkai, N. Kato and I. Hamachi, *Nature Mater.*, 2004, **3**, 58.
20. *DNA Microarrays: A Practical Approach*, ed. M. Schena, University Press, Oxford, UK, 1999; *Biochip Technology*, J. Cheng and L. J Kricka, ed., ISBN: 9057026139, Routledge, 2001; A. Carmen; G. Hardiman, *Biochips As Pathways To Drug Discovery* ISBN: 1574444506, CRC Press, 2006, K. Chakrabarty and Fei Su, *Digital Microfluidic Biochips: Synthesis, Testing and Reconfiguration Techniques*, ISBN: 0849390095, CRC Press, Boca Raton, 2006.
21. P. Hazarika, B. Ceyhan and C.M. Niemeyer, *Small*, 2005, **1**, 844; B. Zou, B. Ceyhan, U. Simon and C.M. Niemeyer, *Adv. Mater.*, 2005, **17**, 1643.
22. V.E. Barsky, A.M. Kolchinsky, Y.P. Lysov and A.D. Mirzabekov, *Molec. Biol.*, 2002, **36**, 437.
23. P.Y.W. Dankers, M.C. Harmsen, L.A. Brouwer, M.J.A. Van Luyn and E.W. Meijer, *Nature Mater.*, 2005, **4**, 568.
24. I. Gill and A. Ballesteros, *Trends Biotech.*, 2000, **18**, 282; V.B. Kandimalla, V.S. Tripathi and H.X. Ju, *Crit. Rev. Anal. Chem.*, 2006, **36**, 73.
25. Review: E. Katz and I. Willner, *Angew. Chem. Int. Ed. Engl.*, 2004, **43**, 6042. See also Smart Nano- and Microparticles, K. Kono, R. Arshady, Eds., London, Kentus, 2006.
26. Reviews: D.G. Anderson, J.A. Burdick and R. Langer, *Science*, 2004, **305**, 1923; S.G. Penn, L. He and M.J. Natan, *Curr. Opin. Chem. Biol.*, 2003, **7**, 609; N.L. Rosi and C.A. Mirkin, *Chem. Rev.*, 2005, **105**, 1547; W. Fritzsche and T.A. Taton, *Nanotechnology*, 2003, **14**, R63.
27. Review: E. Katz and I. Willner, *Chemphyschem*, 2004, **5**, 1085.
28. Review on the use of gold nanoparticles: M.C. Daniel and D. Astruc, *Chem. Rev.*, 2004, **104**, 293.
29. A.K. Boal and V.M. Rotello, *J. Am. Chem. Soc.*, 1999, **121**, 4914.
30. G.M. Credo, A.K. Boal, K. Das, T.H. Galow, V.M. Rotello, D.L. Feldheim and C.B. Gorman, *J. Am. Chem. Soc.*, 2002, **124**, 9036; for a review on the use of hydrogen bonding for the design of new materials see G. Cooke, V.M. Rotello, *Chem. Soc. Rev.,* 2002, **31**, 275.

31. T. Torimoto, M. Yamashita, S. Kuwabata, T. Sakata, H. Mori and H. Yoneyama, *J. Phys. Chem. B*, 1999, **103**, 879.
32. K. Hamad-Schifferli, J.J. Schwartz, A.T. Santos, S.G. Zhang and J.M. Jacobson, *Nature*, 2002, **415**, 152.
33. A. Jordan, R. Scholz, P. Wust, H.F. Wehling and R. Felix, *J. Magn. Magn. Mater.*, 1999, **201**, 413.
34. E.A. Smith, M.J. Wanat, Y.F. Cheng, S.V.P. Barreira, A.G. Frutos and R.M. Corn, *Langmuir*, 2001, **17**, 2502.
35. C.A. Mirkin, R.L. Letsinger, R.C. Mucic and J.J. Storhoff, *Nature*, 1996, **382**, 607.
36. B. Dubertret, M. Calame and A.J. Libchaber, *Nature Biotechnol.*, 2001, **19**, 365.
37. (a) *Polymeric Drugs and Drug Administration*, R.M. Ottenbrite, ed., American Chemical Society, Washington, DC, 1994; (b) *Pulsed and Self-Regulated Drug Delivery*, J. Kost, ed., CRC Press, Boca Raton, Florida, 1990; (c) *Multilayer Hollow Microspheres. MML Series* **1–5** ed. R. Arshady, Citus Books, 2001/2004; (d) M. Staples, K. Daniel, M.J. Cima and R. Langer, *Pharm. Res.*, 2006, **23**, 847; (e) L.Y. Qiu and Y.H. Bae, *Pharm. Res.*, 2006, **23**, 1; (f) G.B. Sukhorukov, A.L. Rogach, B. Zebli, T. Liedl, A.G. Skirtach, K. Kohler, A.A. Antipov, N. Gaponik, A.S. Susha, M. Winterhalter and W.J. Parak, *Small*, 2005, **1**, 194; (g) J. Siepmann and A. Gopferich, *Adv. Drug Deliv. Rev.*, 2001, **48**, 229; (h) K.F. Arndt, T. Schmidt, A. Richter and D. Kuckling, *Macromolecular Symposia*, 2004, **207**, 257; (i) W.M. Saltzman and W.L. Olbricht, *Nature Rev. Drug Discov.*, 2002, **1**, 177; (j) N.A. Peppas, J.Z. Hilt, A. Khademhosseini and R. Langer, *Adv. Mater.*, 2006, **18**, 1345.
38. W. Meier, *Chem. Soc. Rev.* 2000, **29**, 295; S. Forster, M. Konrad, *J. Mater. Chem.*, 2003, **13**, 2671; C. Dennis, *Nature* 2003, **423**, 580; M.E. Akerman, W.C.W. Chan, P. Laakkonen, S.N. Bhatia and E. Ruoslahti, *Proc. Natl. Acad. Sci. USA* 2002, **99**, 12 617; W.C.W. Chan, S. Nie, *Science*, 1998, **281**, 2016; K. Greish, T. Sawa, J. Fang, T. Akaike, H. Maeda, *J. Contr. Rel.*, 2004, **97**, 219; O.J. Plante, E.R. Palmacci and P.H. Seeberger, *Science*, 2001, **291**, 1523.
39. J.M. Anderson and M.S. Shive, *Adv. Drug Deliv. Rev.*, 1997, **28**, 5.
40. A.C.R. Grayson, I.S. Choi, B.M. Tyler, P.P. Wang, H. Brem, M.J. Cima and R. Langer, *Nature Mater.*, 2003, **2**, 767.
41. J.T. Santini, A.C. Richards, R. Scheidt, M.J. Cima and R. Langer, *Angew. Chem. Int. Ed. Engl.*, 2000, **39**, 2397.
42. J.T. Santini, M.J. Cima and R. Langer, *Nature*, 1999, **397**, 335.
43. S. Murdan, *J. Control. Rel.*, 2003, **92**, 1; see also I.C. Kwon, Y.H. Bae and S.W. Kim, *Nature* 1991, **354**, 291; L.L. Miller, *Mol. Cryst. Liq. Cryst.* 1998, **160**, 297.
44. E.A. Moschou, S.F. Peteu, L.G. Bachas, M.J. Madou and S. Daunert, *Chem. Mater.*, 2004, **16**, 499.
45. E.A. Moschou, M.J. Madou, L.G. Bachas and S. Daunert, *Sens. Actuators B-Chem.*, 2006, **115**, 79.

46. M.B. Fox, D.C. Esveld, A. Valero, R. Luttge, H.C. Mastwijk, P.V. Bartels, A. van den Berg and R.M. Boom, *Anal. Bioanal. Chem.*, 2006, **385**, 474; D. Rabussay, N.B. Dev, J. Fewell, L.C. Smith, G. Widera and L. Zhang, *J. Phys. D-Appl. Phys.*, 2003, **36**, 348; J.C. Weaver, *IEEE Trans. Dielect. Electric. Insul.*, 2003, **10**, 754; J. Gehl, *Acta Physiol. Scand.*, 2003, **177**, 437.
47. E. Eljarrat-Binstock and A.J. Domb, *J. Control. Release*, 2006, **110**, 479; M.E. Myles, D.M. Neumann and J.M. Hill, *Adv. Drug Deliv. Rev.*, 2005, **57**, 2063; Y.N. Kalia, A. Naik, J. Garrison and R.H. Guy, *Adv. Drug Deliv. Rev.*, 2004, **56**, 619.
48. V. Labhasetwar, T. Underwood, S.P. Schwendeman and R.J. Levy, *Proc. Nat. Acad. Sci. USA*, 1995, **92**, 2612.
49. M.J. Madou and R. Cubicciotti, *Proc. IEEE*, 2003, **91**, 830.
50. L.M. Low, S. Seetharaman, K.Q. He and M.J. Madou, *Sens. Actuators B-Chem.*, 2000, **67**, 149.
51. J.M. Maloney, S.A. Uhland, B.F. Polito, N.F. Sheppard, C.M. Pelta and J.T. Santini, *J. Control. Rel.*, 2005, **109**, 244.
52. A. Gutowska, J.S. Bark, I.C. Kwon, Y.H. Bae, Y. Cha and S.W. Kim, *J. Control. Rel.*, 1997, **48**, 141; L.Y. Qiu and K.J. Zhu, *Int. J. Pharm.*, 2001, **219**, 151; J.H. Kim, J.Y. Kim, Y.M. Lee and K.Y. Kim. *Appl. Polym. Sci.*, 1992, **44**, 1923; L. Lu, C.A. Garcia and A.G. Mikos, *J. Biomed. Mater. Res.*, 1999, **46**, 236, and references cited therein.
53. C. Giacomelli, L. Le Men, R. Borsali, J. Lai-Kee-Him, A. Brisson, S.P. Armes and A.L. Lewis, *Biomacromolecules*, 2006, **7**, 817.
54. H.L. Jiang and K.J. Zhu, *Int. J. Pharm.*, 2000, **194**, 51.
55. F. Fischelghodsian, L. Brown, E. Mathiowitz, D. Brandenburg and R. Langer, *Proc. Nat. Acad. Sci. USA*, 1988, **85**, 2403.
56. G.P. Misra and R.A. Siegel, *J. Control. Rel.*, 2002, **81**, 1.
57. T. Okano, Y.H. Bae and S.W Kim in *Pulsed and Self-Regulated Drug Delivery*, J. Kost, ed., CRC Press, Boca Raton, Florida, 1990, p. 17–45; J. Kost and R. Langer, *ibid.* 3, A. Gutowska and S.W. Kim, *Macromol. Symp.*, 1997, **118**, 545.
58. D. Kuckling, T. Schmidt, G. Filipcsei, H.J.P. Adler and K.F. Arndt, *Macromol. Symp.*, 2004, **210**, 369; see also J.X. Zhang, L.Y. Qiu, Y. Jin and K.J. Zhu, *J. Biomed. Mater. Res. A*, 2006, **76**, 773.
59. J.X. Zhang, L.Y. Qiu, Y. Jin and K.J. Zhu, *J. Biomed. Mater. Res. A*, 2006, **76A**, 773.
60. C. Aschkenasy and J. Kost, *J. Control. Rel.*, 2005, **110**, 58; J. Kost, J. and R. Langer, in *Pulsed and Self-Regulated Drug Delivery*, J. Kost, ed., CRC Press, Boca Raton, Florida, 1990, 3–16; J. Kost, K. Leong and R. Langer,. *Proc. Natl. Acad. Sci. USA*, 1989, **86**, 7663.
61. E. Mathiowitz and M.D. Cohen, *J. Membr. Sci.*, 1989, **40**, 67; I.C. Kwon, Y.H. Bae and S.W. Kim, *Nature*, 1991, **354**, 291.
62. A.G. Skirtach, A.M. Javier, O. Kreft, K. Kohler, A.P. Alberola, H. Mohwald, W.J. Parak and G.B. Sukhorukov, *Angew. Chem. Int. Ed. Engl.*, 2006, **45**, 4612.

63. Reviews: J. Dobson, *Drug Development Res.*, 2006, **67**, 55; J. Kost and R. Langer, *Pharm. Int.*, 1986, **7**, 60.
64. T.J. Yoon, J.S. Kim, B.G. Kim, K.N. Yu, M.H. Cho and J.K. Lee, *Angew. Chem. Int. Ed. Engl.*, 2005, **44**, 1068.
65. T.Y. Liu, S.H. Hu, D.M. Liu and S.Y. Chen, *Langmuir*, 2006, **22**, 5968.
66. J. Yang, H. Lee, W. Hyung, S.B. Park and S. Haam, *J. Microencaps.*, 2006, **23**, 20; see also M. Arruebo, M. Galan, N. Navascues, C. Tellez, C. Marquina, M.R. Ibarra and J. Santamaria, *Chem. Mater.*, 2006, **18**, 1911; J. Chen, L.M. Yang, Y.F. Liu, G.W. Ding, Y. Pei, J. Li, G.F. Hua and J. Huang, *Macromol. Symp.*, 2005, **225**, 71; J.J. Cheng, B.A. Teply, S.Y. Jeong, C.H. Yim, D. Ho, I. Sherifi, S. Jon, O.C. Farokhzad, A. Khademhosseini and R.S. Langer, *Pharm. Res.*, 2006, **23**, 557, and references cited therein.
67. S.J. Son, J. Reichel, B. He, M. Schuchman and S.B. Lee, *J. Am. Chem. Soc.*, 2005, **127**, 7316.

Subject Index

abacus 92–5
abrupt shape transition 288–90
acidity, generative force
    characteristics 197–8
acoustic control 243–5
actin 152
    see also F-actins; polymer–actin
        complexes
actin gels 465–7
actuation
    basics of 397–9
    electrically induced robotic 130–2,
        133
    metal-hydride-actuation principle
        387, 389–94
    natural muscles 152–3
    polyacrylonitrile 194, 196–8, 203
    shape-memory effect in polymers
        308–11
    virtual two-way actuation 323–4
actuators
    artificial 460–1
    carbon nanotubes as 499
    design 405–6
    diaphragm actuators 411–12, 413
    dielectric elastomer 396–421
    energy-to-volume ratio 332
    ferroelectric relaxor polymers as
        256–79
    framed 414–15
    linear 411
    magnetic shape-memory alloy 496
    performance capabilities 201–3
    piezoelectric ceramic actuators
        233–5, 236, 242, 247, 407
    polymeric 153–4, 155–6, 163
    polymeric microactuator 273–8
    as products 169–70
    see also artificial muscles
adaptive structural shape control 246
adaptive structures 331–3
aeronautical applications 320–1, 332–3, 334–6
air, muscles working in 165, 167
air valve 324–5
aircraft couplings 320–1
aluminium/titanium fibre system
    482–4, 488
amino acids 117–18
amplification, signal 56–9
amplifiers, unimolecular 223
AND logic gate 62–3, 112
anion effects
    generative force characteristics
        196–8
    polymer gels 105
antennas
    cellular phone 327
    for light harvesting 53–6
aqueous solution, conducting
    polymers in 143–6
aromatic effectors sensitivity 453
arrays
    diaphragm actuators for 411–12, 413
    enhanced-thickness mode arrays
        412–14

Subject Index

artificial actuators 460–1
artificial muscles 153–70, 181
  for biomimetic robots 409–10
  chitosan-based 460–1
  comparison of actuator technologies 407–8
  contraction mimicked by ferrogels 294–5, 296
  contraction/stretching of 77–8, 79
  control of pseudomuscular contraction 295–9
  ferroelectric relaxor polymers as 256–79
  ionic polymer metal nanocomposites as 126–40
  magnetic polymeric gels as 282–99
  metal hydrides as 386–94
  modified PAN as 191–203
  structures and designs of 157
  see also actuators
asymmetrical monolayers 157–8
ATP-driven gel machine 464–75
azobenzene polymers 424–38
  photodeformable azo materials 434–7
  photoswitchable azo materials 430–2

basic conducting polymers 143
batteries 172–4, 181
bending structures 157–8
  combination of 161, 162
bilayers 158, 159, 164
biochips 508–9
biogenic materials 506–17
biological machines 3–4
biological materials 507
biological motors 3–4
biomedical applications *see* drug delivery systems; medical applications
biomimetic robots 409–10
biomimicking properties 151–2
bionanoparticles 509–11
biosensors 508–9
  nanobiosensors 511–12
bipolarons 150, 170, 171
bismacrocycle transition metal complexes 86–8

brakes, rotary 373–4
breakdown properties, dielectric elastomers 420–1
Brownian motion 2, 3
buckyballs 497–9

cables, extension 53, 54
carbon nanotubes 497–9
cardiovascular stents 329–31
catenanes 12, 16
  controlling rotational motion in 24–9
  reversible rotary motor 28–9
  rotary motor 27–8
  ruthenium-containing 91–2, 93
  transition metal complexed 83–9
cellular-phone antennas 327
ceramics *see* piezoelectric ceramics
CFRP/aluminium laminate 481–3, 488
chemical reactions, molecular motions from 77–83
chemical transduction 151, 152, 174, 175
chemomechanical polymers 100–20
  concentration profiles 110–11
  cooperativity and logical gate functions 111–13
  particle-size effects and kinetics 107–8, 109
  and pH 100, 101–7
  selectivity by covalent interactions 118–19
  selectivity with organic effector molecules 114–17
  ternary complex formation for amino acids and peptides as effectors 117–18
  water uptake and release 108–10
chitin 448, 449
chitosan 102–7
  biological properties 449
  biomedical applications 458–61
  expansion kinetics 107–8
  intelligent properties 450–6
  physical and chemical properties 448–9
  solvent and solubility 449–50
  volume expansions 105–6, 115–16, 450–3

chitosan-based hydrogels 447–61
chitosan-based intelligent materials 456–8
chromophores 150, 170, 171, 181, 425–7, 432–5
circumrotation, directional 27–8
clutches 373–5
colour of light 56
commercial products *see* industrial applications; medical applications
compact hybrid actuators 247
complexation/decomplexation process 78–82
compliant electrodes 400–1
composites 143, 478, 479
   shape-memory polymers and particles 308
   see also intelligent composites; ionic polymer metal nanocomposites
concentration profiles 110–11
condensed phases 32–8
conducting polymers
   devices based on electrochemical properties of 153–74
   electrochemical behaviour in aqueous solution 143–6
   electrochemical properties 149–51
   materials 143
   membranes and electron/chemical transducers 174, 175
   multifunctional and biomimicking properties 151–2
   nonstoichiometric, soft and wet materials 147–9
   theoretical models 174, 176–9, 180
conductive particle networks 400
configurational changes
   controlling 9–12
   inducing net positional change 19–20
conformational control 4–9
connectors, electrical 322–3
conscious systems, primitive 164–5, 166, 167
contacts 207–9

contraction
   mimicked by ferrogels 294–5, 296
   of a muscle-like rotaxane dimer 77–8, 79
   pseudomuscular contraction 295–9
controlled motion
   in solution 30–2
   on surfaces, in solids and condensed phases 32–8
controlling configurational changes 9–12
controlling motion
   in catenanes 24–9
   in covalently bonded molecular systems 4–12
   in mechanically bonded molecular systems 12–29
cooperativity 111–13
copolymers 143
copper-complexed catenane 83–6
copper-complexed rotaxane 88–9
coulomb-blockade device 222
couplings 320–1
covalent bonds
   and molecular shuttling 19
   stimuli-induced conformational control around 4–7, 8–9
covalent interactions 118–19
covalently bonded molecular systems 4–12
covalently crosslinked shape-memory polymers 306–7

dampers 363–73
damping 333–4
decomplexation process 78–82
deformation
   ferrogels 284–8
      interpretation of 288–90
   nonhomogeneous, of ferrogels 290–3
   photodeformable azo materials 434–7
deformation ratio 289–90
degenerate molecular shuttle 78–82
dendrimers
   as fluorescent sensors 56–9
   for light harvesting 53–6
   for a multiple use of light signals 59–61

Subject Index

dental procedures 328
design principles, molecular motors and machines 1–4
dethreading/rethreading of pseudorotaxanes 65–6, 67
diaphragm actuators 411–12, 413
dielectric elastomers 396–421
  actuator design 405–6
  applications 406–17
  basics of actuation 397–9
  compliant electrodes 400–1
  implementation challenges 417–18
  materials development for 418–21
  pre-stress bias 399
  theory and modeling 401–5
dielectric properties 420
differential resistance devices 222
diodes 212–21
directional circumrotation 27–8
distributed nanosensing 132–6
divinylbenzene/styrene-based polymers 129, 130, 131
DNA 507, 509–12
drug delivery systems 102–3, 119, 458–9, 512–17
drug release systems 458–9
dual stimuli-responsive polymers 456–7
dynamic yield stress 351–2

effector molecules
  amino acids and peptides as 117–18
  concentration 107–8, 109, 110
  selectivity with 114–17
  solvation 108–10
effectors sensitivity 453
elastic properties 419–20
  see also superelasticity
elastomers
  liquid-crystal 491–3, 494–5
  magnetorheological 344–7
  see also dielectric elastomers
electric-current effect on force generation 199–201
electrical connectors 322–3
electrical measurements 209–10

electrical responsive polymers 458
electrically driven PAN actuator 194, 196, 197
electrically induced robotic actuation 130–2, 133
electroactive materials 500
electrochemical basic molecular motors 156–7
electrochemical models 177–8
electrochemical properties, conducting polymers 149–51
  in aqueous solution 143–6
  devices based on 153–74
electrochemically controllable artificial muscle 191–203
electrochemically driven machine 88–9
electrochemically induced motions 83–91
electrochemomechanical muscles 153–5, 407
  volume variation 155–6
electrochemomechanical properties 150, 152, 453–4
electrochromic devices 170–2, 181
electrochromic properties 150, 152
electrodes
  compliant electrodes 400–1
  contacts between 207–9
electrokinetic artificial muscles 154
electromagnetic devices 407
electromechanical devices 273–8
electromechanical muscles 153–5
electromechanical response
  of HEEIP 262–6
  in P(VDF-TrFE)-based terpolymers 266–8
electron acceptors 206–7
electron/chemical transduction 151, 152, 174, 175
electron donors 206–7
electron/neurotransmitter 152
electronic devices
  molecular solid-state 32–5
  unimolecular 205–23
electrons, and molecular shuttling 15–17
electrorheological fluids 339, 340, 500–1

electrorheological response 454–5
electrosensitivity 453–5
electrostatic devices 407
electrostrictive polymers 407
energy-to-volume ratio 332
engineering applications *see* industrial applications
enhanced-thickness mode arrays 412–14
entropy-driven molecular shuttling 23–4
excited states 20–2
expansion *see* swelling
extension cables 53, 54
eyeglass frames 326–7

F-actins 465–7
  motility assay 470–1
  polarity 472–4
  polymorphism 467–9
fast-moving electrochemically driven machine 88–9
ferroelectric phase 256–7
ferroelectric relaxor behaviour 268–73
ferroelectric relaxor polymers 256–79
ferrogels 283–8
  abrupt shape transition 288–90
  control of pseudomuscular contraction 295–9
  muscle-like contraction mimicked by 294–5, 296
  nonhomogeneous deformation of 290–3
fibre-reinforced metal 485–7
field effect transistors 222
films with metal supports 160–1
flash memory devices 222
flexible semiconductors 147
fluids *see* electrorheological fluids; magnetorheological fluids
fluorescent sensors with signal amplification 56–9
force generation 194, 196–8
  electric-current effect on 199–201
force–strain behaviour 194, 195
four-probe electrical measurements 209–10
framed actuator 414–15
frames, eyeglass 326–7
fullerenes 497–9

gels *see* ferrogels; hydrogels; polymeric gels
generative force *see* force generation
generators 416–17
glucose 118–19
graft elastomers 407

health monitoring, structural 246–7
heterodinuclear bismacrocycle transition metal complexes 86–8
high-energy electron-irradiated copolymer (HEEIP)
  electromechanical responses of 262–6
  microstructures of 258–62
high-force devices 334
high-performance structural materials 478, 479
home appliances 327
HOMOs 206–7
hybrid actuators 247
hydrogels 101, 102–5
  chitosan-based 447–61
  pH-responsive 456, 457
  water uptake and release 108–10

indirect actuation 308–11
industrial applications 139–40
  IPMNCs 128–30
  linear actuators 411
  piezoelectric 247–52
  shape-memory alloys 331–6
injectable gels 460
intelligent bionanoparticles 509–11
intelligent composites 478–90
  fabricated by new route 481–5
  ionic polymer metals 126–40
  liquid phases for self-repair 485–9
  new route to developing 479–81
intelligent materials
  chitosan-based 456–8
  magnetic polymeric gels as 282–99
  metal hydrides as 386–94
  polyacrylonitrile-derived 191–203
  shape-memory alloys as 331–3
  shape-memory polymers as 301–14
  see also multifunctional materials

intelligent properties 450–6
intelligent soft actuators 256–79
interconnects, organic 223
interfacial properties 37–8
intermittent degenerate molecular shuttle 78–82
interrupter, thermal 325–6
intramolecular complexation/decomplexation process 78–82
intramolecular motion 86–8
ion-metal translocation 82–3
ionic polymer conductor nanocomposites 126, 127, 132, 139
ionic polymer metal nanocomposites (IPMNCs) 126–40
  distributed nanosensing and transduction 132–6
  electrically induced robotic actuation 130–2, 133
  manufacturing methodologies 128–9, 130
  manufacturing steps 129–30
  medical, engineering and industrial applications 139–40
  modeling and simulation 136–8
  smart-product development 138–9
  three-dimensional fabrication of 128, 129
ionic strength sensitivity 452–3

jet-engines 334–6

kinetics 107–8, 109

lanthanum-pentanickel 386–7, 388, 389
lead zirconate titanate (PZT) 233, 234, 235, 237, 239
  applications 245–6
  commercial applications 248
length-change characteristics 194, 196
lenses, molecular 56
light, tuning the colour of 56
light-driven molecular machines 65–9, 91–5
light harvesting 53–6

light-induced shape-memory polymers 311–12
light signals
  multiple use of 59–61
  processing 50–65
lineal movements 158, 160, 164
linear actuators 411
liquid-crystal elastomers 491–3, 494–5
liquid phases in composites 485–9
logical gates 62–5, 111–13
low Reynolds number 3
LUMOs 206–7

machines see molecular machines
magnetic chitosan microsphere 457–8
magnetic polymeric gels 282–99
magnetic shape-memory alloy actuators 496
magnetic shape-memory (MSM) materials 493, 495–7
magnetorheological elastomers 344–7
magnetorheological fluid clutches 373–5
magnetorheological fluid dampers 363–5
  effect of temperature 369–73
  modeling of 365–9
magnetorheological fluid devices 363–76
magnetorheological fluids 339, 341–4
  effects of surface roughness 357–63
  field-induced microstructures 353
  historical perspective 340
  models for shear-yield stress 351–3
  rheological behaviour of 348–50
  rheometry of 354–7, 358
magnetorheological materials 341–63
magnetostrictive actuators 408
manufacturing see industrial applications
materials
  conducting polymers 143
  mechanical characterisation 168, 169
  mechanical properties of 35–7
  nonstoichiometric, soft and wet 147–9
  see also intelligent materials; multifunctional materials
Maxwell stress 396–8, 401–3

mechanical characterisation, materials and devices 168, 169
mechanical properties of materials 35–7
mechanical switches
  and interfacial properties 37–8
  and mechanical properties of materials 35–7
  and optical properties of materials 32
mechanically bonded molecular systems 12–29
mechanochemical polymer/gels 408
medical applications 139–40
  chitosans 458–61
  magnetorheological fluids 375–6
  piezoelectric ceramics 249
  shape-memory alloys 313, 327–31
medical devices, superelastic 328–9
membranes 174, 175, 181
memory devices 222
metal electrodes 400–1
metal-hydride-actuation principle 387, 389–94
metal hydrides 386–94
metal ions
  and molecular shuttling 17–19
  translocation 82–3
metal nanocomposites 126–40
metal supports 160–1
microdevices 161
microelectromechanical devices 273–8
microstructure
  field-induced 353
  HEEIP 258–62
  P(VDF-TrFE)-based terpolymers 268–73
microtools 161
modeling
  dielectric elastomers 401–5
  ionic polymer metal nanocomposites 136–8
  magnetorheological fluid dampers 365–9
  metal hydride actuators 390–2, 393
  piezoelectric ceramics 235–42
  shear-yield stress 351–3
  theoretical, of conducting polymers 174, 176–9, 180

molecular abacus 92–5
molecular devices
  photochemically controlled 48–65
  processing light signals 50–65
  unimolecular electronic 205–23
molecular dynamics treatment 179, 180
molecular electronic devices 32–5
molecular extension cables 53, 54
molecular lenses 56
molecular machines
  controlled motion in solution 30–2
  controlled motion on surfaces and in solids 32–8
  current challenges 29–30
  design principles for 1–4
  electrochemically driven 88–9
  light-driven 65–9, 91–5
  metal-ion translocation 82–3
  photochemically controlled 48–9, 65–9
  transition metal complex-based 76–96
molecular motions
  from a chemical reaction 77–83
  electrochemically induced 83–91
  see also controlled motion; controlling motion
molecular motors
  design principles for 1–4
  electrochemical basic 156–7
  reversible rotary 28–9
  rotary 27–8
  sunlight-powered 66–9
molecular recognition 82
molecular shuttles
  adding and removing covalent bonds 19
  adding and removing electrons 15–17
  adding and removing metal ions 17–19
  adding and removing protons 13–15
  changing configuration and 19–20
  complexation/decomplexation process for making 78–82
  entropy-driven 23–4
  and environment change 24
  stimuli-responsive 13–24
  via excited states 20–2

molecular switches
  photochemically controlled 51
  translational 13–24
  unimolecular 221
molecular systems
  covalently bonded 4–12
  mechanically bonded 12–29
motility assays 470–2
motors *see* molecular motors
MR fluids *see* magnetorheological fluids
MSM materials *see* magnetic shape-memory (MSM) materials
multifunctional materials
  azobenzene polymers as 424–38
  chitosan-based hydrogels as 447–61
  electrorheological fluids as 500–1
  fullerenes and carbon nanotubes as 497–9
  intelligent composites 489
  liquid-crystal elastomers as 491–3, 494–5
  piezoelectric ceramics as 231–52
  shape-memory alloys as 317–38
multifunctional polymers with shape-memory effect 312–13
multifunctionality 151–2, 312–13
  magnetic shape-memory materials 496–7
muscle-like contractions 294–5, 296
muscle-like rotaxane dimer 77–8, 79
muscle potential 163–5, 166–7
muscles 152–3, 408
  sensing 163–4, 165
  tactile muscles 164–5, 166, 167
  working in air 165, 167
  see also artificial muscles
myosin gel 469–71

nanobiosensors 511–12
nanocomposites 126–40
nanomotors 66–9
nanoparticles 509–11
nanosensing 132–6
nanotubes 497–9
natural muscles 152–3, 408
negative differential resistance devices 222

net positional change 13–15, 15–20
NiTi alloy 317, 320–1, 326–30
  engineering applications 333–8
  virtual two-way actuation using 323–4
nonbiased safety devices 324–5
nonhomogeneous deformation of ferrogels 290–3
noninterlocking systems 89–91
nonionic polymer gels 500
nonstoichiometric materials 147–9
nucleic acids 507

optical properties 32
optics, framed actuator for 414–15
organic effector molecules 114–17
  ternary complex formation 117–18
organic effectors sensitivity 453
organic interconnects 223
organometallic systems 8
oriented myosin gel 469–71
orthodontics 328

PAN *see* polyacrylonitrile (PAN)
paraelectric phase 256–7
particle-size effects 107–8, 109
peptides as effectors 117–18
percolating conductive particle networks 400
perfluorinated sulfonic ionic polymers 127–8, 129, 130–1
performance capabilities, actuators 201–3
pH
  and chemomechanical polymers 100, 101–7
  and cooperativity effects 112
  hydrogels 456, 457
  PAN fibres 191–4, 196
  sensitivity 450–2
  and water uptake 108–10
photo-orientation 432–4
photochemically controlled
  molecular abacus 92–5
  molecular devices 48–65
  molecular machines 48–9, 65–9
  molecular switches 51
photodeformable azo materials 434–7

photoinduced decoordination 91–2, 93
photoisomerisation 426–7
photomechanical effects 437
photomechanical materials 424–38
photoresponsive azo materials 432–4
photoswitchable azo materials 430–2
piezoelectric ceramic actuators 233–5,
    236, 242, 407
    compact hybrid actuators 247
piezoelectric ceramics 232–3
    applications 242–7
    commercial products 247–52
    modeling 235–42
        actuators 242
        sensors 239–41
piezoelectric polymers 407
piezoelectric single crystal 407
piezoelectricity 232
plug/socket systems 52–3
polarity 472–4
polyacrylonitrile (PAN) 191–203
    actuation properties 194, 196–8, 203
    force–strain behaviour 194, 195
    PAN bundle muscle crane 201
    performance capabilities 201–3
    performance of PAN bundle
        artificial muscle 198–201
    volume change 193–4, 195
    work performance 201, 202
polyelectrolytes 100
polymer-protein complexation 464–75
polymer science 176
polymer–actin complexes 465–7
    motility assay 471–2
    polarity of actin in 472–4
    polymorphism of 467–9
polymeric actuators 153–4, 155–6, 163
polymeric batteries 172–4, 181
polymeric blends 143, 145, 146
polymeric electrochemomechanical and
    electromechanical muscles 153–5
polymeric gels 100, 102–7, 282–3
    ATP-driven 464–75
    magnetorheological 342–3
    nonionic 500
    see also magnetic polymeric gels

polymeric microactuator 273–8
polymers see azobenzene polymers;
    chemomechanical polymers;
    conducting polymers
polymorphism 467–9
polypeptides 465
polypyrrole 144–5, 148, 149
porosity 150–1, 152, 181
porous devices 336–8
positional switches 24–7
pre-stress bias 399
primitive conscious systems 164–5,
    166, 167
protein-polymer complexation 464–75
protons, and molecular shuttling 13–15
pseudomuscular contraction 295–9
pseudorotaxanes 65–6, 67
pumps 411–12
PVDF polymer 256–7
P(VDF-TrFE)-based terpolymers 407
    electromechanical response in 266–8
    microstructure and ferroelectric
        relaxor behaviour of 268–73
P(VDF-TrFE) copolymer 256–66

rectifiers 212–21
redox-driven translocation 89–91
relaxation models 178–9
relaxor ferroelectric polymers 256–79
resistance devices 222
resistors 210–12
responsive gels 283–8
rethreading of pseudorotaxanes 65–6
reversible rotary motors 28–9
Reynolds number 3
rheological properties 341–2, 348–50
rheometry 354–7, 358
robotic actuation 130–2, 133
rotary brakes 373–4
rotary motors 27–9
rotational motion in catenanes 24–9
rotaxanes 12–24
    complexation/decomplexation
        process for making 78, 80
    molecular abacus 92–5
    muscle-like dimer 77–8, 79

Subject Index 531

pseudorotaxanes 65–6, 67
sunlight-powered nanomotor 66–9
technology applications 31, 33–7
transition metal complexed 83–9
rotor-blade flap 245–6
ruthenium-containing catenane 91–2, 93

sacrificial mechanism 94–5
safety devices 324–5
salt effects in polymer gels 105–6
scale effects 2
seals 321–2
selectivity
   covalent interactions 118–19
   organic effector molecules 114–17
self-repair 485–9
sensing 489
   magnetic shape-memory materials 496–7
   muscles 163–4, 165
   unparalleled simultaneous 151
sensors 239–41, 415–16
   fluorescent 56–9
   see also biosensors
shape control 246
shape distortion, ferrogels 284–90
shape-memory alloys 317–38, 407
   applications 320–7
   engineering applications 331–6
   medical applications 313, 327–31
   thin-film and porous devices 336–8
shape-memory effect 303–4
   multifunctional polymers with 312–13
   thermally induced 308–11
shape-memory materials, magnetic 493, 495–7
shape-memory particles 308
shape-memory polymers 301–14
shear flow 469–70
shear-yield stress 351–3
shock absorbers 363–73
shuttles *see* molecular shuttles
signal amplification 56–9
silicone dielectric elastomers 407
simulation *see* modeling
single covalent bond 4–7

single-electron transistor 222
smart fin 243–4
smart materials *see* intelligent materials; multifunctional materials
smart-product development 138–9
socket/plug systems 52–3
soft actuators 256–79
soft materials 147–9
solid-state molecular electronic devices 32–5
solids, controlled motion in 32–8
solutions, controlled motion in 30–2
static yield stress 351, 352
stents 329–31
stimuli-induced conformational control
   around a single covalent bond 4–7
   around several covalent bonds 8–9
   in organometallic systems 8
stimuli-responsive molecular shuttles 13–24
strain–force behaviour 194, 195
stress
   Maxwell stress 396–8, 401–3
   shear-yield stress 351–3
stretching, rotaxane dimer 77–8, 79
structural damping 333–4
structural health monitoring 246–7
structural shape control 246
structured metal electrodes 400–1
styrene/divinylbenzene-based polymers 129, 130, 131
sunlight-powered nanomotor 66–9
supercapacitors 174
superelastic medical devices 328–9
superelasticity 317–20, 327
surface mass transport 434–7
surface properties 434
surface roughness 357–63
surfaces, controlled motion on 32–8
swelling 101, 102–7
   chitosan-based hydrogels 105–6, 450–3
   and concentration profiles 110–11
   and cooperativity 111–13
   glucose-driven 118–19

and kinetics 107–8, 109
organic effector molecules 114–17
in ternary complexes 117–18
and water effects 108–10
switches *see* mechanical switches; molecular switches

tactile muscles 164–5, 166, 167
temperature, magnetorheological fluids and 369–73
ternary complex formation 117–18
theoretical models, conducting polymers 174, 176–9, 180
thermal interrupter 325–6
thermal recoordination 91–2, 93
thermally induced shape-memory polymers 303–11
thermodynamics 176
thermoplastic shape-memory polymers 304–5
thermoresponsive polymers 456–7
thermosensitivity 455–6
thin-film devices 336–8
three-dimensional fabrication of IPMNCs 128
three-probe electrical measurements 209–10
three-stage catenanes 83–6
three-way catenane positional switches 24–7
tissue engineering 460
titanium fibre/aluminium system 482–4, 488
transduction 132–6
electron/chemical 151, 152, 174, 175
transistors 222
transition metal complexes
catenanes and rotaxanes 83–9

heterodinuclear bismacrocycle 86–8
molecular machines 76–96
translational molecular switches 13–24
translocation
metal-ion 82–3
redox-driven 89–91
triple layers 154, 158, 160, 164, 166
tubes with metal supports 160–1
two-probe electrical measurements 209–10
two-way catenane positional switches 24–7
two-way devices 323–4

unimolecular amplifiers 223
unimolecular electronic devices 205–23
unimolecular rectifiers 213–21
unparalleled simultaneous sensing 151

valves 324–5
thin-film 336–7
vibration/acoustic control 243–5
virtual two-way actuation 323–4
volume variation
electrochemomechanical muscles 155–6
PAN fibre gel 193–4, 195
see also swelling

water uptake and release 108–10
wet materials 147–9
wires 50
work performance 201, 202

XNOR logic gate 63–5
XOR logic gate 63–5

yield stress 351–3